Springer Proceedings in Physics

Volume 252

Indexed by Scopus

The series Springer Proceedings in Physics, founded in 1984, is devoted to timely reports of state-of-the-art developments in physics and related sciences. Typically based on material presented at conferences, workshops and similar scientific meetings, volumes published in this series will constitute a comprehensive up-to-date source of reference on a field or subfield of relevance in contemporary physics. Proposals must include the following:

- name, place and date of the scientific meeting
- a link to the committees (local organization, international advisors etc.)
- scientific description of the meeting
- list of invited/plenary speakers
- an estimate of the planned proceedings book parameters (number of pages/articles, requested number of bulk copies, submission deadline).

More information about this series at http://www.springer.com/series/361

Iker Zuriguel · Angel Garcimartín ·
Raúl Cruz Hidalgo
Editors

Traffic and Granular Flow 2019

Springer

Editors
Iker Zuriguel
Departamento de Física y Matemática
Aplicada
Universidad de Navarra
Pamplona, Navarra, Spain

Angel Garcimartín
Departamento de Física y Matemática
Aplicada
Universidad de Navarra
Pamplona, Navarra, Spain

Raúl Cruz Hidalgo
Departamento de Física y Matemática
Aplicada
Universidad de Navarra
Pamplona, Navarra, Spain

ISSN 0930-8989 ISSN 1867-4941 (electronic)
Springer Proceedings in Physics
ISBN 978-3-030-55972-4 ISBN 978-3-030-55973-1 (eBook)
https://doi.org/10.1007/978-3-030-55973-1

This Springer imprint is published by the registered company Springer Nature Switzerland AG
The registered company address is: Gewerbestrasse 11, 6330 Cham, Switzerland

Preface

These are the proceedings of the 13th Traffic and Granular Flow (TGF) Conference, which was held in Pamplona, Spain, from July 2 to July 5, 2019. TGF started in 1995 at Forschungszentrum Jülich (Germany) as a single time event, and has grown in 25 years to a conference series that is a reference in the field of pedestrian dynamics.

This time, more than 120 international researchers from different fields ranging from physics to computer science and engineering came together to discuss the state-of-the-art developments. The chosen dates (just before the worldwide known bull-running festivals of Sanfermín) and the variety of social events prepared by the organizers gave rise to a stimulating atmosphere that facilitated many discussions and the establishment of lots of new research connections.

In this edition, the hosting group, The Granular Lab of the University of Navarra, made an effort to encourage the participation of several members of the granular community, a discipline that was very present at the beginning of the conference series, but had practically disappeared from the main topics of the previous events. It was interesting to see that the connections between the behavior of granular media and other systems such as pedestrian dynamics are more tangible than ever. At the same time, there were several contributions from people working on emergent-related subjects such as self-propelled particles, data transport, swarm behavior, intercellular transport, and collective dynamics of biological systems.

This conference would not have been possible without the trust of Andreas Schadschneider and the rest of the TGF scientific committee. Of course, we are very grateful to those who shared with us the responsibility of organizing this conference: Daniel R. Parisi, Luis A. Pugnaloni Diego Maza, and César Martín-Gómez. A special thank goes to Silvia Olza, whose effort and dedication allowed the success of the conference. We would also like to state our recognition for the amazing performance of Iñaki Echeverría, Jose Ilberto Fonceca, Diego Gella, Bruno Guerrero, and Dariel Hernández, who were all Ph.D. students at the Granular Lab. Finally, we would like to acknowledge the logistic support of the Faculty of Sciences of the University of Navarra.

We are happy to contribute to the tradition of Springer proceedings of the Traffic and Granular Flow conference, and we are looking forward to the next edition, organized by K. Ramachandra Rao in New Delhi, India. Exceptionally, due to COVID19 pandemic and the delay of Pedestrian Evacuation Dynamics conference to the year 2021, the next Traffic and Granular Flow conference will take place in 2022.

Pamplona, Spain

Iker Zuriguel
Angel Garcimartín
Raúl Cruz Hidalgo

Contents

Contributors

Juliane Adrian Forschungszentrum Jülich GmbH, Jülich, Germany

A. P. F. Atman Programa de Pós-Graduação em Modelagem Matemática e Computacional, PPGMMC, CEFET-MG, Belo Horizonte, Brazil;
Departamento de Física, CEFET-MG, Belo Horizonte, Brazil;
Instituto Nacional de Ciência e Tecnologia de Sistemas Complexos, INCT-SC, CEFET-MG, Belo Horizonte, Brazil

Stefania Bandini Research Center for Advanced Science and Technology, The University of Tokyo, Tokyo, Japan;
Complex Systems and Artificial Intelligence Research Center, University of Milano-Bicocca, Milan, Italy

Benjamin Sebastian Bergner Faculty of Urban and Environmental Planning, Urban Sociology, University of Kaiserslautern, Kaiserslautern, Germany

Stefan Bittihn Institut für Theoretische Physik, Universität zu Köln, Köln, Germany

Maik Boltes Institute for Advanced Simulation IAS-7: Civil Safety Research, Forschungszentrum Jülich GmbH, Jülich, Germany

Ernst Bosina SBB AG, Bern, Switzerland

Marek Bukáček Faculty of Nuclear Sciences and Physical Engineering, CTU in Prague, Prague, Czech Republic

Hans-Joachim Bungartz Department of Informatics, Technical University of Munich, Garching, Germany

Antonio J. Caamaño Rey Juan Carlos University, Móstoles, Spain

Yang Cao Department of Automation, University of Science and Technology of China, Hefei, China

Laura C. Carpi Programa de Pós-Graduação em Modelagem Matemática e Computacional, PPGMMC, CEFET-MG, Belo Horizonte, Brazil;
Departamento de Física, CEFET-MG, Belo Horizonte, Brazil;
Instituto Nacional de Ciência e Tecnologia de Sistemas Complexos, INCT-SC, CEFET-MG, Belo Horizonte, Brazil

Tao Chen Tsinghua University, Beijing, China

Mihaela I. Chidean Rey Juan Carlos University, Móstoles, Spain

Mohcine Chraibi Civil Safety Research, Institute for Advanced Simulation, Forschungszentrum Jülich GmbH, Jülich, Germany

Daniel Clarke Department of Chemical and Process Engineering, University of Canterbury, Christchurch, New Zealand

Richard D. Connors University of Leeds, Leeds, UK

Jakob Cordes Institut für Theoretische Physik, Universität zu Köln, Cologne, Germany

Luca Crociani Complex Systems and Artificial Intelligence Research Center, University of Milano-Bicocca, Milan, Italy

Winnie Daamen Department of Transport and Planning, Technical University of Delft, Delft, The Netherlands

Azita Dabiri Department of Transport and Planning, Technical University of Delft, Delft, The Netherlands

Irene de la Torre - Arenas SENSEable City Laboratory, Massachusetts Institute of Technology, Cambridge, MA, USA

Eduardo del Arco Rey Juan Carlos University, Móstoles, Spain

Felix Dietrich Whiting School of Engineering, Johns Hopkins University, Baltimore, USA;
Princeton University, Princeton, USA

Dorine C. Duives Department of Transport and Planning, Delft University of Technology, Delft, The Netherlands

I. Echeverría Facultad de Ciencias, Dpto. de Física y Matemática Aplicada, Universidad de Navarra, Pamplona, Spain

Alonso Espinosa Mireles de Villafranca University of Bristol, Bristol, UK

Claudio Feliciani School of Engineering, Department of Aeronautics and Astronautics, The University of Tokyo, Bunkyo-ku, Tokyo, Japan;
Research Center for Advanced Science and Technology, The University of Tokyo, Meguro-ku, Tokyo, Japan

Yan Feng Delft University of Technology, Delft, The Netherlands

Rainer Fischer Munich University of Applied Sciences, Munich, Germany

Zhijian Fu School of Transportation and Logistics, Southwest Jiaotong University, Chengdu, PR China

Akihiro Fujita School of Engineering, Department of Aeronautics and Astronautics, The University of Tokyo, Bunkyo-ku, Tokyo, Japan

Taku Fujiyama Department of Civil, Environmental and Geomatic Engineering, University College London, London, UK

Luke Fullard School of Fundamental Sciences, Massey University, Palmerston North, New Zealand

Angel Garcimartín Facultad de Ciencias, Dpto. de Física y Matemática Aplicada, Universidad de Navarra, Pamplona, Spain;
Dep. Física y Mat. Apl., Fac. Ciencias, Universidad de Navarra, Pamplona, Spain

Vicente Garzó Departamento de Física and Instituto de Computación Científica Avanzada (ICCAEx), Universidad de Extremadura, Badajoz, Spain

František Gašpar Faculty of Nuclear Sciences and Physical Engineering, Czech Technical University in Prague, Prague 1, Czechia

Alexandra Gavriilidou Delft University of Technology, Delft, The Netherlands

Juan I. Giribet Dto. Ing. Electrónica, Facultad de Ingeniería, Universidad de Buenos Aires, C. A. de Buenos Aires, Argentina;
Dto. de Matemática, Facultad de Ingeniería, Universidad de Buenos Aires, C. A. de Buenos Aires, Argentina;
Instituto Argentino de Matemática "Alberto Calderón" CONICET, C. A. de Buenos Aires, Argentina

Marion Gödel Munich University of Applied Sciences, Munich, Germany;
Technical University of Munich, Garching, Germany

Rubén Gómez González Departamento de Física, Universidad de Extremadura, Badajoz, Spain

Juan F. González-Saavedra Departamento de Física and Instituto de Computación Científica Avanzada (ICCAEx), Universidad de Extremadura, Badajoz, Spain

Andrea Gorrini Complex Systems and Artificial Intelligence Research Center, University of Milano-Bicocca, Milan, Italy

B. V. Guerrero Dep. Física y Mat. Apl., Fac. Ciencias, Universidad de Navarra, Pamplona, Spain

Milad Haghani Institute of Transport and Logistics Studies—Business School, The University of Sydney, Darlington, NSW, Australia

Samer H. Hamdar The George Washington University, Washington, D.C., USA

Flurin Hänseler Delft University of Technology, Delft, The Netherlands

Anna Belle Harada Department of Physics, Mount Holyoke College, South Hadley, MA, USA

Stephan Heule SBB AG, Bern, Switzerland

R. C. Hidalgo Dpto. de Física y Matemática Aplicada, Facultad de Ciencias, Universidad de Navarra, Pamplona, Spain

Daniel Holland Department of Chemical and Process Engineering, University of Canterbury, Christchurch, New Zealand

Serge P. Hoogendoorn Department of Transport and Planning, Delft University of Technology, Delft, The Netherlands

Pavel Hrabák Faculty of Information Technologies, Czech Technical University in Prague, Prague 6, Czechia

Kai Huang Experimental Physics V, University of Bayreuth, Universitätsstraße 30, Bayreuth, Germany;
Division of Natural and Applied Sciences, Duke Kunshan University, Kunshan, Jiangsu, China

Yoshinari Inomata Oita University, Oita, Japan

Samuel Irvine School of Fundamental Sciences, Massey University, Palmerston North, New Zealand

Xiaolu Jia Department of Advanced Interdisciplinary Studies, School of Engineering, The University of Tokyo, Meguro-ku, Tokyo, Japan

Takayuki Kanda Kyoto University, Kyoto, Japan

Yu Kang State Key Laboratory of Fire Science, University of Science and Technology of China, Hefei, China;
Department of Automation, University of Science and Technology of China, Hefei, China

Ioannis G. Kevrekidis Whiting School of Engineering, Johns Hopkins University, Baltimore, USA;
Princeton University, Princeton, USA

Basma Khelfa Division for Traffic Safety and Reliability, University of Wuppertal, Wuppertal, Germany

Megan M Khoshyaran ETC Economics Traffic Clinic, Paris, France

Janine Klein Institute for Advanced Simulation IAS-7: Civil Safety Research, Jülich, Germany

Benedikt Kleinmeier Department of Computer Science and Mathematics, Munich University of Applied Sciences, Munich, Germany;
Department of Informatics, Technical University of Munich, Garching, Germany

Victor L. Knoop Department of Transport and Planning, Delft University of Technology, Delft, The Netherlands

Gerta Köster Department of Computer Science and Mathematics, Munich University of Applied Sciences, Munich, Germany

Tobias Kretz PTV Group, Karlsruhe, Germany

Valentina Kurtc Peter the Great St. Petersburg Polytechnic University, St. Petersburg, Russia

Pierre-Yves Lagrée Sorbonne Université, CNRS, Institut Jean le Rond d'Alembert, Paris, France

Sylvain Lassarre GRETTIA, COSYS, IFSTTAR, Gustave Eiffel University, Champs-sur-Marne, France

Jean-Patrick Lebacque UGE-IFSTTAR-COSYS-GRETTIA, Marne-la-Vallée, France;
GRETTIA, COSYS, IFSTTAR, Gustave Eiffel University, Champs-sur-Marne, France

Daniel Lehmberg Department of Computer Science and Mathematics, Munich University of Applied Sciences, Munich, Germany;
Department of Informatics, Technical University of Munich, Garching, Germany

Hongliu Li State Key Laboratory of Fire Science, University of Science and Technology of China, Hefei, China;
Department of Architecture and Civil Engineering, City University of Hong Kong, Hong Kong, China

Meng Li Tsinghua University, Beijing, China

Tao Li Faculty of Geosciences and Environmental Engineering, Southwest Jiaotong University, Chengdu, China

Zhihua Li Tsinghua University, Beijing, China

Hans van Lint Department of Transport and Planning, Delft University of Technology, Delft, The Netherlands

Dan Liu Department of Physics, University of Hartford, West Hartford, CT, USA

Hezi Liu Tsinghua University, Beijing, China

Long Liu Tongji University, Shanghai, China

Zhigang Liu College of Urban Railway Transportation, Shanghai University of Engineering Science, Shanghai, People's Republic of China

Siuming Lo City University of Hong Kong, Hong Kong, People's Republic of China

Miguel A. López-Castaño Departamento de Física and Instituto de Computación Científica Avanzada (ICCAEx), Universidad de Extremadura, Badajoz, Spain

Thomasin Lynch School of Fundamental Sciences, Massey University, Palmerston North, New Zealand

Jian Ma School of Transportation and Logistics, National Engineering Laboratory of Integrated Transportation Big Data Application Technology, Southwest Jiaotong University, Chengdu, China

Qing Ma State Key Laboratory of Fire Science, University of Science and Technology of China, Hefei, China

Rafael F. Martin Instituto Tecnológico de Buenos Aires (ITBA), C. A. de Buenos Aires, Argentina

C. Martín-Gómez Dpto. de Construcción, Instalaciones y Estructuras, Escuela Técnica Superior de Arquitectura, Universidad de Navarra, Pamplona, Spain

Ignacio Mas Dto. de Matemática, Instituto Tecnológico de Buenos Aires (ITBA), CONICET, C. A. de Buenos Aires, Argentina

Diego Maza Dpto. de Física y Matemática Aplicada, Facultad de Ciencias, Universidad de Navarra, Pamplona, Spain

David Méndez-Esteban Dpto. de Física y Matemática Aplicada, Facultad de Ciencias, Universidad de Navarra, Pamplona, Spain

José Méndez Omaña Beuth-Hochschule für Technik Berlin, Berlin, Germany

Inmaculada Mora-Jiménez Rey Juan Carlos University, Móstoles, Spain

Gerhard Müller Department of Physics, University of Rhode Island, Kingston, RI, USA

Hisashi Murakami Research Center for Advanced Science and Technology, The University of Tokyo, Meguro-ku, Tokyo, Japan

Akihito Nagahama Ritsumeikan Global Innovation Research Organization, Ritsumeikan University, Kusatsu, Shiga, Japan

Hari Hara Sharan Nagalur Subraveti Delft University of Technology, Delft, The Netherlands

Hana Najmanová Faculty of Civil Engineering, CTU in Prague, Prague, Czech Republic

Alexandre Nicolas Institut Lumière Matière, CNRS and Université Claude Bernard Lyon 1, Villeurbanne, France

Katsuhiro Nishinari Research Center for Advanced Science and Technology, The University of Tokyo, Tokyo, Japan;
School of Engineering, Department of Aeronautics and Astronautics, The University of Tokyo, Tokyo, Japan

Kerstin Nordstrom Department of Physics, Mount Holyoke College, South Hadley, MA, USA

Tobias Oberholzer SBB AG, Bern, Switzerland

Izabela M. Oliveira Departamento de Matemática, Centro Federal de Educação Tecnológica de Minas Gerais, CEFET-MG. Av. Amazonas, Belo Horizonte, MG, Brazil;
Programa de Pós-Graduação em Modelagem Matemática e Computacional, PPGMMC, CEFET-MG, Belo Horizonte, Brazil

Laura Pardini Susacasa Delft University of Technology, Delft, The Netherlands

Daniel R. Parisi Instituto Tecnológico de Buenos Aires (ITBA), CONICET, Buenos Aires, Argentina;
Consejo Nacional de Investigaciones Científicas y Técnicas, Buenos Aires, Argentina

Shinkyu Park SENSEable City Laboratory, Massachusetts Institute of Technology, Cambridge, MA, USA;
Department of Mechanical and Aerospace Engineering, Princeton University, Princeton, NJ, USA

J. M. Pastor Focke Meler Gluing Solutions S.A., Orkoien, Navarra, Spain

Germán A. Patterson Instituto Tecnológico de Buenos Aires (ITBA), CONICET, Buenos Aires, Argentina

Veronika Pešková Faculty of Civil Engineering, CTU in Prague, Prague, Czech Republic

Jana Pick Institute for Advanced Simulation IAS-7: Civil Safety Research, Jülich, Germany

Claudio D. Pose Dto. Ing. Electrónica, Facultad de Ingeniería, Universidad de Buenos Aires, C. A. de Buenos Aires, Argentina

Shunzhi Qian School of Civil and Environmental Engineering, Nanyang Technological University, Singapore, Singapore

Carlo Ratti SENSEable City Laboratory, Massachusetts Institute of Technology, Cambridge, MA, USA

Nagarjun Reddy Delft University of Technology, Delft, The Netherlands

Giulia Reggiani Department of Transport and Planning, Technical University of Delft, Delft, The Netherlands

Xiangxia Ren State Key Laboratory of Fire Science, University of Science and Technology of China, Hefei, China

Álvaro Rodríguez-Rivas Departamento de Física and Instituto de Computación Científica Avanzada (ICCAEx), Universidad de Extremadura, Badajoz, Spain

Paolo Santi SENSEable City Laboratory, Massachusetts Institute of Technology, Cambridge, MA, USA;
Instituto di Informatica e Telematica, CNR, Pisa, Italy

Majid Sarvi Department of Infrastructure Engineering, The University of Melbourne, Parkville, VIC, Australia

Andreas Schadschneider Institut für Theoretische Physik, Universität zu Köln, Cologne, Germany;
Institut für Physikdidaktik, Universität zu Köln, Cologne, Germany

Tobias Schrödter Forschungszentrum Jülich, Institute of Advanced Simulation, Jülich, Germany

Stefan Seer SENSEable City Laboratory, Massachusetts Institute of Technology, Cambridge, MA, USA;
Austrian Institute of Technology, Vienna, Austria

Sebastian Seriani Faculty of Engineering and Applied Sciences, Universidad de los Andes, Santiago, Chile

Armin Seyfried Forschungszentrum Jülich GmbH, Jülich, Germany;
Bergische Universität Wupperal, Wuppertal, Germany

Dongdong Shi School of Transportation and Logistics, National Engineering Laboratory of Integrated Transportation Big Data Application Technology, Southwest Jiaotong University, Chengdu, China

Kenichiro Shimura Research Center for Advanced Science and Technology, The University of Tokyo, Meguro-ku, Tokyo, Japan

Anna Sieben Forschungszentrum Jülich GmbH, Jülich, Germany;
Ruhr-University Bochum, Bochum, Germany

Weiguo Song State Key Laboratory of Fire Science, University of Science and Technology of China, Hefei, People's Republic of China

Martijn Sparnaaij Department of Transport and Planning, Delft University of Technology, Delft, The Netherlands

Bernhard Steffen Civil Safety Research, Institute for Advanced Simulation, Forschungszentrum Jülich GmbH, Jülich, Germany

Toshiya Takami Oita University, Oita, Japan

Alexander G. Tatashev MADI and MTUCI, Moscow, Russia

Jasmin Thurau SBB AG, Bern, Switzerland

Antoine Tordeux Division for Traffic Safety and Reliability, University of Wuppertal, Wuppertal, Germany;
Institut für Sicherheitstechnik, Bergische Universität Wuppertal, Wuppertal, Germany

Matrin Treiber Technische Universität Dresden, Dresden, Germany

Bart van Arem Transport & Planning, Delft University of Technology, Delft, The Netherlands

Arjan van Binsbergen Transport & Planning, Delft University of Technology, Delft, The Netherlands

Francisco Vega Reyes Departamento de Física and Instituto de Computación Científica Avanzada (ICCAEx), Universidad de Extremadura, Badajoz, Spain

Simeon Völkel Experimental Physics V, University of Bayreuth, Universitätsstraße 30, Bayreuth, Germany

Cornelia von Krüchten Institut für Theoretische Physik, Universität zu Köln, Cologne, Germany

Takahiro Wada College of Information Science and Engineering, Ritsumeikan University, Kusatsu, Shiga, Japan

Ke Wang School of Civil and Environmental Engineering, Nanyang Technological University, Singapore, Singapore

Qiao Wang University of Science and Technology of China, Hefei, Anhui, People's Republic of China;
City University of Hong Kong, Hong Kong, People's Republic of China

Marie-Jette Wierbos Delft University of Technology, Delft, The Netherlands

R. Eddie Wilson University of Bristol, Bristol, UK

Qiancheng Xu Institute for Advanced Simulation, Forschungszentrum Jülich GmbH, Jülich, Germany

Daichi Yanagisawa School of Engineering, Department of Aeronautics and Astronautics, The University of Tokyo, Bunkyo-ku, Tokyo, Japan;
Research Center for Advanced Science and Technology, The University of Tokyo, Meguro-ku, Tokyo, Japan

Lizhong Yang University of Science and Technology of China, Hefei, People's Republic of China

Marina V. Yashina Moscow Automobile and Road State Tech. Univ. (MADI) and Moscow Tech. Univ. of Communications and Informatics (MTUCI), Moscow, Russia

Rui Ye State Key Laboratory of Fire Science, University of Science and Technology of China, Hefei, People's Republic of China

Yuji Yoshimura SENSEable City Laboratory, Massachusetts Institute of Technology, Cambridge, MA, USA;
Research Center for Advanced Science and Technology, the University of Tokyo, Meguro-ku, Tokyo, Japan

Kai Yuan Department of Transport and Planning, Delft University of Technology, Delft, The Netherlands

Zeynep Yücel Okayama University, Okayama, Japan;
ATR, Kyoto, Japan

Kwok Kit Richard Yuen Department of Architecture and Civil Engineering, City University of Hong Kong, Hong Kong, China

Francesco Zanlungo ATR International, Kyoto, Japan

Jun Zhang State Key Laboratory of Fire Science, University of Science and Technology of China, Hefei, People's Republic of China

Yuan Zhao Shanghai Shentong Metro Group Co., Shanghai, China

Yifan Zhuang College of Urban Railway Transportation, Shanghai University of Engineering Science, Shanghai, People's Republic of China

Benedikt Zönnchen Department of Computer Science and Mathematics, Munich University of Applied Sciences, Munich, Germany;
Department of Informatics, Technical University of Munich, Garching, Germany

Xiaolei Zou Tongji University, Shanghai, China

Iker Zuriguel Facultad de Ciencias, Dpto. de Física y Matemática Aplicada, Universidad de Navarra, Pamplona, Spain;
Dep. Física y Mat. Apl., Fac. Ciencias, Universidad de Navarra, Pamplona, Spain

Boudewijn Zwaal Department of Transport and Planning, Delft University of Technology, Delft, The Netherlands

Part I
Pedestrian Dynamics

Chapter 1
Influence of Corridor Width and Motivation on Pedestrians in Front of Bottlenecks

Juliane Adrian, Maik Boltes, Anna Sieben, and Armin Seyfried

Abstract We present experiments on the behaviour of participants in front of a bottleneck. The waiting area was limited by a corridor of barriers leading straight to an entrance gate. We varied the width of the corridor and the degree of motivation of the participants to rush through the gate. We show that the density in front of the bottleneck grows with increasing corridor width and increasing motivation. Therefore, the velocity of individuals in the jam in front of the bottleneck is slower for wider corridors. Furthermore, it is shown that the flow through the bottleneck is controlled by the bottleneck width itself but is independent from the corridor width. The degree of motivation has, however, an influence on the flow rate through the bottleneck.

1.1 Introduction

Understanding the behaviour of crowds is important in order to draw up or to adapt safety regulations for buildings and event areas. People confronted with spatial bottlenecks either follow the social norm of queuing, they compete, or show a mixture of both behaviours. A competing behaviour leads to a high density of persons per square meter which can result in fatalities. A typical bottleneck situation, in which pushing might occur, is at the entrance gates to concert areas or events.

Sieben et al. [1] investigated the effect of different geometries of barriers in an entrance scenario. The experiments included two scenarios: an open waiting area without barriers and a corridor of barriers leading perpendicularly to the gates. In the first scenario, the participants arranged themselves in a semicircle around the gate. As soon as the gates were opened, they contracted until all interspaces were filled leading to high densities. In case of an upstreamed corridor, the participants arranged themselves in loose lanes before the gates were opened. After opening the gates, the

J. Adrian (✉) · M. Boltes · A. Sieben · A. Seyfried
Forschungszentrum Jülich GmbH, Jülich, Germany
e-mail: j.adrian@fz-juelich.de

A. Sieben
Ruhr-University Bochum, Bochum, Germany

I. Zuriguel et al. (eds.), *Traffic and Granular Flow 2019*,
Springer Proceedings in Physics 252,
https://doi.org/10.1007/978-3-030-55973-1_1

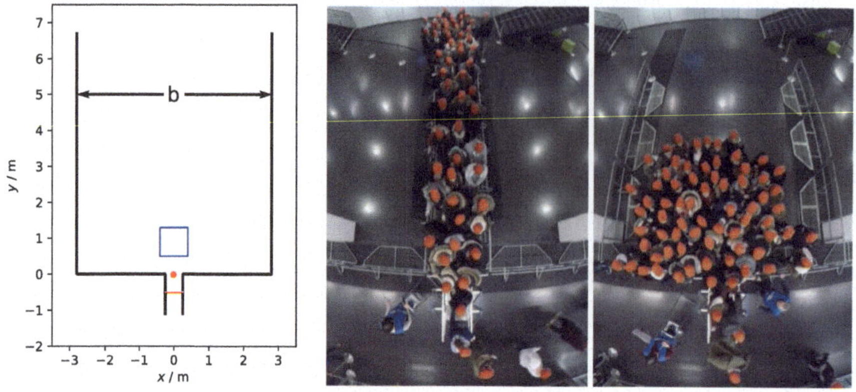

Fig. 1.1 Left: Sketch of the experimental setup with varying corridor width b. The red dot indicates the target at $x = 0$, $y = 0$, the red line is the target line for flow measurements, the blue square is the density measurement area. Centre, right: Screenshots of overhead video recordings at $t = 5$ s with high motivation of a run with $b = 1.2$ m (centre) and with $b = 5.6$ m (right)

densities were lower than in the first scenario. According to a questionnaire study, in which participants were asked to rate the scenarios after a presentation of video recordings of the experiments, the open space is perceived as less safe and more unjust than the corridor.

The results lead to the question to what extent the width of a corridor influences the participants behaviour in terms of physically measurable quantities, such as the density, and socio-psychological factors, such as the perception of social norms.

In this work, bottleneck experiments are presented focussing on the behaviour of persons in front of a bottleneck while their degree of motivation and the width of a corridor leading to the bottleneck is varied.

1.2 Methods

The presented experiments were carried out at the University of Wuppertal, Germany. As participants, university students were recruited at the end of their lectures. Due to the recruiting strategy, the number of participants varied. In the following, we concentrate only on runs with more than 40 participants.

The experimental setup is sketched in Fig. 1.1. A typical entrance gate with a width of 0.5 m served as bottleneck and simulated the entrance to a concert hall. The height of the gate was 1.18 m. Therefore, shoulder parts could be protruded over the structure. The waiting area was limited by barriers forming a corridor leading straight to the entrance. The corridor width were systematically varied from 1.2 m to 5.6 m.

All participants were asked to imagine that they are waiting in front of the entrance to a concert of their favourite artist or band. Each group of participants had to perform

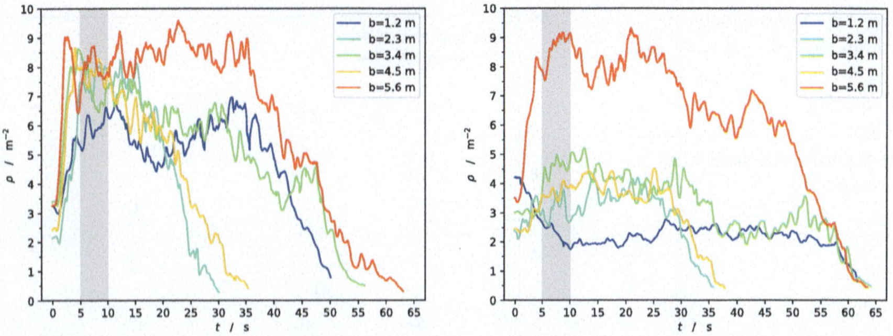

Fig. 1.2 Time series of the mean density ρ within the measurement area (see Fig. 1.1-left) for one experimental run for each corridor width. Left: high motivation, right: low motivation

two experimental runs in the same corridor width. In the first run they were told that, if they stand in the back, they can hardly see the stage or only on the video screen. So, they absolutely want to be standing next to the stage. Therefore, the motivation to access the concert as fast as possible was high. In the second run, the degree of motivation was reduced by the announcement that this time everybody will have a good view of the stage, but still they should enter the concert quickly. Even though the motivation of participants was different, we could not detect any obvious temporal decrease of motivation during a run (typically 30 to 60 s). However, we could in some cases observe that highly motivated participants pushed and passed the bottleneck earlier than participants with low motivation.

For all experimental runs, the individual head trajectories were automatically extracted from overhead video recordings following [2]. These trajectories are the basis for all data analysis presented in the following. The moment at which the entrance was opened is defined as $t = 0$ s.

1.3 Results

1.3.1 Influence of Corridor Width on Density

The personal space of each participant is defined as the area of the individual Voronoi cell being determined according to [3] using the software `JuPedSim` [4, 5]. Individual density is then given by the inverse of personal space.

Figure 1.2 shows the time series of the mean density within the measurement area (shown in Fig. 1.1) for one representative set of runs for each corridor width. We chose the measurement area to be small enough to fit in all corridor widths and to be in some distance (0.5 m) to the bottleneck in order to cover the area in which we expect the highest densities to occur.

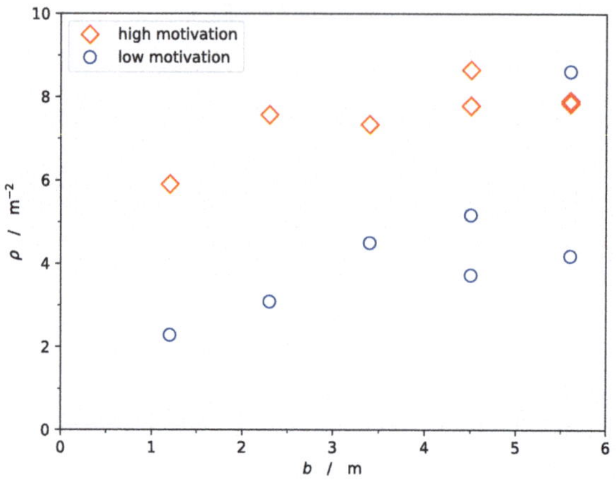

Fig. 1.3 Mean density ρ within the time interval from $t = 5$ s to $t = 10$ s (cf. grey area in Fig. 1.2). For $b = 4.5$ m and $b = 5.6$ m, two pairs of experimental runs were performed

A density increase within the first seconds indicates a contraction phase being most prominent in the highly motivated runs (cf. Fig. 1.2-left). The highest densities are found for the widest corridor, and the lowest densities for the narrowest corridor. Furthermore, the degree of motivation has a significant impact on the maximum densities that are reached.

It should be noted that the differences in runtime are caused by a different number of participants. For a comparison independent of variations in time, we consider the mean density within the 5s-interval from $t = 5$ s being shortly after the contraction phase to $t = 10$ s. The relation of the mean densities and the corridor width is summarized in Fig. 1.3. The data shows two trends: First, the density increases with increasing corridor width and second, there is a gap of 3–4 m^{-2} between the runs with high motivation and the corresponding runs with lower motivation. However, there is one outlier (for $b = 5.6$ m), in which the low motivated run results in a higher density than the highly motivated run.

1.3.2 Waiting Time and Velocity in Target Direction

To describe the movement towards a target, studying the individual progress towards that target is useful. Waiting time plots are characteristic for each geometry and motivation. An sample waiting time plot is displayed in Fig. 1.4-left for a run with the widest corridor and high degree of motivation. Each dot indicates the initial distance to the target (being located within the entrance gate, see red dot in Fig. 1.1) and the complete runtime until the target is reached. Each line represents the individual relation between target distance and remaining runtime during the experimental run.

After a contraction phase which takes about 5s, all interspaces are filled, the density is high and the participants are stuck in a jam in which only low velocities are possible. The velocity in jam v_{jam} in the target direction is derived from the inverse

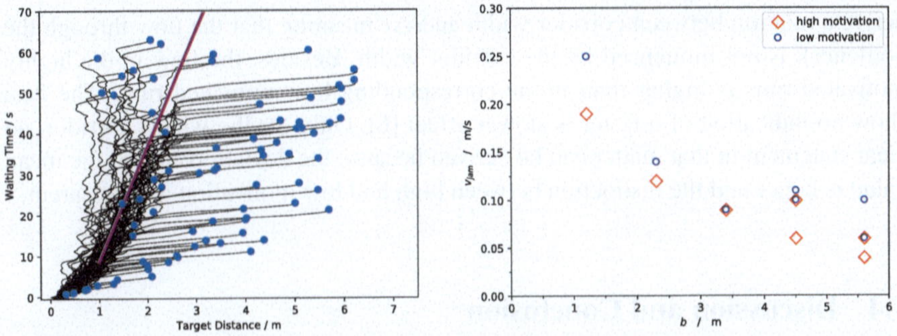

Fig. 1.4 Left: Waiting time plot for a run with highly motivated participants in the widest corridor ($b = 5.6$ m). Blue dots indicate initial target distance and the complete runtime of each person. Right: Velocity in jam v_{jam} versus corridor width. For $b = 4.5$ m and $b = 5.6$ m, two pairs of experimental runs were performed

of the slope of a linear fit to the waiting time plots (see pink line in Fig. 1.4-left). Because we are only interested in the intermediate phase, in which the jam occurs, the initial phase (first 5 s) and the final phase being the outflow through the gate (last meter to target) are omitted.

Figure 1.4-right summarizes the velocities in jam v_{jam} for different corridor widths and degrees of motivation. It is shown that v_{jam} is reduced with increasing corridor width and with increasing motivation.

1.3.3 Flow Through the Bottleneck

We examine the flow through the bottleneck by the time gaps Δt between two consecutive persons crossing the target line in the entrance gate (see Fig. 1.5) meaning that the longer the time gaps, the smaller the flow through the bottleneck. We cannot

Fig. 1.5 Boxplot representation of the time gaps Δt between consecutive persons crossing the target line (red line in Fig. 1.1-left). Grey circles: mean value, black rhombus: outliers

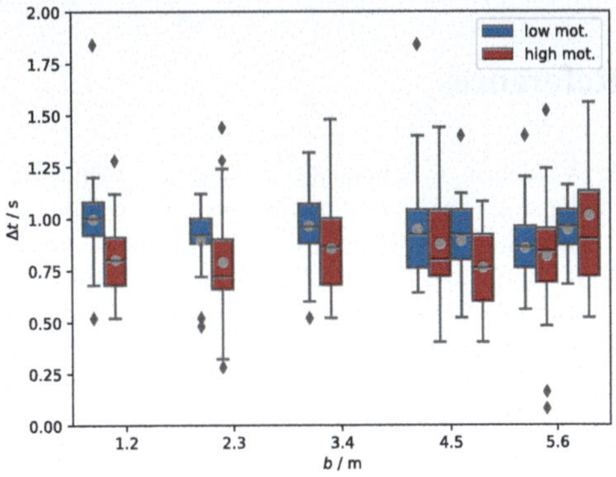

derive a relation between corridor width and Δt meaning that the flow through the bottleneck is not influenced by the corridor width. Besides, the flow in the highly motivated runs is higher than in the corresponding low motivated runs. The data show no indication of a faster is slower effect [6]. Only for the widest corridor, no clear statement in that matter can be derived because the deviation around the mean value is larger and the distinction between high and low motivation is unapparent.

1.4 Discussion and Conclusion

When people are confronted with a bottleneck, they can follow two opposite strategies in order to reach their target. Either they follow certain social rules and show a queuing-like behaviour or they show a rather competitive pushing behaviour. However, a mixture of both strategies is also possible. We suggest that, in contrast to a queuing behaviour, the pushing behaviour can lead to high densities, high initial velocities (together with a rapid contraction), low velocities in jam and collective transversal movements. We have shown that these factors are facilitated by widening the width of the corridor leading to the bottleneck and by increasing the degree of motivation.

We have seen that the velocity in jam is always slow but we measure slower velocities in wider corridors. When people cannot feel any progress towards the target or when they think that a situation is unfair, people can get frustrated because they assess that it will take longer to reach a target than in comparable situations in which they are slow but feel a progress. Then, the frustration might trigger a pushing behaviour which makes the situation less safe.

Dangerous situations in which a pushing behaviour is provoked and/or the density is too high can be avoided by narrowing the width of barrier systems. It is worth noting that, according to our findings, the flow through the bottleneck itself is independent of the width of the corridor leading to the bottleneck. Therefore, safety can be improved without compromising the performance of a facility.

References

1. A. Sieben, J. Schumann, A. Seyfried, Collective phenomena in crowds—where pedestrian dynamics need social psychology, Chialvo DR, editor. PLoS ONE **12**(6), e0177328 (2017)
2. M. Boltes, A. Seyfried, Collecting pedestrian trajectories. Neurocomputing. **100**, 127–33 (2013)
3. B. Steffen, A. Seyfried, Methods for measuring pedestrian density, flow, speed and direction with minimal scatter. Physica A: Statist. Mech. Appl. **389**(9), 1902–10 (2010)
4. M. Chraibi, A.U.K. Wagoum, C. Liao, E. Andresen, Graf A. JuPedSim, JuPedSim: JuPedSim v0.8.3. Zenodo [Internet]., June 20. Available from: (2018). https://doi.org/10.5281/zenodo.1293784

5. A.U,K. Wagoum, M. Chraibi, J. Zhang, JuPedSim: an open framework for simulating and analyzing the dynamics of pedestrians, in 3rd Conference of Transportation Research Group of India, p. 12 (2010)
6. D. Helbing, L. Buzna, A. Johansson, T. Werner, Self-organized pedestrian crowd dynamics: experiments, simulations, and design solutions. Transp. Sci. **39**(1), 1–24 (2005)

Chapter 2
The Measurement of Stress at Open-Air Events: Monitoring Emotion and Motion Utilizing Wearable Sensor Technology

Benjamin Sebastian Bergner

Abstract The application of wearable sensor technologies offers new insights in emotional states and resulting behaviors of event attendees. The here presented monitoring of emotion and motion shows continuously, when and where negative feelings of stress are experienced. Additional context data enable a more definite stimuli and behavior identification. The gathered information can help event planners to make events more safe and secure. Further, new perspectives on behavior in crowds are given. We conducted a mixed-method approach, consisting of real-time stress monitoring, movement pattern analyses, spatial analyses, and classical empirical methods at two different events with selected study participants ($n = 25$). We discuss the benefits, deficits, and limitations of the methodical approach regarding the identification of resulting individual and crowd behavior. We also show the implications of utilizing stress and motion monitoring at dynamic movements in public space, like political demonstration marches.

2.1 Introduction and Research Objectives

Our research focuses on emotional states of event attendees at open-air events, and primarily on situations, in which people feel stressed, unsafe or insecure with impact on their (social) behavior. The course of an event is endangered, when an as risky perceived situation leads to stress, and further towards e.g. avoidance or aggression behavior. In contrast to classical empirical methods, like quantitative and qualitative interviews, such situations can be captured by a continuous real-time stress and motion monitoring. The influence of cognitive self-perceptions, experience reports, and social desirability are eliminated by using the here presented approach. But only the triangulation with classical empirical and additional external data acquisitions gives a comprehensive picture of the concurrence of stimuli identification, emotional

B. S. Bergner (✉)
Faculty of Urban and Environmental Planning, Urban Sociology, University of Kaiserslautern, Pfaffenbergstraße 95, 67663 Kaiserslautern, Germany
e-mail: benjamin.bergner@ru.uni-kl.de

I. Zuriguel et al. (eds.), *Traffic and Granular Flow 2019*,
Springer Proceedings in Physics 252,
https://doi.org/10.1007/978-3-030-55973-1_2

outcomes, and resulting behavior. We used a sensor wristband in our real-world studies, which allows to collect location-independent, peripheral-physiological data as indicator for emotional stress in real-time. An additional GPS-tracker delivered the exact geo-location of measured stress reactions. The investigated use cases are a city festival and an open-air music festival, where study participants were equipped with respective sensor technologies. In order to identify stress stimuli, further data sources were considered (e.g. spatial analyses and neck-camera data). Standardized questionnaires conducted before and after visiting the event gave insights in the participants' personalities, actual states of anxiety/fear, event affinities, perceived feelings of safety and security, and socio-structural data. Overall, the paper focusses on following basic research questions: (A) How can individual-based sensor technologies be used to capture emotional states of individuals for a better understanding of crowd behavior? (B) What limitations are present in utilizing wearable sensor technology in real-world applications?

2.2 Theoretical Background

Understanding human behavior, especially in situations where people feel unsafe or insecure, includes a deeper knowledge of the interrelation of the situation itself, elicited stress and emotions, and specific resulting behavior. A basic definition of stress is given by Lazarus and Launier [1: p. 296]: "Stress is any event in which environmental or internal demands tax or exceed the adaptive resources of an individual, social system, or tissue system." For Lazarus [2: p. 5] stress can be triggered by internal factors within the person or by external and environmental factors, for example conflicting situations in crowds. The perception of harm, threat, and challenge are antecedent conditions for stress experience [2: p. 5], which go hand in hand with feelings of unsafety and insecurity. Lazarus defines fear and anger as emotional states, which are connected to stressful situations [2: p. 12]. Kleinginna and Kleinginna [3: p. 355] propose a comprehensive definition of emotion: "Emotion is a complex set of interactions among subjective and objective factors, mediated by neural/hormonal systems, which can [...] (c) activate widespread physiological adjustments to the arousing conditions; and (d) lead to behavior that is often, but not always, expressive, goal-directed, and adaptive." Though this definition is in the light of current research [4, 5] incomplete, it shows a linkage between physiological changes and resulting behavior beside other components of emotion. Nearly simultaneous changes in skin conductivity (increase) and skin temperature (decrease) indicate the experience of the negative emotions fear or anger [6, 7]. This physiological pattern figuratively defines a stress reaction in a specific situation, which is transferred to our sensor-based stress monitoring application at events. Next, we focus on possible (social) behavior resulting from stress experience. The term social behavior is basically characterized as behavior which is oriented towards others, and includes interpersonal exchange with other people and in groups or crowds [8: pp. 20 f.]. Social behavior is a most integrative part in understanding human and crowd behavior at events. The effects of emotional states on different types of behavior regarding individuals, in groups, and in crowds are of special interest. These types of behavior are deviant behavior, spatial behavior in due consideration of proxemics, territorial behavior and personal space, and walking behavior. These kinds of behavior have an impact on the overall crowd behavior. Wijermans [9] describes in her work how crowd behavior is composed and which overall mechanisms underlie. Especially the role of the individual as the generating unit of crowd behavior is accentuated. Crowd behavior is characterized as context-dependent and dynamic, and depends on the physical and social context, as well as on the internal state of the cognitive system of the individual [9: pp. 137, 141].

2.3 Methodical Framework

Core element of our research is the real-time monitoring of stress in space, including a simultaneous recording of geo-positions. The stress monitoring describes a psychophysiological monitoring as a method within ambulatory assessment [10: pp. 559 ff.]. It covers the part of physiological changes triggered by a distinctive stimulus in the test person itself or in the environment. Physiological parameters, which are recordable by non-invasive sensor technologies, are cardiovascular, respiratory, electrodermal, and brain activity measures [6, 10: p. 560]. We focus on utilizing electrodermal parameters; to be exact, changes on skin conductivity and skin temperature. "With the help of proper algorithms in appropriate software, the resulting data consists of gradient values for every second" [7]. The analyzation of the parameters is done retrospectively. "The serial nature of ambulatory data also allows identification of the specific time points when abrupt and statistically significant level shifts occur that may be associated with psychological or other events" [10: p. 565]. Thus, we consider only short-termed, high intensity physiological changes in skin conductivity and skin temperature in our stress reaction analysis. Parameter changes resulting from physical body activities (e.g. dancing) are leveled out by this procedure. The measurement device (sensor wristband BMS Smartband) presents a practical solution for not influencing the test person by the measurement instrument itself. It includes further sensors recording outdoor temperature, triaxial acceleration, and electrode contact pressure in real-time [11]. The test people are additionally equipped with GPS-trackers, which are synchronized with the sensor wristband. The GPS-devices offer the possibility to derive movement patterns, walking speeds, directions, as well as rest and activity cycles. The usage of unobtrusive neck-cameras and noise recorders (or other environmental sensors) provides a relevant surplus in showing what the test person visually and acoustically perceives when being stressed. External observations are useful if they are not influencing the test persons' natural behavior. An additional analysis of the spatial environment provides indications if spatial deficient spaces are responsible for experienced stress. The so far listed data via wearable sensor technologies, external observations, and spatial analyses are objectively identifiable and measurable. They offer a good chance for stimulus recognition fastened from an objective perspective.

But how people feel and behave in their immediate environment depends not only on (socio-)spatial or other environmental factors. It is the individual itself which influences how and in what intensity certain situations are perceived. Personal factors like constitution, mental state of mind, experiences with certain situations, attitudes and values, socio-structural characteristics, cultural imprints, and so on describe essential information for understanding the context how a person feels [4]. Standardized questionnaires before and after the study fulfil the purpose to extract this information. Open questions give the test person furthermore the chance to express her/his impressions and emotional experiences. In order to get information on the personality of the test people, we used the Big-Five-Inventory-10 [12]. Another standardized questionnaire applied before and after the case studies was the State-Anxiety-Inventory (STAI-S) [13]. It measures the state anxiety/fear level of a person at the moment of its conduction. Besides a general questionnaire regarding event affinity, perceived sense of safety and social basic data, "the test people also had the chance to give self-assessment regarding stressing situations at the event via tape recorders" [7: p. 2224].

The methodical framework comprises on the one hand objectively gathered data, which can be visualized in maps, videos, and pictures. On the other hand, subjective self-assessments of the test people themselves are included. It is our aim to bring both information threads together in a mixed-method approach.

2.4 Application of Stress and Motion Monitoring—Case Studies

The overall objective of the case studies was the identification of stress hotspots as indications when, where, and why event attendees feel unsafe or insecure. In contrast to former studies, where the individual was most often observed isolated in linkage to material space characteristics, the event setting is more shaped by the social environment, and thus by social behavior. Nevertheless, material space provides the frame where events take place and has to be considered as such.

The city festival "Altstadtfest" 2014 in Kaiserslautern/Germany is a traditional event taking place every year in summer for three days, and has about 100,000 visitors. It is situated in the pedestrian zone in the historic district with its own spatial characteristics. The city festival is free to visit without access controls. In the stretch of three days, 17 people participated in our stress monitoring study. 14 data sets have been evaluable; three datasets have been invalid due to technical issues in the data recording [14: p. 41]. Most of the participants were initially organized in measurement groups consisting of friends or family members, and thus in a well-known social environment. This has influence on the perceived safety and security of the test people themselves and in regard to their group members as well.

In the presentation of the results, we mainly focus on the movement patterns for every day and the aggregated heatmap of stress hotspots for all participants over the complete stretch of the city festival (Fig. 2.1). On all three days, the participants were mostly concentrated on the main axis of the festival area. On Saturday though, it was observed that test people were consciously dodging specific areas, because the density of the crowd was very high. It was especially observable at locations where the pedestrian zone becomes narrower or where music stages were situated. The movement pattern analyses illustrate that the higher the point density, the more frequented was the certain area by the test people. These areas were not only walked-through. The participants stayed for a longer time at these specific places. Reasons were sitting or standing together with friends or family members, ordering food and drinks, or dance in front of the music stages. The shown movement patterns represent an aggregation of the participants over a complete day. Time series analyses offer the chance for a higher resolution of specific time stretches, and heighten the quality of the results. This is especially helpful when the resulting stress hotspots are included in the consideration. Not every marked space is stressful for the whole period of the event (dynamic character: daytime and visitors' number). The aggregated stress hotspots indicate that areas are perceived as most stressful where the density of people was very high. The availability of space was shortened. Personal space was restricted and the possibility to move with the own preferred movement speed was lessened. We observed this especially before the stages, between the stages and the food and drink stands, and at bottlenecks where the path width is reduced [14: p. 41].

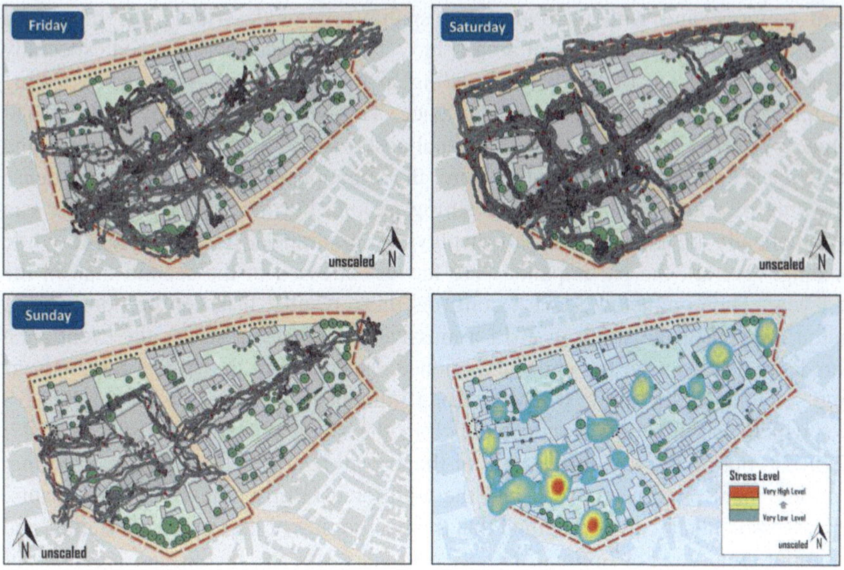

Fig. 2.1 Participants' movement patterns by day, and overall aggregation of stress hotspots over the stretch of all three days [14: p. 42]

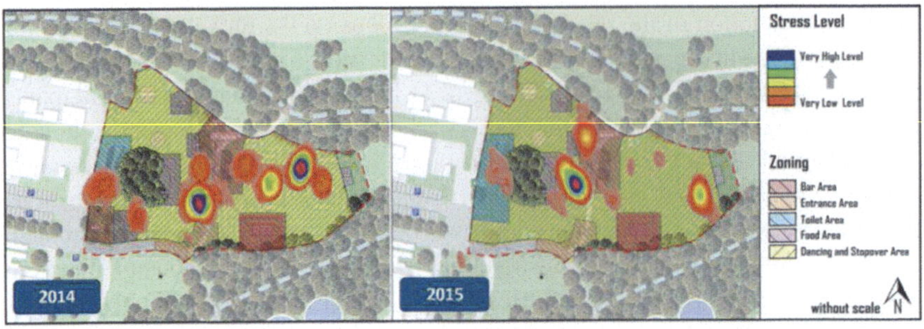

Fig. 2.2 Localization of stress hotspots in comparison 2014 and 2015 [14: p. 40]

The second test setting was the "Back to the Woods Open-Air" in Munich-Garching/Germany in two consecutive years (2014 and 2015). It is a one-day open-air electro-music festival visited by about 5000 people. The Back to the Woods Open-Air is not situated in the open public space. Access controls, ticketing, and a fenced event area are completely opposite conditions in comparison to the city festival. The security measures were consequently predefined and clearly visible for everyone [14: pp. 13 f.]. In the first year, six datasets of the study participants have been evaluable. In the second year the analysis of five datasets was conducted. Equally to the city festival in Kaiserslautern, the test people started the study in social groups consisting of friends and/or partners.

The results show the stress hotspots of the participants in regard to their movement patterns, and with an underlying spatial zoning describing the functional assignment of the specific event area (Fig. 2.2). Stress was experienced especially in the areas in front of the bars. Standing in line for a long time and crowding are the reasons for this negative experience. This was confirmed by participants' subjective self-assessments and by external observations. In both years, the dancing area appears surprisingly stressful. Here, the situation was characterized by dense crowding of dancing, standing, and sitting people. Further reasons for stress can hypothetically be factors like self-displays or hindrances by other people while dancing. Overall, the participants were far less stressed in the second than in the first year. The conducted STAI-S confirms these results. It has to be pointed out, that at the end of the study in 2014 a massive thunderstorm struck the event area. This had an impact on the stress experience, especially at the end of the study as time series analyses show. It had a strong influence on the results of the STAI-S too, which was not only conducted before the study but afterwards as well [7: 2225; 14: pp. 38 ff., 45]. In the light of the overall results of the stress monitoring, the movement pattern analysis, and the additional information from participants' self-assessments, it can't be denied that the positioning of bars and food stands, and the toilet area was not optimal. For the festival 2015, the layout and locations of these areas were accordingly mathematically optimized [14: p. 6]. The stress level (number of stress events/measurement time in seconds) was the highest at the city festival in Kaiserslautern, followed by the Back to the Woods Open Air 2014. The lowest stress level value has the Back to the Woods Open Air 2015, where the layout and locations of the event facilities were optimized and the event overall took place at a regular level without external disturbances [14: p. 45].

2.5 Conclusion and Discussion

The studies show the potential benefits of the application of a continuous real-time stress monitoring in linkage to a synchronized motion recording. We can see when and where people are stressed while attending an open-air event. A high level of stress indicates feelings of unsafety and insecurity, and thus influences the behavior in and of crowds. That fact offers event planners and security forces information which places require closer attentions during which times. If located stress hotspots are available in real-time, it is conceivable to transfer this data immediately to a security headquarter. In this case, stress monitoring could represent a form of an early-warning system and likewise a decision support system for organizing security personnel on-site.

The question is still open which specific kind of behavior is elicited by the measured stress experiences. Event attendees are inclined (or even forced) to let other people into their intimate and personal distance. It is in the nature of things at events, which have the basic characteristics to be crowded, having certain (dynamic) emotional atmospheres, and a shortage of space availability. The effects on stress experiences and behaviors have to be further discussed in this special context, particularly if people are placing themselves into crowds by their own free will. Appropriate, unobtrusive observations are still rare, which give an objective overview of the whole event area at all time. Here, not only organizational but ethical and legal aspects have to be considered (e.g. surveillance and privacy debate). Wearable technologies, like neck-cameras, can help to identify resulting behavior if they meet an overall acceptance. In case of the Back to the Woods Open-Air, the test people refused to use a continuous recording via neck-cameras, because of the perceived intrusion in their sphere of personal privacy. What we can observe is a coherence between movement patterns and localized stress hotspots, further validated by participants' subjective self-assessments. Areas which were very crowded, at bottlenecks, and where people have to stand in line for a longer time in order to get drinks were overall the most stressful locations.

The dialogue between objectively measured data resulting from stress monitoring, environmental data, and spatial analyses on the one hand, and subjective self-assessments evaluated from questionnaires on the other hand, is a fundamental concern of emotion research in real-world environments. The comparison of all gathered information represents an approach of triangulation not only between the methods, but as well between what we can measure objectively and how the human being assesses specific situations by its own perception. The decision to keep the studies as uncontrolled as possible reflects the reality. Natural resulting behavior is thus observable. A secured stimuli identification is affected by uncontrolled real-world settings, too. Information on what possibly elicited stress in certain situations can be evaluated from literature research and highly controlled settings and, in case of real-world studies, from context data (e.g. observations or spatial analyses). Subjective self-reports also give information about reasons for specific stress experiences. A more basic shortcoming of the instrument of stress monitoring is the availability of the used sensor technologies and thus the small number of participants. From our point of view, only indications on stress-eliciting stimuli and on resulting behavior can be made at this point.

When it comes to ethical aspects, the application of a stress monitoring is not practicable in every social environment. In our current research, we investigate emotional aspects and social behaviors in dynamic movements in public space, like parades and political demonstration marches. They imply continuously changing conditions regarding space, emotional atmospheres, and modes of behavior. Of late, we focused on conflicting political demonstrations with counter demonstrations (where police forces were involved). It turned out that the application of continuous real-time monitoring of stress is difficult if not impossible in this context. The stress monitoring depends strongly on the acceptance of the potential study participants as demonstrators. Furthermore, we were confronted with serious and critical concerns about data usage and data protection. The real-time monitoring of stress experience shows its limits of application in controversial situations.

Acknowledgements This research is part of the research project "Support Systems for Urban Events—Multi-criteria Integration for Openness and Safety (MultikOSi)" 2013–2016, and the research project "Organized Pedestrian Movement in Public Spaces: Preparation and Crisis Management of Urban Parades and Demonstration Marches with High Conflict Potential (OPMoPS)" 2017-2020, funded by the German Federal Ministry of Education and Research (BMBF).

References

1. R.S. Lazarus, R. Launier, Stress-related transactions between person and environment, in *Perspectives in Interactional Psychology*, ed. by L.A. Pervin, M. Lewis (Plenum, New York, 1978), pp. 287–327
2. R.S. Lazarus, From psychological stress to the emotions: a history of changing outlooks. Annual Rev. Psychol. **44**(1), 1–21 (1993)
3. P. Kleinginna, A. Kleinginna, A categorized list of emotion definitions, with suggestions for a consensual definition. Motiv. Emotion **5**(4), 345–379 (1981)
4. K.R. Scherer, What are emotions? And how can they be measured? Trends Dev. Res. Emotion **44**(4), 695–729 (2005)
5. K. Mulligan, K.R. Scherer, Toward a working definition of emotion. Emotion Rev. **4**(4), 345–357 (2012)
6. S.D. Kreibig, Autonomic nervous system activity in emotion: a review. Biol. Psychol. **84**(3), 394–421 (2010)
7. B. Bergner, D. Steffen, The role of user-generated emotion data and their optimized visualization for planning decision making, in *Proceedings AESOP 2015 Prague Annual Congress*

2015—Definite Space—Fuzzy Responsibility, ed. by M. Macoun, K. Maier (Prague, 2015), pp. 2216–2230
8. B. Miebach, Dimensionen sozialen Handelns, in *Soziologische Handlungstheorien – Eine Einführung. 2. grundlegend überarbeitete und aktualisierte Auflage*, ed. by B. Miebach (VS Verlag für Sozialwissenschaften|GWV Fachverlage GmbH, Wiesbaden, 2006)
9. F.E.H. Wijermans, *Understanding Crowd Behaviour: Simulating Situated Individuals* (University of Groningen, SOM research school, Groningen, 2011)
10. F. Wilhelm, P. Grossman, Emotions beyond the laboratory: theoretical fundaments, study design, and analytic strategies for advanced ambulatory assessment. Biol. Psychol. **84**, 552–569 (2010)
11. Bodymonitor 2019 Homepage, https://bodymonitor.de/. Last accessed 4 Nov 2019
12. B. Rammstedt, C.J. Kemper, M.C. Klein, C. Beierlein, A. Kovaleva, Eine kurze Skala zur Messung der fünf Dimensionen der Persönlichkeit. 10 Item Big Five Inventory (BFI-10). Methoden Daten Analysen **7**(2), 233–249 (2013)
13. C.D. Spielberger, R.L. Gorsuch, R.E. Lushene, *Manual for the State-Trait Anxiety Inventory* (Consulting Psychologist Press, Palo Alto, 1970, revised 1983).
14. A. Spellerberg, B.S. Bergner, M. Ehbrecht, U. Neumann, H.W. Hamacher, P. Heßler, A. Maier, Unterstützungssysteme für urbane Events: Multikriterielle Vernetzung für Offenheit und Sicherheit (MultikOSi): Teilprojekte: SASKuE und StanLay. Multikosi Endbericht: kurze & eingehende Darstellung (TU Kaiserslautern, Kaiserslautern, 2017).

Chapter 3
Smoothing Trajectories of People's Heads

Maik Boltes, Jana Pick, and Janine Klein

Abstract This paper compares three methods for smoothing trajectories from controlled experiments to determine the main movement direction: central moving average, spline interpolation of inflection points, and moving convex hull. The often-used method of averaging needs an adaptation due to the velocity but nevertheless is less accurate than the others. The spline method gives best results for high to moderate velocities. The newly introduced method using convex hulls is very robust and offers especially for low velocities good results. Smoothing has a large influence on the calculation of pedestrians' velocity and thus on the course of the fundamental diagram.

3.1 Introduction

Controlled experiments give the possibility to extract trajectories of the head of each individual person with high accuracy also in dense crowds as long as the head is visible to overhead cameras. These trajectories allow a detailed analysis of the movement, provide a basis for quantifications in legal regulations, guidelines and manuals for the construction of pedestrian facilities and enables the design, calibration and verification of microscopic models.

Caused by swaying and up and down bobbing of human walking these trajectories do not directly show the main movement direction (MMD) also called principal movement. The path of the head is influenced not only by the MMD but also by its bipedal gait and the associated change of the centre of mass (CM) [1]. Physiologically walking is characterized by an inverted pendulum movement in which the CM vaults over a stiff leg with each step [2] so that the path of the CM does not describe the MMD.

M. Boltes (✉) · J. Pick · J. Klein
Institute for Advanced Simulation IAS-7: Civil Safety Research, Forschungszentrum Jülich, 52428 Jülich, Germany
e-mail: m.boltes@fz-juelich.de
URL: https://www.fz-juelich.de/ias/ias-7/

I. Zuriguel et al. (eds.), *Traffic and Granular Flow 2019*,
Springer Proceedings in Physics 252,
https://doi.org/10.1007/978-3-030-55973-1_3

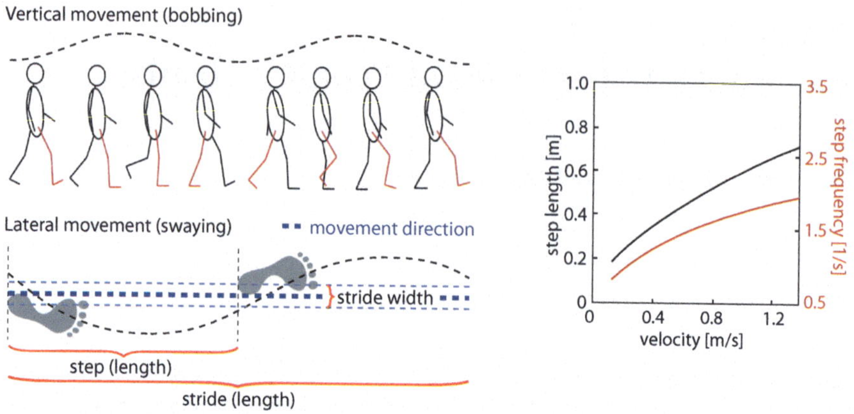

Fig. 3.1 Sketches of: top left: the vertical movement (bobbing; dashed black line) for one stride consisting of two steps; bottom left: bipedal gait from one foot to the other leads to a lateral movement (swaying; dashed black line) of the head; the thick blue dashed line shows the MMD; right: the dependency between step length and frequency of the gait cycle and velocity

The top left of Fig. 3.1 sketches the vertical movement (bobbing) resulting in a height variation for one stride consisting of two steps. In normal walking, the amplitude of vertical CM displacement increases linearly with walking speed from about 0.015 m at low (0.3 m/s) to 0.034 m at high speed (1.6 m/s) [3].

The bipedal gait from one foot to the other with a stride width of 0.05 m to 0.13 m [4] leads also to a lateral movement (swaying) of the head (dashed curved line at the bottom left of Fig. 3.1). The peak-to-peak amplitude (PPA) of the swaying of the head is around 0.05 m for normal walking in free flow condition [5]. For low velocity we see in our data a PPA of up to 0.3 m.

The right of Fig. 3.1 sketches the dependency between step length or frequency of the gait cycle and velocity [6, 7]. Due to the bipedal gait the velocity of body parts is not homogeneous. But for normal walking the head only has a low speed variation of 0.02 m/s [8].

The knowledge of the full body motion enables the understanding of the physical part of pedestrian dynamics as a whole, but for some measures and analysis only derived values lead to a correct understanding. For instance using directly the positions along the trajectory of the head for calculating the velocity of a person leads to an overestimated velocity [9]. Also, lane formation is more difficult to identify, if directly the location of the peoples' head is chosen than the path of the MMD. Microscopic models describing individual movement of pedestrians often represent a person only by a trajectory along their MMD. Thus, to compare simulation results with empirical data on a microscopic level the empirical data has to be smoothed.

This paper describes methods smoothing trajectories of the head to determine the MMD and does not deal with outliers, sudden jerks or a jitter along the trajectory for example coming from an imprecise detection method during field studies without the possibility to use markers [10]. Also colored caps used for detection without

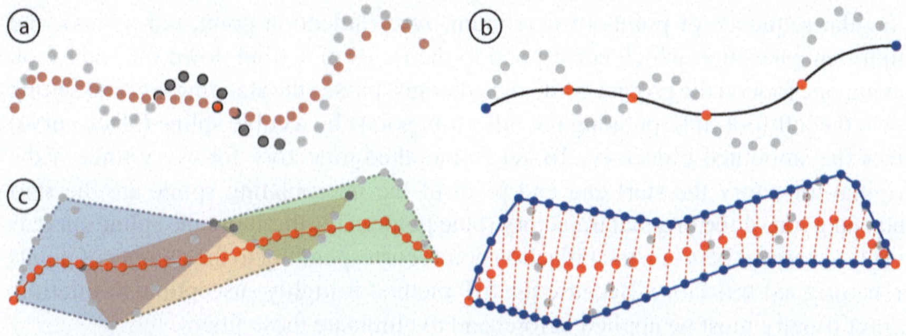

Fig. 3.2 Sketches of the smoothing methods: **a** CMA method forms arithmetic averages (red points) over partial intervals (dark grey points) of the original data (grey points); **b** SIP method interpolates the inflection points (red points) by a cubic spline (black curve); **c** MCH method uses the envelope of the union of moving convex hulls (colored areas) over a shifting interval along the trajectory; the centre (red points) of the dashed red lines between points on the left and right part of the envelope (blue points) belonging to the same time form the smoothed trajectory

structured markers lead to jittering [11]. For these type of errors we refer to methods described in [12], which have to be applied beforehand if necessary.

In the following sections three methods are presented and compared, especially their influence on quantities used for describing pedestrian dynamics.

3.2 Methods

3.2.1 Central Moving Average (CMA)

An often-used method for smoothing data series is the central moving average (see Fig. 3.2a) also called central rolling average. This method forms arithmetic averages (red points) over partial intervals (dark grey points) of the original data (grey points) equally spaced on either side of the point in the trajectory where the mean is calculated. At the beginning and end of the trajectory the equally spaced interval has to be reduced symmetrically so that the start and end point is part of the resulting smoothed trajectory.

3.2.2 Spline Interpolation of Inflection Points (SIP)

The second examined method chooses the inflection points along the trajectory as knots for a spline interpolation (see Fig. 3.2b) [13]. As long as there is substantial forward motion the mode of movement is the swinging of the legs which leads to a regular sequence of points of maximum, zero (inflection point, red points), and minimum curvature, which correspond to the times of setting down the right foot, having one foot on the ground while the other just passes the standing leg, and setting down the left foot. Interpolating the inflection points by a cubic spline (black curve) gives the smoothed trajectory. To get a smoothed trajectory for every time of the original trajectory the start and end point of the interpolating spline are the start and end point of the original trajectory (blue points). Furthermore the spline curve is evenly separated between the inflection points corresponding to the number of points in the original trajectory. Because the SIP method is highly susceptible to jittering, a filter usually must be applied beforehand to eliminate these jitters.

3.2.3 Moving Convex Hull (MCH)

This new approach for smoothing a trajectory is based on a union of moving convex hulls (see Fig. 3.2c) over a shifting interval of a defined size along the trajectory. The resulting smoothed trajectory is located between the partial envelopes connected to a single tube around the trajectory. Figure 3.2c shows on the left in blue, red, and green three overlapping convex hulls of three time steps with an interval of 16 successive points. The method starts and ends with full sized intervals which automatically includes the beginning and end of the original trajectory (blue and red area). On the right of Fig. 3.2c the blue line forms the overall envelope of the union of all convex hulls. The left and right part of the envelope is evenly segmented (blue points) between points on the envelope belonging to the original trajectory. The number of additional points corresponds to the number of intermediate points along the original trajectory. The centre of the line between points on the left and right part of the envelope belonging to the same time (red points on dashed red lines) form the resulting smoothed trajectory.

To determine positions to every time step along the spline curve (SIP method) or envelope of the convex hull (MCH method) one could use a projection of the original points to the curve or envelope as well. This would more accurately reflects the velocity of the head within one step in MMD, but other body parts have different progresses of the velocity so that no universal progress can be given and the SIP and MCH method use an even distribution of smoothed points.

All methods can be improved by a preceding step detection to adapt the parameters of the methods. For the CMA method the size of the interval could be set to the length of a stride so that only the period of one stride is continuously contracted. The SIP method would be more reliable, because one could limit the number of inflection points to one per stride located between two steps. The idea behind the MCH method is the construction of a tube just touching the original trajectory at the maximum swaying amplitude. For this the number of successive points of the original trajectory forming a single convex hull has to be the number of points belonging to one stride.

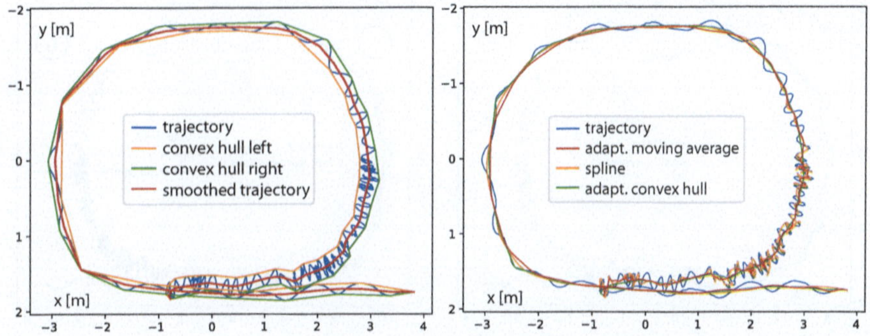

Fig. 3.3 Left: smoothed trajectory (red) from the MCH method and the envelope of the united convex hulls (orange and green); right: smoothing results of the three methods CMA (red), SIP (orange), and MCH (green)

The step detection in video sequences is error prone especially for low velocities. IMU, stereo, or motion capturing systems would give a much more robust detection of steps. Thus, for trajectory extraction from overhead video recordings the approximated velocity can be used to estimate the step length and frequency. The relationship between these variables is sketched on the right of Fig. 3.1. The calculation of this velocity with the CMA method with a fixed large interval covering also low velocities or directly with difference quotients over about one second is reasonable.

Trajectories used for this study have been collected during single file experiments with varying velocity regimes performed with pupils at schools in Germany [14]. The extraction of the data was done automatically from overhead video recordings [15]. For one trajectory (blue) of this data set with varying velocity from stand still to free flow the smoothed trajectory (red) resulting from the MCH method and the appropriate enclosing envelope (green and orange) can be seen on the left of Fig. 3.3. The result of all three smoothing methods for this trajectory is shown on the right of Fig. 3.3.

3.3 Results

All three introduced methods have drawbacks for varying conditions. Figure 3.4 shows some of these constraints.

For the CMA method the length of the interval is crucial. A long interval contracts the whole trajectory to their centre. A short interval cannot smooth out the swaying of slow walking. With an adaptation due to the velocity the error is less but still present (see Fig. 3.3).

The SIP method needs a stable walking behaviour with one inflection point while the person is stepping from one onto the other foot. This is not the case for low velocities so that the SIP method cannot smooth out the swaying of slow walking.

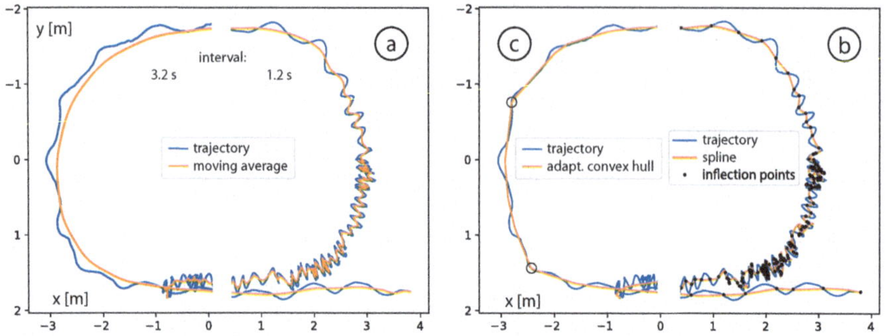

Fig. 3.4 Constraints of the three smoothing methods **a** CMA, **b** SIP, and c) MCH; **a** depending on the length of the interval (here 3.2 s and 1.2 s) the smoothed trajectory contracts or is unable to smooth out swaying; **b** for low velocities inflection points appear beside the one located between two steps; **c** angularity coming from piecewise straight parts of the envelope

The limit of the handling velocity is given in the original work with a minimum of 0.3 m/s.

The largest drawback for the MCH method is the angularity of the smoothed trajectory because of a composition of piecewise straight parts of the envelope so that also the smoothed trajectory is a polygonal chain, a connected series of line segments. The length of the interval has to be at least as long as one stride, but too long intervals can mask real changes of the MMD. The choice of the length is not as crucial as for the CMA method and should rather be chosen slightly larger than one stride.

For an quantitative evaluation of the methods following factors are examined on the full data of [14] with varying velocity regimes:

Distance between a smoothed and original trajectory should be limited and equally distributed on both sides of the trajectory. The maximum distance is most scattered for the SIP method and has an upper quartile of 0.23 m and a 98.5th percentile of 0.3 m. The upper quartile for the CMA and MCH method is 0.15 m and 0.13 m and the 98.5th percentile 0.17 m and 0.14 m respectively and fits to the observed swaying amplitude.

Length difference between smoothed trajectory and the direct path on a straight way should be near zero for single file movement and purposeful walking. The original trajectory shows a median deviation of 0.08 m and all smoothed results approximately 0.02 m, whereby the deviation of the SIP method is a bit more scattered, but the median is the lowest.

Velocity along the smoothed trajectory is an important variable, but as mentioned on page 4 there is no universal truth. For measuring the progress in MMD, for example utilized in the fundamental diagram, it is important to use smoothed trajectories as seen in Figs. 3.5 and 3.6. Otherwise the velocity is overestimated and the swaying has large influence. Despite a typical calculation of the velocity by using a time interval of one second the velocity along the original trajectory is much higher and more

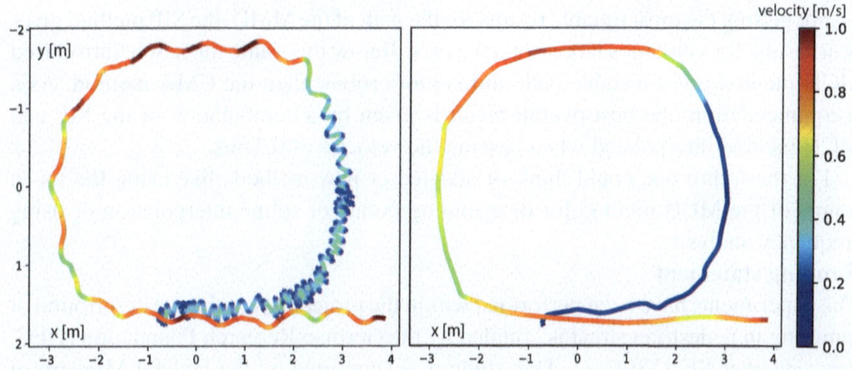

Fig. 3.5 Microscopic progression of the velocity along the original and smoothed trajectory (MCH method)

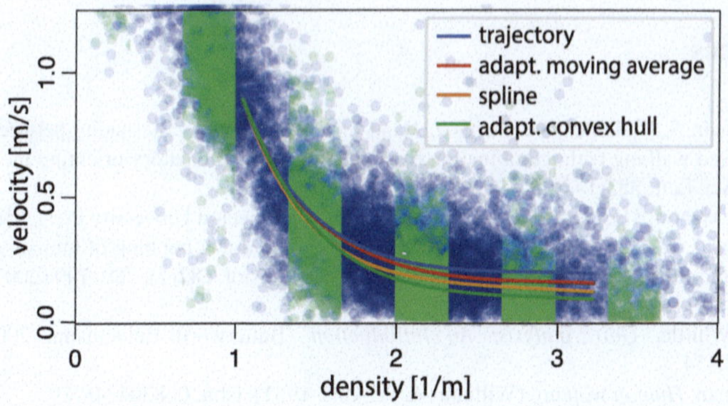

Fig. 3.6 Fundamental diagram as relation between velocity v and density ϕ: the curves are approximations by a cubic function for $\phi \in [1, 3.2]$; for the original (blue) and the trajectories coming from the MCH method (green) individual values are visualized as underlying stripes for better comparison of the distributions

scattered for every density value than for all smoothing methods. For example, the average velocity at a density of 2/m is 46% higher than the velocity in MMD using the smoothed trajectories of the MCH method. This relative error increases strongly for higher densities. The CMA and MCH method have the lowest average velocity.

3.4 Conclusion and Outlook

In summary, all smoothing methods are useful for determining the velocity in MMD. The smoothed trajectories should be the basis for the determination of mesoscopic data: average velocities, densities, and fluxes. The derivative of the smoothed trajectory allows a high resolution in time and is less affected by the swaying of heads.

For getting the most reliable results for the path of the MMD, the SIP method gives best results for velocities larger than 0.3 m/s. Below this value the newly introduced MCH method gives a stable path and is more robust than the CMA method. As a recommendation, the best overall result is given by a combination of the SIP and MCH method interpolated when passing the velocity of 0.3 m/s.

For the future one could think of adapted or new methods like using the touch points of the MCH method for determining points of spline interpolation or using frequency analysis.

Funding statement

The experiments have been performed within the project "Quantitative description of jamming in pedestrian streams" funded by the German Research Foundation (DFG, grant number SE 1789/1-1). This study was supported by the Federal Ministry of Education and Research (BMBF, grant number 13N13950 and 13N14533).

References

1. H. Hicheur, S. Vielledent, A. Berthoz, Head motion in humans alternating between straight and curved walking path: combination of stabilizing and anticipatory orienting mechanisms. Neuro-Sci. Lett. **383**(1–2), 87–92 (2005)
2. T.A. McMahon, *Muscles, Reflexes, and Locomotion* (Princeton University Press, 1984)
3. F. Massaad, T.M. Lejeune, C. Detrembleur. The up and down bobbing of human walking: a compromise between muscle work and efficiency, J. Physiol. **582**(2), 789–799 (2007). https://doi.org/10.1113/jphysiol.2007.127969.
4. M.W. Whittle, *Gait Analysis: An Introduction* (Butterworth-Heinemann, 2001). isbn: 0750600454
5. V.T. Inman, *Human walking* (Williams & Wilkins, 1981). isbn: 068304348X
6. D.W. Grieve, R.J. Gear, The relationships between length of stride, step frequency, time of swing and speed of walking for children and adults, in Ergonomics 9.5 (1966). PMID: 5976536, pp. 379–399. https://doi.org/10.1080/00140136608964399
7. J. Asja et al., Properties of pedestrians walking in line: fundamental diagrams, Phys. Rev. E E **85**(85), 9 (2012). https://doi.org/10.1103/PhysRevE.85.036111
8. P. Jacquelin, Gait Analysis—Normal and Pathological Function. Thorofare, NJ, USA: SLACK (1992). isbn: 978-1-556-42192-1
9. B. Steffen, A. Seyfried, M. Boltes, Reliability issues in the microscopic modeling of pedestrian movement, in Mathematical results in Quantum Physics, pp. 254–259 (2011)
10. M. Boltes et al., T-junction: Experiments, trajectory collection, and analysis, in IEEE International Conference on Computer Vision Workshop on Modeling, Simulation and Visual Analysis of Large Crowds, pp. 158–165 (2011). https://doi.org/10.1109/ICCVW.2011.6130238
11. M. Boltes et al., Influences of extraction techniques on the quality of measured quantities of pedestrian characteristics. Proc. Pedestrian Evacuat. Dyn. **2016**, 500–547 (2016)
12. T. Li et al., Joint smoothing, tracking, and forecasting based on continuous-time target trajectory fitting, in IEEE Transactions on Automation Science and Engineering PP (2017)
13. B. Steffen, A. Seyfried, Methods for measuring pedestrian density, flow, speed and direction with minimal scatter, Physica A **389**(9), pp. 1902–1910 (2010). https://doi.org/10.1016/j.physa.2009.12.015
14. University of Wuppertal and Forschungszentrum Jülich. Single-file motion of pupils (2014). https://doi.org/10.34735/PED.2014.2

15. M. Boltes, A. Seyfried, Collecting pedestrian trajectories, in Neurocomputing, Special Issue on Behaviours in Video 100, pp. 127–133 (2013). issn: 0925-2312. https://doi.org/10.1016/j.neucom.2012.01.036

Chapter 4
Influence of Small-Scale Obstacles on Passenger Flows in Railway Stations

Ernst Bosina, Jasmin Thurau, Tobias Oberholzer, and Stephan Heule

Abstract To improve the situation for passengers at crowded railway stations in a short time perspective, different measures were taken, and their effect was determined using state-of-the-art tracking sensors. In Lenzburg, standing benches were placed on one platform to encourage people to move away from the bottlenecks and thus improve the safety. In Bern railway station the main underpass was first equipped with yellow arrows and then with slightly inclined advertisement boards in order to encourage flow separation. It is shown that the standing benches attract passengers, part of whom were waiting before at the bottlenecks between the platform accesses and the platform edge. Also for the interventions in the underpass in Bern, a higher share of pedestrians walking in the main flow direction close to the center columns was observed. Nevertheless, the effects are relatively small compared to bigger improvements, but they can be useful for the time, until station upgrades are in place.

4.1 Introduction

4.1.1 Background and Objective

The railway network is the backbone of the well-known public transport system in Switzerland. To keep the system reliable and the planned connections on time, the railway stations and especially the pedestrian facilities therein need to be properly designed. Due to the integrated periodic timetable in Switzerland, trains arrive at the main nodes shortly before the node times and allow a passenger exchange between the train lines in a short time. This allows short travel times for connecting passenger but results in high pedestrian peaks within the stations, where short connection times are still required. In addition, pedestrians have to be kept from getting close to the

E. Bosina (✉) · J. Thurau · T. Oberholzer · S. Heule
SBB AG, 65, 3000 Bern, Switzerland
e-mail: Ernst.bosina@sbb.ch

I. Zuriguel et al. (eds.), *Traffic and Granular Flow 2019*,
Springer Proceedings in Physics 252,
https://doi.org/10.1007/978-3-030-55973-1_4

railway tracks for safety reasons and a comfortable walking and waiting situation shall be maintained.

As in most other countries, railway stations in Switzerland are often within a dense city environment. In addition, the continuous improvement of the public transport system and the population rise lead to a significant increase in demand, which is expected to continue in the future. Therefore, several railway platforms and underpasses are close to capacity or experience crowding. Completely rebuilding stations, as it is planned for several big stations, is costly and time consuming. Hence, other interventions are needed to improve the passenger flows on platforms and in underpasses in a short-term perspective.

In this work, two small scale interventions are presented, which shall improve the situation in existing stations in a short time perspective. In Chap. 2, an experiment is presented, where standing benches are placed on a platform to influence the waiting location of passengers. The second study focused on separating the pedestrian flow within the underpass in the railway station Bern (Chap. 3). First, yellow arrows were used to mark a part of the underpass for unidirectional flow. In a later attempt to influence the flow, advertisements boards in the center of the corridor where placed in a way that they also influence the flow separation.

To determine the effect of these interventions, stereo sensors were installed (Manufacturer: Xovis). These sensors allow to track pedestrian in real-life conditions with high accuracy while complying to data privacy regulations [1].

4.1.2 Related Research

Most pedestrians at the observed railway stations are expected to be regular commuters. Hence, they are aware of the specific situation and have their preferred walking paths and waiting locations. In addition, connection times are often short, and some commuters even aim at connections which are below the official connection times. Hence, they optimize for short distances and want to walk fast. For example, passengers tend to wait at their boarding platform at an area, where they can reduce the walking distance at their destination [2]. When choosing a waiting position, also preferred locations, such as close to walls or railings, can be identified [3]. This behavior often leads to uneven distribution of passengers at platforms and in underpasses. Hence, some locations might reach capacity, whereas others would still provide enough and comfortable space. In addition, some of these locations are safety critical bottlenecks at the platform, such as the space between the railings of the platform accesses and the platform edge. There, people might step in the danger zone close to the platform edge, which poses a safety risk. Previous research also showed that more pedestrians step into the danger zone at this location, hence it is desired to move people away from bottlenecks on the platform [4, 5].

Another aspect of pedestrian movement is the distribution of pedestrians within a corridor. In general, pedestrians keep a certain distance away from walls and obstacles and they show a tendency to walk on the right side [6–8]. To account for the wall distance, depending on the wall or obstacle type a certain distance is deducted from the total width [9, 10]. Without the need of external influences, lane formation can be observed, if not obstructed by external influences [11]. If lane formation is not possible, also the capacity is lower. A flow segregation is also beneficial and can theoretically also achieved using specifically shaped corridor walls [12]. For this reason, a layout, which facilitates lane formation or increases areas with unidirectional flow will improve the flow and walking speed.

4.2 Influence of Standing Benches on the Waiting Location

Lenzburg railway station is located on the main line between Zurich and Olten, showing a high number of passing long distance and freight trains at the tracks next to the island platform, where passengers wait for their trains. About 26,800 passengers are using the station every working day [13]. In addition, the platform is narrow, especially between the platform accesses and the platform edge. Therefore, several measures and test setups were made in order to improve the safety and comfort of the users.

One of these measures was the installation of standing benches at the platform (Fig. 4.1). They do not need much space, but they are expected to attract waiting passengers, hence leading to a better pedestrian distribution along the platform. The main goal was to reduce the number of waiting pedestrians at the bottlenecks and therefore the number of people walking in the danger zone next to the platform edge.

To determine the effect of the standing benches, the average density for the platform area, the area of the standing benches and the bottlenecks next to the accesses were determined before and after the installation of the standing benches (Fig. 4.2). Here, a clear increase in the pedestrian density in the area of the standing benches

Fig. 4.1 A standing bench next to a bottleneck at the railway station Lenzburg

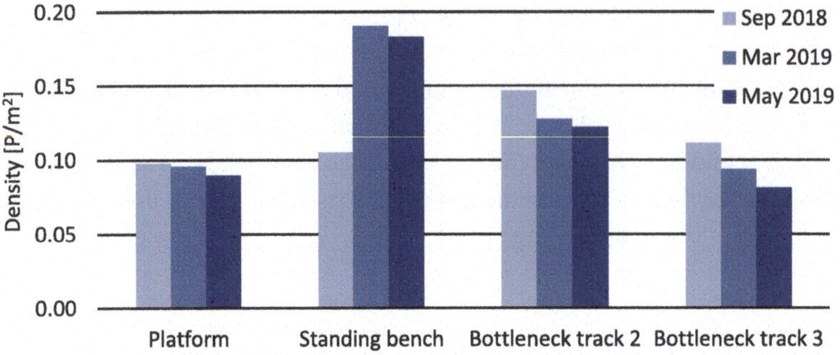

Fig. 4.2 Average density during peak hours at different locations on the island platform in Lenzburg. Platform indicates the total platform area covered by the sensors excluding the danger zone next to the tracks (540 m^2). The area covering the standing bench is 17 m^2 and the two bottleneck areas are about 32 m^2. In September 2018 the standing benches were not present yet

is visible. This contrasts with the bottleneck areas, where a density reduction is observed. Hence, the desired effect is visible, although only a reduction of 13–27% was reached.

4.3 Influencing the Passenger Flow

With about 206,400 passengers every working day, Bern is the second largest railway station in Switzerland and one of the main nodes of the integrated periodic timetable [13]. Especially during the rush-hour, a lot of people use the main underpass below the railway tracks to change between trains or to reach the city. Therefore, different settings were tested in the underpass in order to improve the passenger flow.

At first, yellow arrows were installed in the middle section next to the center columns, dividing the section in four different areas, two per direction (Fig. 4.3). On the outside, next to the platform accesses, no guidance is applied. The yellow arrows

Fig. 4.3 Pedestrian underpass in railway station Bern. Left: yellow arrows, Right: advertisement boards

in the middle shall encourage people to walk in the indicated direction without too much interference from crossing and opposing flow. Still, as platform entrances are present on both sides of the underpass, crossing flow is present, which is expected to reduce the effect of the yellow arrows.

In 2018 the yellow arrows were removed, and a new regime was introduced. Here, advertisement monitors were installed in the middle part of the underpass in an 30° angle to the passenger flow, facing the desired walking direction. Only the side wall of the monitor is visible in the opposite direction. The goal of this adaption is to segregate pedestrians walking in the middle of the underpass as it is assumed that pedestrians do not walk towards the side walls of the monitors. This arrangement is based on a theoretical study which showed theoretically that asymmetrically shaped walls can have a beneficial effect to the flow separation in a corridor [12].

In Fig. 4.4, the distribution of pedestrians along the cross-section of the underpass is shown. Here, the three phases, without measures, with yellow arrows and with advertisement boards, are distinguished. The flow distribution on the outside of the corridor remained quite stable, whereas in the middle, next to the columns, differences between the three phases can be seen. As a more unidirectional flow is desired, the situation with yellow arrows (green line) shows the best effect. But for the advertisement boards, an improvement is visible as well. Nevertheless, a direction separation also existed in the initial situation and the improvements are small. When considering the morning and evening rush-hour separately, generally a stronger direction separation is visible, with similar effects for the two measures.

The tracking data also allow to determine the speed distribution in the corridor. Here, higher walking speeds are observed for the main flows compared to the opposing traffic. These differences are about 0.1–0.4 m/s. Hence, walking in the

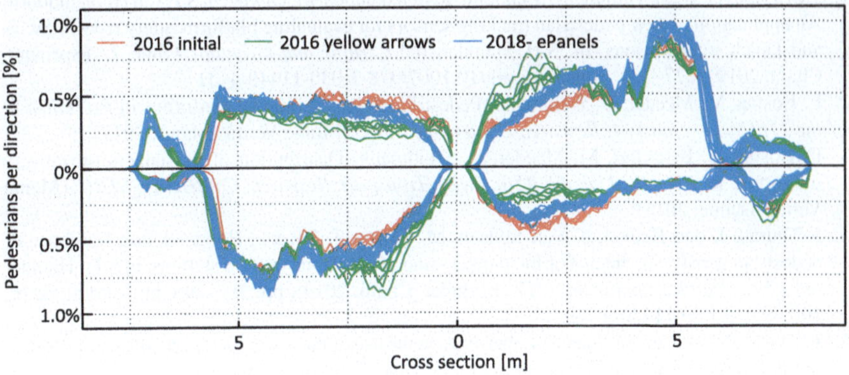

Fig. 4.4 Distribution of pedestrians at a cross-section in the pedestrian underpass in Bern. The lower part shows the direction towards the main hall (~45% of total), the upper part towards Grosse Schanze (~55% of total). The columns are the middle of the cross-section. At around 6.0 m, poles are located to separate people in front of the shops on the outside. A more unidirectional flow is shown by a higher share of pedestrians in the upper right and lower left quarter of the figure (right-hand traffic)

desired direction in the main flow is also beneficiary for the users. However, no differences were found between the three setups here.

The measurements also revealed that the pedestrian density is rarely higher than 0.5 P/m^2, hence the flow is clearly below capacity. Still, the situation is often experienced as crowded, which might result from the present multidirectional flow regime and the desire, to reach the next connection. Comparing the situations also showed no influence on the speed-density relation and therefore on the flow in the underpass.

4.4 Discussion

To improve the existing situation at train stations also in a short time perspective, small scale improvements were tested in two different locations. Both, the installation of standing benches in Lenzburg and the flow guidance in Bern showed a positive effect. As the impact of these measures is small compared to station upgrades, they cannot replace these works, but they can improve the situation until bigger measures are performed.

Real-life data measuring the impact of such interventions is still scarce and limited to a few locations. To be able to generalize the findings it is therefore important, to measure at other locations and to also determine the effect of similar improvements.

References

1. J. den Heuvel, J. van Thurau, M. Mendelin, R. Schakenbos, M. Ofwegen, S.P. van Hoogendoorn, An application of new pedestrian tracking sensors for evaluating platform safety risks at Swiss and Dutch train stations. In: S.H. Hamdar, (ed.) *Traffic and Granular Flow '17* (Springer, Cham, 2019), 277–286. https://doi.org/10.1007/978-3-030-11440-4_31
2. E. Bosina, M. Meeder, U. Weidmann, Pedestrian flows on railway platforms, in *Presented at the 17th Swiss Transport Research Conference STRC* (Monte Verità/Ascona, 2017)
3. E. Bosina, S. Britschgi, M. Meeder, U. Weidmann, Distribution of passengers on railway platforms, in *Presented at the 15th Swiss Transport Research Conference STRC* (Monte Verità/Ascona, 2015)
4. J. Thurau, J. den Heuvel, N. van Keusen, M. Ofwegen, S.P. van Hoogendoorn, Influence of pedestrian density on the use of the danger zone at platforms of train stations, in S.H. Hamdar (ed.) *Traffic and granular flow '17* (Springer, Cham, 2019), pp. 287–296. https://doi.org/10.1007/978-3-030-11440-4_32
5. J. Thurau, N. Keusen, Influence of obstacles on the use of the danger zone on railway platforms, in *9th International Conference on Pedestrian and Evacuation Dynamics (PED2018)* (Lund, in press)
6. E. Bosina, M. Meeder, B. Büchel, U. Weidmann, Avoiding walls: what distance do pedestrians keep from walls and obstacles?, in *Traffic and Granular Flow '15* (Springer, Cham, 2016), pp. 19–26. https://doi.org/10.1007/978-3-319-33482-0_3
7. T.A. Habicht, J.P. Braaksma, Effective width of pedestrian corridors. J. Transp. Eng. **110**, 80–93 (1984). https://doi.org/10.1061/(ASCE)0733-947X(1984)110:1(80)

8. F. Zanlungo, T. Ikeda, T. Kanda, A microscopic, "social norm" model to obtain realistic macroscopic velocity and density pedestrian distributions. PLoS ONE **7**, e50720 (2012). https://doi.org/10.1371/journal.pone.0050720
9. FGSV (ed.), Handbuch für die Bemessung von Strassenverkehrsanlagen (HBS 2015). Forschungsgesellschaft für Straßen- und Verkehrswesen FGSV, Köln (2015)
10. TRB (ed.), Highway capacity manual HCM 2010. Transportation Research Board. Washington, DC. (2010)
11. C. Feliciani, H. Murakami, K. Nishinari, A universal function for capacity of bidirectional pedestrian streams: filling the gaps in the literature. PLoS ONE **13**, e0208496 (2018). https://doi.org/10.1371/journal.pone.0208496
12. C.L.N. Oliveira, A.P. Vieira, D. Helbing, J.S. Andrade, H.J. Herrmann, Keep-left behavior induced by asymmetrically profiled walls. Phys. Rev. X **6**, 011003 (2016). https://doi.org/10.1103/PhysRevX.6.011003
13. SBB AG, Passagierfrequenz, https://data.sbb.ch/explore/dataset/passagierfrequen. Last accessed 21 Oct 2019

Chapter 5
Analysis of Pedestrian Motion Using Voronoi Diagrams in Complex Geometries

Mohcine Chraibi, Bernhard Steffen, and Antoine Tordeux

Abstract Voronoi diagrams are an established method in the analysis of pedestrian motion for constructing a density from two-dimensional positions. It is in turn used to give pointwise values for speed, movement direction, flow etc. The method was first described for high-density situations inside a crowd moving in a simple geometry without considering the influence of walls. However, more complicated distance calculations are needed for more complicated geometries where there are several obstacles or corners. In addition, partially empty spaces also require special treatment to avoid excessively big cells. These problems can lead to estimation errors when not handled correctly in subsequent use. In this work, we give details on how to adapt the calculations of Voronoi diagrams to make them fit for the presence of walls and obstacles in complex geometries. Furthermore, we show how that for persons at the edge of a group the personal space can be reasonably restricted. Based on these modifications, having pointwise values for quantities of interest allows to give average values for arbitrary geometries, not just for lines or rectangles of measurements. However, in order to obtain reasonable measurement values, different quantities may need different kind of averages—arithmetic or harmonic, or weighted with density.

M. Chraibi (✉) · B. Steffen
Civil Safety Research, Institute for Advanced Simulation, Forschungszentrum Jülich GmbH, Jülich, Germany
e-mail: m.chraibi@fz-juelich.de
URL: https://www.fz-juelich.de/ias/ias-7

B. Steffen
e-mail: b_steffen@live.de
URL: https://www.fz-juelich.de/ias/ias-7

A. Tordeux
Division for Traffic Safety and Reliability, University of Wuppertal, Wuppertal, Germany
e-mail: tordeux@uni-wuppertal.de

I. Zuriguel et al. (eds.), *Traffic and Granular Flow 2019*,
Springer Proceedings in Physics 252,
https://doi.org/10.1007/978-3-030-55973-1_5

5.1 Introduction

Only positions of persons can be measured directly, all other values are derived by more (density) or less (speed) involved computations. The density, quantifying the spatial proximity of the pedestrians, is in particular a fundamental variable to assess the safety of a crowd gathering and is considered a classic indicator for the level of service and safety of pedestrian facilities [1–4]. The density is also central when analysing the individual fundamental diagram, a characteristic phenomenological relationship linking the flow rate and the density [5, 6]. Different definitions for the density for pedestrians can be found in the literature. Some of them are based on kernels [6, 7], Voronoi diagram [8], Minkowski fractal dimension [9], features tracks (and the distinction between moving and static features) [7], or background removal and pattern recognition [10]. See [11–13] for a comprehensive comparison and survey on techniques used to estimate the density.

The use of Voronoi diagrams for the analysis of pedestrian density was first developed and described for high density situations in simple geometries without the influence of walls and inside a crowd. The resulting density distribution was used also to calculate averages of quantities of interest over certain regions. However, more complicated geometries need more complicated distance calculations. Partially empty spaces require some modifications, too; an additional mechanism of limiting the (indifinte) size of the Voronoi cells is needed. These problems are only sketched in the original paper [8] and therefore subsequently not always handled correctly. Similarly, the averaging process is not always a trivial choice to make. Depending on the quantity obtained, different averages have to be used for optimal results. In this paper we address these issues and propose new methodologies to solve them.

5.2 Voronoi Diagrams in the Presence of Obstacles

It is obvious that obstacles (walls, barriers, gates) do not belong to the personal space of the person and the Voronoi cell has to be cut with the obstacle. But the effect of an obstacle is possibly more elaborated than simply cutting off the Voronoi cell—the shape of the Voronoi cell changes, too. The distance between points in space in the presence of obstacles should not be the standard Euclidian distance, but the length of the shortest possible path. For an example let us consider the dividing line for the cells of two persons, one standing at $(-1, 0)$ and the other at $(0, -2)$ with a wall between them from (0,0) to $(-100, -100)$, with a cutoff of 4 (see Fig. 5.1). The normal dividing line $(-0.5, -1) + \alpha(1, 2)$ is correct only for $y \geq 0$ (above the end of the wall), for $y \leq 0$ the division line due to the wall is given by the equation

$$\sqrt{x^2 + y^2} + 1 = \sqrt{x^2 + (y + 2)^2},$$

which gives a hyperbola

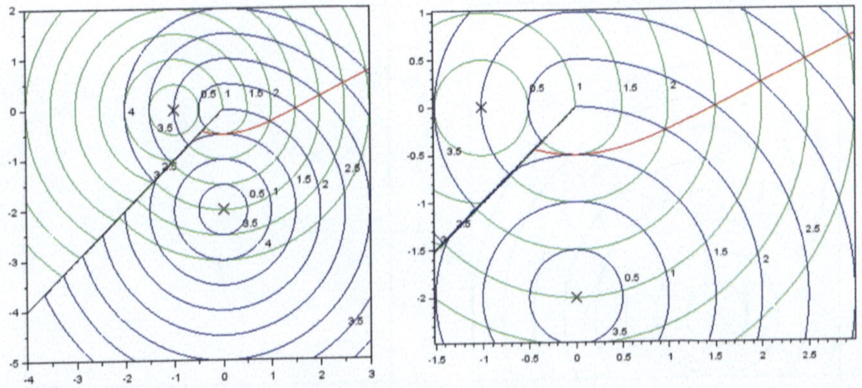

Fig. 5.1 Calculation of Voronoi diagram with an obstacle. Left: all area; right: center part. The black curve is the obstacle while the curve is the dividing line. The green and blue curves are respectively the equidistance lines for the pedestrian located at (−1,0) and the pedestrian located at (0,−2)

$$x = \sqrt{3y^2 + 6y + 2.25},$$

and, above of the wall, all space belongs to the first person or to no one.

This is still a fairly simple situation to compute, where the distance from point 1 is either the Euclidian distance or the Euclidian distance from the endpoint of the wall plus the distance from the agents to this endpoint. In more complicated situations—multi corner, rounded or multiple obstacles—the dividing line may be made up of many pieces of analytic functions or require numerical calculation. This exact dividing line is much more complicated for computation than the line based on Euclidean distance. Fortunately, this will usually be needed only for very few cells. As people have a non-zero extend, it seems appropriate not to allow a path to touch an obstacle, but keep some distance, for instance half shoulder width. In the examples used in this work we set this minimal distance to 0 cm and 15 cm, although the proper value depends on many factors. See also Fig. 5.2.

5.3 Cutoff and Construction of a Density Distribution

Basic proxemics rules show that the personal space should not extend too far out, therefore some kind of a cutoff is necessary to limit the size of the cells. The simplest case is a constant cutoff, e.g. cutting all the cells at 0.5 m from the person. Unfortunately, this distance is variable and depends on the actual situation and density. While 0.5 m is probably sufficient for an evacuation situation with an homogeneous density, it may be too short for a relaxed waiting situation, where 1 m for instance will be more appropriate. A reasonable adaptive cutoff would be cutting at the min-

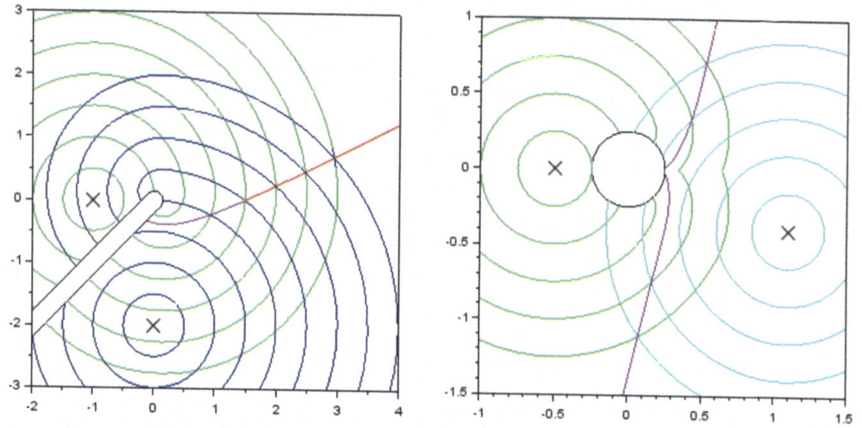

Fig. 5.2 Examples of obstacles with 15 cm distance required. Left: wall from Fig. 5.1; right: vertical pole. See Fig. 5.1 for the legends

imal distance to another person, maximal at 1 m. If the cutoff is required to be the same for all persons, the 25% quantile or the mean of the minimal distance may be reasonable.

It is possible to use a "soft" cutoff, e.g a linear drop of density at the distance 0.4 to 0.6 m, or even a differentiable density, but this leads to complicated computations with little benefit. The cutoff proposed in [14]—restricting the whole space to the convex hull of the persons and additionally for each cell extending beyond the convex hull cutting out the triangle made of the position of the cells person and the two intersection points with the convex hull—is not systematically applicable. A density distribution can also be constructed by attaching a standard density kernel to every person. The Gaussian is the common choice, but others are possible. These functions have the same problems with obstacles, and the choice of the size of the kernel needs the same considerations as the cutoff choice, but is needed everywhere and not just for the edge of a group. A possible combination would be choosing Gaussians with σ such that the corresponding circle around a person has the same area as the Voronoi cell [13].

5.4 Averaging

The density function allows to define pointwise other quantities of interest for the entire area. These are for instance velocity $\mathbf{v}$, speed $v = ||\mathbf{v}||$, flux $J = \mathbf{v} \cdot \rho$. While this can lead over time to a rich data set, for most purposes, a drastic reduction in the amount of data is necessary. This is usually achieved by averaging the obtained data over time or space. In stationary situations , e.g. the neighbourhood of a bottleneck for most of the evacuation time, averages over time [15] can give useful insights into the system. Similarly, averages over space or space and time may contain most of the important information of an experiment. However, there are pitfalls. It is necessary that the averages to allow the same algebra as the initial quantities, at least approximately. This means that the average speed times average density should be average flow, length divided by speed should be time needed (space mean speed, see [16]), etc. In the presence of large fluctuations, this requires care in the choice of the average.

The simplest problem is averaging density, the formulas for average over space D, time $T_2 - T_1$, and space and time are

$$\frac{1}{|D|}\int \rho(x,t)dx, \quad \frac{1}{T_2 - T_1}\int \rho(x,t)dt, \quad \frac{1}{|D|}\frac{1}{T_2 - T_1}\int \rho(x,t)dxdt.$$

Some quantities—like speed, direction are defined only where there is a person, and undefined where $\rho(x) = 0$. Here two different averages are possible, e.g. for speed, over the area occupied (5.1) or over (partial) persons (5.2):

$$\frac{\int v(x)\chi(\rho(x))dx}{\int \chi(\rho(x))dx} \tag{5.1}$$

$$\frac{\int v(x)\rho(x)dx}{\int \rho(x)dx} \tag{5.2}$$

Here $\chi(\cdot)$ is the characteristic function. These averages differ, but both may be used in different contexts (5.1) may be used to predict behaviour if the speed is mostly determined by the floor (e.g. uphill, rough, slippery), (5.2) shows the movement of groups. Using

$$\frac{1}{|D|}\int |v(x)|\chi(\rho(x))dx$$

for an average means setting the undefined speed to zero and is definitely not reasonable. Here the average speed depends on the size of the empty area.

The velocity is important for the calculation of two inherently different quantities—the flux and the required passing time. These two uses require different averages. The average speed along a path is the harmonic average of the local speed (see (5.3))

$$\frac{\int ds}{\int ds/\mathbf{v(s)}} \tag{5.3}$$

$$\frac{\int <\mathbf{v(s)}, \mathbf{n(s)}> ds}{\int ds} \tag{5.4}$$

while the average speed across a line is the arithmetic average of the normal component (5.4). If (5.3) is useful to predict timings, (5.4) gives fluxes across lines.

There is a general problem with the algebra of averages—the average of a product of two values is the product of the averages only if the values are uncorrelated, a condition that is definitely not present for the quantities of interest in pedestrian dynamics unless the dynamics can be mainly described in steady state.

References

1. D. Oeding, Verkehrsbelastung und dimensionierung von gehwegen und anderen anlagen des fussgängerverkehrs. Forschungsbericht 22, Technische Hochschule Braunschweig, in German (1963)
2. J.J. Fruin, *Pedestrian Planning and Design* (Elevator World, 1971)
3. V.M. Predtechenskii, A.I. Milinskii, Planning for foot traffic flow in buildings. Amerind Publishing, New Delhi, *Translation of: Proekttirovanie Zhdanii s Uchetom Organizatsii Dvizheniya Lyuddskikh Potokov* (Stroiizdat Publishers, Moscow, 1978), p. 1969
4. U. Weidmann. Transporttechnik der Fussgänger. In *Schriftenreihe des IVT*, volume 90. Institut für Verkehrsplannung, Transporttechnik, Strassen- und Eisenbahnbau (1993)
5. A. Seyfried, B. Steffen, W. Klingsch, M. Boltes, The fundamental diagram of pedestrian movement revisited. J. Statist. Mech. Theory Exp. **2005**(10), P10002 (2005)
6. D. Helbing, A. Johansson, H.Z. Al-Abideen, Dynamics of crowd disasters: an empirical study. Phys. Rev. E **75**, 046109 (2007)
7. H. Fradi, J.-L. Dugelay, Crowd density map estimation based on feature tracks, in MMSP, 15th International Workshop on Multimedia Signal Processing, September 30-October 2, 2013 (Pula, Italy, Pula, ITALY, 2013), p. 2013
8. B. Steffen, A. Seyfried, Methods for measuring pedestrian density, flow, speed and direction with minimal scatter. Physica A: Statist. Mech. Appl. **389**(9), 1902–1910 (2010)
9. A.N. Marana, L. da Fontoura Costa, R.A Lotufo, S.A Velastin, Estimating crowd density with minkowski fractal dimension, in Acoustics, Speech, and Signal Processing. Proceedings IEEE International Conference, vol. 6, pp. 3521–3524 (1999)
10. M. Jiang, J. Huang, X. Wang, J. Tang, C. Wu, An approach for crowd density and crowd size estimation. J. Softw. **9**(3), 757–762 (2014)
11. J. Zhang, W. Klingsch, A. Schadschneider, A. Seyfried, Transitions in pedestrian fundamental diagrams of straight corridors and t-junctions. J. Statist. Mech. Theory Exp. **2011**(06), P06004 (2011)
12. D. Duives, W. Daamen, S. Hoogendoorn, Quantification of the level of crowdedness for pedestrian movements. Physica A: Statist. Mech. Appl. **427**, 162–180 (2015)
13. A. Tordeux, J. Zhang, B. Steffen, A. Seyfried, Quantitative comparison of estimations for the density within pedestrian streams. J. Statist. Mech. Theory Exp. **2015**(6), P06030 (2015)
14. A. Nicolas, M. Kuperman, S. Ibaez, S. Bouzat, C. Appert-Rolland, Mechanical response of dense pedestrian crowds to the crossing of intruders. Scient. Reports **9**, 105 (2019)
15. J. Liddle, A. Seyfried, B. Steffen, Analysis of bottleneck motion using voronoi diagrams, in R.D. Peacock, E.D. Kuligowski, J.D. Averill (eds.) *Pedestrian and Evacuation Dynamics* (Springer US, Boston, MA. 2011) , pp. 833–836
16. J.H. Wardrop, Some theoretical aspects of road traffic research. Proc. Inst. Civil Eng. **1**(3), 325–362 (1952)

Chapter 6
The Trouble with 2nd Order Models or How to Generate Stop-and-Go Traffic in a 1st Order Model

Jakob Cordes, Andreas Schadschneider, and Antoine Tordeux

Abstract Classical second order models of pedestrian dynamics, like the social-force model, suffer from various unrealistic behaviors in the dynamics, e.g. backward motion, oscillations and overlapping of pedestrians. These effects are intrinsic to the dynamics and the consequence of strong inertia effects that usually appear in second order models. The experimentally observed stop-and-go behavior, which is an important test for any pedestrian model, can be reproduced with a stochastic first order model that does not suffer from the dynamical artefacts resulting from strong inertia. The model provides a new mechanism for stop-and-go behavior which is based on correlated noise. We show that this mechanism exists in a large class of models.

6.1 Introduction

In recent years, growing evidence suggests that models of pedestrian dynamics based on second order differential equations suffer from intrinsic problems that are not related to numerical issues, e.g. insufficient discretization of the differential equations. In many second order models, unrealistic behaviors like unmotivated backward motion, oscillations of the direction of motion or overlapping for pedestrian and other traffic models have been observed [1–3].

In [4], the influence of the type of interaction force on the dynamics has been studied in a one-dimensional single-file scenario. It has been found that unrealistic behavior is related to instability phenomena and stop-and-go waves in these

J. Cordes (✉) · A. Schadschneider
Institut für Theoretische Physik, Universität zu Köln, Cologne, Germany
e-mail: jcordes@thp.uni-koeln.de

A. Schadschneider
e-mail: as@thp.uni-koeln.de

A. Tordeux
Institut für Sicherheitstechnik, Bergische Universität Wuppertal, Wuppertal, Germany
e-mail: tordeux@uni-wuppertal.de

I. Zuriguel et al. (eds.), *Traffic and Granular Flow 2019*,
Springer Proceedings in Physics 252,
https://doi.org/10.1007/978-3-030-55973-1_6

models. However, since stop-and-go waves have been observed in experiments with pedestrians [5–7] their reproduction is a benchmark test for any model of pedestrian dynamics.

Here, we will address this problem from a slightly different perspective. Instead of considering the influence of the specific force used, we will argue that the inertia term in the second order model is responsible for many unrealistic behaviors observed. We propose a minimal first order model, i.e. a model which has no inertia term related to the physical mass, which is able to reproduce the basic properties of stop-and-go waves observed in single-file experiments with pedestrians. The mechanism for the formation of these waves is different than the classical mechanism which is based on the fact that the homogeneous solution becomes unstable. In our model, it is stable for all densities and the stop-and-go waves are triggered by an additive *correlated* noise. Here we show that this model is a special case of a larger class of models and the mechanism is rather general. It can be observed in other models as well.

6.2 The Model

The Optimal Velocity (OV) models are a class of models set by an optimal velocity function $V(\cdot)$ which typically depends on the headway Δx. The simplest OV model is the first order model

$$\dot{x}_n(t) = V(\Delta x_n(t)). \tag{6.1}$$

In order to incorporate more realistic behaviour a reaction time τ_R can be added as a delay [8]

$$\dot{x}_n(t + \tau_R) = V(\Delta x_n(t)). \tag{6.2}$$

A Taylor expansion on the left hand side gives the second order model proposed by Bando et al. [9] with the sensitiviy $a = \frac{1}{\tau_R}$.

Another important process in traffic is anticipation. Agents extrapolate the current situation in order to reduce the reaction time [10]. This helps to avoid collisions and allows a smoother and faster flow. Therefore an anticipation time τ_A is added on the right hand side

$$\dot{x}_n(t + \tau_R) = V(\Delta x_n(t + \tau_A)). \tag{6.3}$$

Obviously (6.3) can be brought into the form of (6.2) with $\tau_R \to \tau_R - \tau_A$ by shifting the time. A specific form of the Full-Velocity-Difference (FVD) model, which has been proposed in [11], can be derived when Taylor expansions are performed independently in τ_A and τ_R:

$$\ddot{x}_n(t) = \frac{1}{\tau_R}\Big[V(\Delta x_n(t)) - \dot{x}_n(t)\Big] + \frac{\tau_A}{\tau_R}\Delta \dot{x}_n(t) V'(\Delta x_n(t)). \tag{6.4}$$

This is a second order model which does not have an inertia term related to the physical mass m. Instead it has an effective inertia which is determined by the two times τ_A and τ_R. The models (6.3) and (6.4) have the same stability condition

$$\tau_R - \tau_A < \frac{T}{2}, \tag{6.5}$$

where a linear OV function $V(d) = (d - l)/T$ with a desired time gap T and a size l of the agents has been assumed for congested states [11, 12]. The case $\tau_A > \tau_R$ is unrealistic because it corresponds to a motion which chooses the velocity optimal according to the situation in the future. $\tau_R = \tau_A$ corresponds to a full compensation of the reaction time. If, in this case, (6.4) is combined with a white noise $\alpha\xi_n(t)$ one gets

$$\begin{aligned} \dot{x}_n(t) &= V(\Delta x_n(t)) + \epsilon_n(t); \\ \dot{\epsilon}_n(t) &= -\frac{1}{\tau}\epsilon_n(t) + \alpha\xi_n(t), \end{aligned} \tag{6.6}$$

where the model has been rewritten as a first order OV model with a correlated truncated Brownian noise $\epsilon_n(t)$ described by the Orstein-Uhlenbeck process [13]. Equation (6.6) is the form in which the model has been proposed in [14]. In accordance with condition (6.5) it is always linearly stochastically stable, since the deterministic model is intrinsically stable while the noise is additive and independent of the agent states. Note that (6.6) is a special case of (6.4) but still a genuine first order equation in the position variables x_n since the second equation is the definition of the noise that does not involve the variables $x_n(t)$.

6.3 Simulations

The model (6.6) is analyzed numerically using the explicit Euler-Maruyama scheme with a time step $\delta t = 0.01$ s. The parameter values are $T = 1.02$ s, $l = 0.34$ m, $a = 0.09$ ms$^{-3/2}$ and $\tau = 4.4$ s, which are the statistical estimates in [14] with the data of [5, 15]. The linear OV function stated above is used. According to the corresponding experimental situation, the length is $L = 27$ m, the boundary conditions are periodic, and simulations are carried out with $N = 28, 45, 62$ pedestrians. For small densities a homogeneous free flow state is observed ($N = 28$), while stop-and-go waves appear at medium densities ($N = 45$, $N = 62$). The comparison with the empirical trajectories shows good agreement (Fig. 6.1).

The autocorrelation of the spacing for 50 agents for different noise parameters τ and α but a constant variance $\sigma = \alpha\sqrt{\tau/2}$ is shown in Fig. 6.2. The system can be supposed to be in the stationary state (simulation time $t > t_s = 2 \cdot 10^5$ s). The period of the autocorrelation remains the same and stop-and-go behaviour is maintained in all cases but is less pronounced for smaller τ, i.e. less correlated noise. For a more detailed analysis of the model we refer to [14].

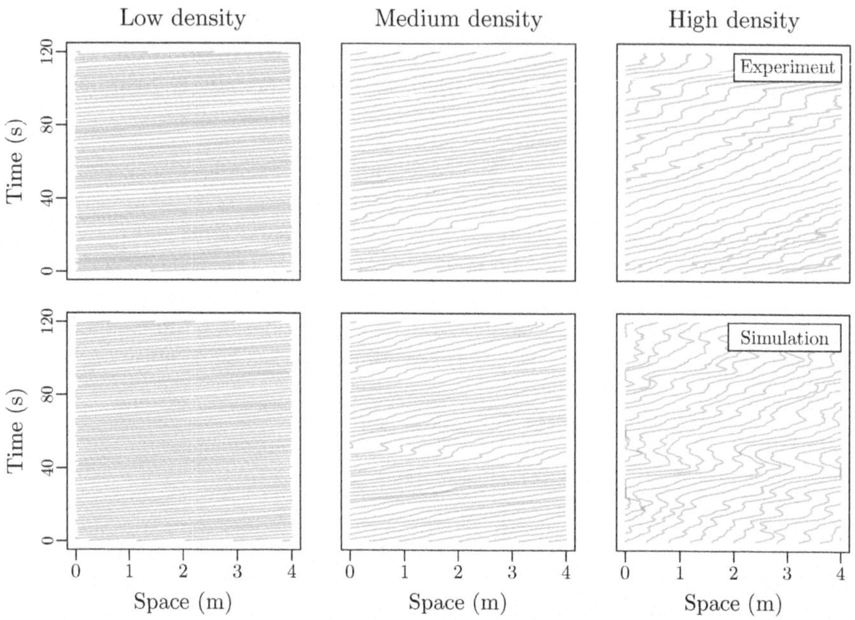

Fig. 6.1 Empirical (top panels) and simulated (bottom panels) trajectories for different densities ($N = 28, 45, 62$ from left to right). The initial configuration is homogeneous, both in the experiments and the simulations

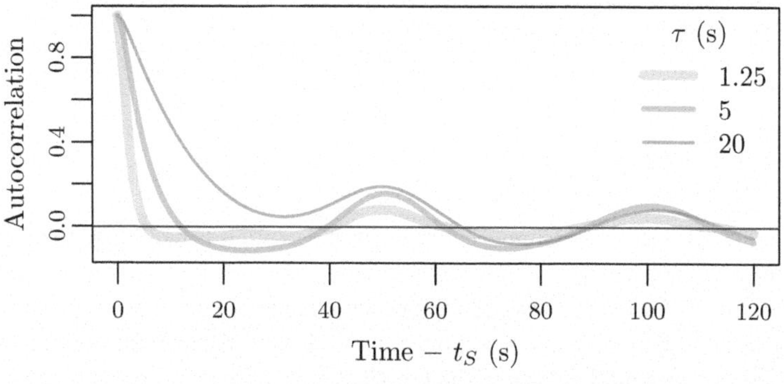

Fig. 6.2 Mean temporal correlation function of the distance spacing in the stationary state of (6.6) for different values of the noise parameter τ. Parameter α is chosen such that the noise amplitude is the same in all cases. The noise parameters do not influence the frequency of the waves, which only depends on N and T

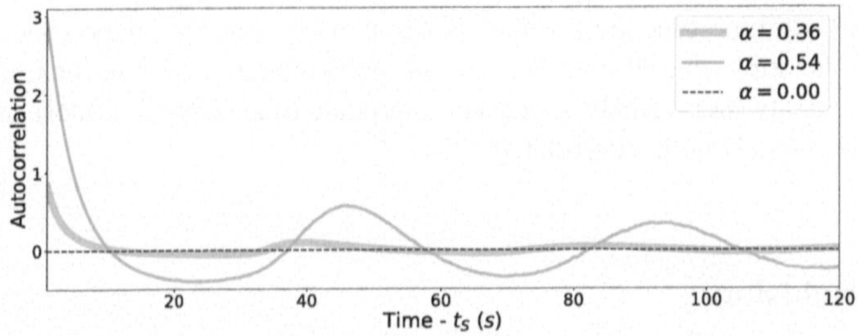

Fig. 6.3 Mean temporal correlation function of the distance spacing in stationary states of the stochastic Gipps model for different values of the noise amplitude α. The oscillations of the autocorrelation arise when a noise is added and get more pronounced with increasing amplitude α. The corresponding parameters are $a = 1.7\frac{m}{s^2}$, $\tau = \frac{2}{3}$ s, $v_{\max} = 30\frac{m}{s}$, $b = 3.0\frac{m}{s^2}$, $\hat{b} = 2.9\frac{m}{s^2}$, $l = 6.5$ m with 50 cars on a ring of length 1000 m

The point to stress is the different mechanism of reproducing stop-and-go behaviour. In most continuous models (e.g. social-force model, deterministic OV models), the formation of a heterogeneous configuration is the consequence of a deterministic instability and can be described as a phase transition. In contrast, in model (6.6) the stop-and-go waves are part of the stationary state of the stochastic system. The system is permanently kicked out of the deterministically stable homogeneous configuration by the noise. This can also be reproduced in the delayed first order OV model, the second order OV model and the FVD model if a white noise is added in the deterministic stable regime. It can be understood in the light of the connections between the models investigated above.

To explore this mechanism further, the effect of noise on a different traffic model, namely the Gipps model, is investigated. In contrast to the other models presented here, Gipps' model is time-discrete and mainly considered for vehicular traffic. The update time τ is also interpreted as the reaction time. It has a maximum acceleration a, maximum deceleration b and an estimated braking capability of the preceding car $\hat{b}$, according to these the cars choose their velocity. The stability of the homogeneous configuration of cars has been investigated by Wilson [16] and strongly depends on the underestimation of the braking ability of the preceding car $\hat{b}$.

Simulations are carried out in the deterministically stable regime with periodic boundary conditions. A white noise with amplitude α is added to the velocity. The system can be assumed to be in a stationary state after waiting for $t_s = 2 \cdot 10^5$ s. In Fig. 6.3 the autocorrelation of the headway for different noise amplitudes is shown. The autocorrelation begins to oscillate when a noise is added. The oscillations are getting more pronounced when α is increased. The stable homogeneous configuration is destabilized by the noise and oscillating behaviour is observed in the stationary state.

Similar observations have been made in the social-force model as defined in [17] when reduced to one dimension. However, it is not reproducible in the first order OV model with a white noise [14]. These findings indicate that the described mechanism is not specific but rather generic for continuous models. It can be reproduced if the model exhibits some inertia-like effect, such as a (inertial) mass (Social-Force model), a relaxation time (Bando-Hasebe and FVD model), or a reaction time, incorporated as a delay (delayed OV model) or an update time (Gipps' model), as well as in first order models with a correlated noise.

6.4 Conclusion

In continuous second order pedestrian models, stop-and-go behaviour is produced if the inertia-related quantity is sufficiently large. However, this seems to lead generically to intrinsic problems, like oscillations, which only can be avoided by choosing fine-tuned parameters values. This has been already indicated by earlier work [4, 18, 19]. The results presented here strongly support this view and give new insights into the mechanism behind this behavior. Generically, the resulting effective inertia in these models is too strong and drives the model into a regime of damped oscillations instead of the overdamped region. We have proposed an alternative approach based on a first order model with a correlated noise that allows to overcome these problems while choosing realistic parameter values. The underlying mechanism seems to be generic and allows to overcome these problems for other models as well.

Acknowledgements Financial support by German Science Foundation (DFG) under grant SCHA 636/9-1 and Bonn-Cologne Graduate School of Physics and Astronomy (BCGS) and the German Excellence Initiative through the University of Cologne Forum "Classical and Quantum Dynamics of Interacting Particle Systems" is gratefully acknowledged.

References

1. L. Davis, Modifications of the optimal velocity traffic model to include delay due to driver reaction time. Physica A **319**, 557–567 (2003)
2. R.E. Wilson, P. Berg, S. Hooper, G. Lunt, Many-neighbour interaction and non-locality in traffic models. Eur. Phys. J. B **39**, 397–408 (2004)
3. D. Helbing, A.F. Johansson, On the controversy around Daganzo's requiem for and Aw-Rascle's resurrection of second-order traffic flow models. Eur. Phys. J. B **69**, 549–562 (2009)
4. M. Chraibi et al., Jamming transitions in force-based models for pedestrian dynamics. Phys. Rev. E **92**, 042809 (2015)
5. A. Portz, A. Seyfried, Analyzing stop-and-go waves by experiment and modeling, in R.D. Peacock, E.D. Kuligowski, J.D. Averill (eds.) *Pedestrian and Evacuation Dynamics* (Springer, 2011), pp. 577–586
6. J. Zhang et al., Universal flow-density relation of single-file bicycle, pedestrian and car motion. Phys. Lett. A **378**, 3274 (2014)
7. M. Boltes, J. Zhang, A. Tordeux, A. Schadschneider, A. Seyfried, Empirical results of pedestrian and evacuation dynamics, 671–699 (2019)
8. G.F. Newell, Nonlinear effects in the dynamics of car following. Oper. Res. **9**, 209 (1961)
9. M. Bando, K. Hasebe, A. Nakayama, A. Shibata, Y. Sugiyama, Dynamical model of traffic congestion and numerical simulation. Phys. Rev. E **51**, 1035 (1995)

10. A. Schadschneider, D. Chowdhury, K. Nishinari, *Stochastic Transport in Complex Systems* (Elsevier, 2011)
11. R. Jiang, Q. Wu, Z. Zhu, Full velocity difference model for a car-following theory. Phys. Rev. E **64**, 017101 (2001)
12. A. Tordeux, M. Roussignol, S. Lassarre, Linear stability analysis of first-order delayed car-following models on a ring. Phys. Rev. E **86**, 036207 (2012)
13. G.E. Uhlenbeck, L.S. Ornstein, On the theory of the Brownian motion. Phys. Rev. **36**, 823–841 (1930)
14. A. Tordeux, A. Schadschneider, White and relaxed noises in optimal velocity models for pedestrian flow with stop-and-go waves. J. Phys. A. **49**, 185101 (2016)
15. Forschungszentrum Jülich and Wuppertal University. www.asim.uni-wuppertal.de/datenbank
16. R.E. Wilson, An analysis of Gipps' car-following model of highway traffic. IMA J. Appl. Math. **66**, 509–537 (2001)
17. D. Helbing, I. Farkas, T. Vicsek, Simulating dynamical features of escape panic. Nature **407**, 487–490 (2000)
18. T.I. Lakoba, N.M. Finkelstein, Modifications of the Helbing-Molnar-Farkas-Vicsek social force model for pedestrian evolution. Simulation **61**, 339 (2005)
19. G. Köster, F. Treml, M. Gödel, Avoiding numerical pitfalls in social force models. Phys. Rev. E **87**, 063305 (2013)

Chapter 7
The Impact of Walking Speed Heterogeneity on the Pedestrian Fundamental Diagram

Dorine C. Duives, Martijn Sparnaaij, and Serge P. Hoogendoorn

Abstract Most works featuring the capacity of pedestrian infrastructures have focussed on unidirectional movement base cases, studied relatively low-density situations and instructed their participants to behave 'normal'. However, during large crowd movements at train stations and events often far higher densities are encountered, the flow situations are more complex, and pedestrians do not always behave 'normal'. The objective of this study is to determine the impact of heterogeneity in walking speed and the flow situation on the global and local dynamics of the crowd. A large pedestrian experiment was performed by Delft University of Technology, coined CrowdLimits, to derive the answer to this question. A preliminary analysis of the participants' movements illustrates that especially the introduction of differences in walking speed and distribution between flows influence the shape fundamental diagram.

7.1 Introduction

The fundamental diagram is often used as an input for the design of pedestrian spaces, as it provides information regarding the capacity of corridors, the unfolding of pedestrian flow scenarios and the jam density. The fundamental diagram has been studied a lot in the last decade. Influential works include the works of [1–6]. An overview by Vanumu et al. [7] of fundamental diagrams presented in the last two decades illustrates that most research considering the pedestrian fundamental diagram features densities up to 4.5 m^2, unidirectional or bidirectional movements and student populations. Consequently, the exact shape of the pedestrian fundamental diagram for higher densities and more complex movement base cases is still uncertain. Moreover, the impact of confounding factors, such as population heterogeneity and goal-orientation, is undetermined.

D. C. Duives (✉) · M. Sparnaaij · S. P. Hoogendoorn
Transport and Planning, Delft University of Technology, 2628 CN Delft, The Netherlands
e-mail: d.c.duives@tudelft.nl

I. Zuriguel et al. (eds.), *Traffic and Granular Flow 2019*,
Springer Proceedings in Physics 252,
https://doi.org/10.1007/978-3-030-55973-1_7

The objective of this study is to determine to the impact of several factors on the fundamental diagram that have not yet been studied thoroughly, namely the flow ratio and speed heterogeneity. A worldwide unique large-scale pedestrian experiment, coined CrowdLimits, has been performed to derive the extent of these impacts.

The remainder of this paper first describes the research methodology in Sect. 7.2. Accordingly, the CrowdLimits experiment is presented in Sect. 7.3. The results with respect of the heterogeneity of goal-orientation and flow situation are presented in Sect. 7.4. This chapter closes with a set of preliminary conclusions and some suggestions for future research considering this topic.

7.2 Research Methodology

This section describes the research methodology. First, Sect. 7.2.1 introduces the CrowdLimits experiment. Accordingly, Sect. 7.2.2 presents the setup of the CrowdLimits experiment. The mathematical definition of the macroscopic flow variables are mentioned in Sect. 7.2.3.

7.2.1 The CrowdLimits Experiment

The CrowdLimits experiment's objective was to study crowd movement dynamics of heterogeneous crowds under higher densities than have been recorded up to this moment and determine the effect of several confounding factors on crowd movement dynamics. Here, this experiment attempted to determine the impact of the three major differences between the current laboratory studies and real-world crowd dynamics in crowded pedestrian spaces simultaneously, namely population heterogeneity, high densities movements and (more) complex movement base cases. Moreover, the impact of two confounding factors were established, namely speed heterogeneity and the flow ratio.

Three variables were operationalised using a set of input parameters that could be varied one-by-one during a set of experimental runs, namely the movement base case, the flow ratio and a set of assignments to vary the speed heterogeneity. Crowd heterogeneity and demand level dynamically varied within each experimental run with time, and as such, were not included in the set of input parameters.

7.2.2 Set-up of the CrowdLimits Experiment

The CrowdLimits experiment studies two movement base cases that often occur at crowded pedestrian spaces, namely bidirectional and intersecting flows under a 90° angle. Figure 7.1 illustrates the layout of the first movement base case, the results of

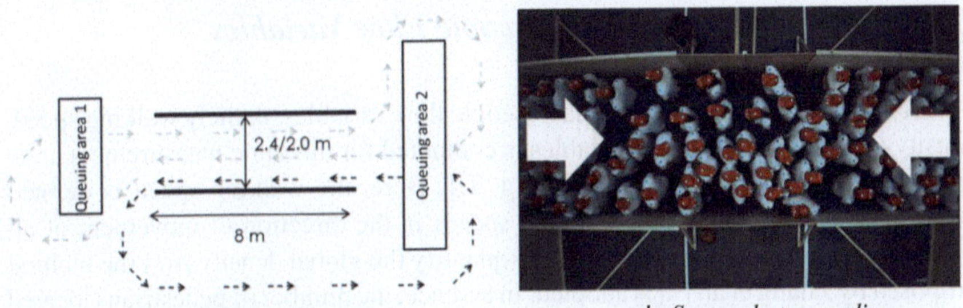

a. Walking directions b. Screenshot recording

Fig. 7.1 Overview setup bidirectional scenario and a screenshot thereof

which will be further analysed in this paper. This movement base case was structured using four movable wooden L-shaped 2 m high wooden panels, which jointly formed a long corridor of 2.4 or 1.8 m wide and 8 m long.

In each run of the experiment, the flow distribution over the entrances was varied. The entrance flow was either distributed 50–50 over both entrances or split in a major flow (80%) and a minor flow (20%). These two particular cases are adopted because the work of Kretz et al. [8] illustrates that these two flow ratio scenarios are the most advantageous (50–50) and disadvantageous (80–20) flow ratio scenarios with respect to the resulting capacity.

Each of the participants received an assignment at registration. Three distinct assignments were handed out: (A) *No assignment*, (B) *Crossing*-designed to decrease the predictability of the crowd movements, (C) *Speeding up*—aimed to increase goal-orientation and thus pressure differences in the crowd. The assignments aimed to increase the walking speed and decrease the predictability of the movement behaviour of a part of the participants. As such, these assignments stimulated the emergence of more 'chaotic' crowd movement dynamics, which increased speed heterogeneity.

In total 24 runs were performed across two days. In this study, the results of 6 runs are analysed, namely (1) 50–50 without assignment; (2) 80–20 without assignment; (3) 50–50 with crossing assignment; (4) 80–20 with crossing assignment; (5) 50–50 with increased speed; (6) 80–20 with increased speed.

During each of the runs, participants were asked to line up in one of the queueing areas. When the traffic light turned green, the first line of participants is asked to start moving through the infrastructure and continue their movement until the end of the run. Over time, the demand within the infrastructure was increases in steps until the maximum flow rate was reached after 5 minutes. Every time, the crowd movements were allowed to stabilize before the additional participants were introduced into each run.

The participant population forms a heterogeneous crowd, in contrast to many other laboratory experiments performed in recent years. That is, the gender distribution is almost 50% and the age of the 139 participants varies greatly between 18 and 70. Their average length is 1.76 m (ranging from 1.40–2.04) and their weight 73.2 kg (ranging from 47 to 140).

7.2.3 Derivation of the Macroscopic Flow Variables

This research makes use of three macroscopic flow variables, namely walking speed, density and flow. Each of the variables is computed for the same measurement area, the location of which is detailed in Fig. 7.1. Here, the walking speed is defined as the average of the effective walking speeds in the direction of movement of all pedestrians in the measurement area. To quantify the global density $\rho(t)$ the method proposed by Zhang et al. [9] is adopted. In essence, the number of pedestrians located inside the measurement area N divided by the summation of the area of the Voronoi cells occupied by those pedestrians, see (7.1).

$$\rho(t) = \frac{\sum_i p_i}{\sum_i A_i}, \forall i \in A \tag{7.1}$$

The flowrate is computed as the multiplication of the average effective speed $v(t)$ times the global Voronoi density $\rho(t)$, see (7.2).

$$q(t) = v(t) \cdot \rho(t) \tag{7.2}$$

7.3 The Impact of Speed Heterogeneity and Flow Ratio

This paper presents an analysis of the fundamental diagrams derived for the 6 runs under investigation. Figure 7.2 illustrates the first analysis, namely a comparison of the fundamental diagrams for the three assignments. This analysis illustrates that the maximum flow rate is roughly similar for all three assignments. Please note that each scenario ran for 4–6 min. At the end of each run, pedestrian demand was allowed to stabilize at the highest flow rate. Even though the team attempted to increase

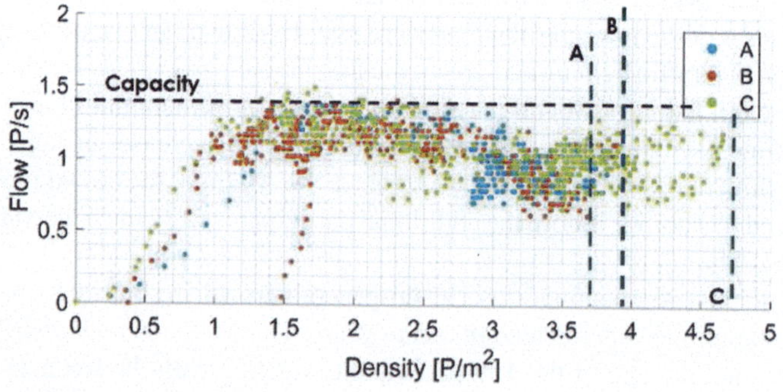

Fig. 7.2 Comparison of the fundamental diagrams for the three assignments for both flow ratios

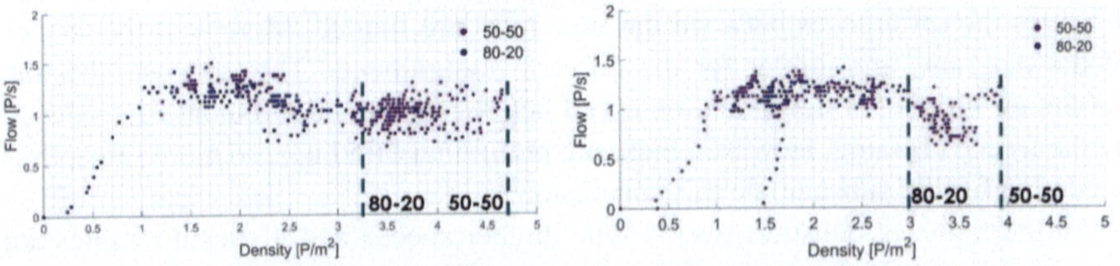

(a) *Assignment B - Forced crossing* (b) *Assignment C – Speeding up*

Fig. 7.3 Comparison of the shape of the fundamental diagram for different flow ratios featuring **a** differences in goal-orientation and **b** speed heterogeneity

throughput, there was no indication that a higher flowrate is to be expected in any of the 6 runs.

Surprisingly, differences in the maximum experienced global density occur. In particular, a 28% higher maximum density is found in case of assignment C. Apparently, the (forced) differentiation of speeds introduces density differences across the infrastructure, which can lead to sudden, relatively local increases of the density.

A second analysis, which compares the differences between the flow ratios 50–50 and 80–20 for the runs with assignments B and C, is visualized in Fig. 7.3. Similar trends are found in both graphs, namely that (a) the maximum flow rate is roughly equal for both flow ratios for both assignments and (b) large differences in the maximum global density occur. For both assignments, the higher maximum densities are found when flows are more equally distributed over both entrances. The authors hypothesize that the increasing densities in the 80–20 case are the result of a lack of opportunities to perform the assignment within the lane that is moving in their direction of movement. Forcing pedestrians to step outside a lane might interrupt flow in the opposite direction, thus creating high-density regimes at the location where the flow was interrupted.

A third analysis was performed to understand to what extend these global differences also arose at the level of an individual. Flow-density and speed-density relations were derived for all individuals in the crowd, the 95th-percentile per density bin is visualized in Fig. 7.4 for each of the runs. These figures illustrate that at low

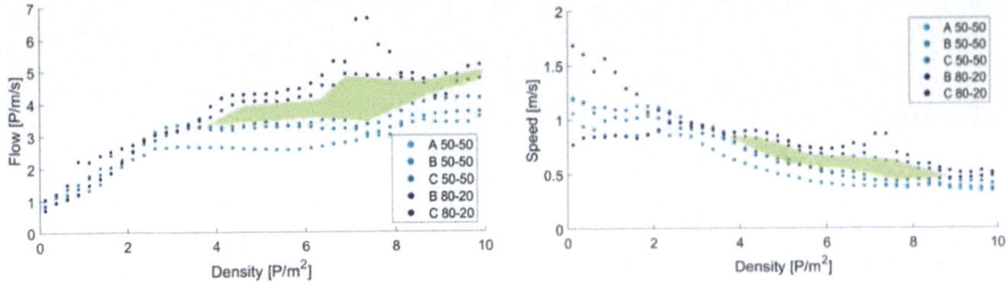

(a) flow – density (b) speed - density

Fig. 7.4 Fundamental relations for the different flow ratios and assignments densities the differences between the flow rates are limited. Moreover, the shapes of the diagrams are similar for the different assignments. This suggests that the maximum flux is not dependent on speed heterogeneity and turbulence introduced by the forced crossing. However, these are preliminary findings, so more research is needed to further substantiate this conclusion.

Furthermore, at densities over 4 P/m^2 higher speeds and higher flow rates are found for the 80–20 flow ratios. It is unclear what causes this trend. Potential causes are that (a) pedestrians have more opportunities to pass in the 80–20 runs, thus their speed is higher or (b) the spatial differences in density in the 80–20 runs result in local differences in speeds, where especially at the transition zones, high speeds at high densities are recorded.

7.4 Conclusions and Implications for Design and Management

This research has attempted to determine the impact of speed heterogeneity and the flow ratio in a bidirectional movement base case on the shape of the fundamental diagram. A part of the data captured during the CrowdLimits experiment was analysed for this purpose, which featured bidirectional crowd movements of a heterogeneous participant population in a corridor.

This study has found variations in the shape of the pedestrian fundamental diagram for 2D flows, namely (1) the maximum flow rate (capacity) is not dependent on the speed heterogeneity and the flow ratio; (2) the maximum global density measured in the infrastructure is dependent on the speed heterogeneity and the flow ratio; (3) the maximum flux is dependent on the flow ratio but not the speed heterogeneity.

These findings suggest that increased speed heterogeneity leads to local increases of the density, but that the impact of these local disturbances seems to dampen out at a global level. This has two implications for design and crowd management operations. Firstly, one can potentially not objectify personal safety by means of macroscopic flow variables, such as area-wide counts and flow rates per minute. Second, when designing pedestrian infrastructures one should also account for heterogeneity in behaviour in order to capture the real experience of pedestrians in a crowd.

References

1. S. Older, Movement of pedestrians on footways in shopping streets. Traffic Eng. Control **10**(4), 160–163 (1968)
2. U. Weidmann, *Transporttechnik der Fussgänger, Transporttechnische Eigenschaften des Fussgängerverkehrs* (Schriftenreihe des IVT, ETH Zürich, 1993)
3. W. Daamen, Modelling passenger flows in public transport facilities. Delft University of Technology (2004)

4. A. Seyfried, B. Steffen, W. Klingsch, M. Boltes, The fundamental diagram of pedestrian movement revisited. J. Stat. Mech. Theory Exp. **2005**(10), P10002 (2005)
5. D. Helbing, A. Johansson, H. Al-Abideen (2007) The dynamics of crowd disasters: an empirical study. Phys. Rev. Part E **75**, 1–7 (2007)
6. J. Zhang, A. Schadschneider, A. Seyfried, Empirical fundamental diagrams for bidirectional pedestrian streams in a corridor, in *Pedestrian and Evacuation Dynamics 2012*, ed. by U. Weidsmann, U. Kirsch, M. Schreckenberg (Springer, Heidelberg, 2014), pp. 245–250
7. L. Vanumu, R.K. Ramachandra, G. Tiwari, Fundamental diagrams of pedestrian flow characteristics: a review. Eur. Transp. Res. Rev. 9–49 (2017)
8. T. Kretz, A. Grunebohm, M. Kaufman, F. Mazur, M. Schreckenberg, J. Stat. Mech. Theory Exp. P10001 (2006)
9. J. Zhang, W. Klingsch, A. Schadschneider, A. Seyfried, Transitions in pedestrian fundamental diagrams of straight corridors and T-junctions. J. Stat. Mech. Theory Exp. **2011**, P06004 (2011)

Chapter 8
Experimental Investigation on Information Provision Methods and Guidance Strategies for Crowd Control

Claudio Feliciani, Hisashi Murakami, Kenichiro Shimura, and Katsuhiro Nishinari

Abstract Recent improvements in crowd sensing and dynamic signage are paving the way for automatic crowd control, in which real-time information is used to steer crowds of people. However, little is known on the impact that automatic information provision has on crowd dynamics. In this study, pedestrians moving in a two-lane fork loop were given different types of information and were guided using different strategies. Our analysis focused on which combination had the best performance and was perceived as comfortable from the participants. Results indicate that a human guidance is preferred when comfort is concerned, but providing real-time traffic information is more efficient in terms of pedestrian flow. It is speculated that the processing of a large volume of information limits the perceived comfort when real-time data are provided. As a consequence, people prefer the less effective human guidance in which decision-making is entrusted into others. With this said, the speculative nature of this conclusion puts emphasis on future studies which will need to investigate more in detail how information provision affects pedestrians' behavior and comfort perception in public spaces.

8.1 Introduction and Background

Pedestrian guidance and crowd control are the key elements to ensure safe and smooth mass events and are also an important aspect in the design of urban infrastructures dealing with a large number of people [3, 6]. In the recent years, advances in pedestrian sensing technology [4] and computational performances are making it possible to simulate large crowds on real-time, thus increasing the volume and the type of information which can be used to guide people.

C. Feliciani (✉) · H. Murakami · K. Shimura · K. Nishinari
Research Center for Advanced Science and Technology, The University of Tokyo, 4-6-1 Komaba, Meguro-ku, Tokyo 153-8904, Japan
e-mail: feliciani@jamology.rcast.u-tokyo.ac.jp

K. Nishinari
Department of Aeronautics and Astronautics, The University of Tokyo, 7-3-1 Hongo, Bunkyo-ku, Tokyo 113-8656, Japan

I. Zuriguel et al. (eds.), *Traffic and Granular Flow 2019*,
Springer Proceedings in Physics 252,
https://doi.org/10.1007/978-3-030-55973-1_8

Nowadays, expected waiting time is often provided at the entrance of security screening in airports and, in the context of vehicular traffic, dynamic routing is increasingly used to avoid congestion. However, there is only little experimental knowledge on which type of information can help efficiently guiding people and to which extent will people follow instructions given.

A number of studies considered the case of a fork road and investigated its dynamics in regard to real-time information in the frame of vehicular traffic, both theoretically [7, 8] and experimentally [9–11]. However, to the knowledge of the authors, no study has been performed to date for the case of pedestrian traffic.

In our previous research [5], we investigated the effect between information provision and compliance by means of simulation and found that information is beneficial to the overall crowd dynamics, but high levels of compliance can be an issue if guidance is not performed in the proper way. However, as the study only considered a simple simulation model, the validity of the conclusions is questionable.

Bode and Codling [1] considered how information affects the way in which people make decisions in virtual evacuations, concluding that stress contributed in making less rational choices and people did not prefer familiar routes. However, also in their study, non-real conditions were tested, thus showing the need to collect experimental data on the relation between information provision and crowd dynamics.

In this study, different methods to guide crowds of people are considered in a supervised experiment, where both efficiency and perceived comfort for different solutions are investigated.

8.2 Experimental and Technical Setup

To study experimentally the above topics a course like the one schematically presented in Fig. 8.1 was built. The course consisted of two lanes, clearly divided by a wall. People were allowed to move from the internal to the external lane (or vice-versa) only in two locations, where an opening of 80 cm was provided. Participants to the experiments were asked to avoid unnecessarily wasting time and had therefore to chose the fastest lane when necessary.

Sensors were installed to measure in real time walking speed in different locations and obtained speed could be displayed on PC displays provided along the course. Speed shown in the display was relative to the average in the inner and outer course (two measuring locations were used to get each average). We also tested the case in which people are guided by staff on-site and rely on their decisions to change lane or not. In this case, guidance staff (students belonging to the group organizing the experiment) was located at each lane-change location and was requested to guide participants showing left/right arrows on a dedicated PC display (different from the ones used for speed). In a first experimental trial, guidance staff had to rely only their local vision (limited to 90 degrees by internal walls), while in a later trial they had access to the real-time video stream of the experiment, entirely showing the course (the video offered the same view to what is seen on Fig. 8.1).

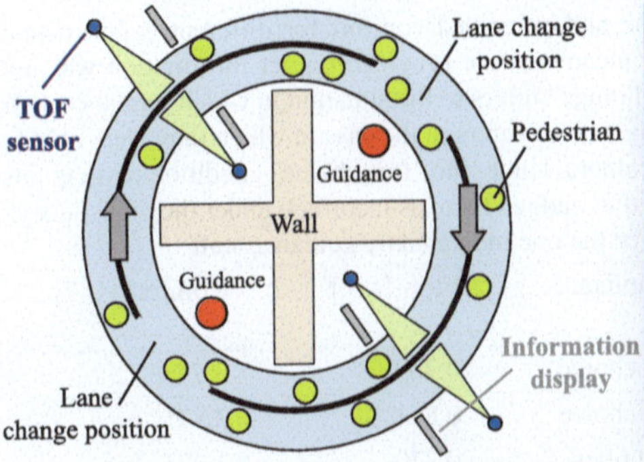

Fig. 8.1 Experimental setup employed: schematic representation (left) and top-view snapshot (right)

Each experiment started with 35 participants in the inner loop and 5 in the outer loop and lasted 3 min (start and stop signal were given by a programmed clock). Three variables were varied to study crowd behavior: availability and timing of information provision (either real-time/present or predicted/future speed), availability and strategy employed for staffed human guidance (direct visual inspection or use of the overhead camera stream) and compliance (participants could choose by their own or had to obey to the instructions/information given them). Predicted (future) speed was obtained by means of a simulation model [5] using the speed measured by sensors as initial conditions. Speed predicted 10 seconds ahead in time was computed in less than one second and promptly displayed on the PC displays along the course.

Videos of the experiments were taken using a camera fixed directly above the center of the experimental course (see Fig. 8.1) and pedestrians' positions were later obtained using PeTrack software [2]. A comparison between the ground truth speed obtained from video recordings and sensor measurements gained during the experiments revealed that sensors' error was about 3% for speeds between 0.5 and 1.2 m/s and generally above 10% for lower and higher speeds. We can thus assume that speed measurement was accurate when both lanes moved at similar speed and differences in speed were difficult to notice by visual inspection. When one lane was slow (and congested) the other one was always much faster (as a consequence of the experimental design), thus making inaccuracy at low and high speed of little concern.

8.3 Results

To evaluate the efficacy and suitability of different strategies of information provision and crowd control, two criteria have been used: the average flow of people transiting over a single lane during each trial and the comfort perceived by the participants. For the latter measure we provided participants with a questionnaire and asked them

Table 8.1 Average pedestrian flow per lane and perceived comfort for different experimental conditions. N/A is used when the indicated mean was not provided (either information was not available or guidance was not given). "Visual range" refers to the situation in which guidance staff only relied on their (limited) vision, "real-time video" refers to the case in which they were able to see the real-time footage from the overhead camera. Under the "free choice" condition participants were allowed to ignore information/orders if they judged them as incorrect; under the "must obey" condition they had to choose the fastest lane or the one indicated by guidance staff

Information provided	Guidance	Compliance	Flow [s^{-1}]	Comfort [1–7]
N/A	N/A	Free choice	0.789	3.775
Present speed	N/A	Free choice	0.851	3.875
Present speed	N/A	Must obey	0.770	3.375
Predicted speed	N/A	Free choice	0.772	4.049
Predicted speed	N/A	Must obey	0.492	2.829
N/A	Visual range	Free choice	0.794	4.436
N/A	Visual range	Must obey	0.769	4.175
N/A	Real-time video	Free choice	0.780	4.100
N/A	Real-time video	Must obey	0.749	3.950

to indicate at the end of each experimental trial the level of comfort using a score varying from 1 to 7 and based on their own perception. We did not give instructions on how to judge the perceived level of comfort, but simply asked to participants to provide a score for it. Results for different experimental conditions are given in Table 8.1.

Results presented in Table 8.1 clearly indicate that the type of information provided, the inspection method used by the staff to guide crowds of people and the degree of compliance among participants clearly affect the overall dynamics of the group. In particular, it seems there is a discrepancy between what is efficient from a physical/engineering perspective and what people perceive as comfortable. While real-time speed information achieved the highest flow, the highest degree of comfort was recorded when human guidance was provided along the course.

Although the reasons for this difference are not completely clear, a few participants interviewed after the experiments stated that a continuous stream of information forced them to make very often decisions on which lane to use, thus creating a sort of psychological stress, related to the constant pressure on taking the right choice. Also, since some participants were elderly in their 60s, they did not perceive the computer system constantly providing numbers as user-friendly. On the other side, human guidance was more accepted and more warmly perceived. Also, since decisions were entrusted into guidance staff, participants simply had to follow them, reducing their pressure on making prompt decisions.

When information timing is considered, real-time information seems to be slightly better than the predicted one. However, on this topic, further investigations will need to assess the accuracy of predicted speed.

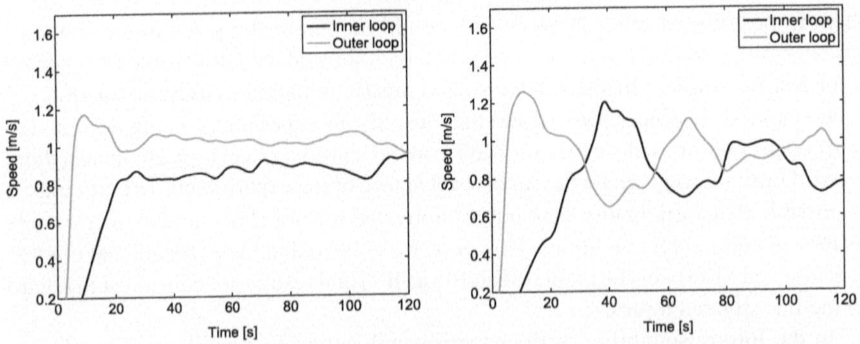

Fig. 8.2 Speed profile for the two cases where real-time speed is shown to pedestrians. Left: signage has only an informative nature, people are able to act or not in regard to the information given (free choice); right: people must compulsory move to the lane which has the highest speed (must obey). Speed shown here is computed from video processing (ground truth)

Finally, it can be seen that differences also appear in regard to compliance. An hypothesis to explain those differences is that in the "must obey" case a sort of resonance may have formed in the control system thus amplifying the fluctuations in speed and worsening the outcome.

The relation with compliance is illustrated better in Fig. 8.2 which provides the speed change in the inner and outer lanes for two selected cases. As it is seen, when people were allowed to use the real-time speed only as an information (which could have been disregarded), both the inner and outer speeds rapidly converged to similar levels and stood almost constant for the length of the experiment. On the other side, the constraint on moving into the faster lane led to strong oscillations. The cause for this oscillations may be related to the slowing down created by a large number of people changing lane and the small distance between the speed signage and the lane change position, thus creating a sort of delay in the control loop.

8.4 Conclusions and Future Work

Effect of information provision on an infinite fork path has been experimentally studied for the pedestrian case. This study showed that type of information provided to people and crowd control strategy can influence the overall dynamics of the crowd. While directly providing information to pedestrians is surely important and was found being the best solution in terms of traffic flow, our results also showed the importance that guidance staff has on the comfort perceived by people.

It is therefore our belief that having an informed and trained staff/guidance personnel is of central importance in crowd management. In this regard, the development of a system able to provide guidance staff and security personnel with relevant information and possibly also predicting changes occurring in the near future, would allow them to perform an efficient work and should be one of the goals in the frame of research on crowd management. As this research showed, benefits would be reflected in pedestrian traffic efficiency, but also and mostly in regard to user satisfaction.

It is nonetheless necessary to add that despite our experiments being designed to reproduce conditions close to reality and to avoid learning effect by participants, there are still limitations given by the supervised nature of the experiments. In particular, it is possible that participants were not as motivated toward choosing the fastest route as they would in real conditions. However, we believe that the experimental context still allowed to answer the main research question and results are consistent in regard to the investigated topics.

In the future, suitability of the experimental setting to describe real conditions should be further studied, possibly conducting similar experiments in a real environment. Additional investigations on the experimental data are also required, considering for example the reliability of the predicted speed and the local relationship between information provided and pedestrian decision-making. It could be also interesting investigating whether there is a relationship between speed profiles and comfort perceived by participants.

Acknowledgements The authors would like acknowledge the help from the members of the Nishinari laboratory in organizing the experiments. This work was supported by JST-Mirai Program Grant Number JPMJMI17D4, Japan.

References

1. N.W. Bode, E.A. Codling, Human exit route choice in virtual crowd evacuations. Animal Behav. **86**(2), 347–358 (2013). https://doi.org/10.1016/j.anbehav.2013.05.025
2. M. Boltes, A. Seyfried, Collecting pedestrian trajectories. Neurocomputing **100**, 127–133 (2013). https://doi.org/10.1016/j.neucom.2012.01.036
3. C. Feliciani, H. Murakami, K. Shimura, K. Nishinari, Efficiently informing crowds-experiments and simulations on route choice and decision making in pedestrian crowds with wheelchair users. Transp. Res. Part C: Emerg. Technol. **114**, 484–503 (2020). https://doi.org/10.1016/j.trc.2020.02.019
4. C. Feliciani, K. Nishinari, Estimation of pedestrian crowds' properties using commercial tablets and smartphones. Transportmetrica B: Transp. Dyn. **7**(1), 865–896 (2019). https://doi.org/10.1080/21680566.2018.1517061
5. C. Feliciani, K. Shimura, D. Yanagisawa, K. Nishinari, Study on the efficacy of crowd control and information provision through a simple cellular automata model, in International Conference on Cellular Automata, pp. 470–480 (Springer, 2018). https://doi.org/10.1007/978-3-319-99813-8_43
6. D. Gibson, *The wayfinding Handbook: Information Design for Public Places* (Princeton Architectural Press, 2009)
7. Y. Hino, T. Nagatani, Effect of bottleneck on route choice in two-route traffic system with real-time information. Physica A: Statist. Mech. Appl. **395**, 425–433 (2014). https://doi.org/10.1016/j.physa.2013.10.044
8. T. Imai, K. Nishinari, Optimal information provision for maximizing flow in a forked lattice. Phys. Rev. E **91**(6), 062818 (2015). https://doi.org/10.1103/PhysRevE.91.062818
9. R. Selten, T. Chmura, T. Pitz, S. Kube, M. Schreckenberg, Commuters route choice behaviour. Games Econ. Behav. **58**(2), 394–406 (2007). https://doi.org/10.1016/j.geb.2006.03.012

10. R. Selten, M. Schreckenberg, T. Chmura, T. Pitz, S. Kube, S.F. Hafstein, R. Chrobok, A. Pottmeier, J. Wahle, Experimental investigation of day-to-day route-choice behaviour and network simulations of autobahn traffic in north rhine-westphalia, in:Human behaviour and traffic networks (Springer, 2004), pp. 1–21. https://doi.org/10.1007/978-3-662-07809-9_1
11. R. Selten, M. Schreckenberg, T. Pitz, T. Chmura, J. Wahle, Experiments on route choice behaviour, in Interface and Transport Dynamics (Springer, 2003), pp. 317–321. https://doi.org/10.1007/978-3-662-07969-0_30

Chapter 9
The Impact of Guidance Information on Exit Choice Behavior During an Evacuation—A VR study

Yan Feng, Dorine C. Duives, and Serge P. Hoogendoorn

Abstract This paper presents a Virtual Reality (VR) experiment to study pedestrian exit choice behavior during evacuations. It investigates whether and to what extent different types of guidance information (i.e. exit signs and directional signs) influence pedestrian exit choice during an evacuation (drill). The analysis focuses on the commonalities and differences in the pedestrians' exit choice behavior between the scenario without any additional guidance information and two scenarios with different types of guidance information. The findings suggest that guidance information of exit signs and direction signs have significant influences on the number of recognized exits. Meanwhile, guidance information might have a relevant effect on final exit choice of participants.

9.1 Introduction

During a building evacuation, the choice of which exit to use to leave a building is vital for pedestrians' survival. Pedestrians usually have a choice of multiple exits and need to decide which one to use [1]. Literature shows that the exit decision is affected by multiple factors, such as the visibility of exits, guidance information provided by signs, lights and sounds [2–4], and the presence of other pedestrians [1, 5, 6]. The relation between those factors has predominantly been investigated using laboratory experiments [3, 4, 7, 8].

Another promising experimental method is Virtual Reality, which also has been used to study pedestrian evacuation behavior. VR experiments can potentially obtain high experimental control to analyze the influence of different factors on pedestrian behavior more precisely. Studies have established that VR is a useful tool to study pedestrian behavior during evacuations [9–11]. Yet to date, few studies have focused on exploring the effect of influential factors on pedestrian exit choice during evacuations with VR. Often, the validity of the results generated from VR experiments has

---

Y. Feng (✉) · D. C. Duives · S. P. Hoogendoorn
Delft University of Technology, Stevinweg 1, 2628 CN Delft, The Netherlands
e-mail: y.feng@tudelft.nl

I. Zuriguel et al. (eds.), *Traffic and Granular Flow 2019*,
Springer Proceedings in Physics 252,
https://doi.org/10.1007/978-3-030-55973-1_9

been questioned. Therefore, there is a need to conduct more exploratory VR experiments to understand how different factors influence pedestrian exit choice behavior and how to implement VR technique to pedestrian study.

This study builds on a method to conduct experiments on pedestrian exit choice behavior in virtual environments, validated in [12]. It investigates to what extent different types of guidance information present in the environment (i.e., visibility of exit signs and directional information) influence pedestrian exit choice behavior during evacuations. The analysis focuses on: (1) analyzing pedestrian exit choice during different evacuation scenarios; and (2) comparing the commonalities and differences in the pedestrian exit choice behavior between the scenario without additional guidance information and the scenarios with different types of guidance information.

9.2 Experimental methods

9.2.1 Experiment design

A field experiment during an unannounced evacuation drill took place in the building of Architecture Faculty at the Delft University of Technology. The experimental area was a workshop space with multiple exits. A video recording of the unannounced evacuation drill of the workshop space was used as the benchmark for the VR experiment. The camera was located at the center of the workshop area, as identified by a circle icon in Fig. 9.1. The video recording provides a 360° view of the space

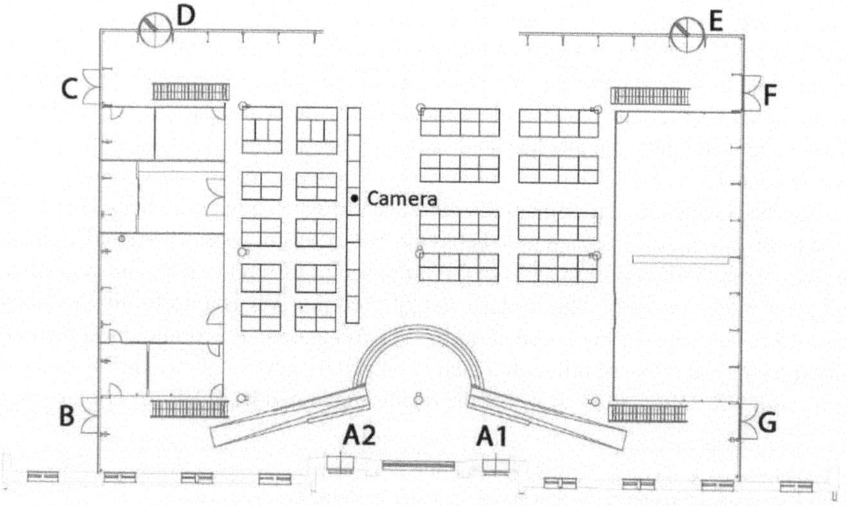

Fig. 9.1 A schematic illustration of the workshop space (exits are marked with A1-G)

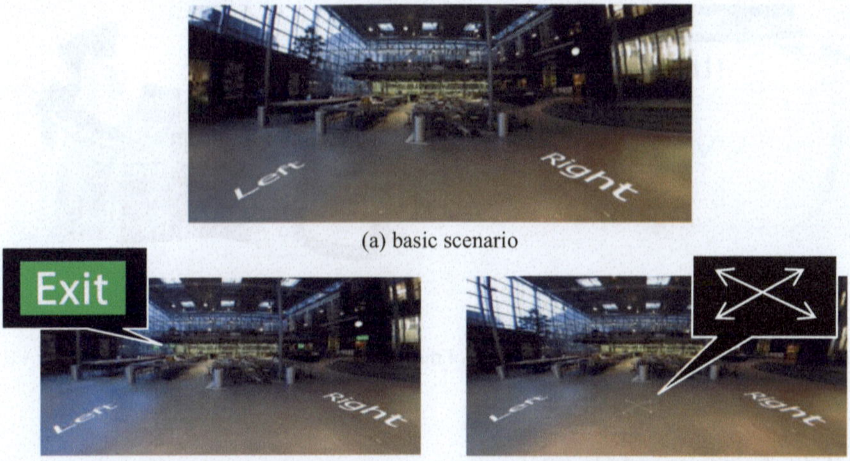

(a) basic scenario

(b) exit scenario

(c) direction scenario

Fig. 9.2 Screenshots of experiment scenarios

including all exits. All exits are easily observable from the central spot at which the camera was located.

In order to investigate and compare the effect of different information on pedestrian exit choice behavior, two variables were tested: exit signs near the exits, direction signs on the floor in front of the participant. In total, three different evacuation scenarios were created: one basic scenario and two experimental scenarios, as shown in Fig. 9.2. The "basic scenario" consisted of the recorded area and an added evacuation alarm sound. For the "exit signs" scenario, eight emergency exit signs were added to the environment, which were located above each of the exits. For the "direction" scenario, four arrows were added on the ground in front of the participant, which point to four exit directions. Besides the three experimental scenarios, a general familiarization scenario was developed to allow participants to become familiar with the navigation in VR and the sensation of VR.

9.2.2 Experiment Apparatus

Participants were immersed in the virtual environment via a VR head-mounted display called VR Pro Virtual Reality Glasses (Fig. 9.3), which has an approximate 90° horizontal and 110° vertical field of view. The virtual environment was presented on a smartphone, the screen was 14 cm length and provided a resolution of 1125 × 2436 pixels for 3D effects.

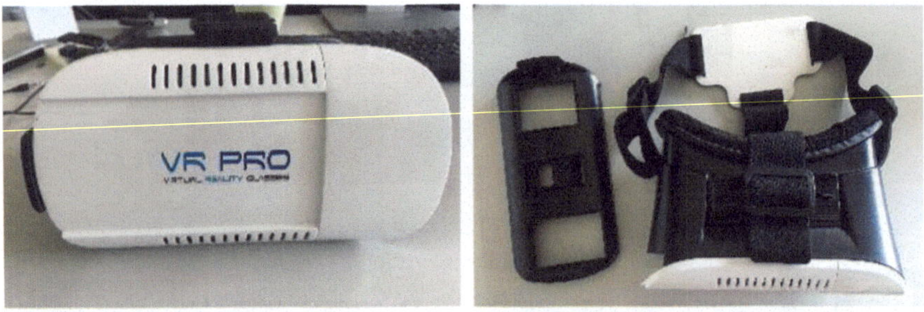

(a) (b)

Fig. 9.3 The front view **a** and the top view **b** of the head-mounted display was used during the VR experiment

9.2.3 Experiment Questionnaire

In order to obtain personal features and experiences of participants regarding the VR experiment, participants were asked to complete a questionnaire immediately after the experiment. The questionnaire contained four sections: (1) participants' information, (2) the Simulator Sickness Questionnaire [13], which determined simulator sickness of participants, (4) the Presence Questionnaire [14], which measured participants' experience presence in the virtual environment(3) the System Usability Scale [15], which assessed usability of the simulator.

9.2.4 Experiment Procedure

The VR experiment was conducted during the International Festival of Technology on 6th, 7th, 8th June 2018 in Delft, the Netherlands. The VR experiment was approved by the Human Research Ethics Committee of the Delft University of Technology. Participants for the VR experiment were recruited in four ways, namely through the festival website, social media, posters distributed at the university, and direct acquisition of visitors during the festival.

The procedure of the VR experiment included the following parts: (1) participants were introduced to the purpose and instructions of the experiment; (2) got familiar with VR environment and the HMD device (Fig. 4a); (3) were presented with the scenario of the evacuation drill and chose one exit; (4) completed the questionnaires (Fig. 4b).

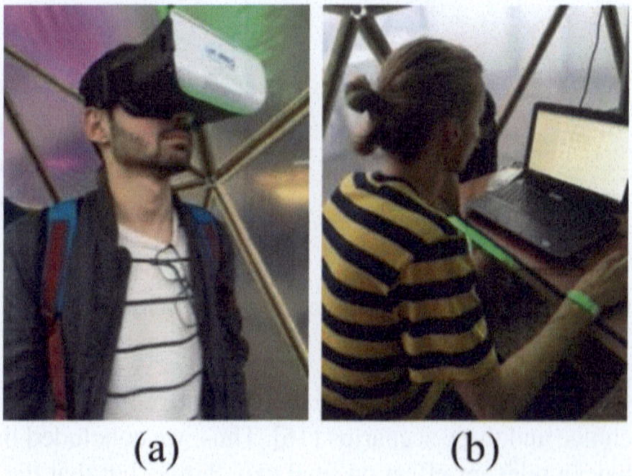

Fig.4 Participants were **(a)** experiencing the virtual experiment; **(b)** filling in the questionnaire

9.3 Findings and Conclusions

In total, sixty-eight participants joined the VR experiment. Every participant was randomly assigned to take part in one scenario. The collected data was two-fold: (1) participants' exit choice (including observed exits and final exit choice), and (2) participants' experience regarding the VR experiment.

Figure 9.5 presents the mean value of the number of exit participants observed in each scenario. A non-parametric Kruskal–Wallis test was conducted to test whether guidance information has a significant influence on the number of observed exits. The results found that there was a statistically significant difference in observed exits between different scenarios, $\chi 2(2) = 28.266$, $p = 0.000$. It showed that the visibility of exits and directional information influenced the number of exits that the participants can recognize and recall during the experiment.

Figure 9.6 presents the participants' final exit choice in each scenario. Although eight exits were available, the usage of the exits was not evenly. Amongst all the exits, exit D, C and A1 were chosen more often than the other exits. It shows the exit choices of pedestrians are asymmetrical. In order to examine whether different types of guidance information had a significant influence on pedestrian final exit choice during evacuations, the Fisher-Freeman-Halton exact test was used because of the relatively small sample size. Fisher's exact test found there was no significant difference of pedestrian final exit choice among three scenarios ($p = 0.057$). However,

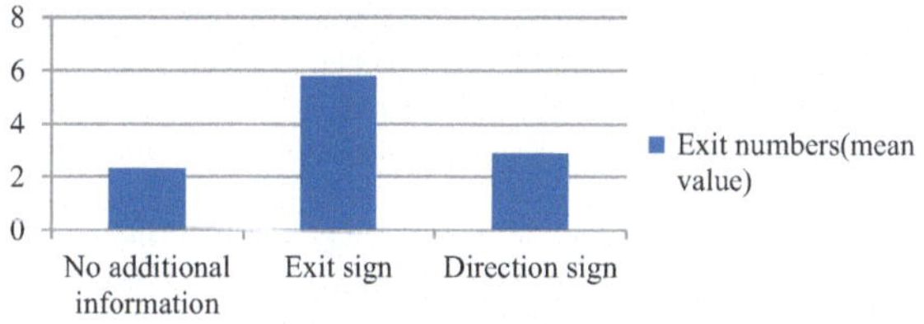

Fig. 9.5 Number of observed exits by participants

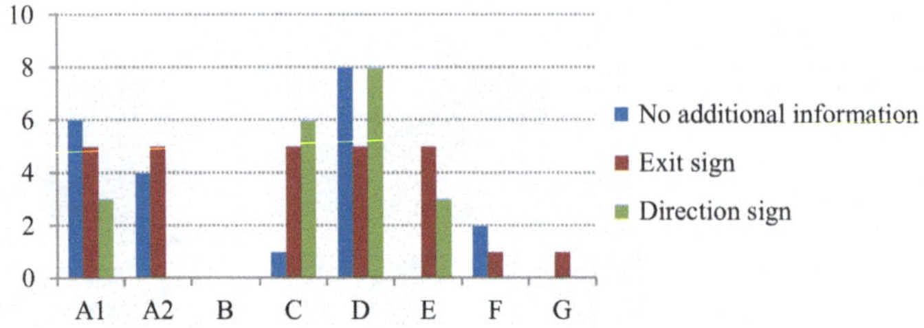

Fig. 9.6 Participant' exit choice of different scenarios

Cramer's V = 0.358 indicated a relatively strong relationship between final exit choice and three scenarios [16]. Thus, we concluded that guidance information may have a relevant effect on final exit choice, but that the sample size was too small to estimate it with sufficient statistical significance.

Combing these findings from the pedestrian exit choice study and the questionnaire, we conclude that the experimental settings and the VR device can be used as a valid method to study pedestrian exit choice during evacuations. In the future, larger samples should be collected to statistically validate the results of this study. Moreover, in order to study pedestrian behavior under more complex situations, the enhancement of experimental setting and better VR equipment are required.

References

1. S. Heliövaara, J.M. Kuusinen, T. Rinne, T. Korhonen, H. Ehtamo, Pedestrian behavior and exit selection in evacuation of a corridor—an experimental study. Saf. Sci. **50**, 221–227 (2012)
2. J.D. Sime, Crowd Psychol. Eng. **21**, 1–14 (1995)
3. E.R. Galea, S.J. Deere, C.G. Hopkin, H. Xie, Evacuation response behaviour of occupants in a large theatre during a live performance. Fire Mater. **41**, 467–492 (2017)
4. L.T. Wong, K.C. Lo, Experimental study on visibility of exit signs in buildings. Build. Environ. **42**, 1836–1842 (2007)
5. C. von Krüchten, A. Schadschneider, C.V. Krüchten, A. Schadschneider, Empirical study on social groups in pedestrian evacuation dynamics. Phys. A Stat. Mech. Appl. **475**, 129–141 (2017)
6. D. Duives, H. Mahmassani, Exit Choice Decisions during pedestrian evacuations of buildings. Transp. Res. Rec. J. Transp. Res. Board. **2316**, 84–94 (2012)
7. K. Fridolf, E. Ronchi, D. Nilsson, H. Frantzich, Movement speed and exit choice in smoke-filled rail tunnels. Fire Saf. J. **59**, 8–21 (2013)
8. M. Kobes, I. Helsloot, B. De Vries, J. Post, Exit choice, (pre-)movement time and (pre-)evacuation behaviour in hotel fire evacuation—behavioural analysis and validation of the use of serious gaming in experimental research. Procedia Eng. **3**, 37–51 (2010)
9. M. Kinateder, M. Müller, M. Jost, A. Mühlberger, P. Pauli, Social influence in a virtual tunnel fire—influence of conflicting information on evacuation behavior. Appl. Ergon. **45**, 1649–1659 (2014)
10. M. Kobes, I. Helsloot, B. de Vries, J.G. Post, N. Oberijé, K. Groenewegen, Way finding during fire evacuation: an analysis of unannounced fire drills in a hotel at night. Build. Environ. **45**, 537–548 (2010)

11. M. Moussaïd, N. Perozo, S. Garnier, D. Helbing, G. Theraulaz, The walking behaviour of pedestrian social groups and its impact on crowd dynamics. PLoS ONE **5**, 1–7 (2010)
12. Feng, Y., Duives, D., Daamen, W., Hoogendoorn, S.: Pedestrian exit choice behavior during an evacuation using virtual reality—a comparison study with a real-life evacuation drill. In: Transportation Research Board 98th Annual Meeting Transporation Research Board (2019).
13. R.S. Kennedy, N.E. Lane, K.S. Berbaum, M.G. Lilienthal, Simulator sickness questionnaire: an enhanced method for quantifying simulator sickness. Int. J. Aviat. Psychol. **3**, 203–220 (1993)
14. J. Brooke, SUS—a quick and dirty usability scale. Usability Eval. Ind. **189**, 4–7 (1996)
15. B.G. Witmer, M.J. Singer, Measuring Presence in virtual environments: a presence questionnaire. Presence. **7**, 225–240 (1998)
16. J. Cohen, Statistical power analysis for the behavioral sciences. Academic Press (2013)

Chapter 10
Experimental Study on Crowds with Different Velocity Composition

Akihiro Fujita, Claudio Feliciani, Daichi Yanagisawa, and Katsuhiro Nishinari

Abstract This study focuses on the difference in walking speed among pedestrians, whereby experiments were conducted asking participants to walk at different speeds. We examined the various factors that affect a pedestrian's individual speed and the macroscopic difference between crowds with the same (homogeneous) velocity and crowds composed of people with different target velocities.

10.1 Introduction

With a rapidly growing global population and advances in transportations, people in such crowds have become more heterogeneous than before. They are different in body size, cultural backgrounds, age, and so forth.

Among these heterogeneities, age is attracting much interest because of the ageing population in several countries. Investigations in the form of observations and experiments have studied the way age difference affects the flow characteristics. In this field, observational studies on pedestrians with different age compositions were conducted [1, 2], and controlled experimental studies that consider age difference have been conducted widely. The subjects of these studies included small children [3], people who were older or less fit [4], and people with disabilities [5, 6]. However, these experimental studies focused mainly on groups of one type of pedestrian (i.e. groups comprising only children, older adults or people with disabilities), and mixed situations (i.e. groups comprising different types of pedestrians walking simultaneously) were rarely considered. One study that did focus on age differences in mixed situations concluded that heterogeneous flow is more likely to cause jams because of

A. Fujita (✉) · C. Feliciani · D. Yanagisawa · K. Nishinari
School of Engineering, Department of Aeronautics and Astronautics, The University of Tokyo, 7-3-1, Hongo, Bunkyo-ku, Tokyo, Japan
e-mail: f-akihiro@g.ecc.u-tokyo.ac.jp

C. Feliciani · D. Yanagisawa · K. Nishinari
Research Center for Advanced Science and Technology, The University of Tokyo, 4-6-1 Komaba, Meguro-ku, Tokyo 153-8904, Japan

I. Zuriguel et al. (eds.), *Traffic and Granular Flow 2019*,
Springer Proceedings in Physics 252,
https://doi.org/10.1007/978-3-030-55973-1_10

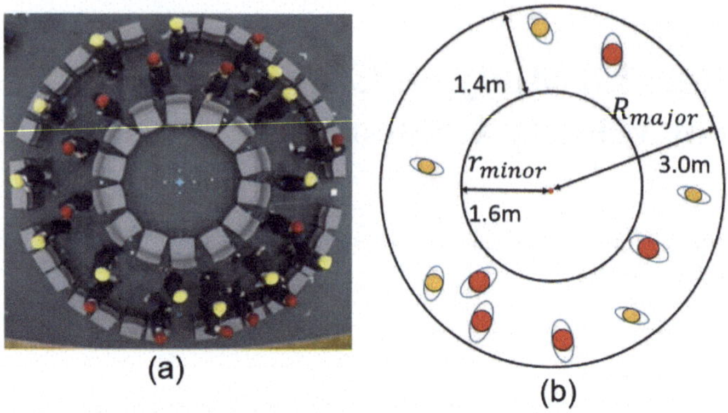

Fig. 10.1 **a** Snapshot of the experiment. **b** Schematic representation

pedestrian interactions [7]. However, that study allowed space for only one lane of pedestrian traffic, thereby permitting only one-dimensional movement and limiting the knowledge obtained from the study.

Even though previous research has focused on age difference, that aspect encompasses several physical heterogeneities such as differences in walking speed and reaction time. Therefore, if we focus on age heterogeneity alone, it is still difficult to obtain generalized idea on what physical quantity affect the flow characteristics.

Accordingly, in the present study, we focus on speed differences among pedestrians and investigate how they influence the walking speed of an individual pedestrian and the macroscopic behaviour of the flow such as mean speed and density distribution. To understand the effect of speed difference and to study two-dimensional movements, we conducted experiments using the circular course shown in Fig. 10.1.

10.2 Experiment Setup

Our experiments aimed to elucidate how speed heterogeneity influences individual pedestrians and the macroscopic flow characteristics. The experiments were conducted in 2017 at the University of Tokyo, Japan. Figure 10.1 illustrates a snapshot of an experiment, all of which were conducted in a ring-shaped corridor, and a sketch of the experimental field, the inner and outer radii of which were set to 1.6 and 3.0 m, respectively. The width of the corridor was thus 1.4 m, and the area of the experimental field was approximately 20.0 m^2. This course was sufficiently wide for two-dimensional behavior to occur. We recruited 32 students as participants, all of whom were male university students aged from 18 to 25 years old. Because the aim of the experiments was to investigate the effects of velocity heterogeneity, crowds with different speed composition were investigated, as shown in Table I. In the case of the slow-mix composition, for example, the participants were divided into two

Table 10.1 Five crowd compositions

Composition	Mixed elements
Fast-mix	Fast and normal
Slow-mix	Slow and normal
Fast (homogeneous)	Fast walkers only
Normal (homogeneous)	Normal walkers only
Slow (homogeneous)	Slow walkers only

groups: slow walkers and normal walkers. The slow walkers were instructed to walk slower than their normal walking speed (the slower speed was measured as 0.49 ± 0.12 m/s), whereas the normal walkers were instructed to walk at their normal speed and maintain that speed. In addition, the normal walkers were allowed to pass the slow walkers to maintain their walking speed (1.07 ± 0.09 m/s), thereby making the experimental conditions resemble those of an actual walkway. Similarly, in the case of the fast-mix composition, the fast walkers were asked to walk faster than their normal speed (the faster speed was measured as 1.71 ± 0.2 m/s), whereas the normal walkers were asked to walk normally as before. In this case, the faster walkers were allowed to pass the normal walkers. We performed several test trials in which the participants were instructed to walk at these three different speeds. Note that these desired speeds were measured under low-density conditions. The participants wore caps whose color indicated the walking speed, as shown in Fig. 10.1a. Normal walkers wore yellow caps, whereas fast and slow walkers wore red caps. In both the slow mix and the fast mix, there were equal numbers of red-cap and yellow-cap participants. We also conducted experiments with homogeneous crowds for reference. The global density (i.e., the mean density of the entire experimental field) was varied from 0.1 to 1.5 m^{-2} in increments of 0.2 m^{-2} by changing the number of walkers (Table 10.1).

10.3 Fundamental Diagrams

In the present study, the speed is calculated as Eq. (10.1).

$$v_t = \frac{|\mathbf{x_{t+5}} - \mathbf{x_{t-5}}|}{10\delta t} \tag{10.1}$$

The δt is the frame interval of the data (1/30 s) and $\mathbf{x_t}$ is the location of each walker at time step t. We calculate local density on the basis of the Voronoi method, as presented in [8].

From Fig. 10.2a, we see that all the plots are in the free-flow region in which the flow increases with local density, so no congested flow was observed in those experiments.

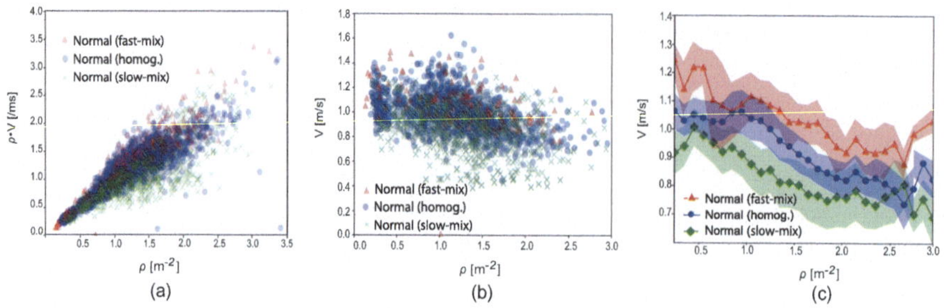

Fig. 10.2 Fundamental diagrams of normal pedestrians in different compositions: blue represents normal walkers walking with other normal walkers (homogeneous runs), and red and green represent normal walkers walking with fast and slow walkers, respectively (heterogeneous runs). **a** flow rate versus density; **b** speed versus density; **c** average speed versus density: the shaded area represents one-sigma error

Contrary to previous research, Fig. 10.2b implies that individual walking speed cannot be determined from the local density alone, especially in a crowd with different speeds. Therefore, to study the effect of the surroundings in more detail, in Fig. 10.2c, we plot the average speed of the normal walkers for each composition and also the variance of that speed.

The average speed of normal walkers grouped with fast walkers is clearly always higher than that of normal walkers in homogeneous flow. Similarly, the speed of normal walkers in homogeneous flow is almost always higher than that of normal walkers in the slow mix composition. Even though each participant tried to walk at his normal speed, his actual walking speed was affected significantly by the speed of those around him (the detailed statistical analysis were conducted in [9]).

10.4 Density Distribution and Mean Speed

As for macroscopic characteristics, no lane formation was observed in these experiments, which resulted in the formation of clusters as shown in Fig. 10.3. These clusters were triggered by the block of slower walkers that prevent normal walkers from walking at their target speed.

In order to get the density heat maps as shown in Fig. 10.3b, the whole experimental field was split into cells (600×600 cells), and each cell was assigned to a continuous density value that represents the number of people in the cell. The detailed information on this density value is summarized in [9]. Thus, clear density gaps can be defined as sharp drops in the density value. In the present study, since the experimental field is ring-shaped, these density values are summed up according to the angle from the center of the experiment field (i.e. density values of different cells are summed up if the cells share the same angle from the center), and the summed density values are stored in the list with the length of 360 (from 1° to 360° with the

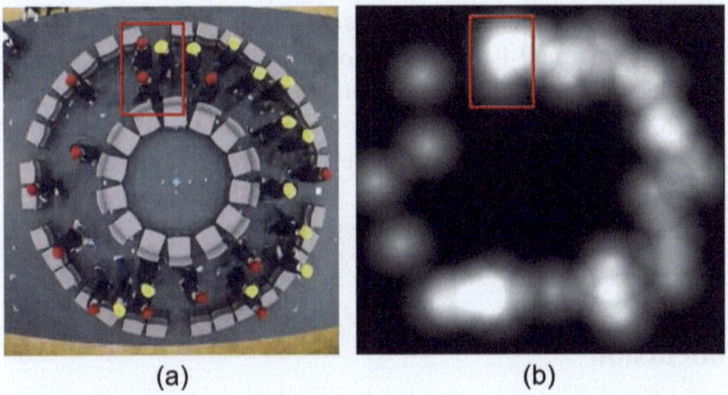

Fig. 10.3 Blocks and density distribution: **a** blocks formed by slower walkers where normal walkers are prevented from walking at their desired speed as represented by the red square area, **b** density heat map corresponding to the snapshot, where brighter area and darker area represents high and low density areas, respectively

increment of 1°). In this study, if there is a drop in density value larger than 1.0 at the angle of θ degree, we define that there is a cluster at angle θ.

By using this definition, we investigated relationship between the existence of blocks and the flow smoothness of the heterogeneous speed crowd. We plotted the speed fluctuation of slow-mix experiments under the different density conditions as

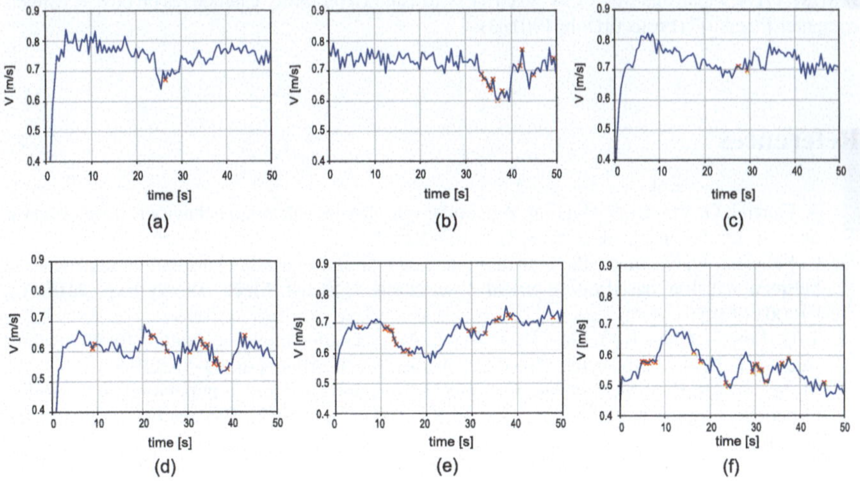

Fig. 10.4 Mean speed of walkers in slow-mix experiments under different global density conditions (from 0.5 m^{-2} to 1.5 m^{-2}): vertical axis and horizontal axis represent mean speed and time from the start of each run, respectively. Red cross markers represent whether there is any block of slower walkers in the experimental field. **a** ρ_g= 0.5 m^{-2} **b** ρ_g= 0.7 m^{-2}; **c** ρ_g= 0.9 m^{-2}; **d** ρ_g= 1.1 m^{-2} (e) ρ_g= 1.3 m^{-2}; (f) ρ_g= 1.5 m^{-2}

shown in Fig. 10.4. Red cross markers represent whether there is any block of slower walkers in the experimental field. The detection of blocks were conducted based on the method introduced in the former paragraph. Interestingly, from Fig. 10.4, it is shown that few crosses are marked when the mean speed is in the process of increasing. On the other hand, however, the mean speed is more or less slow or is about to decrease at the red marks. Namely, it is implied from the result that blocks are the cause of low smoothness of the pedestrian flow.

10.5 Conclusion

In this study, the effect of speed heterogeneity on pedestrian flow was investigated in a series of experiments conducted in a circular corridor. The fundamental diagrams show that local density is not the only factor that determines the speed at which a person walks, and the speed of those walking nearby also has an effect. As for the macroscopic flow characteristics, no spontaneous lane formation was observed despite the speed difference among pedestrians. This triggers the cluster of walkers headed by a block of slower walkers. In addition, it is implied that these blocks decelerate the crowd speed. Therefore, it would be highly advisable to introduce sidewalk markings or lane dividers in pedestrian walkways to promote lane formation and prevent these blocks from being formed.

Acknowledgements This work was partially supported by JST-Mirai Program Grant Number JPMJMI17D4, Japan and MEXT as "Priority Issues and Exploratory Challenges on post-K (Supercomputer Fugaku)" (Project ID: hp190163).

References

1. A. Gorrini, G. Vizzari, S. Bandini, Age and group-driven pedestrian behaviour: from observations to simulations. Collect. Dyn. **1**, 1–16 (2016)
2. C. Feliciani, K. Nishinari, Phenomenological description of deadlock formation in pedestrian bidirectional flow based on empirical observation. J. Statist. Mech. Theory Exp. **2015**(10), P10003 (2015)
3. T.-Q. Tang, L. Chen, R.-Y. Guo, H.-Y. Shang, An evacuation model accounting for elementary students individual properties. Physica A: Statist. Mech. Appl. **440**, 49–56 (2015)
4. M. Spearpoint, H.A. MacLennan, The effect of an ageing and less fit population on the ability of people to egress buildings. Saf. Sci. **50**(8), 1675–1684 (2012). Evacuation and Pedestrian Dynamics
5. M. Manley, Y. Kim, K. Christensen, A. Chen, Modeling emergency evacuation of individuals with disabilities in a densely populated airport. Transp. Res. Record: J. Transp. Res. Board **2206**, 32–38 (2011)
6. J. Koo, Y.S. Kim, B.-I. Kim, Estimating the impact of residents with disabilities on the evacuation in a high-rise building: a simulation study. Simul. Model. Pract. Theory **24**, 71–83 (2012)
7. S. Cao, J. Zhang, D. Salden, J. Ma, C. Shi, R. Zhang, Pedestrian dynamics in single-file movement of crowd with different age compositions. Phys. Rev. E **94**(1), 012312 (2016)

8. J. Zhang, W. Klingsch, A. Schadschneider, A. Seyfried, Transitions in pedestrian fundamental diagrams of straight corridors and t-junctions. J. Statist. Mech. Theory Exp. **2011**(06), P06004 (2011)
9. A. Fujita, C. Feliciani, D. Yanagisawa, K. Nishinari, Traffic flow in a crowd of pedestrians walking at different speeds. Phys. Rev. E **99**(6), 062307 (2019)

Chapter 11
The Effect of an Obstacle Before a Bottleneck: Inert Particles, Sheep, and Persons

Angel Garcimartín, J. M. Pastor, C. Martín-Gómez, I. Echeverría, and Iker Zuriguel

Abstract The placement of an obstacle before the exit has been put forward as a plausible mechanism to alleviate flow interruptions. Whenever a dense flow of discrete particles goes through a bottleneck, arches can form there and restrict the flow. The driving force tends to make the arches more stable. An obstacle before the exit would reduce the load and therefore arches should be easier to break down. We have tested this strategy in different systems. In a two dimensional silo filled with grains, an obstacle can improve by two orders of magnitude the time between clogs. We have also demonstrated that an obstacle can ease the entrance of a sheep herd through a door. Surprisingly, this does not seem to work for people. Although we are still trying to understand the reasons behind this result, some possible explanations are pointed out.

11.1 Introduction

Among of the situations that have recently attracted more attention in the field of pedestrian dynamics, the evacuation through an emergency exit is arguably an outstanding one. Several reasons make it especially appealing: it is undoubtedly a subject worth studying purely for practical reasons, and besides contributions from different viewpoints are likely to help—physics, engineering, psychology, and other

A. Garcimartín (✉) · I. Echeverría · I. Zuriguel
Facultad de Ciencias, Dpto. de Física y Matemática Aplicada, Universidad de Navarra, 31080 Pamplona, Spain
e-mail: angel@unav.es

J. M. Pastor
Focke Meler Gluing Solutions S.A., Pol. Los Agustinos c/G, nave D-43, 31160 Orkoien, Navarra, Spain

C. Martín-Gómez
Dpto. de Construcción, Instalaciones y Estructuras, Escuela Técnica Superior de Arquitectura, Universidad de Navarra, 31080 Pamplona, Spain

I. Zuriguel et al. (eds.), *Traffic and Granular Flow 2019*, Springer Proceedings in Physics 252,
https://doi.org/10.1007/978-3-030-55973-1_11

disciplines, can supply models, ideas and approaches to gain a better understanding about the emergence of flow interruptions and how to avoid them.

One of these insights is due to Helbing and coworkers [1, 2], who proposed to place an obstacle before the exit as an strategy to improve the evacuation. This counterintuitive solution would be beneficial inasmuch pressure would be reduced at the door. In a competitive evacuation, pressure builds up at the door, making clogs more stable and thus longer (this is known as the *Faster-Is-Slower* effect). If the obstacle absorbs part of the pressure, then clogs could be undone more easily, resulting in a lower evacuation time. The authors even performed a mock experiment to demonstrate the notion [2].

11.2 Obstacles with Granular Matter Flows

In principle, this effect should be observed even for inert particles. Some of us performed a series of experiments in a silo filled with granular matter, in both two and three dimensions. The measurements were conclusive by demonstrating how the clogging probability can be reduced by a factor of 100 by the placement of an obstacle before the exit [3]. The obstacle was circular in shape and it was centered with the exit midpoint (Fig. 11.1). This study revealed that the position of the obstacle is crucial to achieve this reduction: a small variation in the distance from the exit to the obstacle caused big variations. Also, in granular matter at least, the flow rate is not changed severely by the obstacle. A small positive variation has been found in numerical simulations if the obstacle is placed at an optimal position [4].

Fig. 11.1 *Clockwise from top left*: outflow of grains (diameter 1 mm) from a 2D silo, sheep entrance to a barn, evacuations carried out by soldiers, and by a group of students

11.3 Observations of a Sheep Herd

In another series of experiments, we recorded the entrance of a sheep herd to a barn were fodder awaited them. We placed an obstacle (a cylindrical column) in front of the door (Fig. 11.1), and measured the time interval between the passage of two consecutive animals. We found that the obstacle reduced the duration of clogs [5]. As in the case of grains, the effect could only be observed if the obstacle was placed at a particular location; the optimal distance was slightly larger than the exit width.

It was also observed that the distribution of clogging durations (the time elapsed between two consecutive animals) displayed fat tails. The obstacle prevented long clogs, thus shortening the tail. Even if the flow is not improved, this effect is notable.

11.4 Evacuation Drills

With the knowledge gained in the experiments with inert grains and with animals, we designed a series of evacuation drills with several groups of persons. The procedures were set with the following desiderata in mind. First, we sought reproducibility and statistical robustness. In many cases the effects can only be observed as an average behaviour (so performing just one evacuation, for instance, will not provide sound quantitative grounds to state that an obstacle is beneficial). In this respect, the number of people in the evacuation drills we carried out was always larger than 70. Indeed, we have found that there is a transient at the beginning and at the end of about a dozen people, so this issue is crucial. Second, we wanted to change the distance from the obstacle to the door by small amounts to check for differences in the distributions of time intervals between consecutive individuals. And third, competitiveness—which is an elusive variable—had to be quantified.

As the evacuation drills have been described elsewhere [6, 8, 9], we will only provide here a short briefing. They were carried out with three sets of young volunteers under different competitiveness. In the first set of experiments, $N = 85$ students were gathered at an indoor gym and had to exit as quickly as possible through a 75 cm wide door. The obstacle was a 300 kg cylindrical column of 100 m diameter, placed at 100 cm. from the door. This obstacle (shown in Fig. 11.1) moved a few cm during the dozen evacuation drills performed, so data are not reported here but are compatible with the findings of the other drills. The second set was a group of $N = 72$ students that performed 10 evacuation drills with a 1 ton column placed at 100 cm from the door (this obstacle did not move). The third group was a set of $N = 165$ soldiers from the Spanish Army that performed several evacuation drills, exiting through a 75 cm wide door with the same 1-ton obstacle in front of it (see Fig. 11.1): 6 evacuations with the obstacle at 50 cm from the door, 6 with the obstacle at 60 cm, 6 with the obstacle at 70 cm, and 7 without an obstacle, all of them in conditions of high competitiveness; besides, 3 evacuations without an obstacle and 3 evacuations with the obstacle at 60 cm were performed in low competitiveness.

The evacuation drills were registered with a video camera hanging from the ceiling and pointing downwards (the field of view covered 6×3 m^2 in the case of the students and 6×3.3 m^2 in the case of soldiers). From these recordings, we tracked the position of every person and then we calculated the individual velocities. Another camera outside the room was used to obtain the times at which each person crossed the door. A pressure sensor at the door jamb measured the local pressure with a sampling rate of 25 Hz.

11.5 The Effect of the Obstacle on a Pedestrian Flow

We will present the measurements corresponding to the soldiers, as they are the largest group and therefore provide the best quality. Let us note once again that all the tests we performed yield similar results. In Fig. 11.2 we represent the boxplots of the time interval Δt_i between consecutive persons (the average flow rate is the inverse of Δt_i). The median and the average are given in Table 11.1.

Two salient features can be noted. First, the mean flow rate is not improved by the presence of the obstacle. The only condition that decreases evacuation time in our experiments is low competitiveness (the well knowm 'Faster-Is-Slower' effect). Second, long clogs (say, those that last longer than 1 s) are not avoided with the obstacle. This is at odds with some previous results (see for instance [2]). Several issues may be at play. We have observed that there is always a short transient at the beginning and at the end of the evacuation of about a dozen individuals. If the group is very small (less than two dozen people) then only these transients will be observed. Besides, the level of competitiveness must be somehow quantified, at least approximatively. (Pressure measurements and forces during the evacuation reported here are reported in [7]). Otherwise, *panic* or *competitiveness* can be empty words.

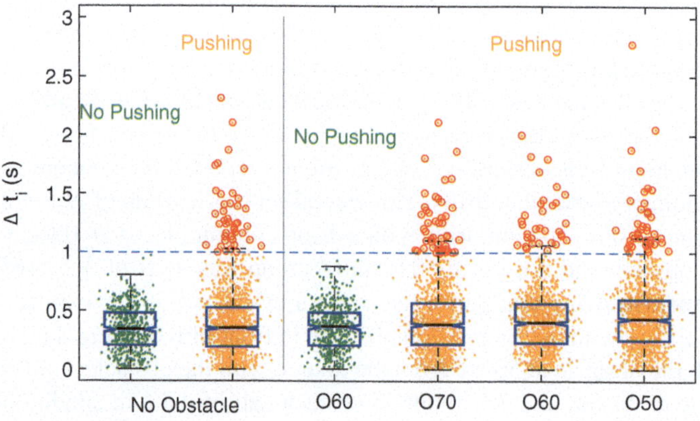

Fig. 11.2 Time intervals between two consecutive persons. Onn stands for the obstacle at nn cm from the door. Red dots mark clogs longer than 1 s

Table 11.1 The mean time interval Δt_i, which is the inverse of the average flow rate, along with the median. Units are seconds. P stands for pushing and NP for no pushing

Case	Mean	Median
NO/NP	0.35 ± 0.02	0.34
NO/P	0.40 ± 0.02	0.35
O60/NP	0.36 ± 0.02	0.37
O70/P	0.42 ± 0.02	0.39
O60/P	0.43 ± 0.02	0.40
O50/P	0.46 ± 0.02	0.44

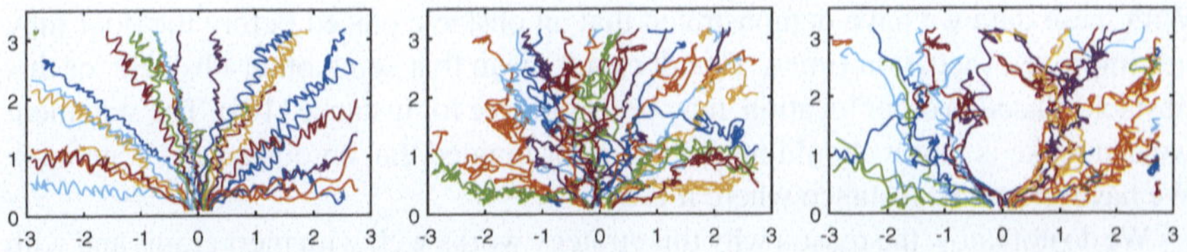

Fig. 11.3 Traces of the paths followed by soldiers: without pushing, pushing, pushing with an obstacle. Only 30 random persons are shown. Units are metres. Note how the obstacle causes paths to be less entangled

Also, maybe the obstacle features are crucial. We have only used a cylindrical column centered in front of the door. Moreover, the column was relatively wide compared to the door. Of course, there can also be other reasons that escape our understanding for the failure of our obstacle to improve the evacuation flow.

However, we have found some benefits from the obstacle. There is one we consider very important: it prevents sideways rushes (Fig. 11.3). These have been observed in several cases of high competitive evacuation. They can easily provoke fallings, which may be as dangerous as clogs because people can be trampled underfoot to dead. So avoiding these rushes is by no means inconsequential.

Fig. 11.4 Pressure at the doorjambs. Each point its an average over the whole area covered by a sensor, sampled every 0.04 s. P stands for pushing, NP for No Pushing, and Onn means that the obstacle was placed at nn cm from the door

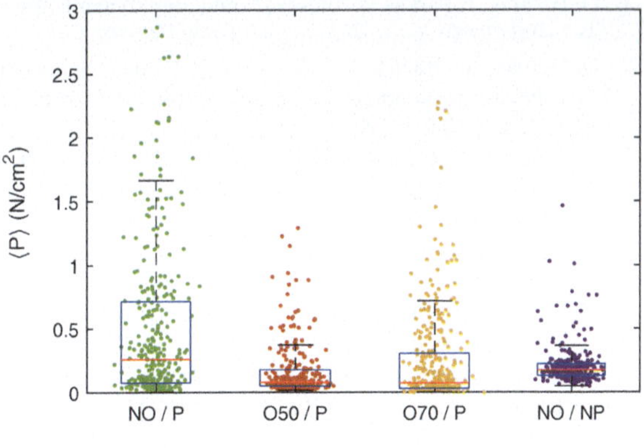

Another benefit of the obstacle is that it reduces pressure at the doorjambs. The average pressure at the doorjambs was sampled every 0.04 s, as explained in [7]. The measurements are shown in Fig. 11.4. As can be seen, evacuation drills performed under high competitiveness and without an obstacle are the ones that give highest pressures at the doorjamb. Placing an obstacle before the door can reduce the pressure to levels comparable to a noncompetitive situation.

11.6 Discussion and Conclusion

With these data we have demonstrated that an obstacle placed before the door may *not* improve evacuation times. This does not mean that some other obstacle, or this obstacle placed in other location, may be useful (see for instance [10]). The statement we can make is that it should not be taken for granted that an obstacle is beneficial. We have shown an instance where it is not.

We do not know the reason why this strategy works well with inert grains and with animals and instead it does not seem to work with people in the same configuration. It is difficult to support the guesses we have with any sound reasoning, and probably more experiments will be needed. In particular, shape, location, size, and an off-centered obstacle can provide meaningful data to understand the role of an obstacle in evacuation dynamics.

Acknowledgements We want to thank L. F. Urrea and J. C. Sánchez for technical help, and R. Cruz Hidalgo and D. Maza for comments and discussions. We also thank the volunteers who participated in the experiments, especially the to the officers and soldiers in the América 66 Regiment of the Spanish Army. This work has been funded by Ministerio de Economía y Competitividad (Spanish Government) through Projects No. FIS2014-57325 and FIS2017-84631-P, MINECO/AEI/FEDER, UE. I.E. acknowledges Asociación de Amigos de la Universidad de Navarra for a grant.

References

1. D. Helbing, I. Farkas, T. Vicsek, Simulating dynamical features of escape panic. Nature **407**, 487–490 (2000). https://doi.org/10.1038/35035023
2. D. Helbing, L. Buzna, A. Johansson, T. Werner, Self-organized pedestrian crowd dynamics: experiments, simulations, and design solutions. Transp. Sci. **39**, 1–24 (2005). https://doi.org/10.1287/trsc.1040.0108
3. I. Zuriguel, A. Garcimartín, C. Lozano, R. Arévalo, D. Maza, Silo clogging reduction by the presence of an obstacle. Phys. Rev. Lett. **107**, 278001 (2011). https://doi.org/10.1103/PhysRevLett.107.278001
4. F. Alonso-Marroquín, S.I. Azeezullah, S.A. Galindo-Torres, L.M. Olsen-Kettle, Bottlenecks in granular flow: when does an obstacle increase the flow rate in an hourglass? Phys. Rev. E **85**, 020301 (2012). https://doi.org/10.1103/PhysRevE.85.020301
5. I. Zuriguel, J. Olivares, J.M. Pastor, C. Martín-Gómez, L.M. Ferrer, J.J. Ramos, A. Garcimartí, Effect of obstacle position in the flow of sheep through a narrow door. Phys. Rev. E **94**, 032302 (2016). https://doi.org/10.1103/PhysRevE.94.032302

6. A. Garcimartín, D. Maza, J.M. Pastor, D.R. Parisi, C. Martín-Gómez, I. Zuriguel, Redefining the role of obstacles in pedestrian evacuation. New J. Phys. **20**, 123025 (2018). https://doi.org/10.1088/1367-2630/aaf4ca
7. I. Zuriguel, I. Echeverría, D. Maza, R.C. Hidalgo, C. Martín-Gómez, A. Garcimartín, Contact forces and dnamics of pedestrians evacuating a rooom: the column effect. Saf. Sci. nn (2019)
8. A. Garcimartín, I. Zuriguel, J.M. Pastor, C. Martín-Gómez, D.R. Parisi, Experimental evidence of the 'Faster Is Slower' effect. Transp. Res. Procedia **2**, 760 (2014). https://doi.org/10.1016/j.trpro.2014.09.085
9. A. Garcimartín, D.R. Parisi, J.M. Pastor, C. Martín-Gómez, I. Zuriguel, Flow of pedestrians through narrow doors with different competitiveness. J. Stat. Mech. **2016**, 043402 (2016). https://doi.org/10.1088/1742-5468/2016/04/043402
10. D. Yanagisawa, A. Kimura, A. Tomoeda, R. Nishi, Y. Suma, K. Ohtsuka, K. Nishinari, Introduction of frictional and turning function for pedestrian outflow with an obstacle. Phys. Rev. E **80**, 036110 (2009). https://doi.org/10.1103/PhysRevE.80.036110

Chapter 12
Towards Inferring Input Parameters from Measurements: Bayesian Inversion for a Bottleneck Scenario

Marion Gödel, Rainer Fischer, and Gerta Köster

Abstract Microscopic crowd simulations can help to enhance the safety of pedestrians in various situations, provided that the simulation yields realistic results. One crucial step on the way toward realistic simulations is calibration. In this paper, we present Bayesian inversion as a systematic method for the calibration of input parameters. We demonstrate how Bayesian inversion works by an example: We infer the mean free-flow speed from Voronoi density using simulated data in a well-investigated bottleneck scenario. Based on the results, we discuss benefits and limitations of applying this technique to crowd simulation.

12.1 Introduction

Crowd simulation is a central part of research on pedestrian dynamics dealing with the prediction of the behavior of a crowd. We focus on microscopic crowd simulation in which the movement path (trajectory) of individual virtual pedestrians (agents) is predicted. Microscopic crowd simulations can be used for a wide range of applications: from identifying potential bottlenecks to testing evacuation concepts, to (hopefully soon) live monitoring and prediction of crowd behavior.

Typically, three layers of models are distinguished: strategical layer, tactical layer, and operational layer. We focus on the operational layer (locomotion models), because we believe that this layer needs to be handled before we tackle more complicated situations that require tactical and strategical layers. Since Bayesian

M. Gödel (✉) · R. Fischer · G. Köster
Munich University of Applied Sciences, Lothstrasse 64, 80335 Munich, Germany
e-mail: marion.goedel@hm.edu

R. Fischer
e-mail: rainer.fischer@hm.edu

G. Köster
e-mail: gerta.koester@hm.edu

M. Gödel
Technical University of Munich, Boltzmannstrasse 6, 85748 Garching, Germany

I. Zuriguel et al. (eds.), *Traffic and Granular Flow 2019*,
Springer Proceedings in Physics 252,
https://doi.org/10.1007/978-3-030-55973-1_12

inversion is independent of the underlying model, it can be applied to all types of locomotion models, such as force-based models, rule-based models or velocity-based models [1].

Simulations can only meet real-world challenges if they correctly predict the movement of a crowd. Therefore, two steps in the development cycle are crucial: to ensure that the software code is correct (verification) and that the results are in line with observations (validation). Verification is typically achieved through software tests, validation by investigating benchmark scenarios, such as the test cases defined by the RiMEA committee [2].

Besides validation and verification of the models, input parameters must be chosen properly, especially when they have a strong influence on simulation outcomes. In this contribution, we investigate the applicability of Bayesian inversion, a method from the field of uncertainty quantification, for parameter calibration.

12.1.1 Parameter Calibration

In scenarios where experiments are available, we want choose the parameter values based on this data. However, measurements may not provide the parameters we need directly. For instance, in a dense crowd, we may have density values, but cannot access free-flow speeds. We need to infer the parameter. The field of uncertainty quantification provides us with a set of systematical methods to investigate the role of parameters. In particular, Bayesian inversion is a likelihood-based method that can be used to infer information about the input parameters from data [3]. We demonstrate how it can be applied to pedestrian dynamics in this work. Other methods for inversion include likelihood-free methods such as Approximate Bayesian Computation [4].

12.1.2 State-of-the-Art

So far, there are few applications of inversion methods to pedestrian dynamics: Bayesian inversion to a force-based model [17], Approximate Bayesian Computation for model calibration and model selection [5], a proof of concept of Bayesian inversion for the optimal steps model in a simple scenario [18]. A more general overview on model fitting and model selection is provided in [6].

12.1.3 Paper Outline

In this work, we focus on the application of Bayesian inversion on a thoroughly investigated scenario in simulations and experiments: a bottleneck setup. At first, we introduce the concept of Bayesian inversion and present how it can be carried out in practice (Sect. 12.2). Then we describe the configuration of the inversion and the simulation scenario (Sect. 12.3). We apply the method to a bottleneck scenario to infer an important input parameter of most microscopic locomotion models, the mean of the free-flow speeds of the virtual pedestrians (agents). In order to test the method, we use simulated data as input for the inversion instead of experimental data. We present the results of inferring one parameter, the mean free-flow speed (Sect. 12.4), and, finally, conclusions and a brief outlook (Sect. 12.5).

12.2 Bayesian Inversion

Figure 12.1 demonstrates the principle of Bayesian inversion: Based on the data d, an initial guess for the distribution of the uncertain parameter (typically referred to as prior distribution) is updated using evaluations of the model V at different parameter guesses x. As a result, one obtains an informed posterior distribution.

12.2.1 Theory Behind Bayesian Inversion

In the Bayesian inverse problem [7]

$$d = V(x) + \eta \tag{12.1}$$

with data d, forward map V, (random) input parameters x, and random (measurement) noise η, one attempts to find the input parameters x that fulfill (12.1). In our case, the forward map V is a simulation of a certain scenario with a simulator. The simulator setup includes a locomotion model, the topography and the methods to calculate the chosen quantity of interest.

The likelihood-based approach requires information about the measurement noise, which is typically not known. This is also the case in our study. Hence, we assume a normal, zero-mean measurement noise $\eta \sim \mathcal{N}(0, \Gamma)$ where covariance Γ is a diagonal matrix meaning that the noise components are not correlated. This assumption leads to the likelihood

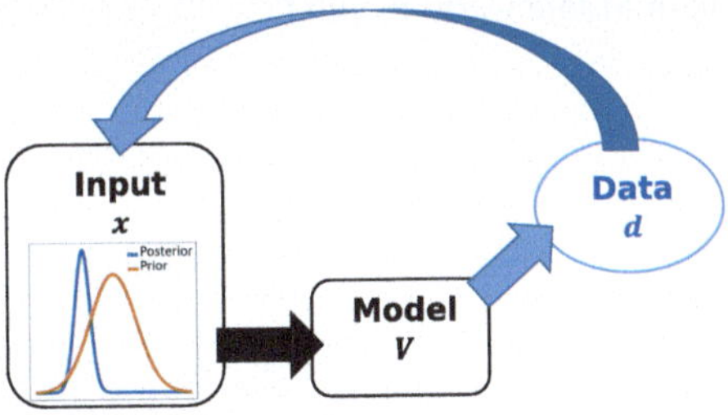

Fig. 12.1 Principle of Bayesian inversion, adapted from [7, p. 3]

$$\rho_{like}(d, x) = \exp(-f_d(x))$$

(see [18, p.84] for more details). The data misfit function

$$f_d(x) = \frac{1}{2} \parallel \Gamma^{-\frac{1}{2}}(d - V(x)) \parallel_2^2 \tag{12.2}$$

measures the difference between data d and model evaluation $V(x)$.

Based on Bayes' theorem, the posterior distribution is proportional to the product of prior distribution ρ_{prior} and likelihood ρ_{like}:

$$\rho_{pos}(d, x) = c_{pos}^{-1}(d) \cdot \rho_{like}(d, x) \cdot \rho_{prior}(x) \propto \rho_{like}(d, x) \cdot \rho_{prior}(x) \tag{12.3}$$

where $c_{pos}(d) = \rho(d)$ is a normalization constant, the so-called evidence.

12.2.2 Bayesian Inversion in Practice

Typically, the posterior cannot be calculated analytically due to high-dimensional integrals that need to be solved. One solution is to sample the posterior instead. There are several sampling methods, a common choice among them for inverse problems are Markov chain Monte Carlo methods [8, 9]. This group of methods constructs a Markov chain, whose stationary distribution is the posterior distribution. Therefore, the samples from the Markov chain are samples from the posterior (after a certain burn-in period). Based on these samples, we can analyze type and modes of the distributions.

Instead of an MCMC algorithm, one could also use other sampling algorithms, e.g. importance sampling or sequential Monte Carlo, or particle filters that may have speed advantages under certain conditions. A drawback of the MCMC approach is that each iteration of the Markov chain requires one simulation of the forward model, i. e. one simulation. As a result, the procedure is computationally expensive. However, since MCMC methods are successfully used in numerous applications of Bayesian statistics, they are often called "the computational workhorse" for Bayesian inversion [9]. Hence, we use this approach for our application. We overcome the high computational effort by replacing the simulation with a simple surrogate model to speed up the computation. We choose to use the Metropolis algorithm [10] as Markov chain Monte Carlo method.

12.3 Parameter Estimation for the Bottleneck Scenario

The concept of Bayesian inversion is now applied to the bottleneck scenario, a simulation scenario that is widely analyzed in experiments and simulations [11, 12]. All simulations are carried out with our simulation framework Vadere and the behavioral heuristics model as locomotion model [13].

12.3.1 Modeling of the Bottleneck Scenario

Figure 12.2 shows a snapshot of the simulation of the bottleneck experiment. We decide to use the simulation scenario setup from [14]. The underlying locomotion model, the behavioral heuristics model (BHM), is based on a set of heuristics which are presented in Fig. 12.3. In [14], a good alignment of the experimental results of Liddle et al. with the tangential evasion heuristic of the BHM is shown.

Instead of experimental data, we demonstrate the functioning of Bayesian inversion based on data retrieved from a simulation. That means we know the actual parameter value that we are trying to infer and are consequently able to test the method. We choose 1.34 m/s as true parameter.

12.3.2 Setup of the Inversion

For all methods from the field of uncertainty quantification, at first, the uncertain parameters and the quantity of interest need to be chosen. We choose the mean free-

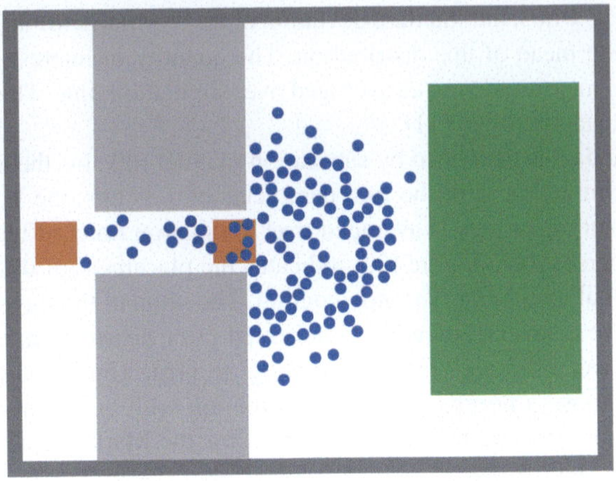

Fig. 12.2 Snapshot of a simulation of the bottleneck scenario after 30 seconds

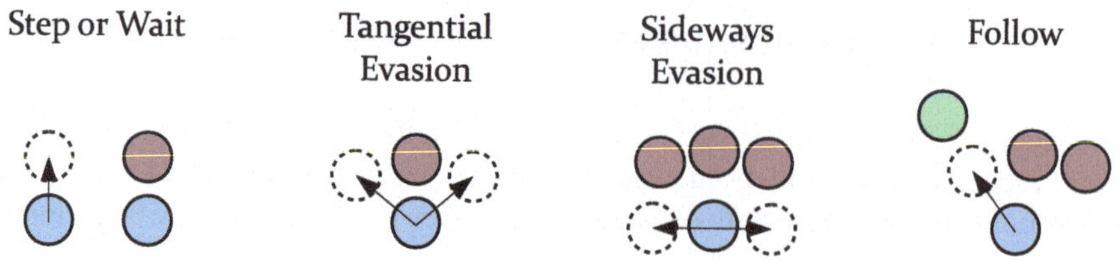

Fig. 12.3 Heuristics of the behavioral heuristics model (BHM) from [14]. In the step-or-wait heuristic, a step in direction of the target is taken if it is free; otherwise, the agent remains. When the evasion heuristics are chosen, the agent can evade other pedestrians by a tangential or by either a tangential or a sideways step. For other scenarios, such as counterflow, the follower heuristic can be used

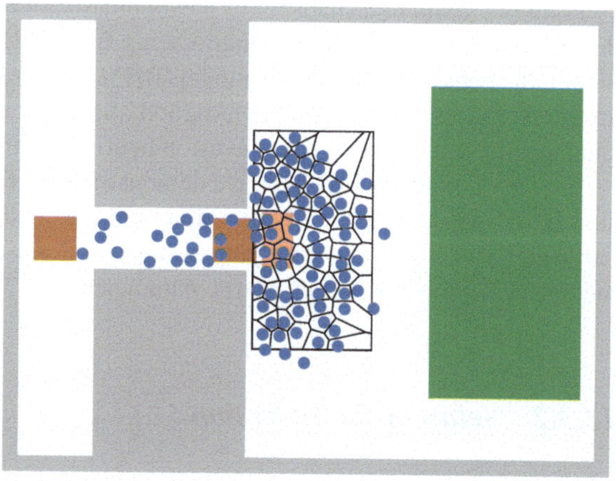

Fig. 12.4 Voronoi diagram of the crowd density. The Voronoi diagram is calculated for a larger area than the measurement area to obtain correct results within the chosen measurement area (red)

flow speed of the agents as uncertain parameter. Within the simulator, each agent is assigned a random free-flow speed from a truncated normal distribution. We set the limits to 0.5 and 2.0 m/s, the standard deviation of this normal distribution to 0.26 m/s [15] and vary the mean of this distribution. The quantity of interest is the Voronoi density within a measurement area averaged over simulation time. The measurement area was placed according to [11].

Note that the data is generated by simulation in order to know the true parameter. Based on the simulation with the true parameter of 1.34 m/s, the Voronoi density within the measurement area is averaged over simulation time, resulting in a mean density of 1.8 persons/m^2. Figure 12.4 indicates the placement of the measurement area and the calculation of the Voronoi diagram. The setup of the Bayesian inversion is as follows: We choose a so-called uninformed prior distribution, i. e., a uniform distribution between 0.5 and 2.0 m/s according to [15]. The initial point is set to 0.75 m/s. We choose a normal proposal distribution with an adaptive jump width regulation [16] to generate the next candidates for the Markov chain and run 10^5 iterations of the Markov chain.

12.4 Applying Bayesian Inversion

In the following, we perform Bayesian inversion for a typical parameter of pedestrian crowd simulations, the mean free-flow speed. In most models, it is assumed that each pedestrian has an inherent speed with which he or she moves when the path is free. Often, free-flow speeds for the agents are drawn as realizations of a (truncated) normally distributed random variable.

12.4.1 *Surrogate Model*

We generate a surrogate model for the inversion for two reasons: First, Bayesian inversion is accelerated by evaluating only the surrogate model instead of performing Vadere evaluations. Generating a surrogate model is costly in itself, but can be easily parallelized (embarrassingly parallel), while the Markov chain within the inversion is, by its nature, a serial procedure and consequently harder to parallelize. Second, the initial state of the simulation is stochastic since the free-flow speeds of the agents as well as the starting positions are chosen in a random manner. Therefore, we want to average over several simulation runs for each parameter configuration. This can be done in the construction of the surrogate model and, since it can be parallelized, it is significantly cheaper than averaging in each step of the Markov chain.

Instead of building a surrogate model for the simulation model itself, we create the surrogate for the data misfit function (12.2). Therefore, we evaluate the data misfit function at 100 equidistant parameter values within the prior distribution. For each parameter value, we average over 10 runs to account for the random setting of initial positions and free-flow speeds. For this test application of the inversion, we choose a simple surrogate: We fit a spline to the averaged data points. The resulting surrogate model is shown in Fig. 12.5 together with all evaluations performed. We observe fluctuations, notably for lower and higher free-flow speeds. Since they are not present in the region of interest, we argue that they do not unduly impact the results of the inversion. Note that the fluctuations are not completely attenuated by the surrogate we chose.

12.4.2 *Results of the Inversion*

The results of the inversion are shown in Table 12.1 and Fig. 12.6. The mean of the candidates is very close to the true parameter value of 1.34 m/s, showing a successful application of Bayesian inversion to calibrate the parameter. Nevertheless, the standard deviation of the result is quite large (0.20 m/s). This indicates that the parameter value could not be inferred with a high confidence. If we take a look at the data misfit function in Fig. 12.5, it becomes clear why: For mean free-flow speeds between

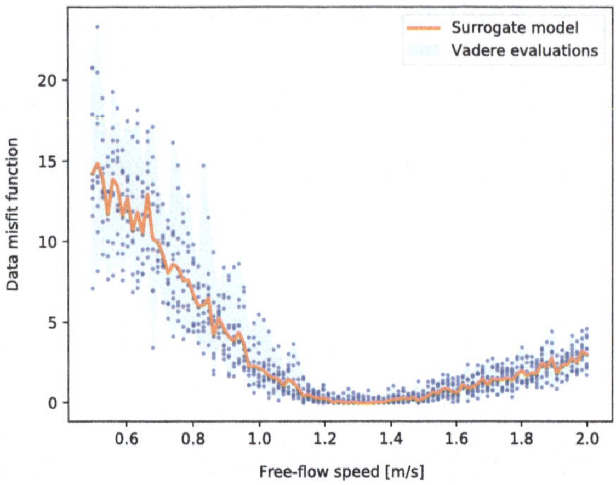

Fig. 12.5 Surrogate model generated from evaluations at 100 parameter values (10 simulations for each mean free-flow speed)

Table 12.1 Results of the inversion applied to the bottleneck scenario to infer the distribution of the mean free-flow speeds

Computation time (excl. surrogate generation)	0.61 min
Mean of samples (without burn-in)	1.38 m/s
Standard deviation of samples	0.20 m/s
Overall acceptance rate	45.49%
Effective sample size	32625.43 (≈ 33%)
Mean jump width	1.00 m/s

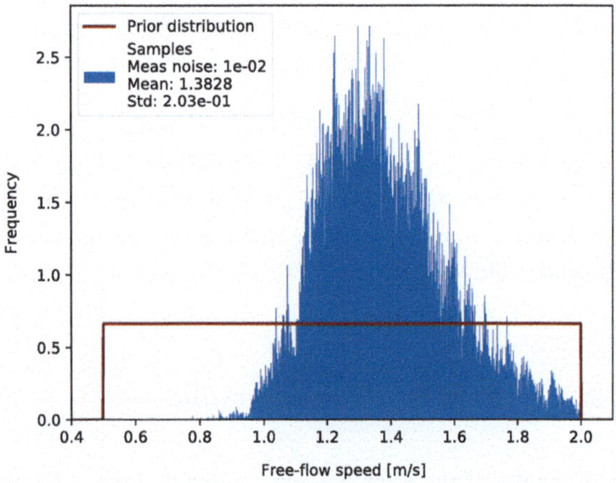

Fig. 12.6 Comparison of prior and posterior distribution. The histogram of the informed posterior samples (mean free-flow speed) is shown together with the uninformed prior distribution

1.2 and 1.4 m/s, the values of the data misfit are almost identical. That means, in this region, the parameter of mean free-flow speed does not have a high impact on the resulting Voronoi density. In addition, the method assumes that the provided measurement data is noisy $\eta \sim \mathcal{N}(0, 0.01)$. Consequently, the method is not able to capture the exact value of the true parameter with a high certainty. Nevertheless, to obtain a density close to the provided data point, an interval of values for the free-flow speed can be chosen. For the modeler, this also means that, in the scenario, the uncertain parameter does not need to be known precisely to obtain predictions close to the experimental results.

12.5 Conclusion and Outlook

We presented a successful application of Bayesian inversion to a typical scenario in crowd dynamics, a bottleneck scenario: The true value of a parameter, the mean free-flow speed of the agents, was inferred correctly based on the Voronoi density calculated from a simulation. While the mean of the samples of the posterior distribution of the uncertain parameters captured the true parameter very well, the standard deviation showed that the confidence in the resulting posterior distribution is lower than we expected. This is due to the weak link between the uncertain parameter and the chosen quantity of interest. This implies that for the density in the bottleneck scenario, a range of free-flow values can be used to obtain similar results. In this setup, the uncertain parameter need to not be known precisely to obtain a meaningful prediction, which is important information for the design of a simulation study.

In the next step, we would like to apply the approach on real data from the experiment carried out in Jülich [11]. In addition, we plan to infer multiple parameters to allow for a calibration of several parameters. Furthermore, we plan to use the results of the inversion as input for forward propagation.

Acknowledgments This work was funded by the German Federal Ministry of Education and Research through the projects S2UCRE (grant no. 13N14464) and OPMOPS (grant no. 13N14562). The authors acknowledge the support by the Faculty Graduate Center CeDoSIA of TUM Graduate School at Technical University of Munich and the research office FORWIN at Munich University of Applied Sciences.

References

1. F. Martinez-Gil, M. Lozano, I. García-Fernández, F. Fernández, ACM Comput. Surv. **50**, 5 (2017). https://doi.org/10.1145/3117808
2. RiMEA, *Guideline for Microscopic Evacuation Analysis*. RiMEA e.V., 3rd edn. (2016). http://www.rimea.de/
3. A.M. Stuart, Acta Numerica **19**, 451 (2010). https://doi.org/10.1017/s0962492910000061
4. T. Toni, D. Welch, N. Strelkowa, A. Ipsen, M.P. Stumpf, J. Royal Soc. Interf. **6**(31), 187 (2009). https://doi.org/10.1098/rsif.2008.0172

5. N. Bode, arXiv 2001.10330(v1) (2020). https://collective-dynamics.eu/index.php/cod/article/view/A68 https://doi.org/10.17815/CD.2020.68
6. N.W.F. Bode, E. Ronchi, Nr. 9 or Nr. 13 *Collective Dynamics* (2019). https://doi.org/10.17815/CD.2019.20
7. A.M. Stuart, *Uncertainty Quantication in Bayesian Inversion* (2014). http://homepages.warwick.ac.uk/~masdr/TALKS/stuart_SIAMUQ.pdf
8. S. Brooks, A. Gelman, G.L. Jones, X.L. Meng, *Handbook of Markov Chain Monte Carlo*, 1st edn. (Chapman and Hall/CRC, 2011)
9. P.G. Constantine, C. Kent, T. Bui-Thanh, SIAM J. Scient. Comput. **38**(5), A2779 (2016). https://doi.org/10.1137/15M1042127
10. N. Metropolis, A.W. Rosenbluth, M.N. Rosenbluth, A.H. Teller, E. Teller, J. Chem. Phys. **21**, 1087 (1953). https://doi.org/10.1063/1.1699114
11. J. Liddle, A. Seyfried, B. Steffen, W. Klingsch, T. Rupprecht, A. Winkens, M. Boltes, arXiv 1105.1532(v1) (2011). arXiv.org/abs/1105.1532
12. A. Schadschneider, W. Klingsch, H. Klüpfel, T. Kretz, C. Rogsch, A. Seyfried, *Encyclopedia of Complexity and Systems Science*, ed. by R.A. Meyers (Springer, New York, 2009), pp. 3142–3176. https://doi.org/10.1007/978-0-387-30440-3_187
13. B. Kleinmeier, B. Zönnchen, M. Gödel, G. Köster, Collect. Dyn. **4** (2019). https://doi.org/10.17815/CD.2019.21
14. M.J. Seitz, N.W.F. Bode, G. Köster, J. Royal Soc. Interf. **13**(121), 20160439 (2016). https://doi.org/10.1098/rsif.2016.0439
15. U. Weidmann, *Transporttechnik der Fussgänger*, *Schriftenreihe des IVT*, vol. 90, 2nd edn. (Institut für Verkehrsplanung, Transporttechnik, Strassen- und Eisenbahnbau (IVT) ETH, Zürich, 1992). https://doi.org/10.3929/ethz-a-000687810
16. A. Gelman, G.O. Roberts, W.R. Gilks, Efficient Metropolis jumping rulesJournal: Bayesian Statistics**5**, 599–607 (1996)
17. A. Corbetta, A. Muntean, K. Vafayi, *Mathematical Biosciences and Engineering*. Parameter estimation of social forces in pedestrian dynamics models via a probabilistic method **12**(2) (2015). https://doi.org/10.3934/mbe.2015.12.337
18. M. Godel, R. Fischer, G. Koster, in UNCECOMP 2019, 3rd ECCOMAS Thematic Conference on Uncertainty Quantication in Computational Sciences and Engineering Applying Bayesian Inversion with Markov Chain Monte Carlo to Pedestrian Dynamics (2019). https://doi.org/10.7712/120219.6322.18561
19. R.C. Smith, Uncertainty Quantication: Theory, Implementation, and Applications. Computational Science and Engineering (Society for Industrial and Applied Mathematics) (2014)

Chapter 13
Spatially Dependent Friction—A Way of Adjusting Bottleneck Flow in Cellular Models

Pavel Hrabák and František Gašpar

Abstract The spatially dependent friction is introduced into cellular floor-field model of pedestrian flow as a possible way of proper adjustment of flow through individual bottlenecks/exits in a rather complex structure containing multiple consecutive bottlenecks of different width. Next to the bulk friction, which is responsible for the conflict-induced delay while walking within the crowd, the local friction is defined individually for each exit-like bottleneck, and is responsible for the clogging in front of the exit. The local friction enables to adjust the flow through given bottleneck according to the observed or estimated value without necessity to change the global parameters. Such method enables to model flow through bottlenecks of similar but different width within one simulation scenario.

13.1 Introduction and Motivation

There is a wide range of models simulating pedestrian dynamics. An important role among them play cellular models [1]. Their advantage are relatively simple rules and thus low computational requirements of the simulation. Moreover, some of the models can be treated analytically by means of statistical physics [2]. Most of the investigated cellular models of pedestrian dynamics are based on the floor-field idea [3–5] that the probabilistic hopping of particles is influenced by the floor-field values of individual cells. Despite the relative simplicity of the rules, the floor-field model is able to reproduce key phenomena of complex collective motion of pedestrians [6].

P. Hrabák (✉)
Faculty of Information Technologies, Czech Technical University in Prague, Thákurova 9, 160 00, Prague 6, Czechia
e-mail: pavel.hrabak@fit.cvut.cz

F. Gašpar
Faculty of Nuclear Sciences and Physical Engineering, Czech Technical University in Prague, Břehová 7, 115 19, Prague 1, Czechia
e-mail: gaspafra@fjfi.cvut.cz

I. Zuriguel et al. (eds.), *Traffic and Granular Flow 2019*,
Springer Proceedings in Physics 252,
https://doi.org/10.1007/978-3-030-55973-1_13

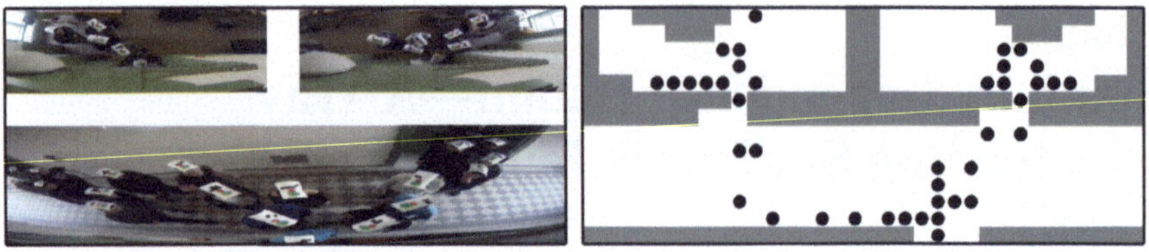

Fig. 13.1 Snapshot from the Prague experiment (left) and its simulation (right) by floor-field model

On the other hand, the cellular structure of the space in cellular models causes problems when calibrating and validating the cellular model against real data [7, 8]. In 2016 two experiments on evacuation of a rather complex structure with multiple consecutive bottlenecks were conducted in Prague and Krakow—see Fig. 13.1 for illustration. The aim of the collaborative research was to model and to simulate such scenarios by means of advanced cellular automata models [9], namely the Social Distances Model [10] and Bonds Floor Field Model [11].

While calibrating the models in the multiple-bottleneck scenario, the most problematic task was to capture the flow through individual bottlenecks of different width (70 cm, 85 cm) and geometry (wall width, presence of niche). The flow is assumed to depend linearly on the bottleneck width [12]. This is however problematic to achieve by cellular automata once the bottleneck width is relatively small with respect to the cell size (e.g. when simulating 70 cm wide exit using cells of the size 40 cm). It was observed that one cell wide bottleneck was too narrow, but two cell wide bottleneck was too wide. It is not usual that there are two independent flows through a door of the width 80 cm.

In floor-field based models, it is common that the behaviour of the particles in a dense crowd is influenced by the friction mechanism of conflict solution [13, 14]. The idea introduced in this paper is to change the value of the friction parameter in the cells preceding the bottlenecks in order to reproduce the desired bottleneck flow without changing any global values of the parameters.

13.2 Model Description

A simple parallel-update driven floor-field cellular model [3–5] with static field S and friction-function mechanism of conflict-solving $\phi(k, \zeta)$ [13, 14] is considered.

The playground of the model is represented by an orthogonal lattice $\mathbb{L}$ consisting of square cells (see Fig. 13.2 for illustration). Cell $x \in \mathbb{L}$ is represented by the position of its center $x = (x_1, x_2) \in \mathbb{Z}^2$, i.e. the neigbouring site to the right is $(x_1 + 1, x_2)$, etc.

Every cell is either empty or occupied by one particle. Particles are moving along the lattice by hopping to one of the neighbouring cells y from the von-Neumann neighbourhood $N(x) = \{y \in \mathbb{L}\,;\ |x_1 - y_1| + |x_2 - y_2| \le 1\}$ i.e. consist-

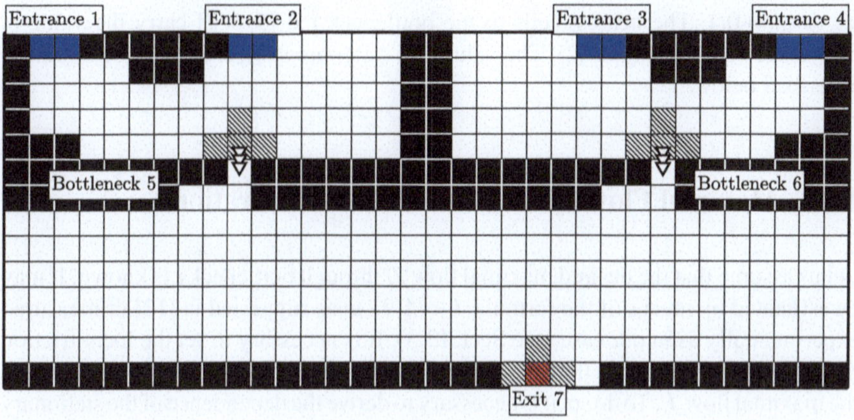

Fig. 13.2 Visualisation of the lattice used for simulations in this paper

ing of the nearest cells in the orthogonal direction. The choice of the particle's next target cell is probabilistic: the particle occupying cell x chooses as its target cell the cell $y \in N(x)$ with probability

$$\Pr(x \to y) \propto \exp(-k_S \cdot S_y) \cdot \mathbf{1}_{\{y \text{ is empty } \vee\ y=x\}}\,, \tag{13.1}$$

where S_y is the value of the static potential field S, $k_S \in [0, +\infty)$ is the sensitivity parameter to the potential, $\mathbf{1}_A$ the identification of the event A. The static field S plays the role of the potential. The value S_x is defined as the number of hops necessary to reach the main exit from the cell x.

Particles are entering the lattice through devoted entrance cells (in our case the entrances 1, 2, 3, 4). The simulation is data-driven, i.e. new particles enter the entrance cell accordingly to the experimental data (after rescaling the time to steps of the algorithm, see Sect. 13.3). Particles leave the lattice through the main exit cell (in our case the main exit 7), i.e. the particle occupying the exit cell leaves the system in next algorithm step with probability one.

The model is driven by fully-parallel update, i.e. all particles choose their target cells simultaneously. This can lead to the conflict, i.e. two or more particles trying to enter the same cell. For our investigation, the friction function mechanism adopted from [14] has been chosen, i.e. if k particles are trying to enter one cell, with probability

$$\phi(k, \zeta) = 1 - (1 - \zeta)^k - k(1 - \zeta)^{k-1} \tag{13.2}$$

none of the involved particles moves; with complementary probability, one particle is randomly chosen to enter the common target cell. Here $\zeta \in [0, 1]$ is the friction parameter – the higher ζ, the higher probability of blocking the motion. The friction parameter is spatially dependent, i.e, $\zeta = \zeta(x)$. More precisely, the bulk value is set to $\zeta_0 = 0.1$. The nearest cells to the bottleneck $i \in \{5, 6, 7\}$ carry the value ζ_i (see dashed cells in Fig. 13.2). The values $\zeta_5, \zeta_6, \zeta_7$ are subject of model calibration described below.

13.3 Maximal Flow and Local Friction Calibration

Let us assume that the desired maximal flow J_i through bottleneck i is known. It may be estimated by means of the formula $J = 1.9 \cdot w$ as suggested in [12] or measured experimentally as in our case (see Sect. 13.3). It is necessary to set the local friction ζ_i such that the stationary flow through the bottleneck i in the model corresponds to the maximal flow J_i. To do so, it is necessary to derive the dependence of the stationary congestion flow on given parameters, i.e.

$$J = J(k_S, \zeta)\,. \tag{13.3}$$

In order to avoid extensive and costly computer simulations, an approximation method based on Markov chain theory for the stationary flow derivation has been used. The method is inspired by the study [15], where the authors approximated the model behaviour by means of the nearest neighbourhood of the exit cells and thus reduced the investigated state space to 16 states in the case of one-cell-wide exit. The congestion flow was obtained assuming that the next-nearest cells to the exit are always occupied. This method has been extended to general values of k_S and wider exit cells in [16]. The approximated dependence of the flow on the parameters k_S and ζ are plotted in graphs in Fig. 13.3. The formula can be reached via [17]. Here we note that the flow $J(k_S, \zeta)$ represents the average number of particles hopping through the exit during one step of the algorithm; further scaling to time units, e.g. seconds, is necessary.

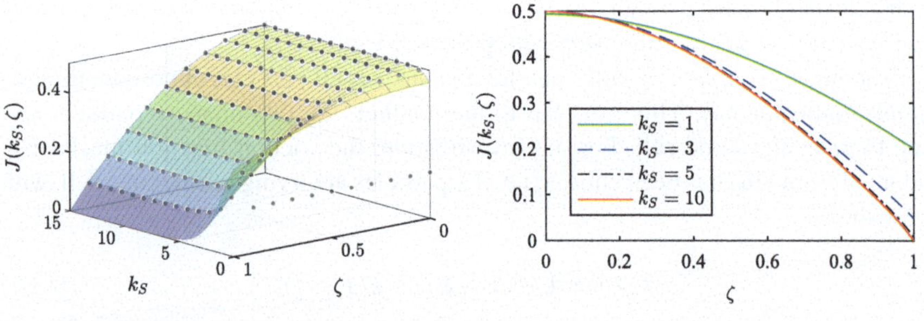

Fig. 13.3 Dependence (13.3) of flow J on parameters k_S and ζ. Left: comparison with Monte-Carlo simulation results (black dots). Right: proper cuts used for local friction calibration

Let us denote by Δt the time-scaling constant, i.e. one algorithm step corresponds to Δt seconds in reality. Similarly, let Δx represent the distance between centres of two neighbouring cells. As usual, we consider $\Delta x = 0.4$ m.

The model calibration is performed in two steps. Firstly, the sensitivity k_S and the time step Δt are set to reproduce the free flow behaviour, i.e. average and variance of the free flow velocity. As mentioned in [5], the effective velocity of floor-field agents can be expressed as probability of forward motion towards the exit as

$$q(k_S) := \Pr(\text{move towards the exit}) = \mathrm{e}^{k_S} \cdot \left(3 + \mathrm{e}^{k_S} + \mathrm{e}^{-k_S}\right)^{-1} . \qquad (13.4)$$

Average free velocity v [m/s] with standard deviation s [m/s] is given by equations

$$v = \tfrac{\Delta x}{\Delta t} \cdot q \,, \qquad s = \tfrac{\Delta x}{\Delta t} \cdot \sqrt{q(1-q)} \,. \qquad (13.5)$$

The task of calibrating the model to fit the bulk behaviour means to solve the system of equations (13.5) in unknown variables q and Δt. The solution has the form $q = \left[1 + (s/v)^2\right]^{-1}$, $\Delta t = \Delta x \left[(1 + (s/v)^2)\right]^{-1}$. The sensitivity to potential is obtained by expressing k_S from the equation (13.4). For considered literature data (see e.g. [18]) $v = 1.58$ m/s and $s = 0.61$ m/s the obtained values are $k_S = 3.00$ and $\Delta t = 0.22$. The global friction was set to $\zeta_0 = 0.1$.

In second step, the local friction ζ_i is set for each bottleneck $i \in \{5, 6, 7\}$, such that the model flow corresponds to the measured flow J_i, i.e.

$$J_i \cdot \Delta t = J(k_S, \zeta_i) \,. \qquad (13.6)$$

using the approximative equation (13.3).

The flow J_i for individual bottlenecks is obtained from the data using the robust least-square linear fit on the time-evolution of the number N of egressed pedestrians through the bottleneck i. As the experimental value of J_i the slope J_{reg} of the linear fit $N(t) = b_0 + J_{\text{reg}} \cdot t$ is taken. The results are: $J_5 = J_6 = 1.56$ ped/s, $J_7 = 1.51$ ped/s; this gives the friction values $\zeta_5 = \zeta_6 = 0.55$ and $\zeta_7 = 0.57$. The comparison of the experimental and model egress times through main exit 7 is plotted in Fig. 13.4.

It is important to note that the measured maximal flow J_i through bottlenecks 5 and 6 are higher than through main exit 7 despite the fact that the width of 5 and 6 was 70 cm and the width of 7 was 85 cm. The reason lies in the geometry of the bottlenecks: there was a niche behind the bottlenecks 5 and 6 enabling faster spreading of pedestrians, while there was a narrow corridor behind bottleneck 7. This illustrates that not only the door width, but even the surrounding geometry has strong impact on the flow.

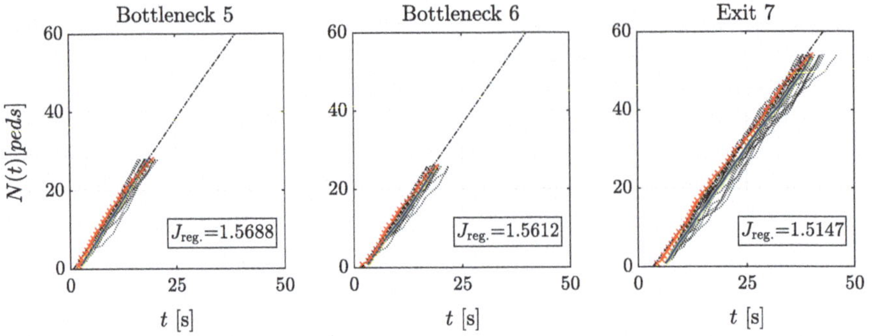

Fig. 13.4 Number of egressed pedestrians $N(t)$ at individual bottlenecks. Experiment (red crosses), linear fit (dashed line), simulation runs (dotted lines), average of simulations (blue line)

13.4 Conclusions

A concept of spatially dependent friction has been introduced into the floor-field cellular model of pedestrian flow. We have shown that this concept enables an effective calibration of this simple model to reproduce the evacuation through a complex system containing multiple consecutive bottlenecks. The main advantage consists in the possibility to adjust properly the bottleneck flow through the bottlenecks of different width even if they are relatively narrow with respect to lattice cell size.

To avoid extensive computer simulations, the individual flow is calibrated by means of analytical dependence (13.3) of the flow J on parameters k_S and ζ. The applicability of this approach is illustrated comparing the model and experimental data.

Acknowledgments The research was supported by the internal grant of Czech Technical University in Prague SGS18/188/OHK4/3T/14.

References

1. Y. Li, M. Chen, Z. Dou, X. Zheng, Y. Cheng, A. Mebarki, Physica A: Statist. Mech. Appl. **526**, 120752 (2019)
2. A. Schadschneider, D. Chowdhury, K. Nishinari, *Stochastic Transport in Complex Systems: From Molecules to Vehicles* (Elsevier Science B. V, Amsterdam, 2010)
3. C. Burstedde, K. Klauck, A. Schadschneider, J. Zittartz, Physica A: Statist. Mech. Appl. **295**(3–4), 507 (2001)
4. A. Kirchner, A. Schadschneider, Physica A: Statist. Mech. Appl. **312**(1–2), 260 (2002)
5. A. Kirchner, H. Klupfel, K. Nishinari, A. Schadschneider, M. Schreckenberg, Physica A: Statist. Mech. Appl. **324**, 689 (2003)
6. A. Schadschneider, A. Seyfried, in *Pedestrian Behavior—Models, Data Collection and Applications*, ed. by H. Timmermans (Emerald Group, Bingley, 2009), pp. 27–43
7. R. Lubaś, J. Porzycki, J. Was, M. Mycek, J. Cellular Automata **11**(4), 285 (2016)

8. C. Dias, R. Lovreglio, Phys. Lett. A **382**(19), 1255 (2018)
9. P. Hrabák, J. Porzycki, M. Bukáček, R. Lubaś, J. Was, in *Cellular Automata*, *Lecture Notes in Computer Science*, vol. 9863, ed. by S. El Yacoubi, J. Was, S. Bandini (Springer International Publishing, 2016), pp. 396–404
10. J. Was, R. Lubaś, Neurocomputing **146**, 199 (2014)
11. P. Hrabák, M. Bukáček, J. Comput. Sci. **21**, 486 (2017)
12. A. Seyfried, O. Passon, B. Steffen, M. Boltes, T. Rupprecht, W. Klingsch, Transp. Sci. **43**(3), 395 (2009)
13. A. Kirchner, K. Nishinari, A. Schadschneider, Phys. Rev. E **67**, 056122 (2003)
14. D. Yanagisawa, A. Kimura, A. Tomoeda, R. Nishi, Y. Suma, K. Ohtsuka, K. Nishinari, Phys. Rev. E **80**, 036110 (2009)
15. T. Ezaki, D. Yanagisawa, K. Nishinari, J. Cellular Automata **8**(5–6), 347 (2013)
16. P. Hrabák, F. Gašpar. Stationary flow approximation in floor-field models using markov-chain theory (2018). https://www.tudelft.nl/citg/over-faculteit/afdelingen/transport-planning/news-agenda/conferences-courses/matts/matts-2018/programme/. Presentation at MATTS 2018, Delft
17. P. Hrabák, F. Gašpar. Equation J(kS,zeta) (2019). https://doi.org/10.6084/m9.figshare.10259999.v1. URL https://figshare.com/articles/Equation%20J%20kS%20zeta%20/10259999/1
18. S.P. Hoogendoorn, W. Daamen, in *TRB 85th Annual Meeting Compendium of Papers CD-ROM* (2006)

Chapter 14
Experimental Study on the Congestion-Sharing Effect of Obstacle on Pedestrian Crowd Egress

Xiaolu Jia, Claudio Feliciani, Daichi Yanagisawa, and Katsuhiro Nishinari

Abstract Many present studies have illustrated the influence of obstacle on the egress efficiency of the crowd. In this paper, we have explored the influence of obstacle on both the egress efficiency and congestion status of pedestrians in a corridor with fixed width. A series of experiments with variable pedestrian number and obstacle size have been conducted with a wall-shaped obstacle set in the middle of the corridor. Results showed that the egress time would linearly increase with the pedestrian number while was not apparently affected by obstacle size. However, the variation of obstacle size would affect the congestion status of the crowd. To be specific, we have observed the congestion-sharing effect, which indicates the phenomenon that increasing the obstacle size could help share the congestion before the exit. Numerical analysis showed that the congestion-sharing effect would only be apparent when the pedestrian number is large enough. Furthermore, we have estimated the optimal obstacle size with the most balanced congestion status in the walking environment, which is expected to help with the actual design of obstacle for better congestion status.

14.1 Introduction

In walking facilities such as subway stations, sports venues and commercial buildings, pedestrian movement is often affected by obstacles like walls, pillars and interior furnishings. Therefore, exploring the influence of obstacles on pedestrian movement

X. Jia (✉)
Department of Advanced Interdisciplinary Studies, School of Engineering, The University of Tokyo, 4-6-1 Komaba, Meguro-ku, Tokyo 153-8904, Japan
e-mail: xiaolujia@g.ecc.u-tokyo.ac.jp

C. Feliciani · D. Yanagisawa · K. Nishinari
Research Center for Advanced Science and Technology, The University of Tokyo, 4-6-1 Komaba, Meguro-ku, Tokyo 153-8904, Japan

D. Yanagisawa · K. Nishinari
Department of Aeronautics and Astronautics, Graduate School of Engineering, The University of Tokyo, 7-3-1 Hongo, Bunkyo-ku, Tokyo 113-8656, Japan

I. Zuriguel et al. (eds.), *Traffic and Granular Flow 2019*,
Springer Proceedings in Physics 252,
https://doi.org/10.1007/978-3-030-55973-1_14

is significant to the actual design of obstacles in order to guarantee a more comfortable and efficient walking environment.

Research on the influence of obstacle began to attract more attention since it was proposed that placing an obstacle before the exit could help improve the evacuation efficiency [1]. Both simulation studies based on agent-based models and experimental studies have been widely used to emulate the influence of obstacle on pedestrian evacuation (a review can be seen in [2]). Although many simulation studies have reproduced the merits of obstacle before the exit in improving the evacuation efficiency [3–6], the lack of experimental evidence has made the simulation results unconvincing to some extent. Therefore, experimental studies are essential in reflecting the influencing mechanism of obstacle on pedestrian flow and providing evidence for modeling and simulation.

Experiments on animals such as sheep [7], ants [8] and mice [9] have given some hints for pedestrian experiments. Taking advantage of the survival instinct of animals to seek food or avoid danger, the animals were forced to be competitive that led to clogs before the exit. Placing an obstacle in this case has resolved the clogs and reduced the conflicts among the animals, thus improving the evacuation efficiency.

Different from animal experiments, it is difficult to implement panic or very competitive experiments due to safety and ethical reasons. It was shown that compared with normal walking conditions where the competition is not so serious, the obstacle could increase the evacuation efficiency under slow running conditions where the competition is more serious [10]. In some experiments where pedestrians were motivated to leave while pushing were rarely observed, it was proposed that the efficiency efficiency would be improved if the obstacle was shifted from the exit center [11, 12]. Other experiments showed that in relatively more competitive conditions where pedestrians have pushed others, placing two obstacles before the exit would contribute to a higher evacuation efficiency compared with one obstacle or no-obstacle case [13]. On the other hand, in some experiments where pedestrians were extremely competitive and the pushing among pedestrians were serious, placing an obstacle before the exit failed to improve the evacuation efficiency [14].

Despite the results of pedestrian experiments were still controversial, we have observed that most of the experiments have managed to increase the conflicts among pedestrians more or less. Besides, the obstacles were mostly pillar-shaped and placed before the exit, and the evacuation time was always their focus. In contrast, we believe that normal walking conditions where pedestrians are not competitive are also very common, and wall-shaped obstacles like partitions, walls and fences have also been widely applied at places not limited to near the exit. Besides, except for egress efficiency, the overall congestion status of the crowd is also significant considering that heavy local congestion could induce psychological and physical discomfort of pedestrians. Therefore, a more generalized study on the influence of wall-shaped obstacle on pedestrian egress efficiency as well as congestion status should be conducted.

14.2 Experiment

Experiments have been performed to explore the influence of obstacle size on the egress of crowd pedestrians under normal conditions. The experiment was conducted on November, 2017 at the lecture hall of the RCAST Building 4, The University of Tokyo, Japan. The experiment scenario was set as a corridor with a wall-shaped obstacle placed at the horizontal middle axis. Totally 30 males aged from 20 to 24 years old have participated in our experiments.

We would detect the locations of pedestrians within the detection region using the coordinates system in Fig. 14.1, where the corridor and the obstacle were built by cardboard boxes. We define the number of boxes used to build the obstacle as box, and the width of each box in y direction is 0.42 m. Totally 20 tests have been conducted with pedestrian number N being 10, 15, 20, 25, 30 and box being 1, 2, 3 and 4. Before each test, N participants were randomly chosen among all the 30 participants and were required to stand in a phalanx before the start line. During the experiments, pedestrians would walk from the start line, pass through the obstacle and finally pass by the exit. Pedestrians were required to traverse the corridor normally and no obvious physical contact has been observed.

A camera was set above the horizontal axis of the corridor and fixed 6 meters above the ground. Recordings of the camera was adjusted to full HD mode (1920 × 1080 pixel) with a frame rate of 30 fps. With the videos of the experiments as rough data, the recognition and tracking of pedestrians could be achieved using PeTrack software [15]. Pedestrians were required to wear colored caps so that their positions at each video frame could be detected.

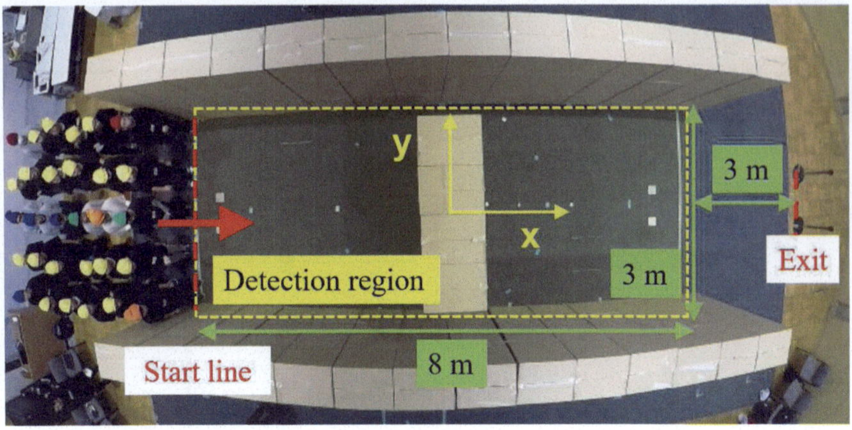

Fig. 14.1 The geometrical layout of experimental setting

14.3 Variation of Egress Time

For a certain test, we define the time from the beginning to when the last pedestrian passed by the exit as egress time T. Because the detection region shown in Fig. 14.1 failed to cover the exit, we have obtained the egress time through manual counting based on video data. The variation of T with pedestrian number N and obstacle size *box* can be seen in Fig. 14.2. The four types of solid dots represent the T under four values of *box* and the (red) hollow rhombic dots represent the average T. The variation of average T with N has been fitted by the (red) linear curve to show the variation trend. It is indicated that T would linearly rise with N, which means the increase of pedestrian number could linearly increase the egress time.

On the other hand, the influence of obstacle size *box* could be evaluated through calculating the CV, i.e. coefficient of variation. Under a certain pedestrian number N, assuming the average value and standard deviation of T under the four types of obstacle sizes as $\overline{T}$ and σ, the CV can be calculated as Eq. 14.1.

$$CV = \frac{\sigma}{\overline{T}} \tag{14.1}$$

Through calculation, the values of CV under $N = 10, 15, 20, 25, 30$ are respectively 0.0525, 0.0461, 0.0438, 0.0397 and 0.0254. Except for the case when $N = 10$, all the values of CV are smaller than 0.05, which means the *box* would not apparently affect the egress time. In the case when $N = 10$, the pedestrian number is small and pedestrian movement is probably more affected by random factors such as the heterogeneity of participants who were randomly chosen in each test. Therefore, we presume that the obstacle size would not apparently affect the egress time.

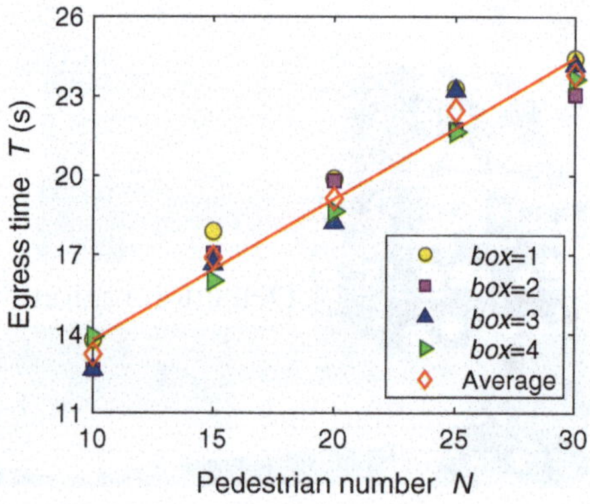

Fig. 14.2 Variation of egress time with pedestrian number

14.4 The Congestion-Sharing Effect

Despite that the obstacle size would not apparently affect egress time, we have discovered that the obstacle size would affect the congestion condition in the walking space. To be specific, we presume that the congestion before the exit could be gradually shared by the increase of obstacle size, which we define as the congestion-sharing effect.

14.4.1 Qualitative Description

For illustration, as shown in Fig. 14.3, we have depicted the pedestrian trajectories under $N = 30$ with the color indicating the local velocity. We presume the velocity could indicate the local congestion level and would analyze the influence of obstacle size *box* on the congestion status.

When the obstacle width is small ($box = 1$), as shown in Fig. 14.3 a, the trajectory color under $x \in [-4, 0]$m appears orange while the color under $x \in [0, 4]$m appears blue. It is indicated that pedestrians walked faster before passing by the obstacle while slower between the obstacle and the exit. We presume the reason is that when the obstacle size is too small to obstruct pedestrians, the pedestrians would cluster before the exit, thus causing a slower velocity and a heavier congestion status before the exit.

By contrast, when the obstacle width is large ($box = 4$), as shown in Fig. 14.3 d, the trajectory color appears blue under $x \in [-4, 0]$m while orange under $x \in [0, 4]$m. It is indicated that pedestrians walked slower before passing by the obstacle while faster between the obstacle and exit. We presume the reason is that when the obstacle

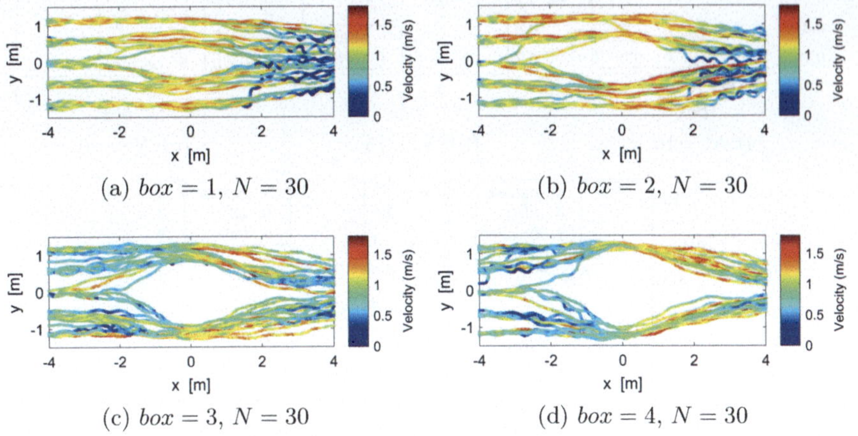

Fig. 14.3 Trajectories and the heat map of pedestrian velocity

size is large enough to obstruct pedestrians, the pedestrians would cluster before the obstacle. As a result, the pedestrian density between the obstacle and exit is reduced and a higher walking velocity is achieved.

Furthermore, with the increase of obstacle size, it can be observed that the congestion before the obstacle becomes heavier while the congestion after the obstacle is gradually reduced. In this way, the congestion before the exit has been transferred by the obstacle and the congestion-sharing effect occurs.

14.4.2 Numerical Analysis

For numerical validation of the congestion-sharing effect under different pedestrian number N, we have calculated the average local velocity of pedestrians before and after the obstacle. The variation of average velocity v against box under different N can be seen in Fig. 14.4. The hollow circular and triangle dots represent the experimental data while the linear curves represent the linear fitting of these dots.

As shown in Fig. 14.4 a, b, when the pedestrian number is small, the v before the obstacle shows a decreasing trend with the rise of box. However, the increasing trend of v after the obstacle is not apparent. We therefore presume that when the pedestrian number is small, though the obstacle would cause more congestion, the congestion after the obstacle could not be shared apparently.

On the other hand, as shown in Fig. 14.4 c, d, e, when the pedestrian number is large, the v before the obstacle would decrease while the v after the obstacle would

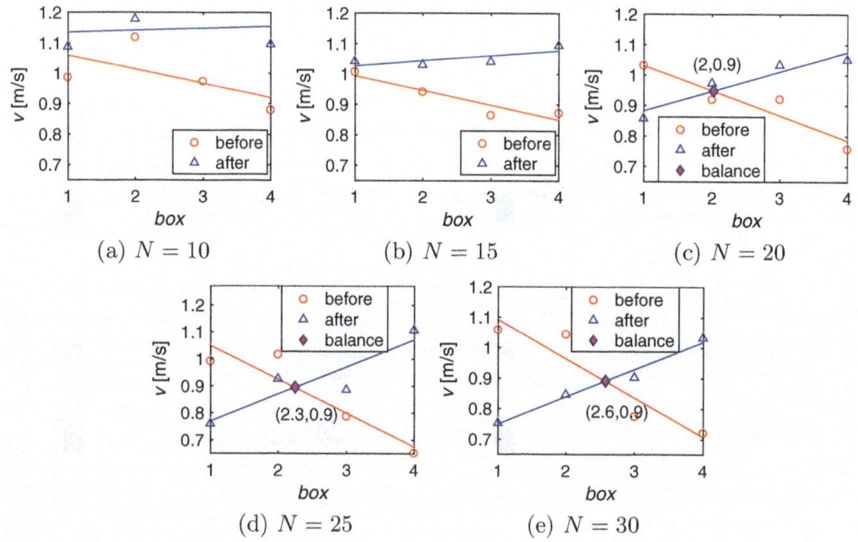

Fig. 14.4 Variation of average velocity before and after the obstacle

increase with the rise of *box*. In this case, it is indicated that with the increase of obstacle size, the congestion after the obstacle is transferred to the region before the obstacle, which means the congestion-sharing effect is more apparent when the pedestrian number is large.

Furthermore, when the congestion-sharing effect is more apparent, we could estimate the optimal obstacle size that corresponds to the most balanced congestion status. In Fig. 14.4 c, d, e, the rhombic dot corresponds to the most balanced obstacle size under which the average velocity as well as the congestion status before and after the obstacle are the same.

Please note that the congestion-sharing effect is due to the combined effect of the obstacle and the width of the corridor. The obstacle and the walls have formed bottlenecks that have been quite narrow when the obstacle size is large. As a result, the narrow bottlenecks have led to lower pedestrian flow rate, making it more congested before the obstacle. Correspondingly, the pedestrian density before the exit would be decreased and the congestion-sharing effect emerged. The performance of congestion-sharing effect might change under a wider corridor, which deserves further experimental exploration.

We presume the most balanced obstacle size is essential to the design of actual walking facilities in avoiding local congestion. In real life, heavy local congestion tends to increase the mental stress of pedestrians under non-emergency conditions and even acts as the inducement of stampede accident under emergency conditions. Therefore, it is significant to keep pedestrians walking smoothly with the help of most balanced obstacle size for a more comfortable and safe walking environment.

14.5 Conclusion

Experiments have been conducted to explore the influence of a wall-shaped obstacle on pedestrian egress. Both the pedestrian number and obstacle size have been adjusted in our experiments to explore the corresponding influence.

Results showed that the egress time would linearly increase with the rise of pedestrian number. On the other hand, despite the variation of obstacle size have not shown apparent influence on egress time, we have discovered its influence on the congestion status.

To be specific, we have observed the congestion-sharing effect that is induced by the combined effect of the obstacle and the corridor. It has been observed that the increase of obstacle size could share the congestion before the exit to the region before the obstacle. Furthermore, through numerical illustration, it is shown that the congestion-sharing effect is only apparent when the pedestrian number is large enough. Taking advantage of the congestion-sharing effect, we have estimated the optimal obstacle size with the most balanced congestion status before and after the obstacle. We presume that in actual design of walking facilities, the congestion-sharing effect could be taken into advantage to assure a more comfortable and safe walking environment.

Acknowledgments This work was partially supported by JST-Mirai Program Grant Number JPMJMI17D4, Japan and MEXT as "Priority Issues and Exploratory Challenges on post-K (Supercomputer Fugaku)" (Project ID: hp190163). The first author would like to thank the Chinese Scholarship Council (CSC) for the financial support during the PhD courses.

References

1. A. Kirchner, K. Nishinari, A. Schadschneider, Friction effects and clogging in a cellular automaton model for pedestrian dynamics. Phys. Rev. E **67**, 056,122 (2003)
2. D.C. Duives, W. Daamen, S.P. Hoogendoorn, State-of-the-art crowd motion simulation models. Transp. Res. Part C: Emerg. Technol. **37**, 193–209 (2013)
3. Y. Zhao, M. Li, X. Lu, L. Tian, Z. Yu, K. Huang, Y. Wang, T. Li, Optimal layout design of obstacles for panic evacuation using differential evolution. Physica A: Statist. Mech. Appl. **465**, 175–194 (2017)
4. G. Frank, C. Dorso, Room evacuation in the presence of an obstacle. Physica A: Statist. Mech. Appl. **390**(11), 2135–2145 (2011)
5. N. Shiwakoti, X. Shi, Z. Ye, A review on the performance of an obstacle near an exit on pedestrian crowd evacuation. Saf. Sci. **113**, 54–67 (2019)
6. V. Karbovskii, O. Severiukhina, I. Derevitskii, D. Voloshin, A. Presbitero, M. Lees, The impact of different obstacles on crowd dynamics. J. Comput. Sci. (2018)
7. Zuriguel, I., Olivares, J., Pastor, J.M., Martín-Gómez, C., Ferrer, L.M., Ramos, J.J., Garcimartín, A.: Effect of obstacle position in the flow of sheep through a narrow door. Phys. Rev. E **94**, 032,302 (2016)
8. S. Nirajan, S. Majid, Enhancing the panic escape of crowd through architectural design. Transp. Res. Part C: Emerg. Technol. **37**, 260–267 (2013)
9. P. Lin, J. Ma, T.Y. Liu, T. Ran, Y.L. Si, F.Y. Wu, G.Y. Wang, An experimental study of the impact of an obstacle on the escape efficiency by using mice under high competition. Physica A: Statist. Mech. Appl. **482**, 228–242 (2017)
10. X. Shi, Z. Ye, N. Shiwakoti, D. Tang, J. Lin, Examining effect of architectural adjustment on pedestrian crowd flow at bottleneck. Physica A: Statist. Mech. Appl. **522**, 350–364 (2019)
11. D. Yanagisawa, A. Kimura, A. Tomoeda, R. Nishi, Y. Suma, K. Ohtsuka, K. Nishinari, Introduction of frictional and turning function for pedestrian outflow with an obstacle. Phys. Rev. E **80**, 036,110 (2009)
12. D. Yanagisawa, R. Nishi, A. Tomoeda, K. Ohtsuka, A. Kimura, Y. Suma, K. Nishinari, Study on efficiency of evacuation with an obstacle on hexagonal cell space. SICE J. Control Measur. Syst. Integ. **3**(6), 395–401 (2010)
13. J. Li, L. Jingyu, S. Chao, Y. Sicong, H. Zhangang, Obstacle optimization for panic flow reducing the tangential momentum increases the escape speed. Plos One **9**, 1–15 (2014)
14. A. Garcimartín, D. Maza, J.M. Pastor, D.R. Parisi, C. Martín-Gómez, I. Zuriguel, Redefining the role of obstacles in pedestrian evacuation. New J. Phys. **20**(12), 123,025 (2018)
15. M. Boltes, A. Seyfried, Collecting pedestrian trajectories. Neurocomputing **100**, 127–133 (2013). Special issue: Behaviours in video

Chapter 15
Experimental Setups to Observe Evasion Maneuvers in Low and High Densities

Benedikt Kleinmeier and Gerta Köster

Abstract Crowd simulations depend on empirical evidence as basis for model development. However, for many scenarios with high practical impact such evidence is still scarce. There are compelling reasons for this: Experiments involving human participants are expensive, labor intensive, and they carry a high risk for bias. This applies to both sides, experimenters and participants. In this contribution we present two experiment setups to observe pedestrian motion through high and low densities. We focus on the measures we take to avoid observer bias, undue influence on the participants and learning effects. In the first experiment, a waiting crowd of 13 participants is passed by a walking proband. In the second experiment, a waiting dyad is passed by a walking proband. Our experiment designs ensure that we can provide the scientific community with reliable data on a crowd phenomenon where evidence is still missing: single pedestrians maneuvering through a crowd.

15.1 Introduction

There are many pedestrian experiments to get a better insight into pedestrian dynamics: Experiments to replicate uni- and bi-directional flows [1–3], bottlenecks [4], stop-and-go waves [5], junctions [6]. Most of them focus on macroscopic quantitative aspects like fundamental diagrams, speeds or evacuation times [7, 8]. These aspects are useful to calibrate and validate pedestrian stream simulators.

On the other hand, some basic day-to-day situations are not yet covered by experiments. Imagine, a single pedestrian passing through a waiting crowd at a railway

B. Kleinmeier (✉) · G. Köster
Department of Computer Science and Mathematics, Munich University of Applied Sciences, 80335 Munich, Germany
e-mail: benedikt.kleinmeier@hm.edu

G. Köster
e-mail: gerta.koester@hm.edu

B. Kleinmeier
Department of Informatics, Technical University of Munich, 85748 Garching, Germany

I. Zuriguel et al. (eds.), *Traffic and Granular Flow 2019*,
Springer Proceedings in Physics 252,
https://doi.org/10.1007/978-3-030-55973-1_15

station. How does the waiting crowd behave and how does the walking pedestrian behave? The authors are aware of only one experiment that (partially) addresses these research questions [9]. Reference [9] compares pedestrian movement at varying densities to granular flow. For this, they observe the mechanical response of a crowd to an intruder. Since, there is no adequate empirical evidence to draw upon, current pedestrian stream simulators lack suitable behavioral models for motion through a dense crowd. Cooperative behavior, which can be observed in real life, is missing in current computer models. As a result, agents simply get stuck in high-density scenarios.

This scientific gap motivated us to design experiments that can provide first evidence. **We describe two experimental setups to observe evasion maneuvers in high and low densities**. Our goal in this is to obtain data from which reliable conclusions can be drawn. Since we are working with humans who react to the way we communicate with them, we need to eliminate undue influence on their behavior through our instructions. Telling them to behave "naturally" would already be such an influence which, as we will show, can be avoided through careful experiment design. We also need to avoid or at least reduce learning effects. **This contribution focuses on the experiment setup and the measures we take to achieve our goal**. Section 15.2 introduces the two experiments. Section 15.3 explains the measures we take to avoid observer biases and participant biases. It also describes the data collection. We conclude with an outlook on the data analysis.

15.2 Methods and Materials: Experiment Description

We developed two experimental setups to observe evasion maneuvers and human behavior in high and low densities.

15.2.1 High-Density Experiment: Motion Through a Dense Crowd

In the high density experiment, called "motion through a dense crowd", we investigate qualitative and quantitative aspects. Qualitatively, we want to observe and identify characteristics of pedestrians motion through a dense waiting crowd. We want to frame these characteristics as requirements for models. Quantitatively, we measure speeds to obtain measures against which we can validate such new models.

Experiment Setup. We use crepe tape to mark a starting line of length 1 m to ensure that all probands start at the same position. The target point for the probands is a tall tree which can easily be seen by all probands. At halfpoint between start and target point we mark a waiting area of 1.55 m $\times$ 1.70 m (2.6 m^2 where 13 participants are placed as waiting crowd. This results in a density of 4.93 peds/m^2 without the

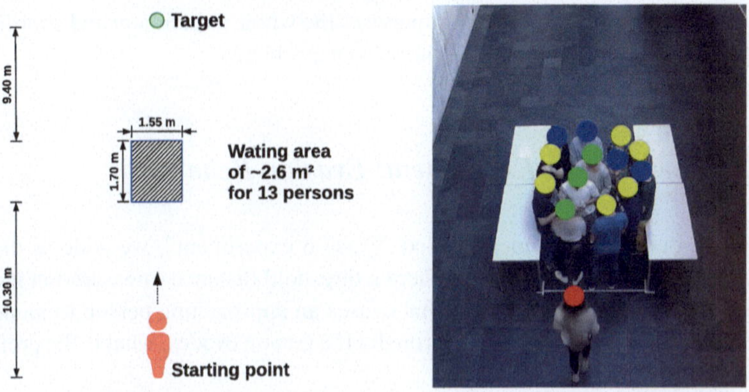

(a) Schematic setup (b) Real setup

Fig. 15.1 Setup of the high-density experiment called "motion through dense crowd"

walking proband and 5.31 peds/m^2 when the walking proband passes through the crowd. The waiting area is delimited with tables on the left- and on the right-hand side to avoid that the waiting participants leave the area unintentionally. The distance between start and target point is 21 m. This ensures that there is enough space in front of and behind the crowd to measure the free-flow velocity of the walking probands. The complete experiment setup is depicted in Fig. 15.1.

Experiment Procedures. We place all participants in a waiting room where they are entertained by an experiment assistant throughout the experiment. We carry out following steps:

1. Choose 13 participants as waiting crowd from 59 available participants and ask them to wait in the delimited area without any further instructions.
2. Choose a proband from the remaining participants to walk through the waiting crowd.
3. Place the walking proband at the starting point and ask them to make their way to the target point.
4. At the target point, ask the walking proband to return to the waiting room.
5. After each run, the waiting participants are asked to leave the waiting area, they are shuffled and then they are asked to return to the waiting area. After five runs, six participants of the waiting crowd are replaced with participants from the available participants. These measures are applied to avoid learning effects for the waiting crowd.

This experiment procedure takes approximately 60 s per run and requires seven or eight experiment assistants. We need four assistants for the experiment steps (1) to (4). Procedure (3) and (5) can be carried out by the same assistant. We prefer to assign an assistant to operating the camera to record the experiment, even if this could be automated. Another assistant entertains the remaining participants in the waiting room. The last assistant supervises the whole experiment and coordinates the experiment assistants in case of unforeseen problems.

15.2.2 Low-Density Experiment: Evasion Behavior

For the low density experiment, called "evasion experiment", we address the following two connected questions: Is there a threshold distance (measured in meters) between a waiting dyad of persons that causes an approaching person to maneuver around the dyad instead of passing through? If a person evades, what is the preferred distance to the dyad?

Experiment Setup. The experiment setup is similar to Sect. 15.2.1. Instead of the waiting area, we place two experiment assistants as waiting dyad halfway between starting and target point. While the dense crowd in the first experiment enforces a minimum of cooperative behavior, walking probands in the second experiment can choose to evade the "conflict". The waiting dyad stands with their backs to the participants to avoid eye contact. We made this choice to facilitate reenactment in a VR-environment. In the course of the experiment, we vary the distance d between the waiting dyad. To achieve a sufficient sample size per distance with a very limited number of probands we need to restrict the measurements to very few, carefully selected distances. Reference [10] observe significant shoulder rotations for humans if d between a waiting dyad is less or equal 1.7 of the participant's shoulder width. If the waiting dyad is replaced by two poles, a significant shoulder rotation is already observed at 1.3 of the participant's shoulder width. We conclude that humans change their evasion behavior if the aperture size falls below a certain threshold of their shoulder width. We assume a typical shoulder width of 0.40 m based on [11] who reported a shoulder width of 355 ± 18 cm for women and 388 ± 19 cm for men (for 268 subjects in total: 201 women and 67 men). Therefore, we choose the distances d: [0.50; 1.70; 0.75; 0] m which corresponds to the following multiples of shoulder width: [1.25; 4.25; 1.88; 0.00]. We expect a significant number of participants to evade if the distance falls below $d <= 0.75$ m. The experiment setup is depicted in Fig. 15.2. The unconventional order of the distances is deliberate and helps to suppress learning effects.

Experiment procedures. We place all participants in a waiting room where they are entertained by an experiment assistant during the experiment. Then, we carry out following steps. For each dyad distance d:

1. Place two experiment assistants at the specified distance d.
2. Choose a proband randomly from the waiting room.
3. Place the proband at the starting position and ask them to make their way to the target point. After at least 40 runs, change to a new dyad distance and start over with step (1).

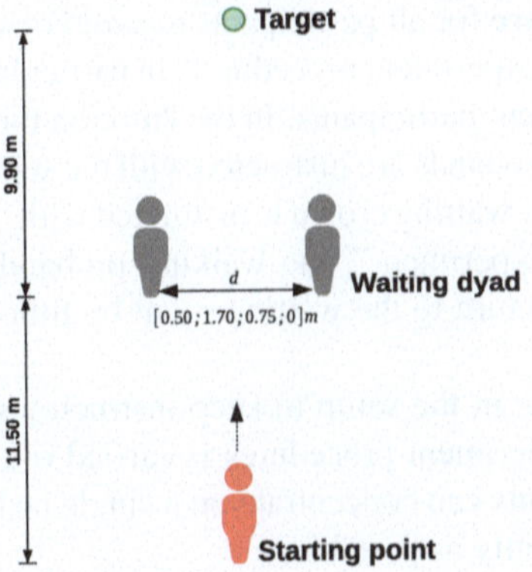

(a) Schematic setup

(b) Real setup

Fig. 15.2 Setup of the low-density experiment called "evasion behavior"

This experiment procedure takes approximately 50 s per run and requires seven experiment assistants. Two assistants as waiting dyad, two assistants for step (2) and (3), one assistant as camera operator, one to entertain the waiting participants and another assistant who supervises the whole experiment and coordinates the experiment assistants in case of unforeseen problems.

15.3 Discussion

In this section, we address the measures we take to avoid observer biases and participant biases to obtain reliable data. Additionally, we describe how the data is collected during the experiment.

15.3.1 Obtain Reliable Data: Measures to Avoid Biases

Researchers from psychologists emphazise the need for careful planning in experiments with human participants [12] to obtain reliable data. Humans are very sensitive to how we communicate with them. For example, it makes a difference whether we instruct them by a command like "Pass the crowd!" or by a polite request "Please, make your way to the other side". With a command, participants tend to hurry. The second option, a kind request, promotes a more relaxed participant behavior. To complicate matters, the experimenters are also humans. Therefore, it also necessary to mitigate the so called observer bias, that is, an error induced by the personal motives and expectations of the viewer [12, p. 22]. We take following counter-measures:

- **Use a standardized experiment procedure** for all participants to avoid observer biases. The procedure is described under "experiment procedure". In particular, we use consistent instructions for all experiment participants. In the "motion through a dense crowd" experiment, the walking probands are instructed with the sentence "Go to the tree by crossing the crowd". The waiting crowd is instructed with "Wait in the delimited area". In the "evasion experiment", the walking probands are instructed with "Go to the tree and then return to the waiting room by following the arrows".
- **Employ assistants at all strategic points** in the setup to keep instructions to a minimum. Each step in the described experiment procedures is carried out by a dedicated assistant. That means, participants can concentrate on a single task at a time and can perform this task as consistently as possible.
- **Entertain the waiting participants** so that they do not discuss the experiment. Since the participants were all fresh first-year students it was easy to engage their full attention in discussing their new student life with senior students (assistants).
- **Exchange and re-shuffle of the participants** as much as possible to avoid learning effects. For instance, in the "motion through dense crowd" experiment, this keeps the waiting crowd from forming a corridor through which the walking proband can pass.
- **Carry out a trial run** with an independent set of participants to identify unexpected and unwanted effects. For example, the trial run revealed training effects for the "evasion experiment" when we increased or decreased the dyad distance d monotonously. Therefore, we choose then following sequence for the dyad distance d: [0.50; 1.70; 0.75; 0] m. That means, we started with $d = 0.50$ m, where we expect most of the participants to evade, because of the shoulder rotations observed in [10]. Then, we increased d to 1.70 m where we expect many probands to pass between the dyad. As third distance we choose $d = 0.75$ m, and as last distance $d = 0$ m, that is, the assistants in the dyad stand shoulder to shoulder.

15.3.2 Data Collection

We conducted the experiment in October 2018 in the foyer of the Munich University of Applied Sciences with 59 first-semester students, aged 18–66. We recorded both experiments from above at an angle of 60°. For the recording, we used a Sony Handycam HDR-PJ780VE camcorder set to a resolution of 1280 pixel × 720 pixel and 25 frames/s. The "motion through a dense crowd" experiment yielded 73 min of video footage and the "evasion behavior" experiment yielded 92 min.

Each walking proband carried a printed id card which was recorded by the camera. Each id card holds a unique number and the following information about the proband: gender, age, height and shoulder width. All shoulder widths were measured by the same experiment assistant. We recorded the shoulder width because we know from [10] that the shoulder width can be used as predictor for the particpant's decision to evade or to pass through the waiting dyad. Refernce [10] observed whether or not participants rotated their shoulders when "squeezing" through two obstacles.

Ethical correctness is another essential requirement for experiments with humans. The experiment was approved by the ethical review committee of Technical University Munich (https://www.ek-med-muenchen.de/). All participants signed an informed consent before they took part in the experiment.

As we are writing this contribution, the video footage is analyzed visually and statistically according to the following procedure: (1) Cut the video material into pieces using the software `ffmpeg` and a self-written `Python` script so that we get a single video per experiment run. (2) Use the software `Tracker` to load each video, eliminate the perspective distortion and extract participant's trajectories by using `Tracker's` "Auto-Tracking" feature. (3) Write `IPython` notebooks to visualize the extracted trajectories. For the "motion through a dense crowd" experiment, generate heatmaps, plot trajectories and evaluate the speed of the walking participant within and outside the crowd. For the "evasion behavior" experiment, focus on the evasion distance if a participant evaded.

15.4 Outlook

In this contribution, we presented two experiment setups to observe pedestrian motion through high and low densities. In the first experiment "motion through a dense crowd", probands pass through a waiting crowd of 13 participants. In the second experiment "evasion behavior", a proband either evades or passes through a waiting dyad of assistants.

We discussed in detail which measures we took to obtain reliable data. We carried out a trial run with an independent set of participants. The trial run revealed unexpected and unwanted effects and helped to optimize the experiment setup and the instructions for the participants. Another useful measure was to exchange and reshuffle the participants to suppress learning effects as much as possible. We employed a large number of assistants at strategic points. This helped to run the experiment smoothly and to keep instructions for participants to a minimum. On the other hand, it bound a lot of resources and required a lot of detailed planning.

We argue that through all these measures, we obtained data that reflects true human behavior with respect to evasion and passing maneuvers in dense crowds ("motion through a dense crowd" experiment). We intend to analyze the data and frame qualitative observations as requirements for a model of cooperative pedestrian behavior in a dense crowd. We strive to identify the quantitative measures against which to validate the models. Thus our data will help to close a scientific gap, since agents in current pedestrian stream simulators get stuck when the crowd density gets too high. The data from the "evasion experiment" will help to answer another open question: When will pedestrians walk through a waiting dyad of persons, when will they evade? And which distance will they keep from the dyad? With measures of the evasion distance we will be able to better validate existing locomotion models.

After the analysis, anonymized experiment data, mostly trajectories, will be made freely available to the scientific community.

Acknowledgments We thank all participants for taking part in the experiments. Additionally, we thank the student team consisting of Karim Belkhiria, Fabian Flach, Jonas Goltz, André Heinrich, Alexandra Mayer, Maximilian Niedermaier, Amir Schnell, Philipp Schuegraf for their help as experiment assistants and for pre-analyzing the video material. Last but not least, we thank the research office (FORWIN) of the Munich University of Applied Sciences and the Faculty Graduate Center CeDoSIA of TUM Graduate School at Technical University of Munich for their support.

Authors Contributions. B. K. organized and conducted the experiments and drafted the article. G. K. advised, assisted with the experiments, critically revised the article and gave final approval for publication.

Funds. B. K. is supported by the German Federal Ministry of Education and Research through the project OPMoPS to study organized pedestrian movement in public spaces (grant no. 13N14562).

References

1. X. Zhang, W. Weng, H. Yuan, J. Chen, Physica A: Stat. Mech. Appl. **392**(12), 2781 (2013). https://doi.org/10.1016/j.physa.2013.02.019
2. J. Zhang, W. Klingsch, A. Schadschneider, A. Seyfried, J. Stat. Mech.: Theory Exp. **2012**(02), P02002 (2012). https://doi.org/10.1088/1742-5468/2012/02/P02002
3. T. Kretz, A. Grünebohm, M. Kaufman, F. Mazur, M. Schreckenberg, J.Stat. Mech.: Theory Exp. **2006**(10), P10001 (2006). https://doi.org/10.1088/1742-5468/2006/10/P10001
4. J. Adrian, M. Boltes, S. Holl, A. Sieben, A. Seyfried, in *Proceedings of the 9th International Conference on Pedestrian and Evacuation Dynamics (PED2018)*, Lund, Sweden, 21–23 August 2018, arXiv:1810.07424
5. A. Portz, A. Seyfried, *Pedestrian and Evacuation Dynamics*, vol. 1 (Springer, 2011), pp. 577–586. https://doi.org/10.1007/978-1-4419-9725-8_52
6. J. Zhang, W. Klingsch, A. Schadschneider, A. Seyfried, J. Stat. Mech.: Theory Exp. **2011**(06), P06004 (2011). https://doi.org/10.1088/1742-5468/2011/06/P06004
7. A. Jelić, C. Appert-Rolland, S. Lemercier, J. Pettré, Phys. Rev. E **85**(3), 036111 (2012). https://doi.org/10.1103/PhysRevE.85.036111
8. A. Jelić, C. Appert-Rolland, S. Lemercier, J. Pettré, Phys. Rev. E **86**(4), 046111 (2012). https://doi.org/10.1103/PhysRevE.86.046111
9. A. Nicolas, M. Kuperman, S.B. Santiago Ibañez, C. Appert-Rolland, Nature Scientific Reports (2019)
10. A. Hackney, M. Cinelli, J. Frank, Acta Psychologica **162**, 62 (2015). https://doi.org/10.1016/j.actpsy.2015.10.007
11. L. Hanson, L. Sperling, G. Gard, S. Ipsen, C. Olivares Vergara, Appl. Ergon. **40**(4), 797 (2009). https://doi.org/10.1016/j.apergo.2008.08.007, http://www.sciencedirect.com/science/article/pii/S0003687008001324
12. R.J. Gerrig, *Psychology and Life*, 20th edn. (Pearson, 2013)
13. A. Sieben, J. Schumann, A. Seyfried, PLoS One **12**, 6 (2017). https://doi.org/10.1371/journal.pone.017732

Chapter 16
How to Change the Value of Social Force Model's Relaxation Time Parameter with Desired Speed Such that Bottleneck Flow Remains Unchanged

Tobias Kretz

Abstract When simulating pedestrian dynamics with the Social Force Model or one of its variants one may commonly be faced with the situation to have calibrated the model such that the model reproduces capacity flow values at bottlenecks which have been observed or which are required by a guideline or regulation, but that then the need arises to change desired speeds. The question is then how the values of the other model parameters need to be adjusted to recover the original flow value. In this contribution it is shown that to achieve this it often suffices to adjust the value of paramter τ and how the new value can possibly be calculated.

16.1 Motivation

It can be shown analytically that in simulations of single-file pedestrian dynamics with the Social Force Model [1] the steady-state capacity flow J_c depends on the value of desired speed v_0 [2] (Eqs. 52–54). This is also true for bottleneck flow at various widths. From the analytical result one could conclude that if v_0/τ [3] is constant also capacity flow remains unchanged. However, again a few quick simulations can show that this is not true. The reason for this being that at bottlenecks the system is not in a steady state since pedestrians accelerate from (nearly) stand still to free speed in a short time when passing through the bottleneck.

This leaves modelers in an unpleasant situation, since calibration for capacity flow is an essential part of a calibration process and desired (free) speeds are an important input value. If a model is calibrated for a capacity flow with a given (distribution of) desired speeds, either obtained from observations or required by regulations, a modification of desired speeds appears to require to completely redo the calibration process. Is there really no way to just calculate new values for parameter τ (or other parameters) to keep capacity flow constant when one must change desired speeds?

The answer of this contribution is: "under certain circumstances, approximately". To achieve this an assumption is made which allows to compute the value for τ

T. Kretz (✉)
PTV Group, 76131 Karlsruhe, Germany
e-mail: Tobias.Kretz@ptvgroup.com

I. Zuriguel et al. (eds.), *Traffic and Granular Flow 2019*,
Springer Proceedings in Physics 252,
https://doi.org/10.1007/978-3-030-55973-1_16

analytically which keeps capacity unchanged when desired speed changes. Values of τ which are obtained in this way are then tried in simulations and results are compared to get an idea when good results could be expected with this method. In other words, an ex post justification for the approximation is attempted.

16.2 Assumptions and Consequences

We consider narrow bottlenecks where only one person at a time can pass. It is assumed that a pedestrian begins to move from stand still after a time $1/J_c$ after the preceding pedestrian has started moving. It is furthermore required that at that time—that is $1/J_c$ after starting to move—the same speed is reached like in the original case with original desired speed and the original value for τ. The hypothesis is that in this case capacity flow takes the same value as in the original case. It is assumed that forces between pedestrians can be neglected—to be precise that forces from the preceding and the following pedestrian(s) cancel to a degree that they can be neglected compared to the inertia encoded in the driving force term against instantaneous velocity adjustments.

With this, pedestrians accelerate according to

$$\dot{v}(t) = \frac{v_0 - v(t)}{\tau} \tag{16.1}$$

this solves as

$$v(t) = v_0 \left(1 - e^{-\frac{t}{\tau}}\right) \tag{16.2}$$

We are interested in time $t = 1/J_c$:

$$v(1/J_c) = v_0 \left(1 - e^{-\frac{1}{J_c \tau}}\right) \tag{16.3}$$

For the modified parameters v_0' and τ' the result is accordingly

$$v'(1/J_c) = v_0' \left(1 - e^{-\frac{1}{J_c \tau'}}\right) \tag{16.4}$$

Above we required

$$v(1/J_c) = v'(1/J_c) \tag{16.5}$$

that is

$$v_0 \left(1 - e^{-\frac{1}{J_c \tau}}\right) = v_0' \left(1 - e^{-\frac{1}{J_c \tau'}}\right) \tag{16.6}$$

which can be resolved for τ':

$$\tau' = -\frac{1}{J_c \ln\left(1 - \frac{v_0}{v_0'}\left(1 - e^{-\frac{1}{J_c\tau}}\right)\right)} \tag{16.7}$$

16.3 Application

For a bottleneck with a width of 0.75 m and a length of 9.5 m Eq. (16.7) is applied. The base value for τ is $\tau = 0.4$ s in all cases. For desired speed three base cases are tried: 0.8, 1.3, and 2.0 m/s. Figure 16.1 shows the values for τ' for these three base desired speeds. Simulations are carried out using PTV Viswalk [4] with default parameters (except for v_0 and τ, obviously). 100 pedestrians are set into a starting area of extent $10 \times 10\,\text{m}^2$ located immediately adjacent (and symmetric) to the bottleneck and immediately begin walking towards the bottleneck. Time measurement begins when the first pedestrian passes (enters) the bottleneck and it ends when the last does so. Figure 16.2 shows the resulting capacity flows.

It can be seen that the capacity flow remains nearly constant if desired speed is increased when Eq. (16.7) is used to compute the new value for τ. If the value is reduced, the method fails in most cases already since Eq. (16.7) does not have a real solution (Fig. 16.3).

For wider bottlenecks, with a width w large enough that more than one person can pass at a time, the reasoning above does not apply since the inverse of the flow is not the headway time between two subsequent pedestrians. Specific capacity $j_c = J_c/w$ must be scaled with a lane width l, such that Eq. (16.7) is modified to

$$\tau' = -\frac{w}{lJ_c \ln\left(1 - \frac{v_0}{v_0'}\left(1 - e^{-\frac{w}{lJ_c\tau}}\right)\right)} \tag{16.8}$$

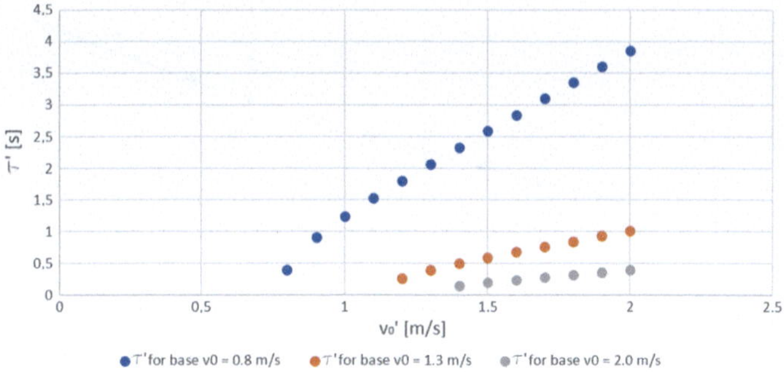

Fig. 16.1 Values for parameter τ calculated—according to Eq. (16.7)—for desired speeds between 0.8 and 2.0 m/s, related to base desired speeds 0.8, 1.3, and 2.0 m/s

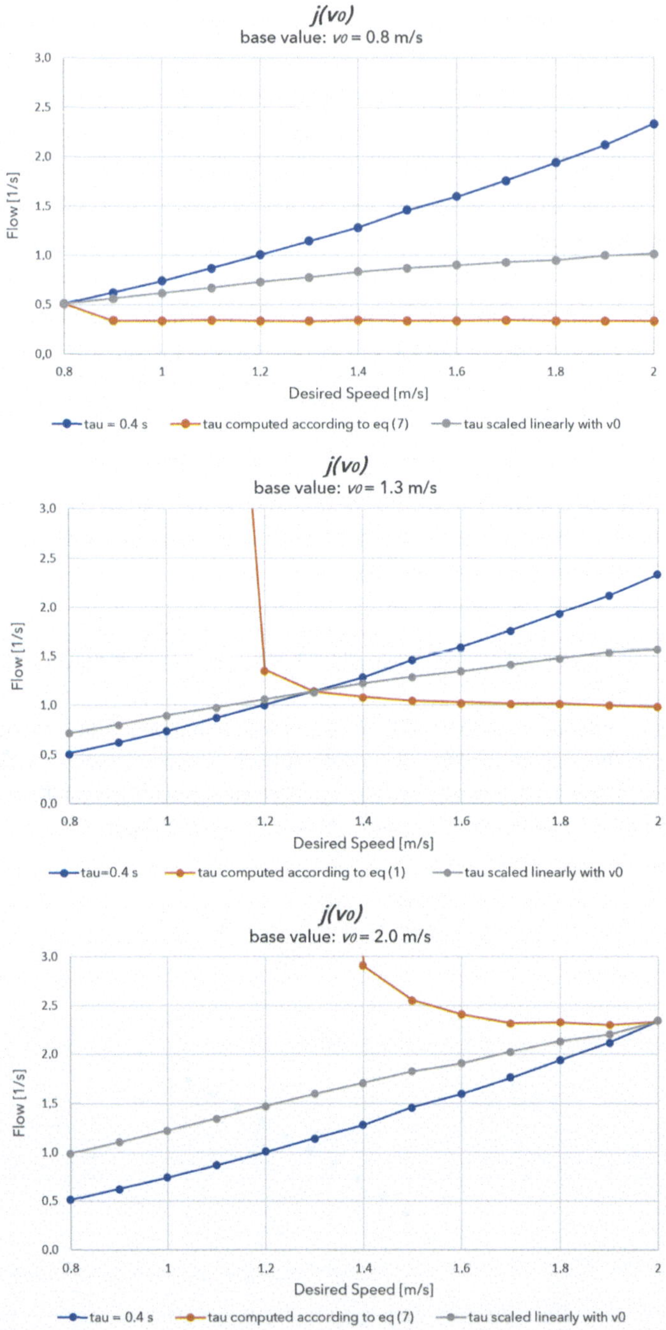

Fig. 16.2 Resulting capacity flows when τ is kept constant (blue), τ is calculated according to Eq. (16.7) (orange), and τ is scaled linearly with desired speed ($v_0/\tau =$ const, gray), for base desired speed 0.8 m/s (top), 1.3 m/s (middle), 2.0 m/s (bottom). Where Eq. (16.7) does not result in a real number $\tau = 0.05$ s has been used

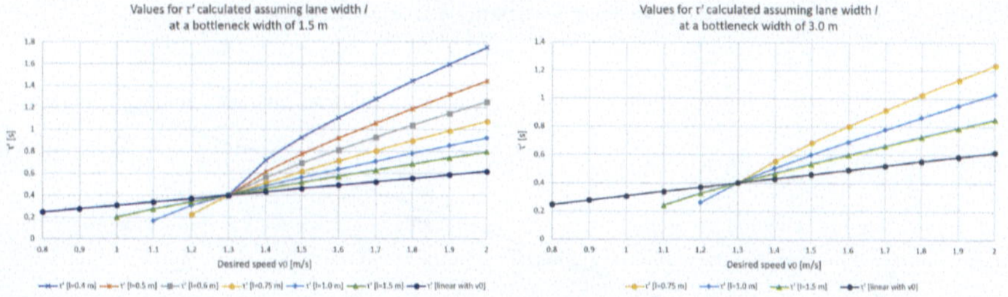

Fig. 16.3 Values for τ' calculated with Eq. (16.8) assuming different lane width values l at bottlenecks 1.5 and 3.0 m wide

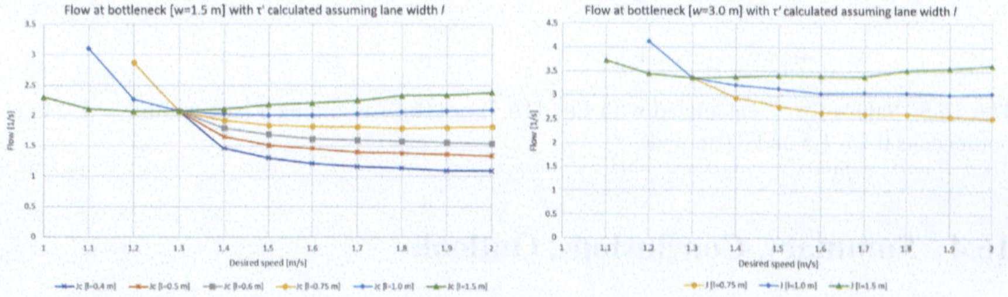

Fig. 16.4 Flow values in the simulation resulting from values for τ' calculated with Eq. (16.8) assuming different lane width values l at bottlenecks 1.5 and 3.0 m wide

The value of the idea proposed in this contribution depends heavily on the answer on the question if l is constant for varying w in the sense that (about) the same value for l produces values for τ with Eq. (16.8) that J_c remains constant for varying desired speeds. For this reason, simulations with bottleneck widths of 1.5 and 3.0 m are done. Flow J_c is measured with $v_0 = 1.3$ m/s and $\tau = 0.4$ s as average of 10 simulation runs. With the result values for τ' are calculated using Eq. (16.8) for some values for l (namely 0.4, 0.5, 0.6, 0.75, 1.0, and 1.5 m). Figure 16.3 shows the values of τ side by side of bottleneck widths w = 1.5 m and w = 3.0 m. With these values flows J_c result as shown in Fig. 16.4. It can be seen that in both cases (w = 1.5 m and w = 3.0 m) a value of l = 1.0 m (or slightly larger) gives the most constant flow for desired speeds larger than 1.3 m/s. It can be seen that these are (mutually) similar if a comparison with linear scaling with desired speed is taken as a reference Fig. 16.5.

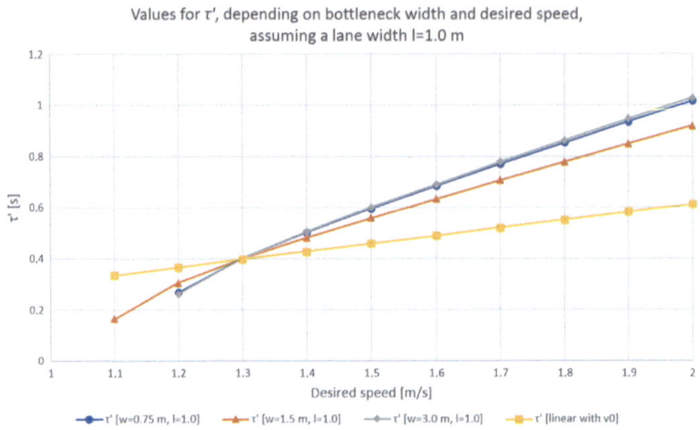

Fig. 16.5 Values for τ' calculated with Eqs. (16.7) or (16.8) assuming a lane width $l = 1.0$ m at bottlenecks 0.75, 1.5 and 3.0 m wide

16.4 Summary, Conclusions, Outlook

A method was proposed that allows to directly calculate a new value for parameter τ of the Social Force Model when desired speed is increased (and in some cases also when it is decreased) such that the flow at bottlenecks of various widths remains nearly unchanged, without having to change further parameter values. For wide bottlenecks the method includes a parameter l which had to be gauged and which can be interpreted as lane width. The resulting value of about $l = 1.0$ m appears a little bit large, albeit it works for the intended purpose. Why it takes such a large value remains an open question. Furthermore, it remains to be shown if the method is still applicable if the other parameters of the Social Force Model have different values.

References

1. A. Johansson, D. Helbing, P.K. Shukla, ACS **10**(supp02), 271 (2007)
2. T. Kretz, J. Lohmiller, P. Sukennik, Transp. Res. Rec. **2672**(20), 228 (2018)
3. F. Johansson, D. Duives, W. Daamen, S. Hoogendoorn, Transp. Res. Procedia **2**, 300 (2014)
4. T. Kretz, F. Reutenauer, F. Schubert, in *92nd Annual Meeting of the Transportation Research Board* (2013). On CD: 13-1943

Chapter 17
An Analytical Solution of the Social Force Model for Uni-Directional Flow

Tobias Kretz

Abstract A function for the dependence of flow on pedestrian density is derived analytically from the Social Force Model (SFM) for the case of a homogeneous population walking in the same direction and being in steady state. Assuming that only nearest Voronoi neighbors effectuate forces the resulting function matches a variety of very different fundamental diagrams that were found empirically.

17.1 Motivation

The motivation for this study comes from different sources. First, the variety of fundamental diagrams of pedestrian dynamics which were reported from field and laboratory—as shown in Fig. 17.1—studies calls for an explanation. Is it measurement methods or evaluation techniques that causes these large differences? Is even the quality of a majority of these studies at question? Is the notion of a fundamental diagram for pedestrians—dissenting with [1]—simply not meaningful? Or is the variation a consequence of varying circumstances which then should be possible to be modeled with parameter value variation within one single model. This is emphasized when comparing with fundamental diagrams discussed for vehicular traffic flow—for an overview see [2]—where there is also a bandwidth of shapes, but it is clearly not as wide as for pedestrian dynamics. Is this because the longer lasting research in vehicular dynamics has led to more agreement or can a reason be identified why pedestrian dynamics would yield different fundamental diagrams than vehicular dynamics?

Second, it was possible before to solve the SFM analytically for single file movement [13, 14] which allowed to derive equations with which model parameter values can be computed from observables [15]. The latter is a decidedly valuable result since it allows direct calibration of the model (with all limitations following from approximations made as part of the analysis).

T. Kretz (✉)
PTV Group, 76131 Karlsruhe, Germany
e-mail: Tobias.Kretz@ptvgroup.com

I. Zuriguel et al. (eds.), *Traffic and Granular Flow 2019*,
Springer Proceedings in Physics 252,
https://doi.org/10.1007/978-3-030-55973-1_17

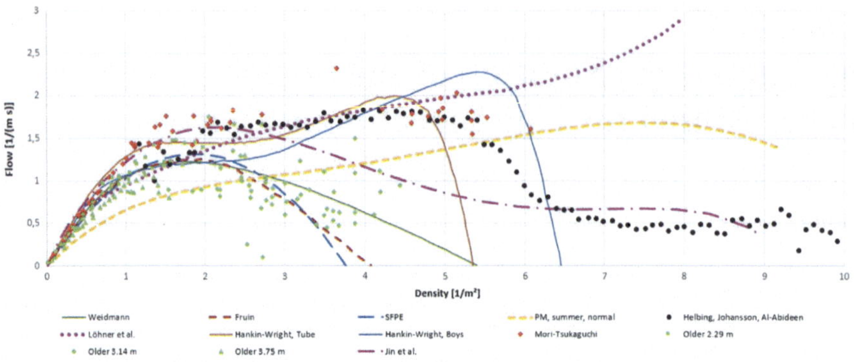

Fig. 17.1 Empirical pedestrian fundamental diagrams as reported in [3–12]. Some of the data is publicly available at https://www.asim.uni-wuppertal.de/de/datenbank/data-from-literature/fundamental-diagrams.html. Where (mean value) functions are shown they are taken from literature, where no average function is given, the scatter plot is shown

Third, an analytical solution of a simulation model allows to formulate an expectation for simulation results. This can be a helpful element for the verification of a software implementation.

17.2 Setting, Assumptions, and Approximations

For this purpose it is assumed that …

- all pedestrians desire to walk in the same direction. This mainly defines the setting for which the obtained function for a fundamental diagram can (approximately) be valid. To a lesser degree it is also an approximation.
- the system is infinitely large or that there are periodic boundary conditions in both dimensions (walking on a torus/"doughnut") This is to avoid having to consider the effect from borders and walls and it implies that the solution will be a better approximation for large than for small systems.
- all pedestrians are identical in the sense that they are described with identical parameter values in the SFM, most importantly this means that the desired speed is set identical for all. This is essential for the analytical treatment since a discussion that includes parameter variations would be much more difficult. At the same time it is unrealistic and one has to be aware that even small variations may disturb and destroy a steady-state.
- only nearest Voronoi neighbors affect a pedestrian. Compared to the original SFM this is a strong limitation since there all pedestrians in a system produced a force on any other. The latter may be largely unrealistic, but it is plausible that in reality beyond nearest also next to nearest and even further neighbors can have a direct effect.

- the system is in a steady-state, all forces on a pedestrian cancel to zero. It is assumed that a steady-state exists and is stable.
- pedestrians walk in a certain formation, namely on the grid points of a triangular lattice, which is not necessarily regular, but can be stretched or quenched. It is not clear to which degree this is an approximation. With the exception of [16] there is little research on walking formation or just the number of Voronoi neighbors.
- density changes affect only longitudinal but not lateral spacing. This is an approximation, but the tendency appears to exist. A bounding box which includes nearest or next to nearest neighbors in data from a laboratory experiment [17] typically has a higher width to length ratio for higher than for lower densities and this makes sense since with speed stride length grows which extends space requirements in walking direction, but not transverse.
- the analytically derived function for steady-state matches (approximately) the average function of empirical data. This is not relevant for the analytical treatment as such, but for the comparison of analytical function and empirical average it is of course important to bear in mind that strictly speaking two different properties are being compared.

17.3 Analysis and Results

Together with these assumptions a highly symmetric walking formation is assumed as shown in Fig. 17.2. The task is now to compute a speed for each density based on the geometry of the walking formation and some appropriate model of pedestrian dynamics, here the SFM serves for this purpose, to be precise circular specification or elliptical specification II (either will work) as of [18].

As a first step it is interesting to note that regarding Voronoi neighbors there are two regimes: one for low and one for high densities as Fig. 17.3 visualizes.

This is a consequence from the Voronoi cells not simply stretching and quenching along with the walking formation, but its edges transform conformal with connecting lines between nearest neighbors as can be seen in Fig. 17.3. This alternative transformation behavior of the cells is neutral with respect to area content and thus has no effect on how density is computed, namely as

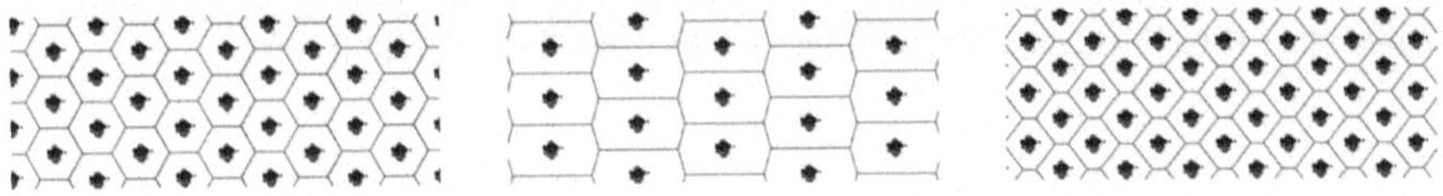

Fig. 17.2 Hypothetical walking formation at medium density, when Voronoi cells are regular hexagons (left), lower density (center) and higher density (right). The lateral spacing is the same in all three cases (i.e. the number of pedestrians in a row remains the same). Walking direction is →

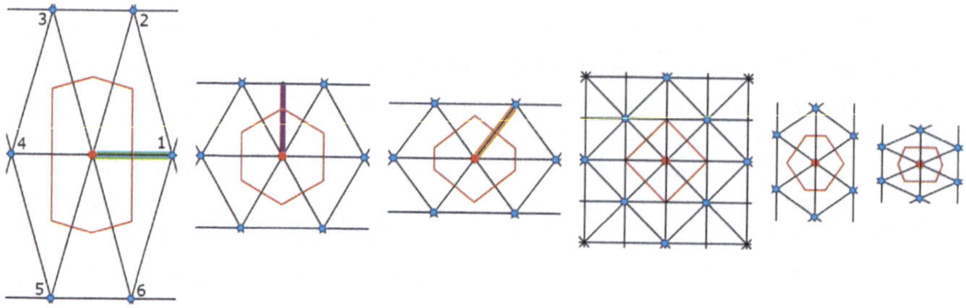

Fig. 17.3 Voronoi cells from lower (left) to higher (right) densities. Walking direction is ↑. Comparing the first and the third figure one can easily see that the Voronoi cell is not simply stretched and quenched with the grid of the walking formation. This has an important consequence: at lower densities a pedestrian has only lateral and diagonal neighbors, whereas above a density of $\rho = 2/b^2$ (4th subfigure) a pedestrian has diagonal and longitudinal neighbors. In the left subfigure neighbors are numbered for use in Eq. (17.4) and the green line marks the lateral spacing b. In the second subfigure the height h of the Voronoi cell is highlighted in magenta and in the third subfigure the 'diagonal' d is wrapped orange

$$\rho = \frac{1}{bh} \tag{17.1}$$

with b and h defined as depicted in Fig. 17.3.

It is clear that forces from lateral neighbors (numbers 1 and 4 in Fig. 17.3) cancel and so do—as a consequence of the high symmetry—pairwise all contributions to the lateral force component. This makes the lateral spacing b an input parameter to the model. The actual walking direction matches the desired walking direction and there is no component of velocity or acceleration orthogonal to it. Thus, we are faced essentially with a 1d equation of motion.

The full model (circular specification)

$$\ddot{\mathbf{x}}_i(t) = \frac{\mathbf{v}_{0,i} - \dot{\mathbf{x}}_i(t)}{\tau_i} + \tilde{A}_i \sum_j w(\mathbf{x}_i(t), \mathbf{x}_j(t), \dot{\mathbf{x}}_i(t), \lambda_i) e^{-\frac{|\mathbf{x}_j(t)-\mathbf{x}_i(t)|-R_i-R_j}{B_i}} \hat{e}_{ij} \tag{17.2}$$

$$w(\mathbf{x}_i(t), \mathbf{x}_j(t), \dot{\mathbf{x}}_i(t), \lambda_i) = \lambda_i + (1-\lambda_i)\frac{1+\cos(\theta_{ij}(\mathbf{x}_i(t), \mathbf{x}_j(t), \dot{\mathbf{x}}_i(t)))}{2} \tag{17.3}$$

can be simplified and re-arranged as a consequence of the high symmetry and the steady-state if only nearest Voronoi neighbors contribute to the force:

$$\dot{x} = \mathbf{v}_0 + \tau A \left(\sum_{j=5,6} \cos(\theta_j) w(\lambda, \theta_j) e^{\frac{-d}{B}} - \sum_{j=2,3} \cos(\theta_j) w(\lambda, \theta_j) e^{\frac{-d}{B}} \right) \tag{17.4}$$

$$\dot{x} = \mathbf{v}_0 - 2\tau A(1-\lambda)\frac{h^2}{d^2} e^{\frac{-d}{B}} \tag{17.5}$$

and get as speed-density relation and fundamental diagram

$$j = v_0\rho\left(1 - 2(1-\lambda)A\frac{\tau}{v_0}\frac{1}{1+\frac{\rho^2 b^4}{4}}e^{-\frac{1}{2}\frac{b}{B}\sqrt{1+\frac{4}{b^4\rho^2}}}\right) \tag{17.6}$$

This holds for the geometry as depicted in the first three subfigures of Fig. 17.3 which is the low density range. For the high density range the analysis a term from the longitudinal neighbor (the one between and ahead of neighbors number 2 and 3 in Fig. 17.3) must be added which can be calculated accordingly.

Finally, the two ranges have to be connected such that a steady function follows. This comes down to the question if the longitudinal neighbor should have a full effect if it shares only one point or a very short edge with the pedestrian for whom forces are computed. A straightforward idea is to weigh forces with the length of the shared edge relative to the total circumference of the Voronoi cell. With this, the function for the fundamental diagram over the full density range results as

$$j(\rho) = \rho v_0\left(1 - \alpha\left(2g_{diag}\frac{1}{1+k^2}e^{-\frac{1}{2}\frac{b}{B}\sqrt{1+\frac{1}{k^2}}} + g_{long}e^{-\frac{b}{B}\frac{1}{k}}\right)\right) \tag{17.7}$$

$$g_{diag} = \frac{4l_{diag}}{4l_{diag}+2l_{long}+2l_{lat}};\; g_{long} = \frac{2l_{long}}{4l_{diag}+2l_{long}+2l_{lat}} \tag{17.8}$$

$$l_{diag} = \frac{b}{2}\sqrt{1+k^2}\min\left(1, \frac{1}{k^2}\right);\; l_{long} = \frac{b}{2}\max\left(0, 1-\frac{1}{k^2}\right) \tag{17.9}$$

$$l_{lat} = \frac{b}{2}\max\left(0, \frac{1}{k} - k\right) \tag{17.10}$$

$$\alpha = (1-\lambda)A\frac{\tau}{v_0};\; k := \rho\frac{b^2}{2} \tag{17.11}$$

where an l gives the length of the respective edge and a g a weight factor. α is a characteristic value for a parametrization of the SFM since it relates the forward-backward asymmetry $(1-\lambda)$, the base strength of forces between pedestrians A and the base self-propelling acceleration v_0/τ. k is a scaled density which has the

Table 17.1 Parameter values which allow to approximate various empirical fundamental diagrams. It is $\lambda = 0.1$ and $\tau = 0.4$ s for all

	P1	P2	P3	P4	P5
v_0 [m/s]	1.3	1.3	0.9	1.0	0.9
A [m/s^2]	4.2	34	3	35	22
B [m]	2.5	0.23	0.6	0.168	0.18
b [m]	0.5	0.9	0.54	0.62	0.7

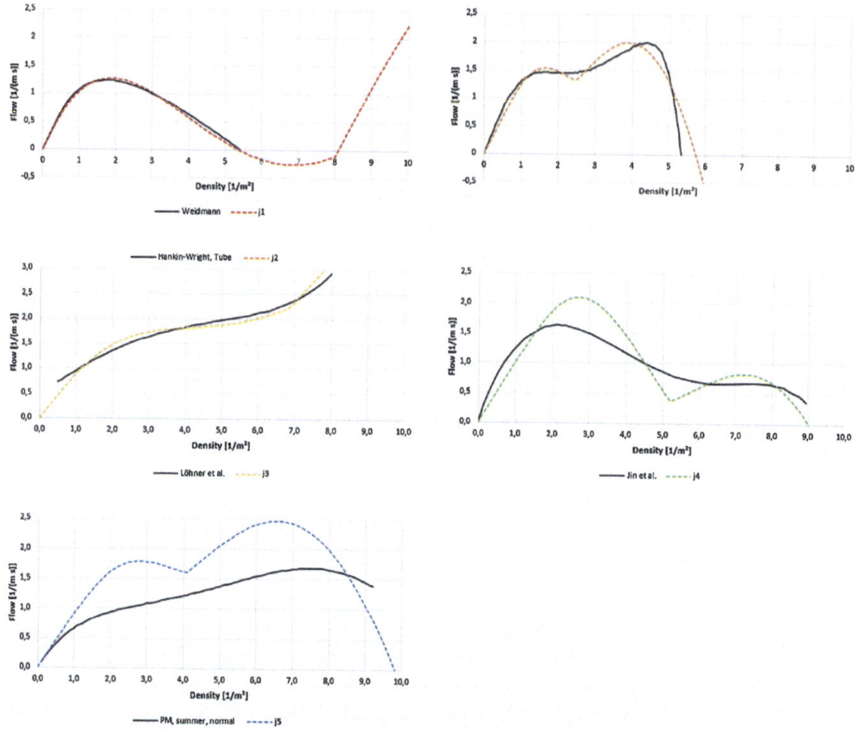

Fig. 17.4 Empirical fundamental diagrams (Weidmann, Hankin-Wright, Löhner et al., Jin et al, Predt.-Mil.) compared to Eq. (17.7) with parameters from Table 17.1

value $k = 1$ at the border of low and high density range. With the parameters from Table 17.1 Eq. (17.7) allows to reproduce fundamental diagrams of all shapes as shown in Fig. 17.4.

17.4 Summary—Conclusions—Outlook

With some simplifying assumptions it was possible to derive analytically from the SFM a function for the fundamental diagram which allows to approximate the entire variety of documented empirical fundamental diagrams by means of parameter variation. This does not guarantee, but it is an indication, that all of the empirical data is valid, despite it being seemingly contradicting, and that the SFM with Voronoi neighborhoods is a realistic model of pedestrian dynamics.

One of the obvious next steps is to implement a variant of the SFM with a Voronoi neighborhood as source of inter-pedestrian forces.

While the starting point was the Social Force Model along the course of the analytical treatment its details seemed less and less relevant. As long as the interaction between pedestrians decreases monotonically with distance and as long as there is some suppression of effects from behind, the main results will not change significantly. One could almost say "the neighborhood is the model". Consequently, an empirical investigation into the structure and geometry of Voronoi cells in real systems appears to be very interesting, particularly into how realistic the assumption of a constant lateral spacing is.

References

1. M.J. Lighthill, G.B. Whitham, Proc. R. Soc. A **229**(1178), 317 (1955)
2. J. Del Castillo, F. Benitez, Transp. Res. B **29**(5), 373 (1995)
3. U. Weidmann, Transporttechnik der Fußgänger (1993)
4. J.J. Fruin, *Pedestrian Planning and Design* (MAUDEP, 1971)
5. P.J. Dinenno et al. (ed.), *SFPE Handbook of Fire Protection Engineering*, 4th edn. (NFPA, 2008)
6. V.M. Predtechenskii, A.I. Milinskii, *Planning for Foot Traffic Flow in Buildings* (Amerind, 1978)
7. D. Helbing, A. Johansson, H.Z. Al-Abideen, Phys. Rev. E **75**(4), 046109 (2007)
8. R. Löhner, B. Muhamad, P. Dambalmath, E. Haug, Collect. Dyn. **2**, 1 (2018)
9. B. Hankin, R.A. Wright, J. Oper. Res. Soc. **9**(2), 81 (1958)
10. M. Mōri, H. Tsukaguchi, Transp. Res. A **21**(3), 223 (1987)
11. S.J. Older, Traffic Eng. Control **10**, 4 (1968)
12. C.J. Jin, R. Jiang, S. Wong, D. Li, N. Guo, W. Wang (2017). arXiv:1710.10263
13. T. Kretz, J. Lohmiller, J. Schlaich (2015). arXiv:1512.01426
14. T. Kretz, J. Lohmiller, J. Schlaich, Collect. Dyn. **1**, 1 (2016)
15. T. Kretz, J. Lohmiller, P. Sukennik, Transp. Res. Rec. **2672**(20), 228 (2018)
16. J. Porzycki, M. Mycek, R. Lubaś, J. Was, Transp. Res. Proc. **2**, 201 (2014)
17. Voronoi density (2018). https://youtu.be/WBBJU2meS34 (contact: M. Chraibi)
18. A. Johansson, D. Helbing, P.K. Shukla, ACS **10**(supp02), 271 (2007)

Chapter 18
A Cognitive, Decision-Based Model for Pedestrian Dynamics

Cornelia von Krüchten and Andreas Schadschneider

Abstract We present a modelling approach that combines several aspects of different model classes. As a basis we use discrete time and continuous space. Principal idea of the model is to reproduce a pedestrian's decision about the direction of motion and the velocity for the next time step. The decision process relies on personal goals and information provided by visual perception of the environment. It is modelled by a stochastic process to represent inaccuracies in the agents' perception and decision. For calibration, simulation results are compared to empirical data qualitatively and quantitatively for different scenarios.

18.1 Introduction

Modelling and model development are valuable sources of information on pedestrian dynamics not only as a supplement to empirical observations but also as tool to understand mechanisms and scenarios that cannot be investigated by experiments or field studies. Therefore, many different approaches have been developed in the course of time. The wide variety of models is mostly categorised by their fundamental properties and structure [1]. Here, we focus on microscopic models which consider each pedestrian and its interactions individually. Microscopic models can be distinguished by the basic approach: acceleration-based models, also referred to as second-order models, consider pedestrians as exposed to extrinsic and intrinsic forces and are generically continuous in both time and space. In contrast, decision-based models are often set in discrete time and space and express the agents' dynamics by a set of

C. von Krüchten · A. Schadschneider (✉)
Institut für Theoretische Physik, Universität zu Köln, Zülpicher Str. 77,
50937 Cologne, Germany
e-mail: as@thp.uni-koeln.de

C. von Krüchten
e-mail: C.vonKruechten@t-online.de

A. Schadschneider
Institut für Physikdidaktik, Universität zu Köln, Gronewaldstr. 2, 50931 Cologne, Germany

I. Zuriguel et al. (eds.), *Traffic and Granular Flow 2019*,
Springer Proceedings in Physics 252,
https://doi.org/10.1007/978-3-030-55973-1_18

pre-defined rules. First-order or velocity-based models determine the velocity of the pedestrians directly, based on the current state of the agent and its environment.

In this contribution, we present an approach that combines aspects of these model classes. As a basis, it is set in continuous space, but discrete time. A pedestrian determines the velocity for the next time step based on its personal goals and information on the environment. In doing so, cognitive abilities like visual perception, decision-making and anticipation are used. At root, the model is a velocity-based approach, but can be expressed in a simple rule-based structure as in decision-based models. Due to the continuous space, numerical and spatial artefacts are prevented. Considering cognitive abilities explicitly, the approach contains elements of operational and tactical level models [1, 2].

The new approach aims at a deeper understanding of human behaviour and decision processes of pedestrians. Stochastic elements in the decision process help at capturing unknown underlying mechanisms and imperfect human decisions and make the modelled dynamics more realistic.

18.2 The Model Approach

Figure 18.1 shows the basic structure of the decision process. In each time step, that can be understood as a reaction time, a pedestrian decides on the velocity for the next step by determining the direction of motion and the absolute value of the velocity successively. Thereby, movement within the environment and the interaction with other pedestrians are considered separately at first and combined at a later stage in the decision process. The main influencing quantity is the distance-to-collision in a given direction which acts as the decision-making basis for both speed and direction of motion.

Starting from the current state and with the personal goals, a pedestrian uses a visual field to gain information on the surrounding. During this perception phase the agent detects the infrastructure (walls and obstacles) as well as other pedestrians. In the latter case, anticipation is used to extrapolate the agents' state in the next time step in order to consider relative motion between the perceiving agent and the other pedestrians.

During the actual decision phase the pedestrian determines two independent angles for collision avoidance ('interaction angle') and the motion within the environment ('target angle'). Considering the perceived infrastructural elements, the pedestrian determines the distance-to-collision into the directions towards the target in order to assess whether the goal is directly accessible. If not, the agent heads for intermediate targets in order to navigate through the geometry. When a final target is selected, the final target angle is determined by using a Gaussian distribution around the optimal (intermediate) target direction.

For collision avoidance the agent determines the distance-to-collision for each pedestrian that has been previously perceived. Combining this information for all pedestrians within the visual field, an angular distribution of the distance-to-collision

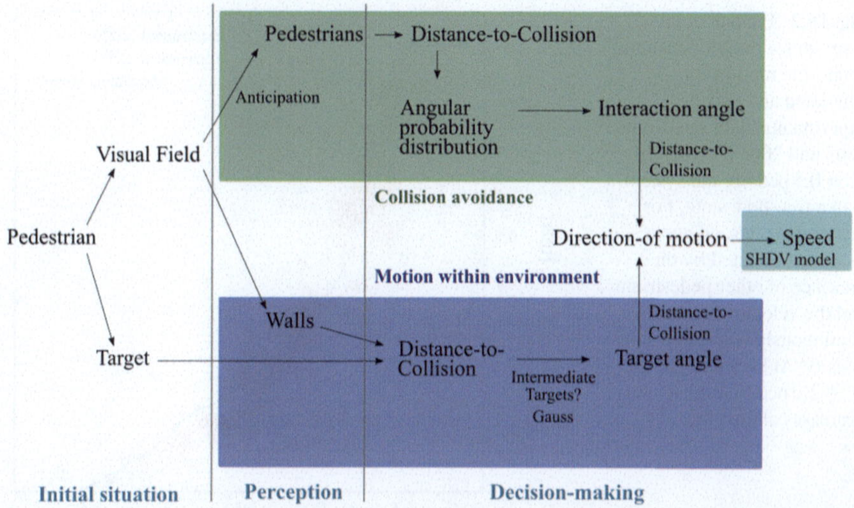

Fig. 18.1 Based on visual information and personal goals, the pedestrian decides on an potential direction for collision avoidance and for reaching the target and then chooses one of these to be the final direction of motion. The distance-to-collision acts as a basis of decision-making for the determination of both direction of motion and speed

can be determined. Whereas free ranges of the visual field are more likely to be chosen for collision avoidance, the likelihood of directions that are covered due to the presence of other agents is reduced accordingly to their relative distance. In doing so, the distance-to-collision is used to determine an angular probability distribution from which the final interaction angle is drawn. Both interaction and target angle are compared in a third step. The agent decides, mostly based on the distance-to-collision, which one of these two angles is chosen as the final direction of motion.

Once this decision has been taken, the distance-to-collision into the direction of motion is used to determine the speed. Referring to the Stochastic Headway Dependent Velocity (SHDV) model [3, 4], the agent's speed for the next time step depends linearly on the headway which is given by the mentioned distance-to-collision.

A more detailed description of the modelling approach is given in [5, 6].

18.3 Modelling Results

In order to calibrate the model parameters and validate the modelled dynamics, simulation results are compared to experimental data for basic scenarios of pedestrian dynamics.

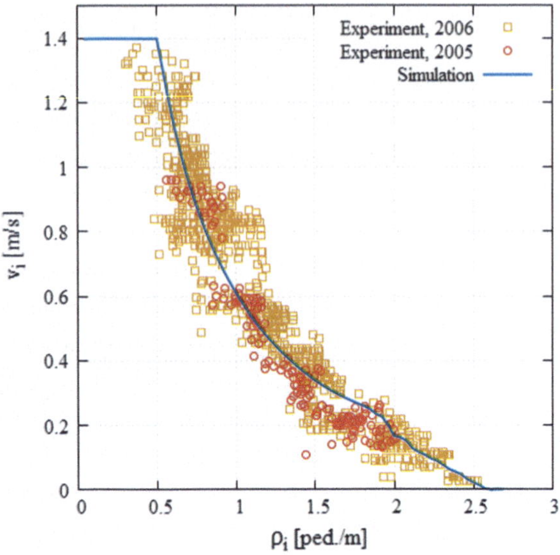

Fig. 18.2 Fundamental diagram for one-dimensional single-file motion for simulated and two experimental data sets from 2005 and 2006 [9, 10]. Up to $\rho_i = 0.5$ ped./m, the system is in a free-flow state. For higher densities, an agent's motion is affected by the presence of other pedestrians and the velocity decreases continuously with increasing density. At high densities, $\rho_i \leq 2.0$ ped./m, the velocity decreases almost linearly

18.3.1 Single-File Motion

In a one-dimensional single-file motion scenario, pedestrians walk on a closed path of length $L = 26.0$ m and width $B = 0.8$ m. After a certain relaxation time, stationary fundamental diagrams were measured. The determination of density and velocity of the experimental data rely on the approaches used in [4, 7, 8]. Figure 18.2 shows the relation of global density and averaged velocity for simulations and experimental results [9, 10]. It can be seen that the simulated data fit well to the experimental ones. At low densities, the agents walk in an undisturbed way at maximum velocity $v_{\max} = 1.4$ m/s. Within the regime 0.5 ped./m $\leq \rho_i \leq 1.8$ ped./m, the pedestrians distribute (on average) homogeneously in the course and walk with reduced speed. At even higher densities, $\rho_i \geq 2.0$ ped./m, the influence of jams and stop-and-go waves becomes significant and the velocity decreases fast with increasing density. To obtain this good quantitative agreement, the modelled dynamics had to be slightly adjusted. For single-file motion, it is crucial that the agents do not overtake or walk next to each other. Therefore, when deciding on the final direction of motion, the urge to avoid the proximity of other pedestrians has to be weakened on favour of maintaining the single-file configuration. Therefore, interaction angles that deviate much from the preferred walking direction towards the target $\left(|\alpha_{\text{interaction}} - \alpha_{\text{target}}| \gtrsim 40°\right)$ are suppressed.

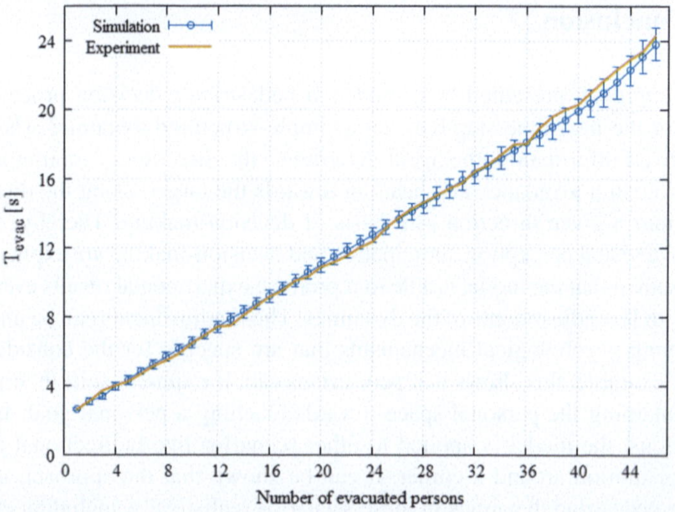

Fig. 18.3 Evacuation times for simulated and experimental [11] evacuation runs. The modelled dynamics reproduces the empirical evacuation behaviour well

18.3.2 Evacuation

Evacuations are important and frequently considered scenarios in pedestrian dynamics. Therefore, simulations are compared to evacuation experiments [11]. The set-up contains 46 agents that are distributed almost uniformly in the centre of a square room of $5 \times 5\,\text{m}^2$ and that leave the room through an exit of width $B = 1.2\,\text{m}$. For the simulations, the initial positions of the pedestrians were taken from the experimental runs in order to increase the comparability. To quantify the progress of the evacuation process, the number of evacuated persons is plotted against the time needed. Since the overall simulation time was short due to the low number of agents, the modelled results for these evacuation times were averaged over 100 simulation runs. Figure 18.3 shows that the model is able to reproduce the experimental evacuation times even quantitatively. The slight deviation of the experimental and simulated results for a larger number of evacuated persons might be caused by a lack of motivation and urge to leave the room of the participants who were last in the experimental set-up. Similar to the single-file scenario, the interaction angle is suppressed for large deviations $\left(|\alpha_{\text{interaction}} - \alpha_{\text{target}}| \gtrsim 60^\circ\right)$ from the target angle. It represents an agent's urge to leave the room as fast as possible and is supported by small anticipation and reaction times and a high urge to follow other agents in front of the exit.

18.4 Conclusion

The new approach presented here models a pedestrian's decision process on the velocity for the next time step built on a simple-structured dynamics. Using visually perceived information, the agent determines the direction of motion as choice between collision avoidance and heading towards the target, using the distance-to-collision into a given direction as a basis of decision-making. Thereby, cognitive abilities like visual perception, anticipation and decision-making are explicitly used. It can be shown that the model is able to reproduce experimental results even quantitatively with few adjustments of the dynamics. These adjustments can be understood as underlying psychological mechanisms that are specific for the considered scenario. For example, this shows that pedestrians can, if required, shift their priorities from maintaining the personal space towards reaching a personal goal. In further investigations, the model is applied to other scenarios like bidirectional flow in a corridor or motion around a corner. It can be shown that the approach is able to reprocude pedestrian dynamics in these scenarios realistically, including characteristic collective phenomena like lane formation. Further information on the modelling approach and the comparison to experimental results is given in [6].

Acknowledgments Financial support by Deutsche Forschungsgemeinschaft (DFG) under grant SCHA 636/9-1, Bonn-Cologne Graduate School of Physics and Astronomy (BCGS) and the German Excellence Initiative through the University of Cologne Forum 'Classical and Quantum Dynamics of Interacting Particle Systems' is gratefully acknowledged. We also thank the Institute for Advanced Simulation: Civil Safety Research by Forschungszentrum Jülich for providing the experimental data used in this contribution [9–11]. The data from 2006 were part of a project funded by DFG under grants KL 1873/1-1 and SE 1789/1-1.

References

1. A. Schadschneider, M. Chraibi, A. Seyfried, A. Tordeux, J. Zhang, Pedestrian dynamics—from empirical results to modeling, in *Crowd Dynamics, Volume 1: Theory, Models, and Safety Problems* ed. by L. Gibelli, N. Bellomo, vol. 63 (Birkhäuser, 2018), p. 102
2. L. Crociani, G. Vizzari, S. Bandini, Adaptive tactical decisions in pedestrian simulation: a hybrid agent approach., in *Traffic and Granular Flow* ed. by V.L. Knoop, W. Daamen, vol. 15 (Springer, 2016), pp. 257–264
3. C. Eilhardt, A. Schadschneider, Stochastic headway dependent velocity model for 1D pedestrian dynamics at high densities. Transp. Res. Proc. **2**, 400–405 (2014)
4. Eilhardt, C.: Computer simulation of pedestrian dynamics at high densities. Dissertation, Universität zu Köln (2014)
5. C. von Krüchten, A. Schadschneider, Concept of a decision-based pedestrian model, in *Pedestrian and Evacuation Dynamics 2018* (2020), p. 316
6. C. von Krüchten, Development of a cognitive and decision-based model for pedestrian dynamics. Dissertation, Universität zu Köln (2019)
7. A. Seyfried, B. Steffen, W. Klingsch, M. Boltes, The fundamental diagram of pedestrian movement revisited. J. Stat. Mech. **2005**, P10002 (2005)

8. A. Portz, A. Seyfried, Analysing stop-and-go waves by experiment and modeling, in *Pedestrian and Evacuation Dynamics* ed. by R.D. Peacock, E. D. Kuligowski, J.D. Averill (Springer, 2011), pp. 577–586
9. Pedestrian Dynamics Data Archive: Single File Movement, Rotunde, http://ped.fz-juelich.de/da/2005singleFile (2005)
10. Pedestrian Dynamics Data Archive: Single File Movement, Caserne, http://ped.fz-juelich.de/da/2006singleFile (2006)
11. Pedestrian Dynamics Data Archive: Evacuation, GymBay, http://ped.fz-juelich.de/da/2014socialGroups (2014)

Chapter 19
Exploring Koopman Operator Based Surrogate Models—Accelerating the Analysis of Critical Pedestrian Densities

Daniel Lehmberg, Felix Dietrich, Ioannis G. Kevrekidis, Hans-Joachim Bungartz, and Gerta Köster

Abstract We apply the Koopman operator framework to pedestrian dynamics. In an example scenario, we generate crowd density time series data with a microscopic pedestrian simulator. We then approximate the Koopman operator in matrix form through Extended Dynamic Mode Decomposition, using Geometric Harmonics on the data as a dictionary. The Koopman matrix is integrated into a surrogate model, which allows to approximate crowd density time series data to be generated, independently from the original microscopic simulator. The evaluation of the constructed surrogate model is orders of magnitude faster, and enables us to use methods that require many model evaluations.

19.1 Introduction

The core of microscopic pedestrian models consists of the update rules for individual agents. This view on a complex system simplifies modeling. Simulating many interacting agents often reveals collective dynamics that are not explicitly modeled,

D. Lehmberg (✉) · G. Köster
Department of Computer Science and Mathematics, Munich University of Applied Sciences, Munich, Germany
e-mail: daniel.lehmberg@hm.edu

G. Köster
e-mail: gerta.koester@hm.edu

D. Lehmberg · H.-J. Bungartz
Department of Informatics, Technical University of Munich, Garching, Germany
e-mail: bungartz@in.tum.de

F. Dietrich · I. G. Kevrekidis
Whiting School of Engineering, Johns Hopkins University, Baltimore, USA
e-mail: felix.dietrich@jhu.edu

Princeton University, Princeton, USA

I. G. Kevrekidis
e-mail: yannisk@jhu.edu

I. Zuriguel et al. (eds.), *Traffic and Granular Flow 2019*,
Springer Proceedings in Physics 252,
https://doi.org/10.1007/978-3-030-55973-1_19

but emerge from the microscopic rules [1]. The dynamic patterns at the collective scale develop and can be observed on a more abstract and macroscopic view. This multi-scale property is common to many complex agent based systems, [2] (and references therein). Emerging patterns relevant for pedestrian dynamics are, for example, pedestrian streams [3] or crowd patterns in high densities [4]. Because macroscopic quantities and their dynamics are relevant for practitioners and model validation, analyzing these emerging patterns is important for microscopic pedestrian models. The difficulty of this analysis is that in most cases, there are no explicit equations for the dynamics of the macroscopic quantities available. The dependence on parameters is thus not immediately clear: How robust are the patterns of a model, given uncertain input parameters and boundary conditions? In what situations are patterns changing? How well does the model approximate crowd patterns observed in reality? To answer these questions we need to *restrict* the microscopic state of the model to the macroscopic quantity of interest. Calibrating a pedestrian model to make it consistent with the macroscopic observations can also require *lifting* the state back to the microscopic agent-based state [5]. In this work, we are computing the macroscopic crowd density evolution from simulated agent trajectories.

19.1.1 Surrogate Model

Sensitivity studies and uncertainty quantification require many simulation runs (often thousands or more) to sample from a set of model input parameters [6]. In order to compute macroscopic observation data the microscopic simulator model state is restricted. However, if a single microscopic model evaluation is computationally expensive this can prevent the analysis [1, 2]. One approach to overcome the computational bottleneck of the microscopic model is to construct a surrogate model, which models the dynamics of the (macroscopic) observations directly [1, 6]. The surrogate model is then based on data which is generated with the original microscopic model in an offline phase. In the context of machine learning, this data is known as "training data" and consists of pairs of input (e.g. parameters we want to analyze in a sensitivity study) and resulting macroscopic output quantities. The surrogate model can be used in an online phase to generate (approximate) macroscopic data, independently from the original model [6]. The evaluation of the surrogate model is much faster than for the original model because the surrogate model fits a specific scenario unlike the more general simulator. In this paper, we consider time-dependent (i.e. dynamic) quantities. The generated example data consist of pairs of initial condition and resulting time series. We aim to construct a model that approximates the example time series data and can be extended to new initial conditions in the vicinity of the sampled data. In our problem setting, detailed in the next section, the time series' initial conditions can be directly inferred from the selected model input parameter which we sample.

19.1.2 Collecting Data from Example Scenario

To explore the proposed method, we set up an example scenario (illustrated in Fig. 19.1). We use the microscopic pedestrian simulation framework Vadere [7] and simulate with the "Optimal Steps Model", which is detailed in [8, 9]. Because our approach is data-driven, the microscopic pedestrian model can easily be replaced.

In the scenario the agents start inside the bus (source), walk through the bus door and station exit, and are removed from the scenario at the target. The only simulation input parameter that we sample is the number of agents starting inside the bus. Here, we take 17 equidistant samples between 10 and 100 agents.

We generate time series by computing the crowd density from the microscopic state, at three observation areas in the scenario, and for every simulated time step. The density data is computed by counting the agents inside of the respective measurement area. Every simulation is 120 s long and we collect data every 0.4 s (simulation time). Because of stochastic elements, we run every parameter setting 10 times and average each time series; see Fig. 19.1 (right) for the final density time series included in the example data. Note that the observation areas "bus" and "bus door" are overlapping (cf. Fig. 19.1, left), which results in a positive density for both quantities at the time series' initial state.

We compile $P = 17$ time series $D_p = \{\mathbf{d}_k\}_{k=0}^{K=300}$ with a single measurement at a given time step $\mathbf{d}_k = (d_{k,\text{bus}}, d_{k,\text{bus door}}, d_{k,\text{station}}, d_{k,\text{combined}})$ into a data matrix $\mathcal{D} = [D_0, D_1, \ldots, D_P]$. We also add a "combined quantity", $d_{k,\text{combined}} = d_{0,\text{bus}} - d_{k,\text{bus}} - d_{k,\text{bus door}} - d_{k,\text{station}}$. The quantity avoids that all time series end up in the origin (see Fig. 19.1 where the quantity is not included). The final example data set $\mathcal{D}$, represented as a matrix, has $N = P \cdot K = 5117$ samples (rows) and $Q = 4$ quantities (columns). Individual samples, such as $\mathbf{d} \in \mathcal{D}$, are treated as column vectors (in bold).

19.2 Koopman-Based Surrogate Model

Our method to construct a dynamic data-driven surrogate model is based on the Koopman operator theory. In this context we view the example data $\mathcal{D}$ as the outcome of a (discrete) dynamical system:

$$\mathbf{d}_{k+1} = F(\mathbf{d}_k) \tag{19.1}$$

with $\mathbf{d}_k \in \mathbb{R}^Q$ (i.e. macroscopic density data) being the system's state at time step k and F as the flow map that pushes the state forward in time. Our aim is to build a model that reflects the dynamics of the system, that is, the unknown and usually non-linear flow map F, by using the example data.

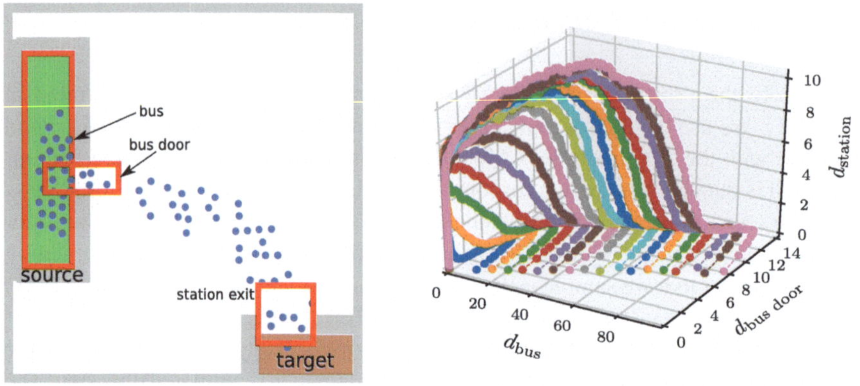

Fig. 19.1 (Left) Example scenario "bus station". We measure the density by counting the number of agents (blue circles) at three rectangular observation areas. All agents exit the bus (source) and move to the target. (Right) Crowd density time series from microscopic simulator for different initial conditions (bottom front points) and each observation area (axis); the color is for visualization only

19.2.1 Koopman Operator

The Koopman operator theory was first developed in the 1930s and regained interest in the last decade with the development of the "Dynamic Mode Decomposition" (DMD) algorithm and subsequent extensions and adaptations [10, 11]. For a detailed introduction we refer to [12]. In the context of pedestrian dynamics the Koopman operator was applied to estimate pedestrian flow in [13]. At the center of the theory is the Koopman operator, which provides an alternative view on the dynamical system compared to its description in Eq. (19.1). The operator $\mathcal{U}$ describes the same fundamental dynamics, but now through the equation

$$[\mathcal{U}\psi]\left(\mathbf{d}_{k+1}\right) = \psi\left(F\left(\mathbf{d}_k\right)\right). \tag{19.2}$$

Instead of acting on the state $\mathbf{d}$ directly, the operator $\mathcal{U}$ acts on functions $\psi \in \mathcal{F}$, the so-called observables, which are elements of a function space $\mathcal{F}$ (a typical choice are square-integrable functions $L^2(\mathbb{R}^Q, \mathbb{C})$). An observable is a *scalar* function defined on the state space of the dynamical system.

The main advantage of changing to the Koopman view is that $\mathcal{U}$ acts linearly on $\mathcal{F}$, while fully capturing the non-linear behavior of the system. Because exact equivalence to the classical view is only possible in the limit, we select a finite subset of observables, referred to a "dictionary" [11]. Using evaluations of the dictionary on the example data, we compute a matrix to approximate the Koopman operator (step 2 in following section).

19.2.2 Construct Surrogate Model

The following three steps describe how we approximate the Koopman operator and integrate it in a dynamic surrogate model. At each step we keep the explanations concise and refer to more detailed literature.

1. **Setting up the dictionary**. Our observables set consists of Geometric Harmonic (GH) functions, which provide a generalized Fourier basis on the data manifold [14]. The GH functions are computed through the solution of an eigenproblem of a kernel matrix consisting of pairwise samples

$$K\Psi = \Psi\Lambda \qquad K_{i,j} = \mathcal{K}(\mathbf{d}_i, \mathbf{d}_j) \tag{19.3}$$

$$\Psi(\mathbf{d}^*) = K^*\Psi\Lambda^{-1} \qquad K^*_{i,j} = \mathcal{K}(\mathbf{d}^*_i, \mathbf{d}_j) \tag{19.4}$$

Here, we set up the radial basis function kernel $\mathcal{K}(\mathbf{d}_i, \mathbf{d}_j) = \exp(-||\mathbf{d}_i - \mathbf{d}_j||_2^2/\varepsilon^2)$, with ε relating to the scale of the example data. We solve for the L largest eigenvectors $\Psi \in \mathbb{R}^{[N\times L]}$ (with respect to the corresponding eigenvalues) and in Eq. (19.4) we extend the eigenvectors via the Nytröm extension for unseen data $\mathbf{d}^* \notin \mathcal{D}$, which fulfills the out-of-sample property of the surrogate model [14].
To also map GH states $\boldsymbol{\psi}$ back to the crowd density data $\mathbf{d}$, we solve the following equation for a matrix C, in a least-squares sense:

$$\Psi C = \mathcal{D}, \text{ with } C \in \mathbb{R}^{[L\times Q]}. \tag{19.5}$$

2. **Compute Koopman matrix**. Using the transformed example data Ψ from Eq. (19.3), we can now approximate the Koopman operator from Eq. (19.2) with a matrix $U \approx \mathcal{U}$. For this task we apply the "Extended Dynamic Mode Decomposition" (EDMD) algorithm described in detail in [11].
The EDMD requires setting up shift matrices $\Psi_{(0:K-1)}$ and $\Psi_{(1:K)}$. The subscript $(0 : K - 1)$ indicates that for each time series we only use the first $K - 1$ samples (i.e. omit the last sample) and omit the first sample for $(1:K)$ accordingly. We then solve a least squares problem, where $\Psi^\dagger$ denotes the Moore-Penrose pseudoinverse:

$$U\Psi^T_{(0:K-1)} = \Psi^T_{(1:K)} \Leftrightarrow U = \Psi^T_{(1:K)}\left(\Psi^T_{(0:K-1)}\right)^\dagger, \text{ with } U \in \mathbb{R}^{[L\times L]}. \tag{19.6}$$

Equation (19.6) corresponds to the Koopman view in Eq. (19.1), using finitely many samples (i.e. $\Psi_{(1:K)}$ is shifted by one time step with the unknown flow map F). Once the Koopman matrix U is set up, it linearly maps observable data at time step k to values at the next time step $k + 1$.

3. **Build and evaluate surrogate model.** With the Koopman matrix from Eq. (19.6) and the coefficient matrix from Eq. (19.5) we can evolve the states $\mathbf{d}_k$ linearly after mapping the initial state $\mathbf{d}_0$ to the dictionary space with the nonlinear GH functions via the Nytröm extension Eq. (19.4):

$$\hat{\mathbf{d}}_k = \left(U^k \cdot \Psi(\mathbf{d}_0^*)\right)^T C = \left(U^k \hat{\boldsymbol{\psi}}_0\right)^T C = \hat{\boldsymbol{\psi}}_k^T C. \tag{19.7}$$

With Eq. (19.7) we can generate new approximate time series $\hat{\mathbf{d}}_k$, with arbitrary values of $0 \leq k \leq K$ and initial conditions $\mathbf{d}_0^*$; in ranges of the example data. In the results below we vary the initial condition with $\mathbf{d}_0^* = (d_{0,bus}^*, 0.08 \cdot d_{0,bus}^*, 0, d_{0,combined}^*)$, where $d_{0,bus}^* \in [10, 100]$ is the simulator parameter setting the agents starting in the bus. The factor 0.08 accounts for the overlapping observation area of the bus door and the interior of the bus (cf. Fig. 19.1).

19.3 Results

After constructing the Koopman-based surrogate model with the above procedure, we can compare the density time series with the original microscopic model. We do not fully benchmark the computational performance between the surrogate model and simulator because to be meaningful this would require both the pedestrian model and surrogate model implementation to be optimized. But as an example, a single time series evaluation of the simulator takes about 2.2–4.2 s, depending on the number of agents. In comparison, the surrogate model takes 11 s in the offline phase (which excludes generating the example data) and 0.03 s for a single time series evaluation, independent of the initial condition.

The quantities d_{bus} and d_{combined} are important parameters for the surrogate model: d_{bus} allows the initial condition to be set for agents starting in the bus and d_{combined} has a positive impact on the error. We leave a full analysis and discussion of these aspects to future work.

In the following results we focus on analyzing how well the surrogate model fits the example data. In Fig. 19.2 we show the reconstructed time series d_{station} from the surrogate model, i.e. we use the same initial conditions from the example data $\mathcal{D}$. The more agents that start in the bus, the longer the duration of (high) density at the station exit. The highest density observed is about 10 agents. The relative error box in Fig. 19.2 shows that data generated by the surrogate model fits the original data. We note that the largest error occurs for the initial conditions of 10–15 agents at times of positive density. At these low densities we expect the relative fluctuations in the time series to be greater.

In Fig. 19.3 we compare the three density time series for the initial setting of 54 agents. All three time series have distinct patterns. They have different peaks of density and a different slope for increasing and decreasing density. For example, the "bus door" time series shows that all agents move immediately towards the

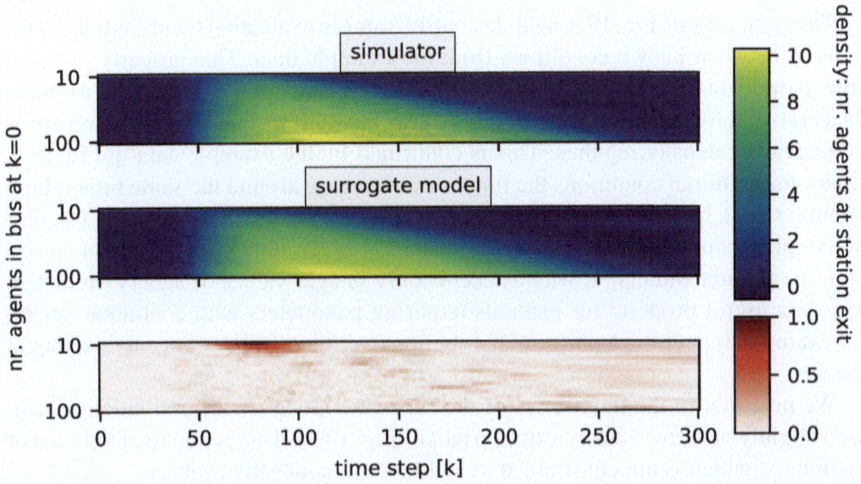

Fig. 19.2 Comparison of the example data d_{station} generated by microscopic simulator (top) with the surrogate model (middle). The color code denotes the measured crowd density. The bottom box shows the error between the data relative to the maximum absolute error (equal to 2.8 agents)

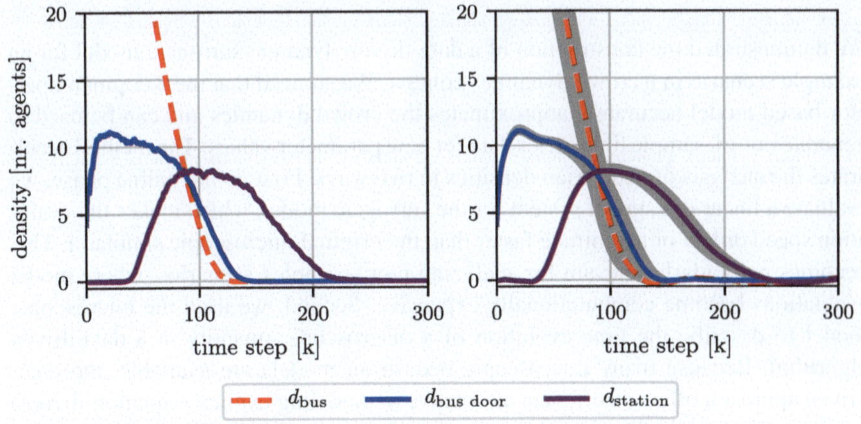

Fig. 19.3 (Left) Crowd density time series with initial condition $d^*_{0,\text{bus}} = 54$ agents from the microscopic simulator (included in the example data). (Right) Time series of surrogate model: Same initial condition to the left and out-of-sample evaluations in gray ($d^*_{0,\text{bus}} = 54 \pm 5$ agents)

measurement area, resulting in the highest density peak shortly after the simulation start. Also, the lower density peaks at the station exit result from a wider exit door, while the narrow bus door marks the bottleneck of the scenario. All patterns are specific to the selected scenario and are a result of its layout. These patterns also change for different initial conditions (for the station exit see Fig. 19.2). Here, we highlight that the constructed (single) surrogate model captures *all* variations of the four time series of the quantities of interest.

The right plot of Fig. 19.3 includes out-of-sample evaluations with initial conditions in between known conditions from the example data. This property of being able to evaluate time series for initial conditions that are not included in the example data is crucial for the surrogate model. The gray band shows that most variation arises in decreasing density regimes. This is confirmed by the example data in Fig. 19.2, where for all initial conditions the time series increase around the same time (about 50 time steps), but the intensity and duration of positive density and varies. The out-of-sample evaluations further include settings that are impossible to parameterize with the original simulator, which accepts only integer values of agents simulated. This is a useful property for methods requiring parameters with continuous input, for example, for setting a uniform distribution over the number of agents starting in the bus.

We note that in the time series of the surrogate model small oscillations occur, with slightly negative density values (right of Fig. 19.3). This is due to our choice of dictionary, which is not constrained to yield only non-negative values.

19.4 Conclusion and Future Work

We demonstrated the construction of a data-driven dynamic surrogate model for an example scenario in a crowd dynamics context. We showed that the Koopman operator based model accurately approximates the crowd dynamics and can be used to generate out-of-sample time series data for new parameter values. The method accelerates the analysis of pedestrian densities in two ways. First, in the online phase, we evaluate a linear dynamical system for the surrogate model, which makes the evaluation speed orders of magnitude faster than the original microscopic simulator. This becomes particularly relevant for analyzing more complex scenarios, where model evaluations become computationally expensive. Second, we used the microscopic model to describe the time evolution of a macroscopic quantity in a data-driven algorithm. Because many microscopic pedestrian models are available, the data-driven approach offers an efficient alternative to modeling explicit (equation-driven) macroscopic models. The data-driven nature of the algorithm requires creating a new surrogate model for every new scenario we wish to analyze. However, the algorithm is generic as it is applicable for a wide range of systems and scenarios.

Acknowledgments This work is supported by the German Research Foundation (DFG) grant no. KO 5257/3-1. The work of I.G.K. was partially supported by the DARPA PAI program. D.L. thanks the research office (FORWIN) of Munich University of Applied Sciences and the Faculty Graduate Center CeDoSIA of TUM Graduate School at Technical University of Munich for their support.

References

1. F. Dietrich, G. Köster, H.J. Bungartz, SIAM J. Appl. Dyn. Syst. **15**(4), 2078 (2016). https://doi.org/10.1137/15M1043613
2. P. Liu, G. Samaey, C. William Gear, I.G. Kevrekidis, Appl. Numer. Math. **92**, 54 (2015). https://doi.org/10.1016/j.apnum.2014.12.007
3. N.W. Bode, M. Chraibi, S. Holl, Transp. Res. Part B: Methodol. **119**, 197 (2019). https://doi.org/10.1016/j.trb.2018.12.002
4. A. Portz, A. Seyfried, in *Pedestrian and Evacuation Dynamics*, vol. 1 (Springer, 2011), pp. 577–586. https://doi.org/10.1007/978-1-4419-9725-8_52
5. I.G. Kevrekidis, G. Samaey, Ann. Rev. Phys. Chem. **60**(1), 321 (2009). https://doi.org/10.1146/annurev.physchem.59.032607.093610
6. F. Dietrich, F. Künzner, T. Neckel, G. Köster, H.J. Bungartz, Int. J. Uncert. Quant. **8**, 175 (2018). https://doi.org/10.1615/Int.J.UncertaintyQuantification.2018021975
7. B. Kleinmeier, B. Zönnchen, M. Gödel, G. Köster, Collective dynamics **4** (2019). https://doi.org/10.17815/CD.2019.21
8. M.J. Seitz, G. Köster, Phys. Rev. E **86**(4), 046108 (2012). https://doi.org/10.1103/PhysRevE.86.046108
9. I. von Sivers, G. Köster, Transp. Res. Part B: Methodol. **74**, 104 (2015). https://doi.org/10.1016/j.trb.2015.01.009
10. P.J. Schmid, J. Fluid Mech. **656**, 5 (2010). https://doi.org/10.1017/s0022112010001217
11. M.O. Williams, I.G. Kevrekidis, C.W. Rowley, J. Nonlinear Sci. **25**(6), 1307 (2015). https://doi.org/10.1007/s00332-015-9258-5
12. J.N. Kutz, Brunton, S.L. Brunton, B.W. Brunton, J.L. Proctor, *Dynamic Mode Decomposition: Data-driven Modeling of Complex Systems* (Society for Industrial and Applied Mathematics, Philadelphia, PA, 2016). https://doi.org/10.1137/1.9781611974508
13. H. Mansour, M. Benosman, V. Huroyan, *International Conference on Sampling Theory and Applications (SampTA)* (2017), pp. 451–455. https://doi.org/10.1109/SAMPTA.2017.8024350
14. R.R. Coifman, S. Lafon, App. Comput. Harmonic Anal. **21**(1), 31 (2006). https://doi.org/10.1016/j.acha.2005.07.005

Chapter 20
Evacuation Characteristics of Students Passing Through Bottlenecks

Hongliu Li, Jun Zhang, Weiguo Song, and Kwok Kit Richard Yuen

Abstract With the increasing of large buildings and public entertainment activities, mass gathering becomes more common. It is of great importance to understand movement characteristics of high motivated running pedestrians passing through bottlenecks in emergency. In this study, a laboratory experiment is performed to study the influence of bottleneck widths on the high motivated pedestrian movement. Voronoi Diagram based methods are applied to obtain the precise density and specific flow. The density in front of the bottleneck is higher than normal motivated pedestrians and the value is up to 8 ped $\cdot$ m^{-2}. Two specific flow peaks appear in the bottleneck. One is near the bottleneck entrance due to the higher density and the other locates around the exit due to the increasing speed in the bottleneck. The findings can help researchers and designers have a better understanding on the movement characteristics of high motivated pedestrians to improve the safety level of pedestrians.

20.1 Introduction

Bottlenecks are common physical geometries in buildings and restrict the pedestrian flow. Congestions often occur when the traffic demand of pedestrian flow exceeds the traffic capacity. When in emergency, the congestion caused by the bottleneck

H. Li · J. Zhang (✉) · W. Song
State Key Laboratory of Fire Science, University of Science and Technology of China, Hefei, China
e-mail: junz@ustc.edu.cn

H. Li
e-mail: hongliul@mail.ustc.edu.cn

W. Song
e-mail: wgsong@ustc.edu.cn

H. Li · K. K. R. Yuen
Department of Architecture and Civil Engineering, City University of Hong Kong, Hong Kong, China
e-mail: Richard.Yuen@cityu.edu.hk

I. Zuriguel et al. (eds.), *Traffic and Granular Flow 2019*,
Springer Proceedings in Physics 252,
https://doi.org/10.1007/978-3-030-55973-1_20

might result in severe disasters like Shanghai stampede in 2014 [1]. To improve safety level, it is of great importance to understand the movement of pedestrian flow passing through the bottleneck. During the past few decades, focusing on this urgent need, several studies have been performed focusing on the movement characteristics of pedestrian flow through bottlenecks [2, 3]. The geometry of bottleneck, like width, length, shape and position have been widely investigated [2]. Researchers propose the relation between flow and bottleneck width based on the laboratory experiments while there is no consensus flow—bottleneck width relation [2, 4–6]. It is pointed out that the decreasing bottleneck length increases the traffic capacity to a certain degree and when the bottleneck length is shorter than a threshold, it shows no impact on the capacity [5]. Appropriate shape improves the traffic capacity and increases the flow apparently [7]. The role of an obstacle near bottlenecks has been studied, but no consensus has been reached. Some researchers propose that the obstacle improves the traffic capacity while some researchers find no specific impact on it [8]. During bottleneck experiment, self-organizing phenomenon are observed, like lane formation and zipper effect in [9, 10]. Besides the impact of the bottleneck geometry, the motivation of pedestrians is considered in the study of pedestrian bottleneck flow [11, 12]. The motivation is usually defined as the intensity of competition between pedestrians. It is proposed that increased competitiveness usually causes clogging, falling pedestrians and paradoxical phenomenon like "fast is slower" effect [3, 11, 13, 14]. While some studies find that the appropriate high motivation increases the speed of pedestrians and improves the traffic capacity [12].

In sum, pedestrian bottleneck flow has been studied in many perspectives. However, the impact of architectural structure like bottleneck width on the high-motivated pedestrians are still lack of study. It is still uncertain about the effective bottleneck width to ensure evacuation efficiency. Less attention are paid on the movement characteristics of pedestrians in emergencies. This increases the difficulties in crowd management.

The lack of study compels us to collect more empirical data to investigate the influence of the bottleneck width on the movement characteristics of running pedestrians (highly motivated). In this study, the distribution of density and specific flow are studied to understand the movement characteristics of high motivated pedestrians. The remainder of the paper is as following. In Sect. 20.2, the experiment setup is briefly described. Results are shown in Sect. 20.3. Section 20.4 summarizes this work.

20.2 Experiment Setup

In this work, a laboratory experiment was performed to study the impact of bottleneck width on the running pedestrians under high motivations.

The bottleneck width was changed from 0.6 to 1.65 at 0.15 m intervals. The bottleneck length was fixed as 1.90 m. A total of 86 students aged about 22 years in a university participated in the experiment. They were asked to run out of the bottleneck

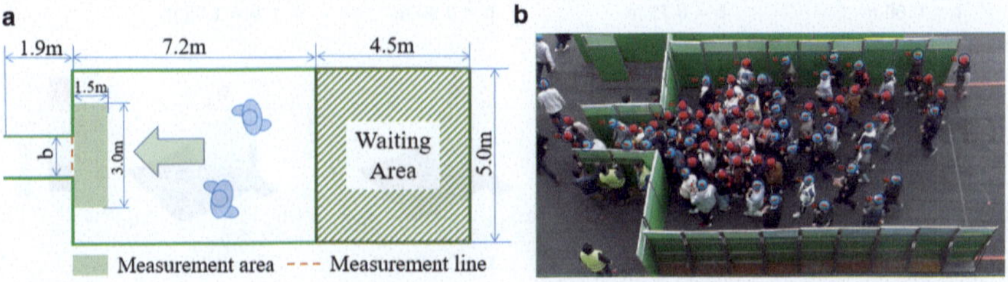

Fig. 20.1 An illustration and a snapshot of the experiment. The bottleneck width b changes from 0.6 to 1.65 at 0.15 m intervals. 86 pedestrians participated into the experiment

quickly while no severe pushing behaviors were allowed. For easy detecting and tracking of their movement afterward, they were asked to wear a hat (red for female and blue for male). At the beginning of each run, the pedestrians stood in the waiting area located 7.2 m away from the bottleneck entrance (see Fig. 20.1). The distance provided adequate space for pedestrians to accelerate. The whole experiment was recorded by a Sony FDR-AX100 camera (resolution: 1920 × 1080 pixels, frame rate: 50 fps) located in the fourth floor. The camera and the whole experiment scenario were calibrated according to PeTrack [15] software instruction to extract high-precision trajectories. Voronoi Diagram based methods [16] are applied to obtain movement characteristics of pedestrians in this study.

20.3 Results and Data Analysis

The influence of bottleneck width on the running pedestrians is investigated in this section. All data analyses following are obtained based on the extracted trajectories combing recorded videos of the experiment.

20.3.1 Density Distributions

Based on the Voronoi Diagram based method [16], the density over small cells (10 cm × 10 cm) is calculated and we acquire the profiles of the density in the experimental areas (see Fig. 20.2). In order to study the movement characteristics of high motivated pedestrians under congested stage, only data in the congestion stage are applied to calculate the density.

The spatial distribution of density is inhomogeneous over space. Transitions of the density are observed in Fig. 20.2, from high density in the middle region to low density in the border area before the bottleneck entrance. Pedestrians mainly distributed in the middle front of the bottleneck entrance. Since pedestrians are eager to pass the bottleneck quickly, they try to reduce the distance from the bottleneck

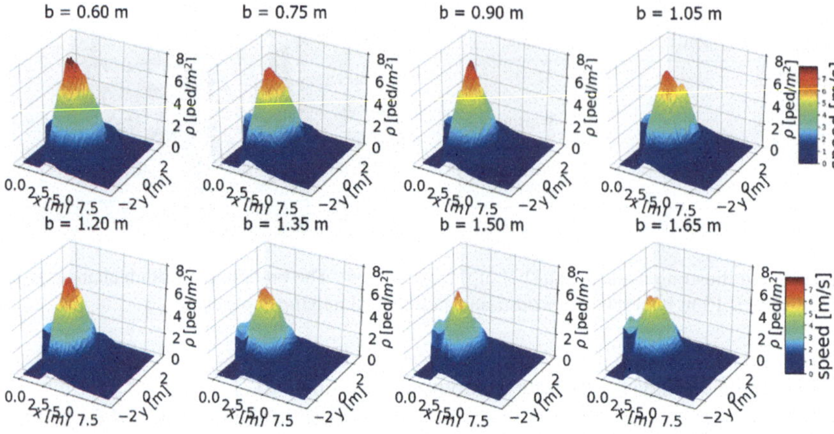

Fig. 20.2 Density distribution of pedestrians during the congested state

entrance and gather around the bottleneck entrance. When the bottleneck width is 0.6 m, the density before the bottleneck entrance can be up to 8 ped/m^2, which is obviously higher than previous study. High motivated pedestrians reduce the distance between pedestrians to pass the bottlenck quickly and this leads to higher density than pedestrians under normal motivation in previous study [2].

Consistent with the existing research results focusing on normal motivated pedestrians, the density decreases with the increasing bottleneck width, since the traffic capacity increases with the broaden bottleneck. The restriction of bottleneck on the pedestrian flow reduces and pedestrians can pass the bottleneck easily without severe congestion before the bottleneck entrance.

20.3.2 Specific Flow Distribution

To study the influence of bottleneck width on specific flow distribution, we plot the specific flow distribution based on Voronoi Diagram based method [16]. The specific flow is obtained by multiplying the density and speed over small cells (10 cm $\times$ 10 cm) in the congested stage. Before the bottleneck entrance, the specific flow increases as the distance from the bottleneck entrance decreases as we expect. While in the bottleneck, the distribution of specific flow is interesting. Two peaks appear in the bottleneck for all studied bottleneck widths. One specific flow peak locates near the bottleneck entrance and the other locates near the exit of the bottleneck. The higher specific flow near the bottleneck entrance mainly result from the high density near the bottleneck entrance. While the higher specific flow near the exit of the bottleneck is due to the higher speed, since pedestrians accelerate in the bottleneck (Fig. 20.3).

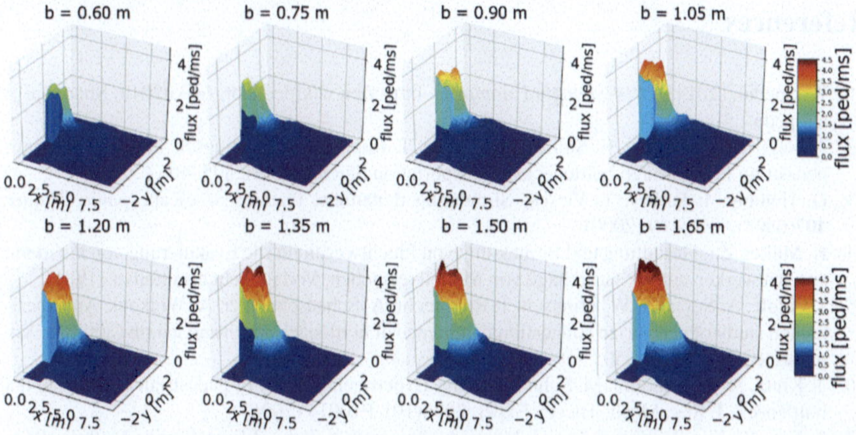

Fig. 20.3 Specific flow distribution of pedestrians during the congested state

With the increasing bottleneck width, the specific flow increases for all studied bottleneck width as we expect. The broaden bottleneck improves the traffic capacity and the specific flow increases. The specific flow is up to 4.5 pedm^{-1} s^{-1} when the bottleneck width >1.35 m.

20.4 Conclusion

In this study, the impact of bottleneck width on the highly motivated pedestrian flow is studied. Density and specific flow are studied to have a better understanding of the traffic capacity.

Compared to normal pedestrian bottleneck flow, the high motivation lead to the higher density around the bottleneck for all studied bottleneck widths as we expected. The increasing bottleneck width increase the traffic capacity and the density decreases with the broaden bottleneck, consistent with current existent studies. Interestingly, the traffic capacity of high motivated pedestrian flow at the entrance and the exit of the bottleneck is higher in this study. Clearly, high motivation lead to higher density and higher specific flow compared to normal motivation.

Acknowledgements The authors acknowledge the foundation support from the National Key Research and Development Program of China (Grant No. 2018YFC0808600), the National Natural Science Foundation of China (Grant No. 71704168, U1933105), from Anhui Provin-cial Natural Science Foundation (Grant No. 1808085MG217) and the Fundamental Research Funds for the Central Universities (Grant No. WK2320000040).

References

1. Wikipedia. (2014). *2014 Shanghai stampede*. https://en.wikipedia.org/wiki/2014_Shanghai_stampede
2. A. Seyfried, O. Passon, B. Steffen, M. Boltes, T. Rupprecht, W. Klingsch, New insights into pedestrian flow through bottlenecks. Transportation Science **43**(3), 395–406 (2009)
3. D. Helbing, I. Farkas, T. Vicsek, Simulating dynamical features of escape panic. Nature **407**(6803), 487–490 (2000)
4. K. Müller, Zur Gestaltung und Bemessung von Fluchtwegen für die Evakuierung von Personen aus Bauwerken auf der Grundlage von Modellversuchen, Verlag nicht ermittelbar (1981)
5. J. Liddle, A. Seyfried, W. Klingsch, T. Rupprecht, A. Schadschneider, A. Winkens, An experimental study of pedestrian congestions: influence of bottleneck width and length (2009), arXiv preprint arXiv:0911.4350
6. T. Kretz, A. Grünebohm, M. Schreckenberg, Experimental study of pedestrian flow through a bottleneck. J. Stat. Mech. Theory Exper. **2006**(10), P10014 (2006)
7. L. Sun, W. Luo, L. Yao, S. Qiu, J. Rong, A comparative study of funnel shape bottlenecks in subway stations. Transp. Res. Part A Policy Pract**98**, 14–27 (2017) (in English)
8. N. Shiwakoti, X. Shi, Z. Ye, A review on the performance of an obstacle near an exit on pedestrian crowd evacuation. Saf. Sci. **113**, 54–67 (2019)
9. X.-D. Liu, W.-G. Song, W. Lv, Empirical data for pedestrian counterflow through bottlenecks in the channel. Transp. Res. Procedia **2**, 34–42 (2014)
10. S.P. Hoogendoorn, W. Daamen, Pedestrian behavior at bottlenecks. Transp. Sci. **39**(2), 147–159 (2005)
11. H.C. Muir, D.M. Bottomley, C. Marrison, Effects of motivation and cabin configuration on emergency aircraft evacuation behavior and rates of egress. Int. J. Aviat. Psychol. **6**(1), 57–77 (1996)
12. A. Garcimartin, J. M. Pastor, C. Martin-Gomez, D. Parisi, I. Zuriguel, Pedestrian collective motion in competitive room evacuation. Sci. Rep. **7**(1), 10792 (2017)
13. J.M. Pastor et al., Experimental proof of faster-is-slower in systems of frictional particles flowing through constrictions. Phys. Rev. E **92**(6), 062817 (2015) (in English)
14. A. Garcimartín, D.R. Parisi, J.M. Pastor, C. Martín-Gómez, I. Zuriguel, Flow of pedestrians through narrow doors with different competitiveness. J. Stat. Mech. Theory Exp. **2016**(4), 043402 (2016)
15. M. Boltes, A. Seyfried, B. Steffen, A. Schadschneider, Automatic extraction of pedestrian trajectories from video recordings,in *Pedestrian and Evacuation Dynamics 2008* (Springer, 2010), pp. 43–54
16. B. Steffen, A. Seyfried, Methods for measuring pedestrian density, flow, speed and direction with minimal scatter. Physica A Stat. Mech. Appl. **389**(9), 1902–1910 (2010) (in English)

Chapter 21
An Efficient Crowd Density Estimation Algorithm Through Network Compression

Meng Li, Tao Chen, Zhihua Li, and Hezi Liu

Abstract Estimating count and density maps from crowd images has a wide range of applications such as video surveillance, traffic monitoring and pedestrian dynamics study. The state-of-the art deep learning approaches generally build a multi-column deep network architecture. Although such architectures perform well, the inference cost is neglected. In this paper, we apply the net-work compression to the convolutional neural network-based crowd density estimation model, to reduce its storage and computation costs. Specifically, we rely on using l1-norm to select unimportant filters and physically prune them. Convolutional filters with small scaling factor value and its corresponding kernels in the next layer will be removed. The model can be trained to improve its regression performance and identify the unimportant filters at the same time. A state-of-the art model, the CSRNet, is tested in the ShanghaiTech dataset. Our method can reduce inference costs for the network significantly (up to approximately 78%) while regaining close to the original accuracy by retraining the networks.

21.1 Introduction

Mass gathering at public places has caused numerous crowd hazards during past years [12]. To prevent deadly accidents, early automatic detection of unusual situations in large scale crowd is highly required. In fact, the spatial distribution of participants is essential for space design and risk prevention in religious festivals, public demonstrations, and sport events. It would certainly assist, as a result, to make appropriate decisions for emergency and safety control by detecting and preventing overcrowd. Therefore, the analysis of crowd dynamics is important in computer vision and safety science. Crowd density estimation technique involves estimating the number of people as well as their spatial distribution. It has its own unique challenges [12] such

---

M. Li (✉) · T. Chen · Z. Li · H. Liu
Tsinghua University, Beijing, China
e-mail: meng-li18@mails.tsinghua.edu.cn

I. Zuriguel et al. (eds.), *Traffic and Granular Flow 2019*,
Springer Proceedings in Physics 252,
https://doi.org/10.1007/978-3-030-55973-1_21

as high clutter, non-uniform distribution of people, non-uniform illumination, scale and perspective, making the problem extremely complicated.

During the past decade, growing number of models have been developed to estimate crowd count and present density maps. Crowd counting, as a computer vision problem, has seen drastic changes in the approaches, from early HOG-based head detections [4, 7, 10] to CNN (Convolutional Neural Networks) regressors predicting the crowd density [1, 2, 6, 8]. The crowd scene analysis methods can be categorized into three schemes [10, 12]: detection-based methods, regression-based methods, and density estimation-based methods. Despite their high performance in congested scene analysis, the application of CNNs in real world is largely hindered by their high computational cost. Since the past few years, in order to achieve stronger representation power, the general trend has been that the networks have grown deeper. However, deeper and larger CNNs are more resource-hungry. In crowd research field, there are some works on the data-hungry problem, but the other problem has not been tackled yet. Consequently, a large CNN may take several minutes to process one single crowd image on a mobile device, making it unrealistic to be adopted for real applications. How to balance between high performance and low resource cost has drawn attention recently in computer vision [3, 5]. The convolutional neural network compression efforts include smaller network architecture design (such as a slight net called the SqueezeNet [5]) and net-work pruning.

As part of long-term research effort aimed at intelligent video surveillance, this paper presents an efficient crowd density estimation algorithm. By doing this, we apply the network compression [9] to the crowd density estimation model, based on convolutional neural network, to reduce its storage and computation costs. Specifically, instead of pruning weights across the network, we focus on filter pruning, which is a naturally structured way of pruning without introducing sparsity and therefore does not require using sparse libraries or any specialized hardware. Specifically, we rely on using $l1$-norm to select unimportant filters and physically prune them. The model can be trained to improve its regression performance and identify the unimportant filters at the same time. A state-of-art model, the CSRNet (Congested Scene Recognition) [8], is tested in the ShanghaiTech datasets. Our method can reduce inference costs for the network significantly (up to approximately 78%) while regaining close to the original accuracy by retraining the networks.

21.2 Methods

21.2.1 The Architecture of Crowd Counting Network

We choose CSRNet because of its strong performance. Unlike the latest works such as [6, 11], CSRNet uses the deep CNN for ancillary, focusing on designing a CNN-based density map generator. The configuration of CSRNet are shown in Fig. 21.1a. It de-ploys the first 10 layers from VGG-16 [13] as the front-end and dilated convolution layers as the back-end to enlarge receptive fields and extract deeper features without losing resolutions (since pooling layers are not used). In addition, it employs the small size of convolutional filters (like 3 × 3) on all layers, for the benefit of reducing the network complexity. By taking advantage of such innovative structure, the network out performs the state-of-the-art crowd counting solutions in ShanghaiTech [14] Part A, Part B, and WorldExpo'10 [14] datasets. Also, it can achieve high performance on the vehicle counting on TRANCOS dataset. However, this network relies on deep networks with 16.26 million parameters. And it needs three days to train the whole model with a NVIDIA GTX 1080 machine. Its high computational cost and long-time of training make it unrealistic to be adopted for real applications. Consequently, the most critical part relays on the balance between high performance and low resource cost.

21.2.2 Network Compression

Aiming to tackle the redundancy and time-consuming problems simultaneously, it is necessary to compress this large crowd counting CNN-model.

As illustrated in Fig. 21.1, the CSRNet deploys 10 VGG-16 layers as feature extractor. The VGG network architecture was designed by VGG group in 2014, which features in its simplicity and high performance. However, it is originally

Configuration of SCRNet
Front-end (fine-tuned from VGG-16)
Conv3-64-1 Conv3-64-1
Max-pooling
Conv3-128-1 Conv3-128-1
Max-pooling
Conv3-356-1 Conv3-356-1 Conv3-356-1
Max-pooling
Conv3-512-1 Conv3-512-1 Conv3-512-1
Back-end (Dilated convolution)
Conv3-512-2 Conv3-512-2 Conv3-512-2 Conv3-128-2 Conv3-64-2
Conv1-1-1

(a)

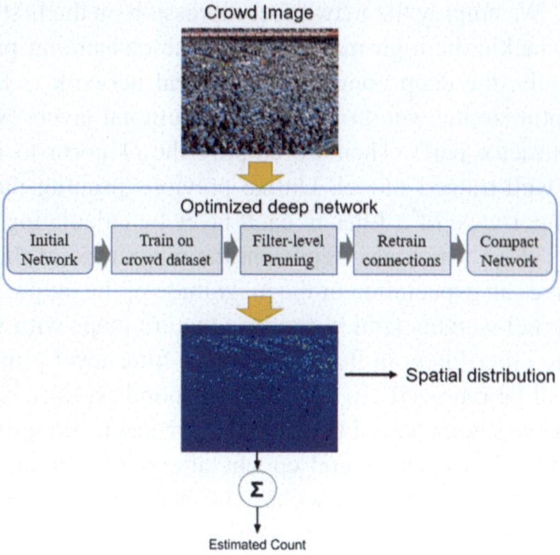

(b)

Fig. 21.1 **a** Configuration of CSRNet. All convolutional layers use padding to maintain the previous size. The convolutional layers' parameters are denoted as "Conv-(filter size)-(number of filters)-(dilation rate)". **b** The illustration of the whole processing pipeline

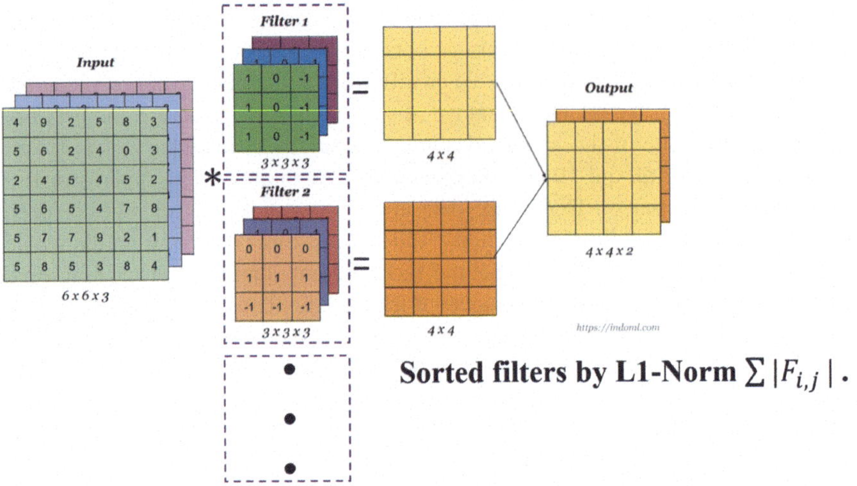

Fig. 21.2 We introduce a new optimizing method that adaptively rank the importance of filters in each layer and zero-out filters with small l1-norms

designed for the ImageNet task, the parameters of the model can be redundant for the crowd dataset. Since the size of the kernel is very small (3×3), and the depth of the network is closely related to the representation ability of the model, the first place to compress is the number of channels.

We employ the network compression on the first ten convolutional layers, aiming to tackle the high-memory and time-consuming problems simultaneously. Specifically, the deep convolutional neural network is first trained on ShanghaiTech to optimize the weights of the convolutional layers (which can be seen as the feature extractor part). Then we employ the $l1$-norm to measure less useful filters from a well-trained model. Unlike previous pruning methods, we measure the relative importance of a filter in each layer by calculating the sum of its absolute weights $\sum |F_{i,j}|$ which also represent the magnitude of the output feature map. This value gives an expectation of the magnitude of the output feature map. Filters with smaller kernel weights tend to produce feature maps with weak activations as compared to the other filters in that layer. In the filter-level pruning step, the unimportant filters will be removed. To avoid the temporal performance degradation due to removing many parameters of the network, we can re-train the pruned network on the ShanghaiTech dataset several epochs afterwards. Finally, a compact network consumes limited resources in test phase is obtained. The main difference between our method and general crowd scene analysis approaches is that we introduce the filter-level pruning step to slim the whole network (Fig. 21.2).

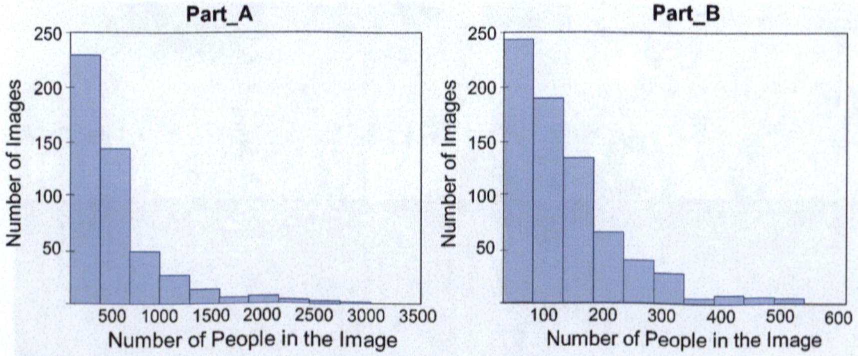

Fig. 21.3 Histogram of crowd number of Shanghaitech dataset

21.2.3 Datasets

In this section, we evaluate our model on two different public datasets. The proposed method not only obtains competitive and even superior estimation performance but also gives significant reductions in the number of parameters and computing operations.

We select Shanghaitech datasets to evaluate our model's performance. The datasets consist of two parts: Part A contains 482 images crawled from the Internet while Part B contains 716 images captured from busy streets in Shanghai. Figure 21.3 plots the histogram of crowd number of this dataset, demonstrating the crowd density difference between the two parts. Both parts are categorized into training and testing sets. For Part A, there are 300 training samples and 182 testing images. For Part B, there are 400 images for training and 316 for testing.

In Fig. 21.4, we have shown the predicted and ground-truth density maps of two exemplar images in ShanghaiTech part A dataset.

21.3 Results and Discussion

In this section, the evaluation metrics are introduced. We show the details of filters of ranked by $l1$-norm, and then evaluate the tested the sensitivity of each layer of CSRNet.

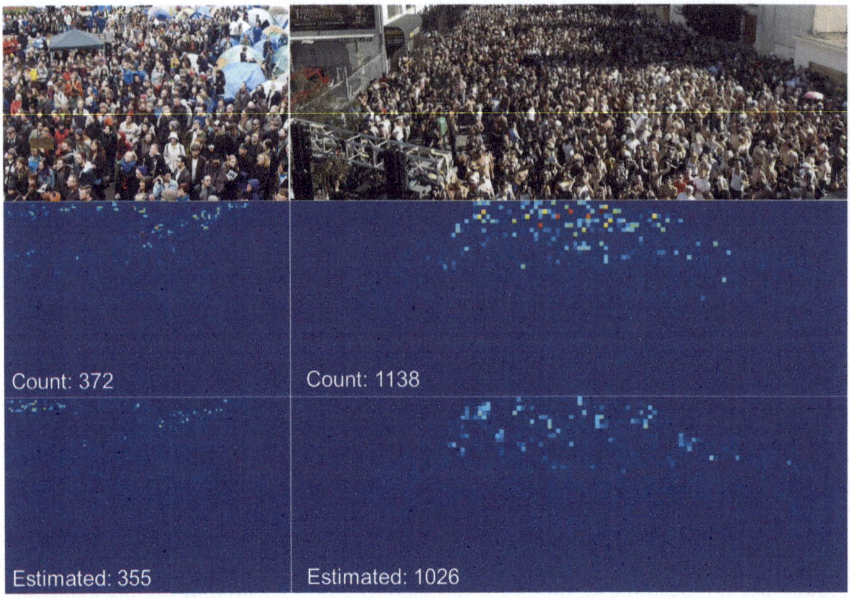

Fig. 21.4 Original images and corresponding crowd density maps in ShanghaiTech Part A dataset. The first row shows the original crowd images. The second row shows the ground-truth density maps. The third row shows the predicted maps by our model

21.3.1 Evaluation Metrics

The MAE and the MSE are used for evaluation which are defined as:

$$\mathrm{MAE} = \frac{1}{\mathrm{N}} \sum_{\mathrm{i}=1}^{\mathrm{N}} \left| \mathrm{C_i} - \mathrm{C_i^{GT}} \right| \tag{21.1}$$

where N is the number of images in one test sequence and $C_i^G T$ is the ground truth of counting. $C_{(i)}$ represents the estimated count which is defined as follows:

$$C_i = \sum_{\mathrm{l}=1}^{\mathrm{N}} \sum_{\mathrm{w}=1}^{\mathrm{w}} \mathrm{Z_{L,w}} \tag{21.2}$$

L and W show the length and width of the density map respectively while $\mathrm{Z_{L,w}}$ is the pixel at(l, w) of the generated density map. C_i means the estimated counting number for image X_i.

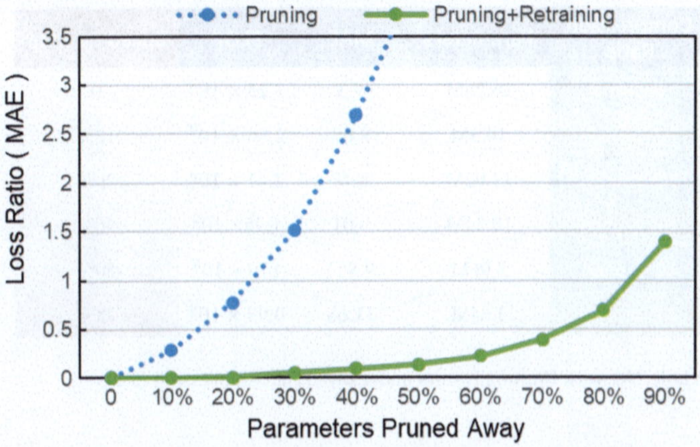

Fig. 21.5 The results of loss increasing ratio with (green line) or without retraining (green dashed line) after pruning

21.3.2 Determining Single Layer's Sensitivity to Pruning

To understand the sensitivity of each layer, we prune each layer independently and evaluate the resulting pruned network's accuracy on the validation set. Layers with flat slops maintain their accuracy as filters are pruned away. On the contrary, layers with relatively steep slops are more sensitive to pruning. For layers sensitivity concerning CSRNet, we observe that layers in the same stage (with the same feature map size) have a similar sensitivity to pruning. For layers that are sensitive to pruning, we prune a smaller percentage of these layers or completely skip pruning them.

We conduct network compression 2 times on the convolutional layers of CSRNet, with and without retraining respectively. The results are shown in Fig. 21.5, that almost 80% of the filters of these layers can be safely removed with retraining.

21.4 Conclusion

Modern CNNs often have high capacity with large training and inference costs. In this paper we reset a method to prune filters with relatively low weight magnitudes to produce CNNs with educed computation costs without introducing irregular sparsity. It achieves about 78% reduction in FLOP for CSRNet (on ShanghaiTech) without significant loss in the original accuracy.

Compression Ratio	Parameters	MAE	FLOP	Speed up Rate
Baseline	16.26M	9.3	3.25×10^8	0
12% pruned	14.3M	8.69	2.86×10^8	12%
26% pruned	12.02M	8.78	2.41×10^8	26%
36% pruned	10.34M	9.01	1.95×10^8	36%
50%pruned	8.06M	9.852	1.61×10^8	50%
78% pruned	3.64M	11.68	0.93×10^8	78%

Fig. 21.6 Whole Network Pruning results under different compression ratios

Instead of pruning with specific layer-wise hyper-parameters and time-consuming iterative retraining, we use the one-shot pruning and retraining strategy for simplicity and ease of implementation. By performing lesion studies on very deep CNNs, we identify layers that are robust or sensitive to pruning, which can be useful for further understanding and improving the architectures (Fig. 21.6).

References

1. L. Boominathan, S.S. Kruthiventi, R.V. Babu, Crowdnet: a deep convolutional network for dense crowd counting, in *Proceedings of the 24th ACM International Conference on Multimedia* (ACM, 2016), pp. 640–644
2. X. Cao, Z. Wang, Y. Zhao, F. Su, Scale aggregation network for accurate and efficient crowd counting, in *Proceedings of the European Conference on Computer Vision (ECCV)* (2018), pp. 734–750
3. Y. Cheng, F.X. Yu, R.S. Feris, S. Kumar, A. Choudhary, S.F. Chang, An exploration of parameter redundancy in deep networks with circulant projections, in *Proceedings of the IEEE International Conference on Computer Vision* (2015), pp. 2857–2865
4. M. Fu, P. Xu, X. Li, Q. Liu, M. Ye, C. Zhu, Fast crowd density estimation with convolutional neural networks. Eng. Appl. Artif. Intell. **43**, 81–88 (2015)
5. F.N. Iandola, S. Han, M.W. Moskewicz, K. Ashraf, W.J. Dally, K. Keutzer, Squeezenet: Alexnet-level accuracy with 50x fewer parameters and <0.5 MB model size. arXiv preprint arXiv:1602.07360 (2016)
6. H. Idrees, I. Saleemi, C. Seibert, M. Shah, Multi-source multi-scale counting in extremely dense crowd images, in *Proceedings of the IEEE Conference on Computer Vision and Pattern Recognition* (2013), pp. 2547–2554
7. V. Lempitsky, A. Zisserman, Learning to count objects in images, in *Advances in Neural Information Processing Systems* (2010), pp. 1324–1332
8. Li, Y., Zhang, X., Chen, D.: CSRNnet: dilated convolutional neural networks for understanding the highly congested scenes. in *Proceedings of the IEEE Conference on Computer Vision and Pattern Recognition* (2018), pp. 1091–1100
9. H. Li, A. Kadav, I. Durdanovic, H. Samet, H.P. Graf, Pruning filters for efficient convnets. arXiv preprint arXiv:1608.08710 (2016)
10. S.A.M. Saleh, S.A. Suandi, H. Ibrahim, Recent survey on crowd density estimation and counting for visual surveillance. Eng. Appl. Artif. Intell. **41**, 103–114 (2015)

11. V.A. Sindagi, V.M. Patel: Generating high-quality crowd density maps using contextual pyramid CNNs, in *Proceedings of the IEEE International Conference on Computer Vision* (2017), pp. 1861–1870
12. V.A. Sindagi, V.M. Patel, A survey of recent advances in CNN-based single image crowd counting and density estimation. Pattern Recogn. Lett. **107**, 3–16 (2018)
13. A. Vedaldi, K. Lenc, Matconvnet: convolutional neural networks for Matlab, in *Proceedings of the 23rd ACM International Conference on Multimedia* (ACM, 2015), pp. 689–692
14. C. Zhang, H. Li, X. Wang, X. Yang, Cross-scene crowd counting via deep convolutional neural networks, in *Proceedings of the IEEE Conference on Computer Vision and Pattern Recognition* (2015), pp. 833–841

Chapter 22
Modelling Pedestrian Social Group Passing Strategy with Expression-Matrix and Social Force

Long Liu, Yuan Zhao, and Xiaolei Zou

Abstract Our experiments and studies in recent years have shown that the complex pedestrian social group has a hierarchical structure. Small pedestrian groups of 2–3 people are simple groups with stable and straightforward formations. The simple groups are also cellular units that act as subgroups and constitute a more extensive complex social pedestrian group. The complex social pedestrian group has multiple patterns, movement and formation change. In this paper, an Expression-Matrix method is used to coordinate the hierarchical relations of the heterogeneous group and the social relations among subgroup members, and to generate global and local strategies for formation changes of complex pedestrian groups. And an extended social force model is proposed to control the dynamics of group members and forms a stable group formation by adding a set of inner subgroup relationship force and a set of inter subgroup relationship force. Finally, trajectories in the bottle-neck scene are compared with the result of the model, which shows the model could accurately describe some of the adoption strategies of complex pedestrian groups.

22.1 Introduction

At present, the research field of pedestrian flow mainly focuses on the characteristics and trajectory analysis in a variety of scenarios based on the single-person model. However, considering that the amount of pedestrian groups in the pedestrian flow is not in the minority [1, 2], and group members have potential impact on the strategy and trajectory of the group in some scenes [3], the primary research and analysis of

---

L. Liu · X. Zou (✉)
Tongji University, 4800# Cao'an Road, Shanghai, China
e-mail: zouxiaolei@tongji.edu.cn

L. Liu
e-mail: liul0121@tongji.edu.cn

Y. Zhao (✉)
Shanghai Shentong Metro Group Co., 909# Guilin Road, Shanghai, China
e-mail: jituanguizhang@126.com

I. Zuriguel et al. (eds.), *Traffic and Granular Flow 2019*,
Springer Proceedings in Physics 252,
https://doi.org/10.1007/978-3-030-55973-1_22

groups behavior in pedestrian flow is essential. The behavior of pedestrians in various scenarios can be interpreted as the individual's response to the external environment. How to define different social attributions, distinguish and accurately describe the behavior of pedestrians under the influence of others in a complex social group are the key objects of this study.

22.2 Methods

In order to give a more accurately description of the location of the pedestrians in the complex group and the structure of the entire group, an Expression-Matrix method is used to record the morphological changes of the pedestrian group structure and relative location. On the basis of social force, the study distinguished different members of different subgroups in one group and figure out their different impact on every single pedestrian according to different social relationships.

22.2.1 Expression-Matrix

Expression-Matrix is a kind of symbolic representation of the group structure and positional statistics based on experimental data. By using Expression-Matrix, this method can express the pedestrian group structure more simply and understandably, and use matrix probability to describe the possibility of pedestrians in the group, which means the probability of distribution of locations. In this method Expression-Matrix, at the level of the individual pedestrian, the pedestrian's heading direction is the Y-axis, and the pedestrian position is arranged from left to right along the X-axis. The distance between the pedestrians determines the positional expression symbol of the pedestrian. On the X-axis, the distance is judged mainly based on the shoulder width of the pedestrian, and different races may affect this value. In the Y-axis, the step distance is mainly used as a key reference for judging the position. This method uses the operands such as "|", " + ", "−", "/", "\" to indicate the relative position of the pedestrian. The rules for "|", " + ", "−", "/" and "\" are detailed in [5]. The matrix is defined as a set of cells that covers all of the group members with side lengths of the standard distance for X-axis and Y-axis. A probability is assigned for each cell to describe the location distribution possibility of group members.

Besides, based on experimental observation data [4], the probability of location distribution of pedestrians in different group sizes is also obtained. Based on Expression-Matrix, Huang [5] proposed the basic pedestrian group composition structure and its symbolic expression. Based on Expression-Matrix, this study selected a five-person group with two cellular groups as the research object and used the extended social force model to simulate the travel strategy of the five-person group.

22.2.2 Extended Social Force Model

The social force model established by Helbing is based on Newton's second law of motion. It is believed that individuals in the crowd are single particles that satisfy Newton's law, and each individual is also affected by the interaction of physical and psychological internal and external factors. These factors are expressed in a forceful manner.

Pedestrian movement meets the basic formula

$$\frac{\mathrm{d}\vec{v}_a}{\mathrm{d}t} = \vec{f}_\alpha(t) + \vec{\xi}_\alpha(t) \tag{22.1}$$

where $\frac{\mathrm{d}\vec{v}_a}{\mathrm{d}t}$ means pedestrian acceleration, $\vec{f}_\alpha(t)$ is the synergy of the social forces that the pedestrians are subjected to, $\vec{\xi}_\alpha(t)$ represents the individual random error term for pedestrian α.

Representatively, for the combined force of the social forces of pedestrians, Helbing proposed that the resultant force $\vec{f}_\alpha$ can be expressed by the following formula

$$\vec{f}_\alpha = \vec{f}_\alpha^{\,0} + \vec{f}_\alpha^{\,wall} + \sum_\beta \vec{f}_{\alpha\beta} \tag{22.2}$$

where $\vec{f}_\alpha^{\,0}$ is the self-driving force of the pedestrian α, $\vec{f}_\alpha^{\,wall}$ is the repulsive force between the pedestrian α and the wall, $\sum_\beta \vec{f}_{\alpha\beta}$ is the force of other pedestrians on the pedestrian α.

This is the result of a single pedestrian as a travel unit. When considering pedestrians traveling in groups, the relationship between pedestrians is not only exclusive but also the attractiveness of distances used to meet communication needs. Based on Helbing's social strength, Moussaïd [2] expands the relationship between pedestrians and pedestrians and adds the concept of a group. Then the social force expression of the pedestrian is

$$\vec{f}_\alpha = \vec{f}_\alpha^{\,0} + \vec{f}_\alpha^{\,wall} + \sum_\beta \vec{f}_{\alpha\beta} + \vec{f}_\alpha^{\,group} \tag{22.3}$$

$$\vec{f}_\alpha^{\,group} = \vec{f}_\alpha^{\,vis} + \vec{f}_\alpha^{\,att} + \vec{f}_\alpha^{\,rep} \tag{22.4}$$

In this formula, Mehdi distinguishes the influence of pedestrians in the group and pedestrians outside the group. The forces of the pedestrians within the group include the self-taking power $\vec{f}_\alpha^{\,vis}$ based on the adjustment of the field of view, the attractiveness, and repulsive force between the members in the group.

Based on the structure of the five-person group, this paper further refines the types of social forces within the group and adds the pedestrian force within the subgroup and the pedestrian force between the subgroups. Based on the different social relationships within the pedestrian group, this paper divides the internal structure of a group of five into three layers, including groups, subgroups, and pedestrians. As shown in the figure, A is the pedestrian group where pedestrian α belongs to, and there is a subgroup inside a and b pedestrian groups A. The pedestrian α belongs to the subgroup a. Then the social force expression of the pedestrian α is

$$\vec{f_\alpha} = \vec{f_\alpha}^{0} + \vec{f_\alpha}^{wall} + \sum_{\beta\epsilon a} \vec{f_{\alpha\beta}} + \vec{f_\alpha}^{group} + \vec{f_\alpha}^{subg} \tag{22.5}$$

where $\sum_{\beta\epsilon a} \vec{f_{\alpha\beta}}$ is the interaction force between pedestrian α and the homogenous group, and $\vec{f_\alpha}^{subg}$ is the interaction force between pedestrian α and other subgroups in the same group. The formula is

$$\vec{f_\alpha}^{subg} = D \cdot exp\left(\frac{\left|d_{\alpha b} - d^*_{\alpha b}\right|}{C}\right) \cdot \vec{e_{\alpha b}} \tag{22.6}$$

where $d_{\alpha b}$ is the real-time distance between pedestrian α and the centroid of the subgroup b, $d^*_{\alpha b}$ is the expected distance between pedestrian α and the centroid of the subgroup b, and $\vec{e_{\alpha b}}$ is the direction of the force. The parameter D is a parameter for judging whether the force is attractive or repulsive according to the distance, and the parameter C is a parameter for controlling the strength (Fig. 22.1).

22.3 Results and Discussion

Pedestrian experimental site is shown as Fig. 22.2.

The width of entrance is defined to ensure the conditions of free flow. And the passage width is gradually reduced from 6 to 1.5 m. The length of limit part is 13 m, and the exit part is 7 m, for a total of 20 m. The experimental participants were college students at Tongji University, and 60 students were asked to be divided into 12 groups of 5 people.

22.3.1 Comparison by Passing Strategies

Based on the previous narrow passages of the five-person group through experiments [6], we found two main strategies that groups choose commonly by summarizing

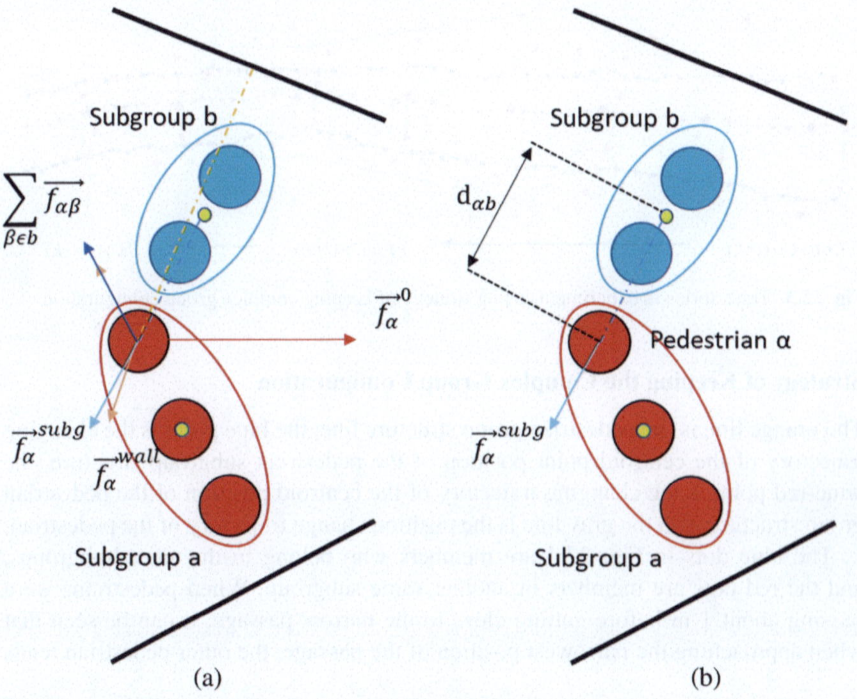

Fig. 22.1 The display of the five-person group under the narrow passage scene. **a** The analysis of pedestrian α in the narrow passage. **b** The influence of subgroup **b** on pedestrian α.

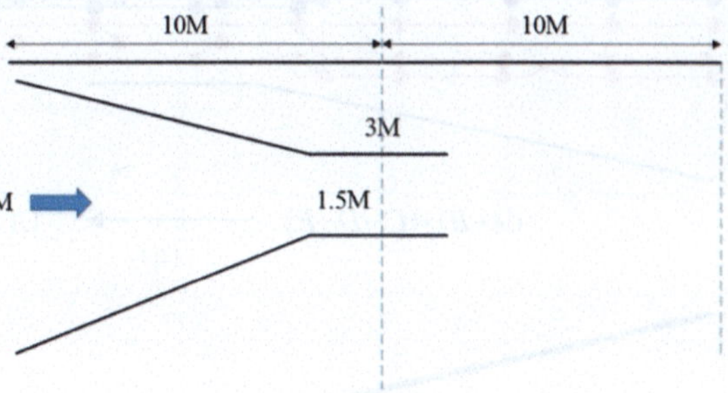

Fig. 22.2 Pedestrian experimental narrow passage

the trajectories of pedestrians. One is keeping the complex group configuration in which the subgroups keep relatively fixed positions to maintain a integrity structure of the complex group. Another is keeping the subgroup configuration in which the pedestrians in a subgroup maintain a relatively fixed structure and the subgroups adjust their relative positions (Fig. 22.3).

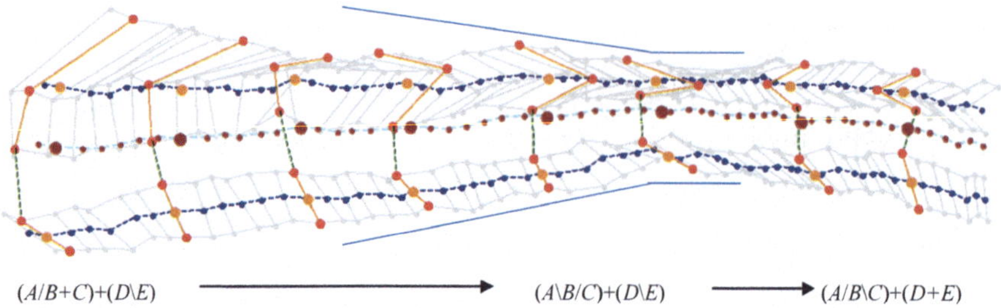

Fig. 22.3 Trajectories in experiments using strategy of keeping complex group configuration

Strategy of Keeping the Complex Group Configuration

The orange line is the pedestrian group structure line, the blue point is the changing trajectory of the centroid point position of the pedestrian subgroup structure, the wine-red point is the changing trajectory of the centroid position of the pedestrian group structure, and the gray line is the position change trajectory of the pedestrian.

The blue dots in Fig. 22.4 are members who belong to the same subgroups, and the red dots are members in another same subgroup. When pedestrians were passing about 1 m before getting close to the narrow passage, it can be seen that when approaching the narrowest position of thc passage, the outer pedestrian tends

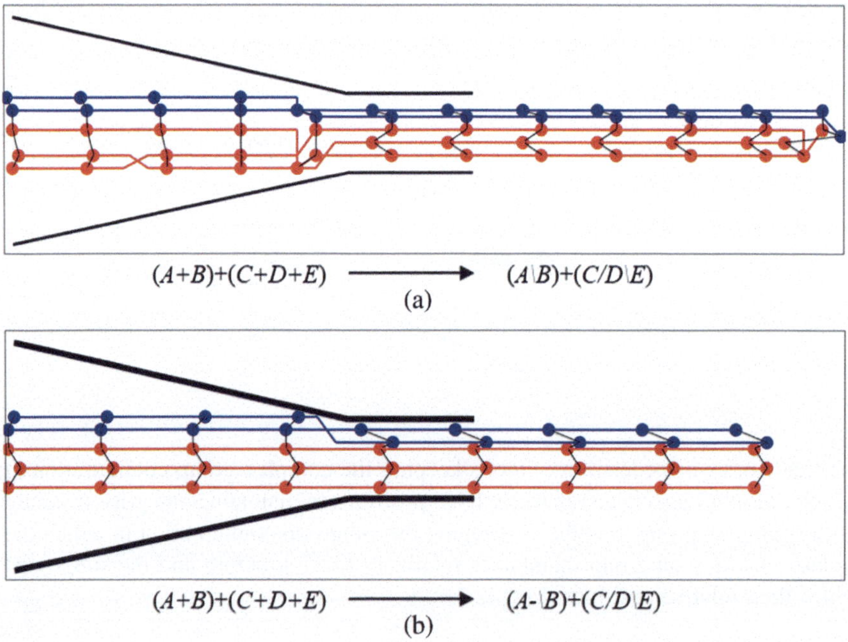

Fig. 22.4 Trajectories in simulation using strategy of keeping complex group configuration

to close inward to compress the space. However, due to space constraints, the position between the individuals become staggered—finally, the whole structure of the heterogeneous groups still in the V-shaped. So in the reorganization area, the strategy in the experiments is consistent with the simulations.

Strategy of Keeping the Subgroup Configuration

In this strategy, the positional relationship between subgroups change, and the left–right relationship becomes the front-rear relationship. While the groups walk from the wide side to the narrow side, as shown in Fig. 22.5, the data [6] shows that 70.2% of the constant groups preferred to keep stable configurations of subgroups, which makes 54.1% of all the complex groups. The simulation result of the passing strategy is shown in Fig. 22.5.

In the simulation, the most influencing factors adopted by the pedestrian group using different strategies are the parameter setting of the wall force and the parameter setting of the social force within the group. When the repulsive force from the wall is as same as the strength of the social force, and even when the social strength of the group is higher, the strategy tends to keep the group structure stable; when the wall strength is higher, the strategy tends to keep the subgroups structure stable. Besides, because the interaction between group members is relatively stable, this study tends to attribute the changing conditions affecting the strategy to the wall

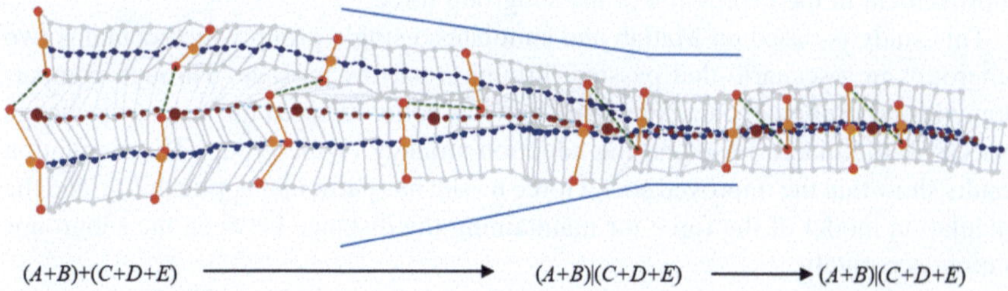

Fig. 22.5 Trajectories in experiments using strategy of keeping subgroup configuration

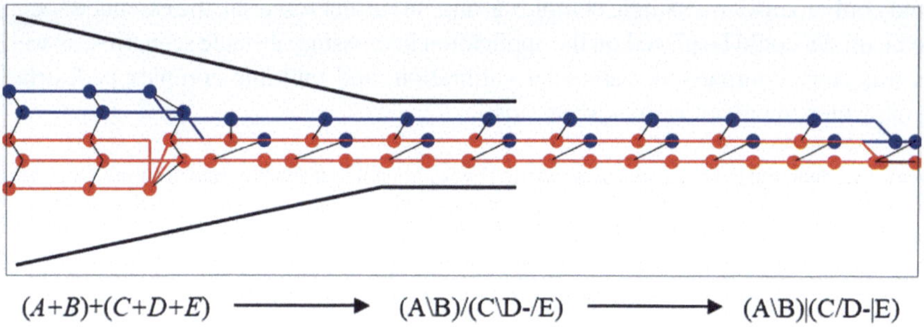

Fig. 22.6 Trajectories in simulations using strategy of keeping subgroup configuration

force. The introduction of random factors in the wall force formula, which controls the probability of occurrence of different strategies, could be helpful to the quantitative analysis and pedestrian experiments. This will be refined in the next step of the study.

22.4 Conclusion

Based on the research basis of the Expression-Matrix and structure of the five-person group, this study attempts to study the heterogeneous groups containing subgroups by using the extended social force model, measure, and record the changing process of group structure by the Expression-Matrix method.

Based on the predecessor model, the extended social force model used in this paper adds the force between other subgroups in the group and the pedestrian individual, by refining the pedestrian relationship in the multiple subgroups. At this point, the initial individual interaction between the pedestrians is subdivided into inter-group forces, inter-subgroup forces, and the interactions between pedestrians within the subgroup based on social relationships. It is worth mentioning that the past formula based on Newton's second law is no longer applicable to the analysis within the group. Because there is a need to maintain a certain distance between the subgroups, and the force contains attraction and repulsion both. Therefore, this requires some improvement in the expression of the subgroup force.

This study is based on Matlab and simulates a single group which contains two subgroups in a scenario that passing through a narrow passage. Through comparison with experimental data, the simulation has better restored the trajectory change process at the macro level and the adoption strategy of the group. The simulation results show that the improved social force model has particular applicability, and the calculation model of the force for maintaining the distance between the subgroups is more accurately.

At the same time, there are still many shortcomings in this study. For a narrow passage scenario of a single complex group, the trajectory is less sensitive to parameters due to the smaller scene with a slight change. Due to the time constraints, this study only focuses on a single complex group. In further research, the extended social force model could be placed on the application in crossing obstacle scenarios, as well as trajectory comparison, parameter calibration, and multiple complex pedestrian groups movements in various scenarios.

Acknowledgements This work was supported by National Key R&D Program of China, Grant No. 2018YFB1201402. This work was also supported in part by Scientific Research Program of Science and Technology Commission of Shanghai Municipality (STCSM), Grant No. 18DZ1201404. Besides, this work was also supported by the "Study on pedestrian limiting strategies during peak hours in metro station" research project from Shanghai Shentong Metro Group Co.

Reference

1. F. Zanlungo, T. Kanda, Do walking pedestrians stability interact inside a large group? Analysis of group and sub-group spatial structure, in *Proceedings of the Annual Meeting of the Cognitive Science Society* (2013), pp. 3847–3852
2. M. Moussaïd, N. Perozo, S. Garnier, D. Helbing, G. Theraulaz, The walking behaviour of pedestrian social groups and its impact on crowd dynamics. PLoS ONE **5**(4) (2010)
3. L. Huang, J. Gong, Social force model-based group behavior simulation in virtual geographic environments. ISPRS Int. J. Geo-Inform. **7**(79), 1–20 (2018)
4. J. Xi, X. Zou, Multi-pattern of complex social pedestrian groups, in *7th International Conference on Pedestrian and Evacuation Dynamics* (Elsevier, Delft, 2014), pp. 60–68
5. J. Huang, X. Zou, A structure analysis method for complex social pedestrian groups with symbol expression and relationship matrix, in *8th International Proceedings on Proceedings* (University of Science and Technology of China, Hefei, 2016), pp. 283–289
6. X. Zou, X. Qu, Experimental study on variation strategies for complex social pedestrian groups in conflict conditions, in *9th International Conference on Pedestrian and Evacuation Dynamics* (Lund University, Sweden, 2018), pp. 123-1–123-8

Chapter 23
Pedestrian Fundamental Diagram in Between Normal Walk and Crawling

Jian Ma, Dongdong Shi, and Tao Li

Abstract Due to the increase of the number and frequency of major events in the past decades, studies on pedestrian dynamics have attracted attentions of researchers of various disciplines. Considering that escape environment, especially the ones under special circumstances including fire, can be greatly affected by factors such as smoke, pedestrian gait is as a result different from normal walking. Thus an experimental study on fundamental diagram of single-file pedestrian movement at different available heights is detailed in the present paper. In total five experimental scenarios with the available heights of 1.0 , 1.2, 1.4, 1.6 and 2.0 m were investigated. The relationship between velocity and density, density and flow, and spatial headway and instantaneous velocity were analyzed. It is found that the pedestrian's free walking speed is the largest at the height of 2 m. At high density scenarios, the speed of pedestrians presents two separated states, i.e., 0.5 m/s at the heights of 2.0 and 1.6 m and 0.25 m/s at the heights of 1.0, 1.2 and 1.4m. It is also found that there are three segments, namely laminar flow state, stop and go phenomenon and jammed state, in the fundamental diagram. The results presented in this paper is helpful in understanding pedestrian evacuation at different heights.

23.1 Introduction

In recent years, pedestrian safety has attracted more and more attentions. It was noticed that there are plenty of interesting phenomena in pedestrian dynamic system, such as herding [1], group behavior [2], stop-and-go [3] and self-organizing [4]. To

---

J. Ma (✉) · D. Shi
School of Transportation and Logistics, National Engineering Laboratory of Integrated Transportation Big Data Application Technology, Southwest Jiaotong University, Chengdu 610031, China
e-mail: majian@mail.ustc.edu.cn

T. Li
Faculty of Geosciences and Environmental Engineering, Southwest Jiaotong University, Chengdu 610031, China

I. Zuriguel et al. (eds.), *Traffic and Granular Flow 2019*, Springer Proceedings in Physics 252,
https://doi.org/10.1007/978-3-030-55973-1_23

understand the origin of such kind of phenomena, researchers from the discipline of physics, computer science and so on have been working on pedestrian system. Pedestrian system is complicated because its performance is affected by pedestrian facility feature and pedestrian characteristics including cultural backgrounds, education level, body sizes, etc.

When emergencies such as fire presents, the pedestrian system becomes even more complex. That is because crowd movement will be restricted by smoke and radiation, pedestrian can only move with a height constraint, which makes the evacuation efficiency much lower [5, 6]. Up to now, researchers have studied evacuation with normal walking and crawling. The most typical kind of response in fire evacuation is that evacuees use the way of crawling to egress. Kade et al. [7] compared individual crawling speed and walking speed. Besides, the influence of gender and individual constitution on crawling evacuation were also investigated. Nagai et al. [8] performed evacuation experiments under two conditions where evacuated people crawled and walked, then they built a simulation model and performed further studies. Typically, people like to evacuate in normal walking ways. However, it is sometime necessary to adopt crawling methods even it leads to long escape times and is particularly vulnerable. To quantify pedestrian movement feature, some researchers conducted a series of well-designed experiments. Seyfried et al. [9] conducted a single-file experiment to study the relationship between speed and density of pedestrians. Jelic et al. [10] carried out a ring experiment with the composition of the inner and outer rings. They studied the fundamental diagram and relationship between spatial headway and speed. It was found that there are three states, i.e., free state, weakly constrained state, strongly constrained state with increasing pedestrian density. Cao et al. [11, 12] conducted a single-file experiment to study the fundamental diagram of different aged pedestrian. The relationship between speed and step width and step length has been explored.

Due to the lack of data for the movement in between normal walking and crawling, it is necessary to perform experiments to quantify this kind of pedestrian movement. In this paper, we conducted a controlled single-file experiment to study the motion characteristics of pedestrians at different heights.

The rest of the article will be organized as follows: the experimental setup will be detailed in the Sect. 2. In Sect. 3, the experimental results will be discussed. The fundamental diagram at different heights will be compared, and the relationship between spatial headway and speed will be discussed. In Sect. 4, we summarize the results of the experiment.

23.2 Experiment Setup

The aim of our experiments is to study the motion characteristics of pedestrian with different height constraints. The experiment scenario and a snapshot of the experiment is shown in Fig. 23.1. The closed ring channel has a center circumference of $C = 23.08$ m, including two 4 m straights and two semicircles with a center radius

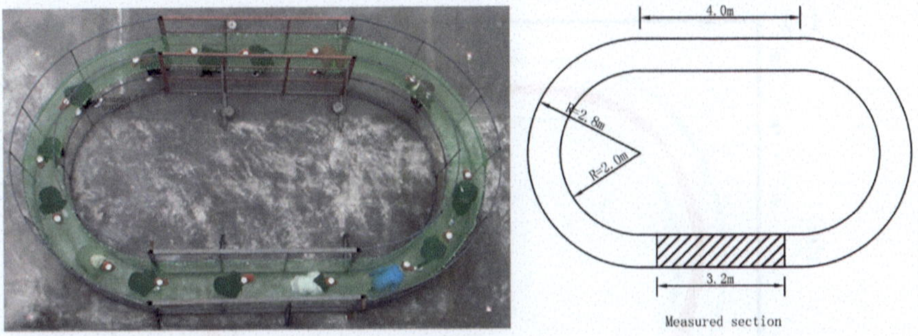

Fig. 23.1 The scenario and a snapshot from the experiment

of 2.4 m. To avoid overtaking, the walkway width is set to 0.8 m. In total 65 young college student volunteers were recruited to take part in the experiment. All of them did not know the purpose of the experiment. They were asked to wear a red hat with a white dot on their heads. They were required to traveling the channel anticlockwise without overtaking. The experiment contains 30 tests to investigate the constraints of five specified heights, i.e., 1.0 m, 1.2 m, 1.4 m, 1.6 m and 2.0 m respectively. For each height constraint, we put N people in the channel, i.e., $N = 5$, 10, 15, 20, 25, and 30, respectively in each test. In this way, the global density $\rho_g = N/C$ is in the range of 0.21 m^{-1} to 1.3 m^{-1}. All participants follow the same instructions in each test. In order to collect enough data, especially to obtain stable data and to ensure repeatable results, they were asked to walk three runs in the channel. Considering that pedestrians will feel tired after a test, we will let the participants have rest time before taking another test.

In the experiment we used a transparent protective net to limit the available heights. The entire experiment was recorded by a camera placed on the top of a laboratory building with a height of approximately 15 m. The recorded video frame rate is 25 frames per second and has a resolution of 1920 × 1080 pixels. The pedestrian trajectory during the experiment was extracted by the software PeTrack. We show the experimental trajectories at different heights in Fig. 23.2. After the extraction, the data is further filtered to avoid extraction errors.

23.3 Results and Discussions

To explore the movement characteristics of pedestrian with different height constraints, longitudinal pedestrian movement in the local region shown in Fig. 23.1 is focused on. The length of the local region has a length of $x = 3.2$ m. In the following section, we first introduce data analysis method and then present the results.

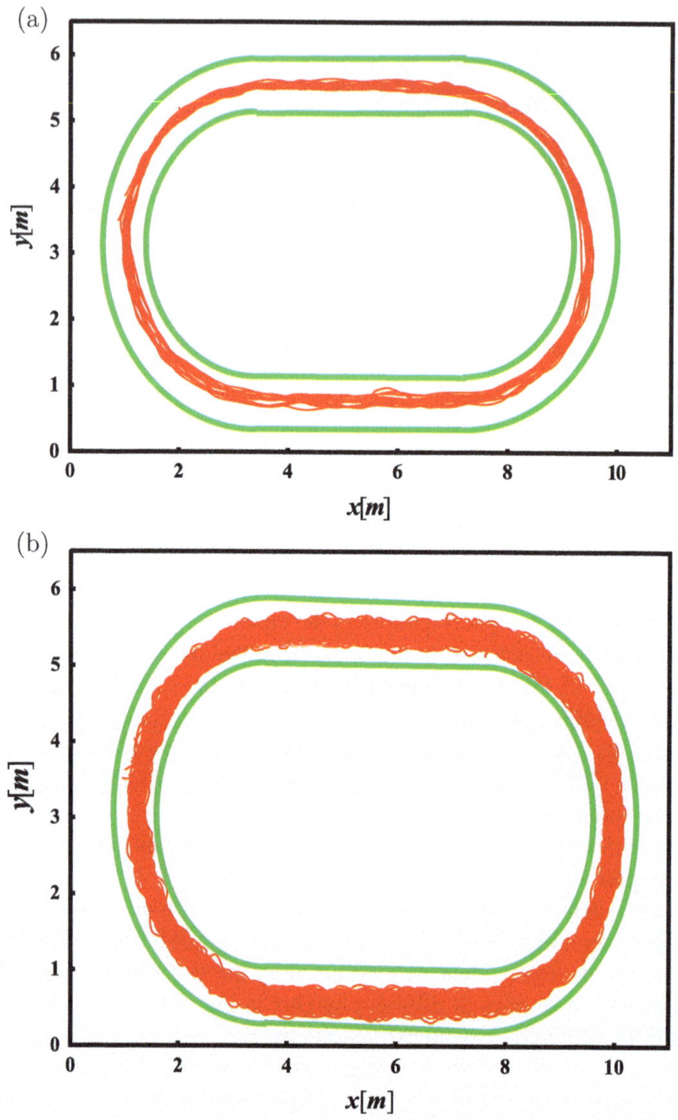

Fig. 23.2 Trajectories of pedestrians in different heights of experiment. **a** Number of student in H = 1.2 m height $N = 5$, **b** number of student in H = 2.0 m height $N = 30$

23.3.1 Method

Local measurement. Measurements of local velocity and density were the same as the ones in [13]. The speed at which person j passes through the measurement area is defined as x divided by the time of his or her pass through this straight area,

$$v_j = \frac{x}{t_j^o - t_j^i} \tag{23.1}$$

t_j^i and t_j^o are the entry and exit times of pedestrians j. In the straight measurement area, the density at time t is calculated by $\rho(t) = \sum_j \phi_j(t)/x$, where $\phi_j(t)$ is the spatial parameter of pedestrians j and followers $j+1$ in the straight area.

$$\phi_j(t) = \begin{cases} \dfrac{t - t_j^i}{t_{j+1}^i - t_j^i} & \text{for } t \in [t_j^i, t_{j+1}^i] \\ 1 & \text{for } t \in [t_{j+1}^i, t_j^o] \\ \dfrac{t_{j+1}^o - t}{t_{j+1}^o - t_j^o} & \text{for } t \in [t_j^o, t_{j+1}^o] \\ 0 & \text{otherwise} \end{cases} \tag{23.2}$$

Instantaneous measurement. For pedestrian j at time t, the headway $d_{h,j}(t)$ is defined as the distance between the centers of him/her and his/her predecessor. Therefore, the instantaneous density $\rho_{h,j}(t)$ and $\rho_{H,j}(t)$ for pedestrian j at time t can be calculated as the inverse of the spatial headway [14]

$$\rho_{h,j}(t) = \frac{1}{d_{h,j}(t)} \tag{23.3}$$

$$\rho_{H,j}(t) = \frac{3}{d_{h,j-1}(t) + d_{h,j}(t) + d_{h,j+1}(t)} \tag{23.4}$$

The individual instantaneous velocity $v_i(t)$ of pedestrian i at time t is calculated as,

$$v_i(t) = \frac{x_i(t + \Delta t/2) - x_i(t - \Delta t/2)}{\Delta t} \tag{23.5}$$

where $x_i(t)$ is lateral coordinate of pedestrian i at time t, Δt is a small time interval and 0.04 s is used in the paper.

23.3.2 Fundamental Diagram

The relationships between velocity and density, flow rate and density considering the influence of pedestrian's available heights are focused on in this section. The entire density is divided into a series of bins per 0.1 m^{-1}. The instantaneous velocity and instantaneous flow value of the pedestrian are averaged by the instantaneous density value per 0.1 m^{-1}, and then one data point is obtained. As shown in Fig. 23.3, although the two density calculation methods are both biased, the biased estimations presence small difference. For the convenient of comparison, the speed and density relations with height constraints of 1.0 and 2.0 m are shown in Fig. 23.3. It can be found that the fundamental diagrams from the two methods show good agreement

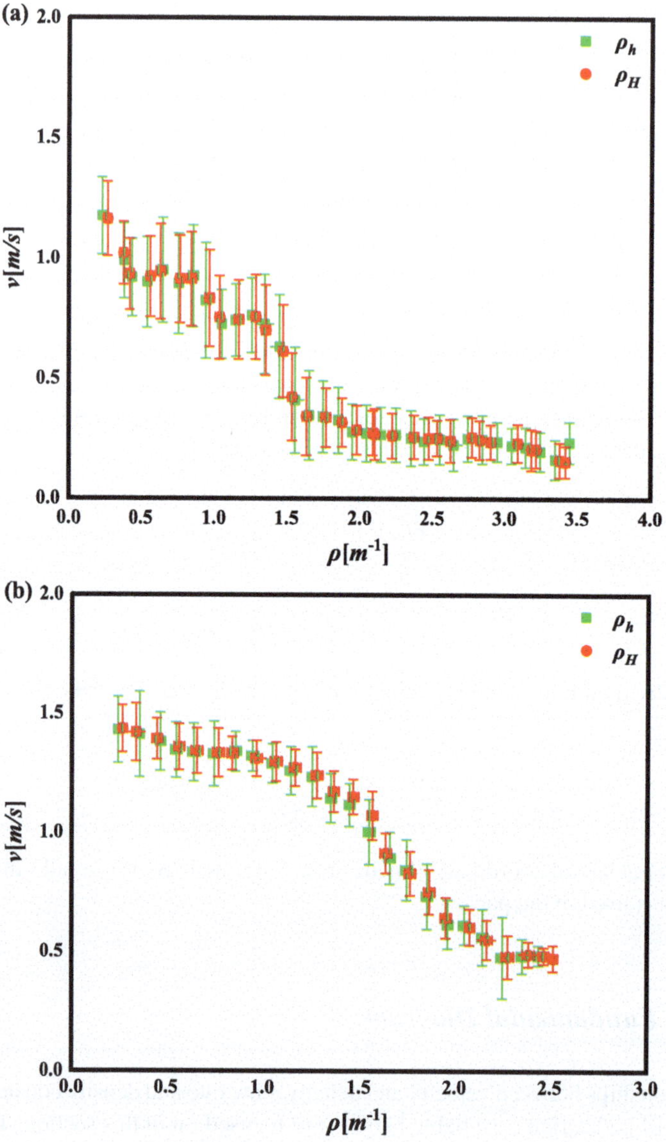

Fig. 23.3 The comparison of two calculational density methods: **a** the height is 1.0 m and **b** the height is 2.0 m

over the whole density range. Thus in the following sections, discussions on base of the density method ρ_h will be detailed.

As shown in Fig. 23.4a, the speed decreases as the density increases. Due to the limited available heights, pedestrians need to bend over and walk, as a consequence,

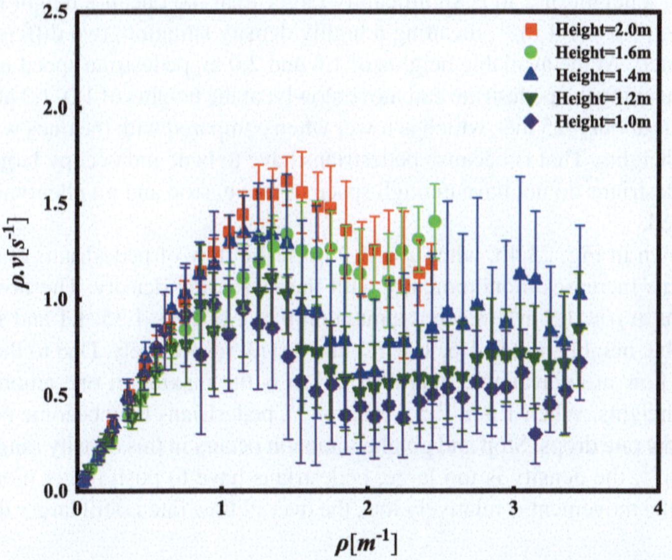

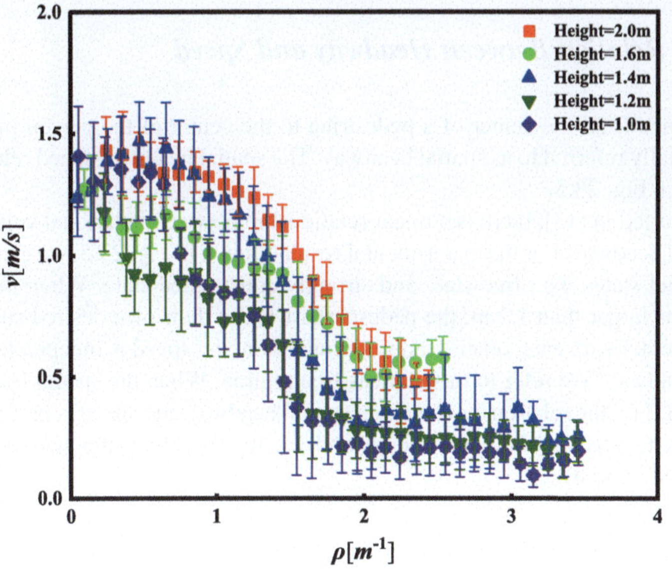

Fig. 23.4 The fundamental diagram: **a** Density-velocity relation, **b** density-flow relation. The purple diamond, green inverted triangle, blue triangle, red circle, gray square stand height 2.0 m, 1.6 m, 1.4 m, 1.2 m, and 1.0 m respectively

their mobility and the corresponding free speed is affected. At low density, i.e., $\rho <$ 0.6 m^{-1}, pedestrians walk at free speed for all available heights. At the heights of 2.0, 1.6, 1.4, 1.2 and 1.0 m, the free speed is 1.4 ± 0.2, 1.1 ± 0.22, 1.35 ± 0.21, 1.2 $\pm$ 0.23, 1.05 ± 0.2 m/s, respectively. It can be found that at a height of 2.0 m, the free speed is the largest. At medium density, ρ is between 0.6 m^{-1} and 1.4 m^{-1}, the velocity at a height of 2 m is significantly larger than the one at a height of 1.0 m. In the case of $\rho >$ 1.4 m^{-1}, meaning a highly density situation, two different states are presented. At the available heights of 1.6 and 2.0 m, pedestrian speed is close to 0.5 m/s, meaning the pedestrian can move slowly; at the heights of 1.0, 1.2 and 1.4 m, the speed is about 0.25 m/s, which is lower when compared with the ones with larger available heights. That is because pedestrians have to bent and occupy larger space. When pedestrians do not have enough space to move, stop and go phenomenon can be observed.

As shown in Fig. 23.4b, when $\rho <$ 1.3 m^{-1}, the flow of pedestrians at all available heights increases monotonically with the increasing density. The flow rate of pedestrians at $\rho =$ 1.3 m^{-1} reaches a maximum of 1.6, 1.35, 1.35, 1.1 and 1.0 s^{-1} at the available heights of 2.0, 1.6, 1.4, 1.2 and 1.0 m, respectively. Due to the heights limit, the flow at an available height of 2.0 m is the maximum one among all the available heights. When 1.3 $m^{-1} < \rho <$ 2.0 m^{-1}, pedestrians flow become congested and the flow rate drops. Stop and go phenomenon occurs in this density range. When $\rho >$ 2.0 m^{-1}, the density is too large, pedestrians have to push to get more space, although the movement is relatively low, the overall flow rate is still larger than zero.

23.3.3 Relation Between Headway and Speed

The distance from the center of a pedestrian to the center of the person in front of him is usually referred to as spatial headway. The spatial headway-speed relation can be found in Fig. 23.5.

As reported in [13], there is a linear relationship between individual velocity and the spatial headway. For the experimental results shown in Fig. 23.5, we can observe two distinct states, i.e., free state and strongly constrained state. When the spatial headway is larger than 1.2 m, the pedestrians walk at their own desired speeds and do not interact with each other. Under this situation, the speed is independent of the spatial headway. We refer to this state as free regime. When the spatial headway is less than 1.2 m, the relationship between spatial headway and speed is more obvious because pedestrians are limited by height and density. We refer to this state as strongly constrained regime.

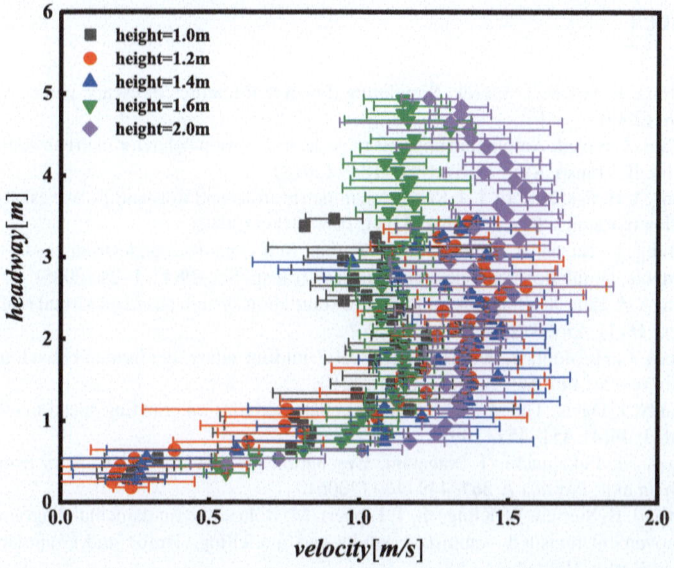

Fig. 23.5 The relation between headway and velocity. The purple diamond, green inverted triangle, blue triangle, red circle, gray square stand height 2.0 m, 1.6 m, 1.4 m, 1.2 m, 1.0 m respectively

23.4 Conclusion

In this paper, the effect of different available heights on the motion characteristics of pedestrians are studied by a single-file pedestrian experiment. Pedestrian trajectories are extracted by the software PeTrack, and the coordinates in the unit of pixel were converted into physical coordinates. All the following analysis is based on these data. It can be seen from the fundamental diagram at different heights that the velocity decreases as the density decreases. Free speed is limited at different heights. When the density is less than 0.6 m^{-1}, the free speed of the person is the highest at the height of 2.0 m. When the density is greater than 1.4 m^{-1}, two different speed states appear. From the relationship between density and flow, three phases can be found. When the density $\rho < 1.3$ m^{-1}, the system is in the laminar phase. When the density 1.3 $m^{-1} < \rho < 2.0$ m^{-1}, the system is in a state of stop and go. When the density 2.0 $m^{-1} < \rho$, the system is in jammed state. In the relationship between speed and headway, only two states, i.e., free state and strongly constrained state can be observed. When the headway is greater than 1.2 m, the pedestrian walks in a free state. When the headway is less than 1.2 m, the pedestrian is in a strongly constrained state.

Acknowledgements The work presented in this paper is fully supported by the National Natural Science Foundation of China (Nos. 71871189 and 71473207) and National Key Research and Development Program of China (No. 2017YFC0804900).

References

1. D. Helbing, I. Farkas, T. Vicsek, Simulating dynamical features of escape panic. Nature **407**, 487–490 (2000)
2. M.H. Zaki, T. Sayed, Automated analysis of pedestrian group behavior in urban settings. IEEE Trans. Intell. Transp. Syst. **19**(6), 1880–1889 (2018)
3. H. Kuang, Y.H. Fan, X.L. Li, L.J. Kong, Asymmetric effect and stop-and-go waves on single-file pedestrian dynamics. Procedia Eng. **31**(4), 1060–1065 (2012)
4. D. Helbing, L. Buzna, A. Johonsson, T. Werner, Self-organized pedestrian crowd dynamics: experiments, simulations, and design solutions. Transp. Sci. **39**(1), 1–24 (2005)
5. C. Chen, A.Z. Ren, X. Zhang, A building fire simulation system based on virtual reality. J. Nat. Disasters **16**(1), 55–60 (2007)
6. M. Kobes, L. Helsloot, B. de Vries, J.G. Post, Building safety and human behaviour in fire: a literature review. Fire Saf. J. **45**(1), 1–11 (2010)
7. R.A. Kady, J. Davis, The effect of occupant characteristics on crawling speed in evacua-tion. Fire Saf. J. **44**(4), 451–457 (2009)
8. R. Nagai, M. Fukamachi, T. Nagatani, Evacuation of crawlers and walkers from corridor through an exit. Physica A **367**, 449–460 (2006)
9. A. Seyfried, B. Steffen, W. Klingsch, T. Lippert, M. Boltes, the fundamental diagram of pedestrian movement revisited—empirical results and modelling. Traffic and Granular Flow'05. Springer, Berlin, Heidelberg (2007)
10. A. Jelić, R.C. Appert, S. Lemercier, J. Pettré, Properties of pedestrians walking in line: fundamental diagrams. Phys. Rev. E **85**(3), 036111 (2012)
11. S. Cao, J. Zhang, W.G. Song, C.A. Shi, R.F. Zhang, Thc stcpping bchavior analysis of pedestrians from different age groups via a single-file experiment. J. Stat. Mech. **2018**(3), 033402 (2018)
12. S. Cao, J. Zhang, D. Salden, J. Ma, C.A. Shi, R.F. Zhang, Pedestrian dynamics in sin-gle-file movement of crowd with different age compositions. Phys. Rev. E **94**, 012312 (2016)
13. J. Ma, D.D. Shi, T. Li, X.F. Li, T.F. Xu, P. Lin, Experimental study of single-file pedestrian movement with height constraints, J. Stat. Mech. **2020**(7), 073409 (2020)
14. A. Tordeux, J. Zhang, B. Steffen, A. Seyfried, Quantitative comparison of estimations for the density within pedestrian streams. J. Stat. Mech. **2015**(6), P06030 (2015)

Chapter 24
Deep Fundamental Diagram Network for Real-Time Pedestrian Dynamics Analysis

Qing Ma, Yu Kang, Weiguo Song, Yang Cao, and Jun Zhang

Abstract The fundamental diagram of pedestrian flow, which describes a relation between pedestrian velocity and crowd density, is an important means forpedestrian dynamics analysis. Some recent work calculates the fundamental diagram of pedestrian flow by tracking each pedestrian in the crowd from video recordings. However, such methods are limited in representation of crowd density and hard to achieve a real-time analysis. To address this problem, this work proposes a novel convolutional neural network-based framework, called deep fundamental diagram network, for real-time pedestrian dynamics analysis. Our proposed framework is consisted of two parts, the multi-scale recursive convolutional neural network (MSR-Net) and an optical flow module, accounting for density distribution estimation and pedestrian motion prediction respectively. Specifically, MSR-Net is presented to learn the direct mapping from the input image of pedestrian flow to the output map of crowd density. Optical flow method is introduced to predict the velocity and direction of pedestrian in real-time. In this way, by aligning the position of pedestrian density map we are able to obtain the fundamental diagram, which shows good agreement with the ones from classical methods but higher computational efficiency. Simultaneously, deep fundamental diagram network can detect anomaly activity of pedestrian (In this work, the anomaly is defined as sudden stop and acceleration, reverse walk.), which is also meaningful for crowd analysis.

Q. Ma · Y. Kang (✉) · W. Song · J. Zhang
State Key Laboratory of Fire Science, University of Science and Technology of China, Hefei, China
e-mail: kangduyu@ustc.edu.cn

Y. Kang · Y. Cao
Department of Automation, University of Science and Technology of China, Hefei, China

I. Zuriguel et al. (eds.), *Traffic and Granular Flow 2019*,
Springer Proceedings in Physics 252,
https://doi.org/10.1007/978-3-030-55973-1_24

24.1 Introduction

Pedestrian dynamics have many practical applications, such as designing evacuation paths, optimizing pedestrian facilities. Experimental studies of pedestrian dynamics date back to 1937 [1], and in decades of development, researchers have carried out a large number of field observation, controlled experiments and animal experiments. Recently, many evacuation dynamics models have been proposed, such as social force model [2, 3], fluid-dynamic models [4, 5] cellular automata [6–8], lattice-gas models [9]. It is worthwhile mentioning that getting the fundamental diagram quickly and accurately is helpful for the study of pedestrian dynamics.

In experimental study, the common method to obtain the fundamental diagram is tracking the experiment video. By tracking each pedestrian in the crowd from video recordings, the velocity and density of pedestrian are calculated and then formed into fundamental diagram. However, this method can only achieve post analysis of pedestrian dynamics, and cannot provide service for onsite decision-making. In this paper, we propose a deep fundamental diagram network for real-time pedestrian dynamics analysis.

Our deep fundamental diagram network is made up of two parts, as shown in Fig. 24.1, a multi-scale recursive convolutional neural network (MSR-Net) is propose for extracting crowd density map. The optical flow module is used for extracting crowd velocity map. Then the fundamental diagram is generated by utilizing the spatial correspondence between density and velocity map. We evaluate the performance of our method on open experiment against tracking based method. The experimental results show that our method can obtain the velocity and density of each frame in real-time, while the generated fundamental diagram is consistent with the tracking method.

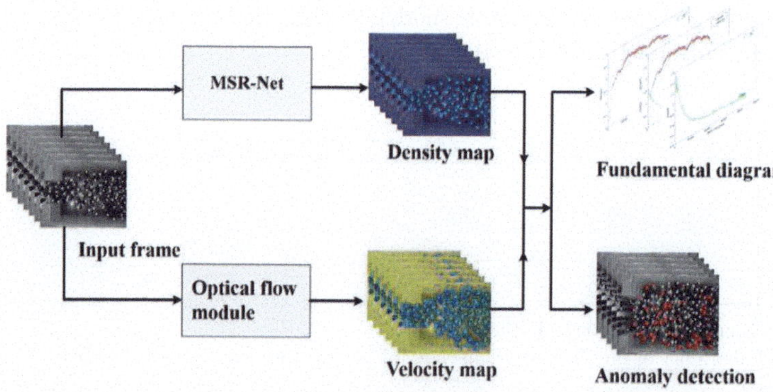

Fig. 24.1 The structure of out proposed deep fundamental diagram network

24.2 Method

This paper aims to obtain crowd density and velocity in real-time. To achieve this, we propose a deep fundamental diagram network. It is composed of two parts: multi-scale recursive convolutional neural network (MSR-Net) and optical flow module. MSR-Net output density map, and optical flow module output velocity map. And the fundamental diagram is obtained by the spatial correspondence of density map and velocity map.

24.2.1 MSR-Net

Motivated by multi-scale recursive convolutional neural network (MRCNN) which has two objectives of estimation of crowd density map and perspective map [10], we propose to use the multi-scale recursive convolutional neural network (MSR-Net) to learn the map from the raw frame image to the density map. Figure 24.2 shows the net-work structure.

Considering that different sizes of the heads in different experimental scenes lead to different number of pixels occupied by the heads, we consider learning the head features at different scales with multi-scale receptive fields. This multi-scale network contains several sub-nets which involve different sizes of filter kernels. Large size of filter kernels capture large size heads and coarse textures, while small size of filter kernels capture small heads and local features. The sub-net is independence, have no impact on each other, and the learning of different scale features depends only on the number of sub-networks. The part of fusion in the MSR-Net aggregate multi-scale feature, and these features processed at various scales again for multi-scale feature fusion and interaction (Fig. 24.3).

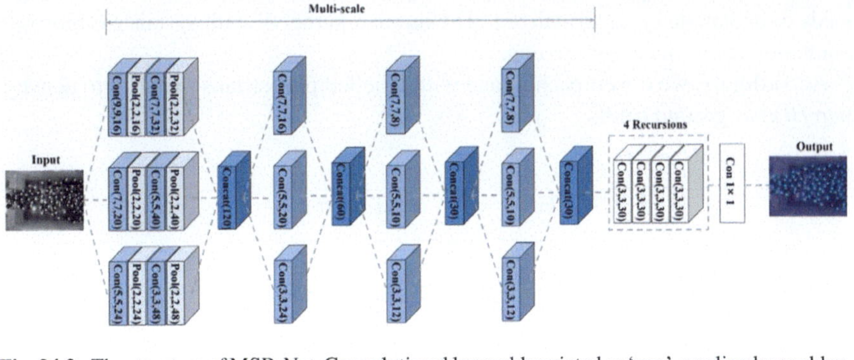

Fig. 24.2 The structure of MSR-Net. Convolutional layer abbreviated as 'con', pooling layer abbreviated as 'pool', concat layer abbreviated as 'concat'. In parentheses, the first two numbers represent the size of the filter kernel, and the last number represents the number of filter kernels. For example, con(9, 9, 16) indicates that the convolutional layer has 16 filters of size 9×9

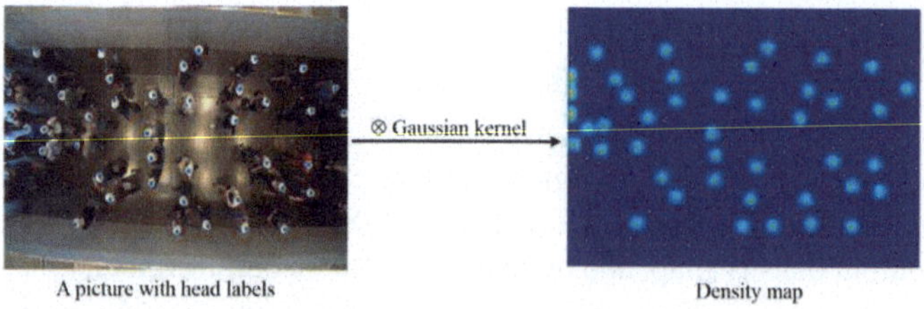

Fig. 24.3 Crowd density map generation process

In our network structure, all pooling layers are use Max Pooling. We use ReLU activation after each convolutional layer. The input of recursive convolution is scale fusion feature which learned mapping relations from multi-scale features to density map. Recursive convolution enlarges receptive field of original input image. Larger receptive field involved more information and interaction that improve network performance. The recursive part has four convolution layers with the same size filters and share the weights of parameters, followed by a filter size of 1 × 1 to map the recursive output to the density map.

In this network, we adopt Euclidean loss functions, which defined as:

$$L_D(\Theta) = \frac{1}{2N}\sum_{i=1}^{N} \| F_d(X_i; \Theta_d) - D_i \|_2^2 \quad (1)$$

where Θ is a set of learnable parameters in the MSR-Net. N is the number of training image. X_i is the input image. $F_d(X_i; \Theta_d)$ stand for density map generated by MSR-Net network. D_i is the ground truth density map. $L_D(\Theta)$ is the loss between density map generate by network and ground truth density map.

The quality of density map affects the effectiveness of the training. In order to generate density map, the position of the pedestrians' head on each frame image needs to be labelled p_i. Normalized 2D Gaussian kernel N with variance σ blurring annotations p_i to make sum to one.

Accordingly, we covert each image with labelled pedestrians' heads into density map *H(x)* as ground truth.

$$\mathrm{H}(x) = \sum_{i=1}^{N} N(x; p_i, \sigma) \quad (2)$$

Thus, the sum of density map represents crowd count C.

$$\mathrm{C} = \sum_{x} H(x) \quad (3)$$

24.2.2 Optical Flow Module

In our optical flow module, we choose LK optical flow algorithm, in which the motion of scene can be described as:

$$\frac{\partial I}{\partial x}V_x + \frac{\partial I}{\partial y}V_y + \frac{\partial I}{\partial t} = 0 \tag{4}$$

The V_x and V_y is the velocity in the x, y, or optical flow called I(x, y, t), $\frac{\partial I}{\partial x}$, $\frac{\partial I}{\partial y}$, $\frac{\partial I}{\partial t}$, is a partial derivative of the image (x, y, t) in the corresponding direction. The relationship between I_x, I_y, and I_t can be expressed in the following equations:

$$I_x V_x + I_y V_y = -I_t \tag{5}$$

Assume that the image content of two adjacent frame move very small, local vectors (V_x, V_y) must meet:

$$I_x(q_i)V_x + I_y(q_i)V_y + I_t(q_i) = 0, i = 1, 2, 3 \ldots n \tag{6}$$

where q_i is the number of pixels in the neighborhood. $I_x(q_i)$, $I_y(q_i)$, $I_t(q_i)$ is the partial derivative of the image at the point q_i and current time to position x, y, and time t. Written in the form of a matrix:

$$Av = b \tag{7}$$

where A $= \begin{bmatrix} I_x(q_1) & I_y(q_1) \\ \vdots & \vdots \\ I_x(q_n) & I_y(q_n) \end{bmatrix}$, $v = \begin{bmatrix} V_x \\ V_y \end{bmatrix}$, $b = \begin{bmatrix} -I_t(q_1) \\ \vdots \\ -I_t(q_n) \end{bmatrix}$.

An approximate solution is obtained by using the least squares method:

$$v = \left(A^T A\right)^{-1} A^T b \tag{8}$$

The pixel real movement distance D_i in real scene conversions as follow:

$$D_i = \frac{d \times x_i}{L} \tag{9}$$

Here d is the number of pixels which represent the distance L in real experiment scene. x is the pixel movement distance. Then, the real velocity of each pixel can be obtained by:

$$u_i = \frac{D_i}{\left(\frac{1}{p}\right)} = D_i \times p \tag{10}$$

where p is the frame rate.

24.2.3 Calculate Fundamental Diagram

In part of 2.1 and 2.2, we put a crowd frame image in the network architecture and optical flow module, obtain the density map and velocity map. The number of pedestrians in any area is equal to the sum of pixels in corresponding area of density map. The density D is expressed as:

$$D = \frac{\sum_{i=1}^{N} d(x, y)}{A} \tag{11}$$

where d (x, y) is the pixel value of the (x, y) position in measuring area. N is the total number of pixels in measuring area. A is the area of measuring area.

Density map shows the positioning of pedestrians in the given frame image. When (x, y) is not 0, this pixel represents pedestrians. Because of the spatial correspondence between the velocity graph and the density map (Fig. 24.4), there are pedestrians in the position of the density map (x, y), then the velocity of this pixel can be obtained in the velocity graph. With such spatial correspondence, we can obtain the average velocity of pedestrian in the measured area through the following formula:

$$V = \frac{\sum v(x, y)}{M} \tag{12}$$

where v (x, y) is the pixel value of the (x, y) position in measuring area. M is the total number of pixels in the measurement area that are not 0 in the density map.

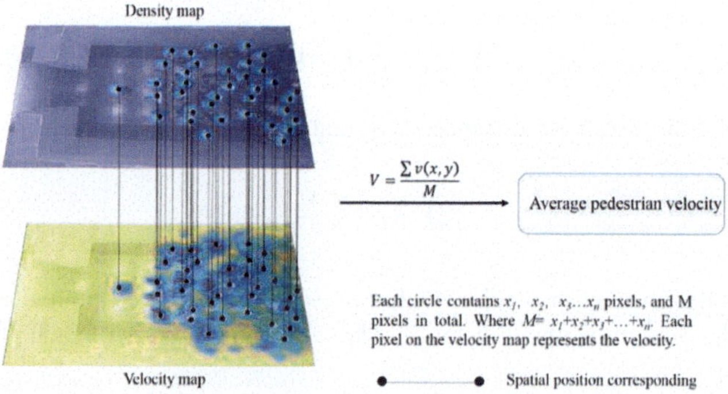

Fig. 24.4 Correspondence between density map and velocity map

Hence, through the two important pedestrian dynamic information of density and velocity, we can draw fundamental diagram.

24.3 Experimental Result and Analysis

In this paper, the network model was fine-tuned using Caffe. (Caffe is an open source deep learning framework developed by the Berkeley Vision and Learning Center created by Jia et al. [11]) Our method have been implemented in MATLAB and run on a PC with 3.60 GHz Intel(R) Core (TM) i7 CPU, 16G RAM. The CNN is implemented on GPU. We used a NVIDIA GTX 1050ti GPU, running Ubuntu 16.04 LTS.

The proposed MSR-Net was trained on the deep learning framework of Caffe [11]. We choose the adaptive learning rate Adam to train the network. Gaussian initialization with standard deviation of 0.01 was carried out of the network, the momentum was 0.9, and the weight decay was 0.0005. The base learning rate was 1e-5. We use the density map to fine-tune the pre-trained network.

We evaluated the performance of our proposed method on the public dataset UNI_CORR_500 [12]. Two terms, Mean Absolute Error (MAE) and Mean Squared Error (MSE), are used in our test.

Table 24.1 shows the quantitative result for pedestrian density and speed. The density of desire area error in sub five experiment is 0.196~0.28%, the average error is 0.22%, and our method work robustly.

We also obtain the fundamental diagram as show in Figs. 24.5 and 24.6.

Table 24.1 Quantitative result for pedestrian density ($1/m^2$) and speed(m/s):

UNI_CORR_500		03	04	05	06	07	Average
MAE	Density	0.0794	0.1116	0.0818	0.0862	0.0784	0.0875
	Speed	0.0618	0.0404	0.0503	0.0312	0.0318	0.0431
MSE	Density	0.0084	0.0162	0.0094	0.0099	0.0101	0.0108
	Speed	0.0149	1.5164	0.0086	1.1659	0.0065	0.54246

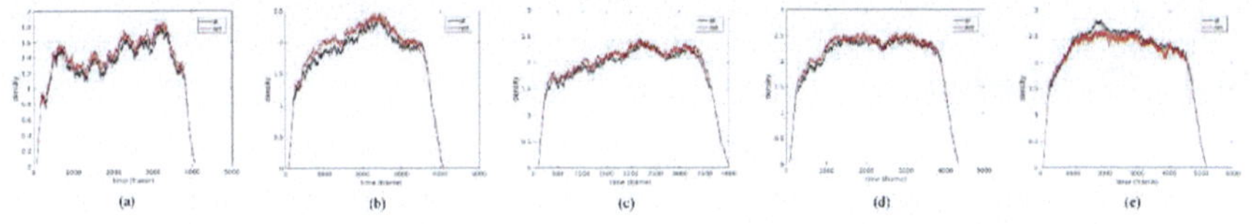

Fig. 24.5 **a, b, c, d, e** is the time evolution of the density of the third to seventh video. The black line represents the ground truth density which obtained from the trajectory file, the red line represents the density obtained from CNN

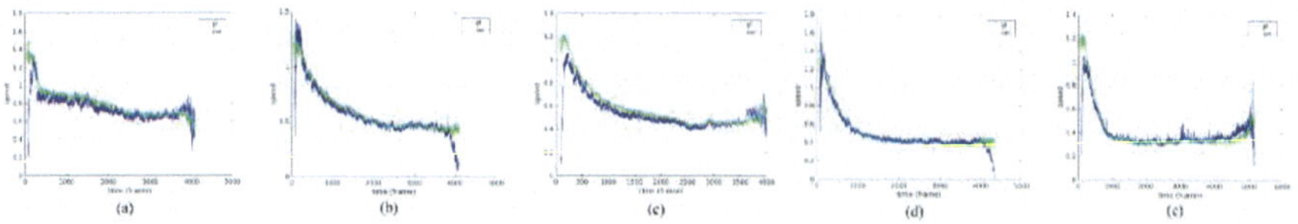

Fig. 24.6 **a**, **b**, **c**, **d**, **e** is the time evolution of the velocity of the third to seventh video. The blue line represents the ground truth speed which calculate from trajectory file. The green line represents the speed obtained from our method

Table 24.2 Obtain pedestrian density and velocity run-time

	Time (second per frame)
Obtain density	0.04
Obtain velocity	0.2

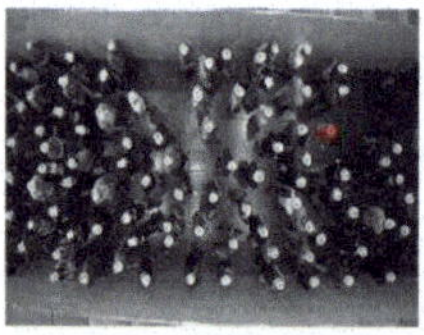

Fig. 24.7 Anomaly pedestrian is marked in red

From a macro perspective, the time-density, time-velocity evolution tends which obtained by our method is basically the same with calculate from trajectory file. In Fig. 24.5, the time-speed curves obtained by our method are smoother and less fluctuating than calculate by trajectory file. We got a relatively accurate pedestrian average speed after a few seconds the experiment in the video begins. However, our method takes very little time and can achieve the density velocity in real time. For the run- time, see Table 24.2

Because of the spatial correspondence between density map and velocity map, we can detect people whose speed and direction are different from other pedestrians. Therefore, we can also get anomaly map (Fig. 24.7), which can be applied to pedestrian anomaly detection. Due to the pedestrian dynamics experiment videos are not for anomaly detection, the dataset lacks of anomaly ground truth.

24.4 Conclusion

In this paper, we propose a method to obtain fundamental diagram in real-time. The multi-scale recursive convolutional neural network is used to estimate density map and we obtain the velocity map by optical flow module. Density map provides crowd density probability and crowd location information, velocity map contains pedestrian movement speed and direction. Through the space correspondence of the two maps, we can get the fundamental diagram in real-time. Experiments have obtained a fundamental diagram consistent with the trajectory tracking method on the open dataset. Meanwhile, the method can be applied to crowd anomaly detection. But, our MSR-Net is suitable for evacuation experiment scenarios with high density and large prediction error for low-density scenarios.

References

1. A. Seyfried, B. Steffen, W. Klingsch, M. Boltes, The fundamental diagram of pedestrian movement revisited. J. Stat. Mech. Theory Exp. **10**, 10 (2005)
2. D. Helbing, I. Farkas, T. Vicsek, Simulating Dynamical Features of Escape Panic. Social Science Electronic Publishing **407**(6803), 487–490 (2000)
3. D. Helbing, P. Molnar, Social force model for pedestrian dynamics. Phys. Rev. E Stat Phys. Plasmas Fluids Relat. Interdiscip. Top. **51.5**, 4282 (1998)
4. L.F. Henderson, On the fluid mechanics of human crowd motion. Transportation Re-Search **8**(6), 509–515 (1974)
5. R.L. Hughes, A continuum theory for the flow of pedestrians. Trans. Res. Part B (Methodological) **36.6**, 507–535 (2002)
6. M. Fukui, Y. Ishibashi, Self-Organized phase transitions in cellular automaton models for pedestrians. J. Phys. Soc. Jpn. **68.8**, 2861–2863 (1999)
7. Y. Imanishi, R. Kuwajima, T. Nagatani, Transition from homogeneous to inhomogeneous flows in a lattice-gas binary mixture of slender particles. Phys. A **387**(10), 2337–2352 (2008)
8. W. Song, Y. Yu, W. Fan, H. Zhang, A cellular automata evacuation model considering friction and re-pulsion. Sci. in China Ser. E Eng. Mater. Sc. **48**(4), 403–413 (2005)
9. S. Marconi, B. Chopard, A Multiparticle Lattice gas Automata model for a crowd, in *Proceedings DBLP Cellular Automata, 5th International Conference on Cellular Automata for Research and Industry, ACRI 2002, Geneva, Switzerland*, 9–11 Oct 2002
10. M. Wei, Y. Kang, W. Song, Y. Cao, Crowd distribution estimation with multi-scale recursive convolutional neural network, in *International Conference on Multimedia Modeling* (Springer, Cham, 2018)
11. Y. Jia et al., "Caffe," in *Presented at the Proceedings of the ACM International Conference on Multimedia—MM '14* (2014)
12. https://ped.fz-juelich.de/db/doku.php?id=corridor5

Chapter 25
Data-Driven Simulation for Pedestrians Avoiding a Fixed Obstacle

Rafael F. Martin and Daniel R. Parisi

Abstract Data-driven simulation of pedestrian dynamics is an incipient and promising approach for building reliable microscopic pedestrian models. We propose a methodology based on generalized regression neural networks, which does not have to deal with a huge number of free parameters as in the case of multilayer neural networks. Although the method is general, we focus on the one pedestrian—one obstacle problem. The proposed model allows us to simulate the trajectory of a pedestrian avoiding an obstacle from any direction.

25.1 Introduction

Recently, we proposed a general framework of pedestrian simulation [1] in which the surroundings of a virtual pedestrian, i.e., obstacles and other noncontacting particles, can only influence its trajectory by modifying its desired velocity.

The basic assumption is that the avoidance behavior can be exerted only by the self-propelled mechanism of the particle itself (usually modeled by the desired velocity).

Under this approach, the problem lies in postulating the heuristics required for computing the variable desired velocity depending on the environment. As in traditional pedestrian theoretical models, any arbitrary heuristic can be proposed (for example, [2, 3]) and then the free parameters could be tuned in order to obtain simulated trajectories that approach experimental micro or macroscopic data.

Instead of this traditional methodology, we can directly use the experimental data so as to compute the desired velocity at each time step. From a set of real trajectories we extract the information for providing a desired velocity to the simulated agent, considering the state of the agent in the simulated and experimental environment.

R. F. Martin (✉) · D. R. Parisi
Instituto Tecnológico de Buenos Aires (ITBA), Lavarden 315, (C1437FBG), C. A. de Buenos Aires, Argentina
e-mail: ramartin@itba.edu.ar

D. R. Parisi
Consejo Nacional de Investigaciones Científícas y Técnicas, Buenos Aires, Argentina

I. Zuriguel et al. (eds.), *Traffic and Granular Flow 2019*,
Springer Proceedings in Physics 252,
https://doi.org/10.1007/978-3-030-55973-1_25

Here we propose a data-driven approach using a nonparametric universal interpolator: the generalized regression neural network (GRNN) [4]. The GRNN needs to have access to the data examples (patterns) when predicting a new output. However, because it has only one degree of freedom (only one free parameter), the number of (input/output) patterns can be relatively low. Also, we postulate that a complete set of (input/output) examples, extracted from experimental trajectories, could be sufficient for simulating and reproducing several configurations. As a starting point, here we present this methodology in the case of one pedestrian avoiding a fixed obstacle.

25.2 The Data-Driven Model

The set of experimental trajectories Because this is a data-driven model, the experimental data are the first ingredient needed. As a case study of the proposed method, we will focus on a simple configuration, considering one pedestrian and one fixed obstacle.

We realize several experiments to obtain real trajectories. Volunteers were instructed to walk from a starting point to a final point. Some of these trajectories have an obstacle in the way.

We choose 13 experimental trajectories represented in solid lines in Fig. 25.1. The lateral swaying of normal walking was smoothed by using a low-pass Fourier filter. Another 13 trajectories were obtained by replicating and rotating extreme trajectories, which do not avoid the obstacle. The 26 trajectories will provide a set of data examples for solving the one pedestrian—one narrow obstacle problem after the following processing.

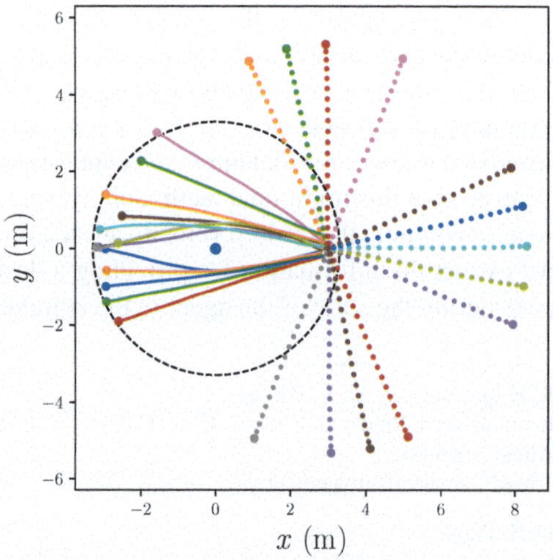

Fig. 25.1 Experimental trajectories (solid lines) and rotated trajectories (dashed lines). The points indicate the initial position of trajectories

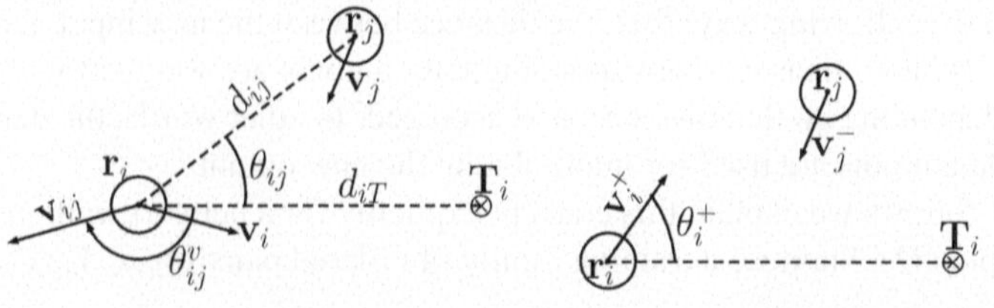

(a) (b)

Fig. 25.2 **a** Basic quantities needed for defining the input vector ($\boldsymbol{\xi}$) where $\mathbf{r}_i$, $\mathbf{v}_i$, $\mathbf{r}_j$, $\mathbf{v}_j$ are the positions and velocities of the pedestrians i and j respectively, $\mathbf{T}_i$ is the target of i and $\mathbf{v}_{ij}$ the relative velocity of j seen from i. **b** Polar coordinates of the future velocity

Input and output We postulate continuous input/state $\boldsymbol{\xi}_{ij}$ and output/reaction $\boldsymbol{\zeta}_i$ vectors given by

$$\boldsymbol{\xi}_{ij} = [\ |\hat{\mathbf{v}}|_i,\ \hat{\theta}_{ij},\ \hat{d}_{ij}\ ,\ \hat{\theta}^v_{ij},\ |\hat{\mathbf{v}}|_{ij},\ \hat{d}_{iT}] \tag{25.1}$$

$$\boldsymbol{\zeta}_i = [\mathbf{v}_i^+,\ \theta_i^+\ ] \tag{25.2}$$

In practice, it is common for NN inputs to be renormalized to values close to 1. Therefore, we define the variables with a hat as rescaled versions of those presented in Fig. 25.2a, for this purpose the two velocity variables were divided by 1.8 m/s, the two distance variables were divided by 4 m saturating its values at 2, $\hat{\theta}_{ij}$ were divided by $\pi/2$ and saturates at -1 and 1 and

$$\hat{\theta}^v_{ij} = \begin{cases} -2(\theta^v_{ij} + \pi)/\pi \text{ if } & \theta^v_{ij} < -\pi/2 \\ -1 & \text{if } -\pi/2 < \theta^v_{ij} < 0 \\ +1 & \text{if } 0 < \theta^v_{ij} \leq \pi/2 \\ -2(\theta^v_{ij} - \pi)/\pi \text{ if } & \theta^v_{ij} > \pi/2 \end{cases}.$$

The theta variables in Fig. 25.2 are angles using as origin the direction to the pedestrian target.

The output $[\mathbf{v}_i^+,\ \theta_i^+\ ]$ are the velocity in the next time step in polar coordinates as show Fig. 25.2b.

The nonparametric neural network We call $\mathbb{E} = \{\boldsymbol{\xi}(t), \boldsymbol{\zeta}(t)\}$ the experimental set of state/action examples having data points for each time step t.

Each one of the two components of the output vector $\boldsymbol{\zeta}(t)$ (Eq. 25.2) will we approximated by one neural network with output ${}^{\mu}O\ :\ \mathbb{R}^6 \rightarrow \mathbb{R}$, where $\mu = 1, 2$ indicates its polar components, i.e., the speed v_i^+ and the angle θ_i^+ respectively.

The neural network we choose is the generalized regression neural network (GRNN) [4], which is a type of radial basis function network [5].

The GRNN is a universal interpolator based on nonparametric regression. The basic idea is that when trying to predict the output for a new input, the data examples are used in the following way: first, the distance between the new input and the data input is calculated, then the corresponding data outputs are weighed with a kernel function, depending on that distance, and averaged. In other words, the data outputs of closer data inputs are used for interpolating the new output.

In what follows we explain this concept explicitly for a network with one dimensional output (O). Suppose a training family of ordered pairs $\{\boldsymbol{\xi}_n, \zeta_n\}_{n \leq N}$, then:

$$O(\boldsymbol{\xi}) = \frac{\sum_{n=1}^{N} \zeta_n K(\boldsymbol{\xi}, \boldsymbol{\xi}_n)}{\sum_{n=1}^{N} K(\boldsymbol{\xi}, \boldsymbol{\xi}_n)} \tag{25.3}$$

where

- $O(\boldsymbol{\xi})$ is the prediction value of an arbitrary input vector $\boldsymbol{\xi}$.
- ζ_n is the output of example n corresponding to the input vector $\boldsymbol{\xi}_n$.
- $K(\boldsymbol{\xi}, \boldsymbol{\xi}_n) = e^{-l_n/2\sigma^2}$ is the radial basis function kernel that weighs the contribution of the n-th output example in order to predict the new output.

Where $l_n = (\boldsymbol{\xi} - \boldsymbol{\xi}_n)^T(\boldsymbol{\xi} - \boldsymbol{\xi}_n)$ is the square distance between data examples $\boldsymbol{\xi}_n$ and the input vector $\boldsymbol{\xi}$.

Once we have a proper set of N patterns, the only degree of freedom in this neural network is the so-called spread σ, which can be taken as a scalar value for all examples and variables of the input vector.

25.3 Simulations

In this section we describe how the spread σ of both GRNN's was calibrated and we present results showing that with the proposed approach we can simulate a configuration of a pedestrian avoiding an obstacle.

Simulation scheme At each time step of the simulation, the input state of the simulated particle is calculated and both GRNN's will provide the speed v^+ and angle θ^+ as the outputs corresponding to the input state, then using $\mathbf{v}_i^s = [v^+ cos(\theta^+), v^+ sin(\theta^+)]$ as the desired velocity we can update the position $\mathbf{r}(t)$ of the particle at time t as follow:

$$\mathbf{r}^s(t + \Delta t) = \mathbf{r}^s(t) + \mathbf{v}^s(t)\Delta t,$$

and this process continues until the simulated particle reach the final target.

As stated in the previous section, there is only one free parameter for each GRNN: the spread (σ). In the next section we specify how this parameter was determined.

Calibrating the GRNN For the determination of the spread, we consider the 13 experimental trajectories $\{\mathbf{r}_1, \mathbf{r}_2, \ldots, \mathbf{r}_{13}\}$ and proceed with a leave-one-out cross-validation. We consider the spread of both GRNN equal and define error functions

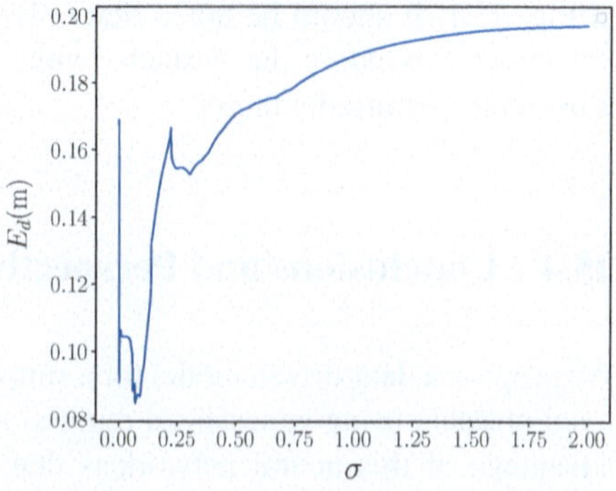

Fig. 25.3 Measures of the error between the simulated and the experimental trajectories as a function of the GRNN parameter σ

between simulated and experimental trajectories based on the minimum distance to the obstacle, $E_d(\sigma)$.

Figure 25.3 shows the results. The optimum spreads found were: $\sigma = 0.074$ with an error $E_d = 0.08$ m.

Results Using the σ obtained in the calibration and the set of 26 trajectories as data we analyze the performance simulating several particles starting around the goal, some of them having to avoid the obstacle.

The system to be simulated consists of a fixed obstacle located and a final target for all particles. Forty-eight new particles were simulated once at a time, with initial positions at 6 m from target as shown in Fig. 25.4a.

In Fig. 25.4b the smoothness and continuity of the trajectories with respect to the initial positions can be seen except for one trajectory that slightly crosses over other neighbors' trajectories. This crossing is also observed in the experiments as is shown

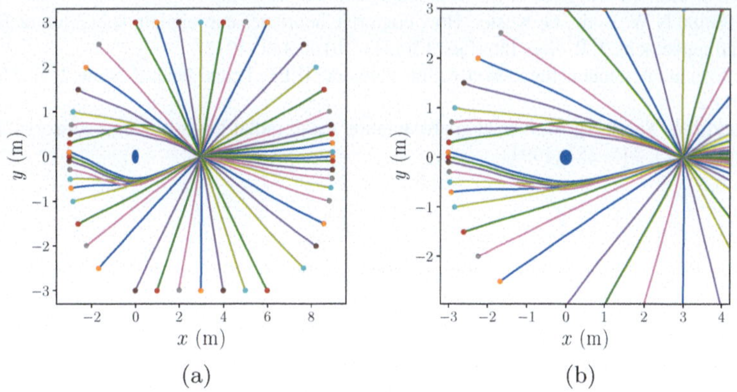

Fig. 25.4 Simulated trajectories with $\sigma = 0.074$. **a** Complete view of the simulated system. **b** Zoom over the avoidance region

in Fig. 25.1. It should be noted that only potentially colliding trajectories produce a detour for avoiding the obstacle, while the rest of the particles describe straight trajectories toward the target.

25.4 Conclusions and Perspective

We propose a data-driven model for a simple configuration of one pedestrian and one fixed obstacle using generalized regression neural network (GRNN). The principal advantage of this neural network is that it only has one free parameter that was calibrated using a leave-one-out cross-validation.

In addition, the method presented allows us to reproduce several configurations with one pedestrian walking freely or avoiding a narrow medium distance obstacle. The method is invariant under rotations.

The presented data-driven method could be extended to simulate more complex configurations considering two or more pedestrians.

Acknowledgments The authors acknowledge financial support via project PID2015-003 (Agencia Nacional de Promoción Científica y Tecnológica, Argentina) and from ITBACyT-2018-42 (Instituto Tecnológico de Buenos Aires).

References

1. R.F. Martin, D.R. Parisi, Pedestrian collision avoidance with a local dynamic goal, in *Proceedings of the PED2018 Conference* (in progress) (2018)
2. M. Moussaïd, D. Helbing, G. Theraulaz, How simple rules determine pedestrian behavior and crowd disasters. Proc. Natl. Acad. Sci. **108**(17), 6884–6888 (2011)
3. M.J. Seitz, N.W. Bode, G. Köster, How cognitive heuristics can explain social interactions in spatial movement. J. R. Soc. Interface **13**(121), 20160,439 (2016)
4. D.F. Specht, A general regression neural network. IEEE Trans. Neural Netw **2**(6), 568–576 (1991)
5. J. Park, I.W. Sandberg, Universal approximation using radial-basis-function networks. Neural Comput. **3**(2), 246–257 (1991)

Chapter 26
Entropy, Field Theory and Pedestrian Dynamics: Prediction and Forensics

José Méndez Omaña

Abstract The aim of this paper is to develop a path or trajectory designing tool for pedestrian dynamics, as a function of time, based on Lewin's concept of psychological field and applying Shannon's concept of entropy as a measure of uncertainty. The geometrical objects we use are Bézier cubics. From the point of view of cybernetics, developing a model of human behaviour should include a discussion of the developing process itself. We present a short discussion based on ideas of von Förster. The use of time variables suggests the analysis in two directions: prediction and forensics. A case study will illustrate the method.

26.1 Introduction

The choice of a path is a basic question in the context of modelling pedestrian dynamics. For instance: applications of CA-Models (see [8]), say developments of the Nagel-Schreckenberger-Algorithm, require a velocity reduction and a subsequent path choice. Path intersections (conflicts) are also important (see [16]). Helbing's Social Force Model of Pedestrian Dynamics [6] presents a discussion of the path choice depending on attraction and repulsion forces. The Gradient Navigation Model for Pedestrian Dynamics [2] uses a navigational vector to describe the individual direction. Out from the system of equations which determine the individual position it is possible to determine paths. Models using some kind of tesellation, say of Voronoi type (see [7]), allow as well path designing. Some other models focusing pedestrian flow [19] present an implicit discussion on the topic. The macroscopic model of Predtechenskii and Milinski [14] focuses on laminar flows and the item path choice does not play an important role. A path choice analysis in the context of pedestrian dynamics includes, necessarily, a psychological part and uncertainty. The aforementioned social force model and a model which considers social reactions presented by von Sievers et al. [18] are examples of approaches with explicit discus-

J. Méndez Omaña (✉)
Beuth-Hochschule für Technik Berlin, Berlin, Germany
e-mail: mendez@beuth-hochschule.de

I. Zuriguel et al. (eds.), *Traffic and Granular Flow 2019*,
Springer Proceedings in Physics 252,
https://doi.org/10.1007/978-3-030-55973-1_26

sions of some psychological aspects of modelling human behaviour. In all cases, the probabilistic aspects of the theory lead to non-deterministic algorithms and a practical way to cope this difficulty is to set a priori conditions. Nevertheless, uncertainty plays a role always. The question is: Are we able to quantify uncertainty? In practice, it is often necessary to find a time estimate for a person or a group of persons walking on some path using a simple but not too elementary setting. The aim of this paper is to develop a path or trajectory designing tool for pedestrian dynamics as a function of time. Besides introduction and conclusions, the paper additionally consists of four sections: Sect. 26.2 presents the geometrical objects: Bézier cubics. The first topic of the next section is Lewin's Field Theory (see [9, 10, 13]). Lewin's Theory is the same underlying Helbing's Social Force Model of Pedestrian Dynamics. The second topic is cybernetics. Shannon's concept of entropy, as introduced in his mathematical theory of communication [17] is the topic of Sect. 26.4. The purpouse of introducing this concept is to offer an answer to the question about quantifying uncertainty. At this place seems to be proper to remark: the concept of entropy we are using is not the same as the thermodynamical one, nevertheless, there are obvious analogies. Last but not least, Sect. 26.5 includes excerpts of a case study. The conclusions focus on the limits of the path developing tool in general and particularly concerning two questions: if we have a model at the time t_0, can we tell anything for $t < t_0$ and for $t > t_0$? The first question concerns forensics, the second one prediction.

26.2 Geometry

There are several mathematical ways to represent a path for pedestrian dynamics, depending on the model, the implementation or resources. The ideas we will present work either applying the direction field of an ordinary first order differential equation and initial/boundary values or if we set up polygonal paths. A disadvantage of the direction field is the lack of flexibility if changes in direction and orientation are required. In the case of a polygonal paths based set up the precision increases with the number of vertices. In this paper we use Bézier cubics because, although they are not as simple as polygonal settings, they allow the construction of suited paths or path segments using only four points for each segment and only one parameter domain, the compact set [0, 1]. Bézier cubics on the plane are oriented curves having a parameter representation $\mathbf{c}(t) = (x(t), y(t)),\ t \in [0, 1]$ with

$$x(t) = (1-t)^3 x_1 + 3t(1-t)^2 x_2 + 3t^2(1-t)x_3 + t^3 x_4,$$
$$y(t) = (1-t)^3 y_1 + 3t(1-t)^2 y_2 + 3t^2(1-t)y_3 + t^3 y_4,$$

where $P_i(x_i, y_i),\ i = 1, \ldots, 4$ are four points, usually called control points, determining the curve. The initial point is $P_1(x_1, y_1)$ and the boundary one is $P_4(x_4, y_4)$. The other two determine the shape (see: [1]). The expressions for $x(t)$ and for $y(t)$ are quite similar. Bézier cubics are non-singular curves on the restricted, open parameter domain $]0, 1[$. The tangent vector and the arc length element are found on the usual way and we are able to calculate the distance between the initial point and the boundary one on this path. Due to this fact, combined with a suited pedestrian velocity, we may associate a time to each path. We implemented the curve definition, the tangent vector an the arc length element in the open source software for numerical computation Scilab 6.0.1 [15] and calculated the arc length by numerical integration. A pseudorandomization of all of the control points, within some constrains, might also be implemented.

26.3 Field Theory and Cybernetics

Lewin's Field Theory is an attempt to describe, on a formal way, the behaviour of persons or groups of persons as a function of the situation of this person or group of persons at a given time (see: [11]). A sketch of this method, which we call diagnostic test has been presented in [13] and shall be specified in the example. Setting the right values of the variables is a difficult task and both experience and interdisciplinary work are needed. Moreover, the psychological field has a dynamic nature. The hodological space represents the psychological space of one person or of one group. In Fig. 26.1. We show a hodological space of a group of persons. The subset P represents a group of persons and includes individual fields information. The subset G represents the goal. Each one of this subsets is included in one region and the number of regions to be crossed from the one, where P is included, to the one including G is the distance. Regions in between represent a kind of psychological barriers. Behaviour is represented by a movement in the hodological space. The individual hodological space is similar, although, it includes a boundary region called psycho-motoric region, connecting the psychological and the physical world. Corresponding to this region we have data concerning temporal or definitive particular characteristics when interacting with the physical world, say wheel chair is needed, or the effect of substances having influence on the perception or motoric. About the cybernetic of observing systems: Lewin was aware of the difficulty involved in achieving knowledge out from empirics due to the effect of the psychological world of the observer. Because of our approach, particularly the use of entropy as well as the lack of space in this paper, we just point to the opinion of von Förster:

'I submit that the cybernetics of observed systems we may consider to be first order cybernetics; while second-order cybernetics is the cybernetics of observing systems'

'From this it appears to be clear that social cybernetics must be a second order cybernetics—a cybernetics of cybernetics in order that the observer who enters the system shall be allowed to stipulate his own purpose: he is autonomous' [4].

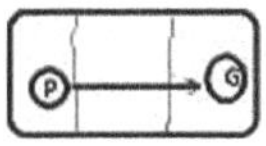

Fig. 26.1 A hodological space of a group of persons

Table 26.1 Entropy of the discrete random variable Y

$P(Y = 0)$	$P(Y = 1)$	$P(Y = 2)$	H(Y)	Measure of uncertainty (%)
0.16	0.16	0.68	0.7725	77.25
0.04	0.01	0.95	0.2035	20.35
0.01	0.01	0.98	0.1019	10.19

26.4 Entropy

C. E. Shannon stated that for a given set of probabilities $\{p_1, \ldots, p_N\}$ quantities of the form $H = -K \sum_{i=1}^{N} p_i \log_b p_i$ play a central role in information theory as measures of information, choice and uncertainty [17]. K is a real constant corresponding to a choice of the logarithmic base b in connection to the choice of a unit for the measures of information. In this paper the logarithmic base is 2 and $K = \frac{1}{log_2(N)}$ leading to $H \in [0, 1]$. The maximal value of H corresponds to the uniform distribution. In practice we are working with random variables and for a random variable X we have $P(X = x_i) = p_i$. $H(X)$ denotes the entropy of the random variable X. If we are given two random variables X, Y the entropy of the joint probabilities is defined as $H(X, Y) = -K \sum_{i,j} p_{i,j} \log p_{i,j}$. In case of two consecutive paths with measures of uncertainty H_1 and H_2 we consider the measure of uncertainty $H_{1,2} = max(H_1, H_2)$. In the Table 26.1. we illustrate the idea. The first line shows a kind of normal distribution.

26.5 Example

The example is part of the modelling of an urban beach. About the topology: The area of the open air ground surface amounts about 1032 m^2. The surface shows an irregular height: the minimum amounts 35.02 m and the maximum 35.56 m above sea level. The mean is 35.38 m. Because of some installations, trees, plants and furnitures the areas of the open surfaces and pedestrian ways amount about one third of the mentioned area. The floor on which pedestrians walk consists of sand or wooden surfaces. In spite of precise measurements, modelling is difficult because of the irregular shapes and materials of the obstacles to be found in reality.

The path design: We design three paths with a common initial point $P_1(13, 23)$ and a common boundary point $P_4(0, 3.5)$. More than one path are required because of the presence of obstacles Fig. 26.2. Shows the oriented Bézier cubics with their corresponding control points and initial tangent vectors. P_1, P_4 and $P_2(12, 20)$, $P_3(1, 4)$ are the control points of the first curve $\mathbf{c_1}(t)$ from the bottom. The common points P_1, P_4 and $P_2(10, 22)$, $P_3(1, 5)$ correspond to the second one $\mathbf{c_2}(t)$. The curve $\mathbf{c_3}(t)$ on the top has the control points P_1, $P_2(3, 23)$, $P_3(0, 19)$, P_4 (Table 26.2).

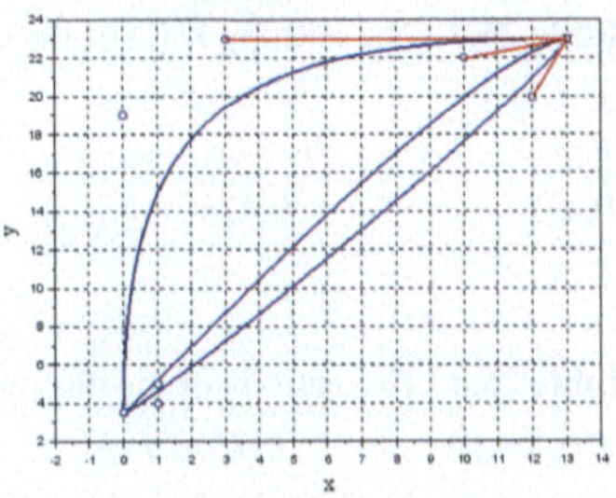

Fig. 26.2 The Bézier cubics $\mathbf{c}_1(t)$, $\mathbf{c}_2(t)$ and $\mathbf{c}_3(t)$

Table 26.2 Arc lengths and estimated pedestrians times on the three paths

Curve	Arc length [m]	Velocity [m/s]	Time [s]
$\mathbf{c}_1$	23.49	[0.4, 0.9]	[26.1, 58.73]
$\mathbf{c}_2$	23.58	[0.4, 0.9]	[26.2, 58.95]
$\mathbf{c}_3$	27.48	[0.4, 0.9]	[30.53, 68.7]

The velocity interval [0.4, 0.9] m/s (own measurements) includes velocities of two persons walking together on the sandy floor. The ground might be considered as the same for all three paths. Lastly, we point out one fact: the initial walking direction (see tangent vector) and the sight direction are not always the same. The last one might be oriented directly towards the target P_4.

The psychological field: The population size ranges between 5 and 500 visitors at a given time intervall. At six different days, each day for one hour, we obtained an impression of the behaviour of the visitors, collecting data of 473 persons merely by observation. There was a high, registered visitors fluctuation, hence, density and potential behaviour fluctuated too. The average visitors number was 79 persons/day, the minimum 20 and the maximum 157. Concerning the age we can tell: at a given time interval the average age was estimated to amount 25 years but during another time interval a range between 1 and 70 years has also been observed. The visitors are, almost always, part of a group (clustering). The groups sizes range between 2 and 12 persons. The scenario we are working on is false alarm. We assume the management will react immediatly keeping all persons informed. A panic scenario would require more variables and would be too extensive to be presented in this paper. We set up the field of two persons, who might walk on the paths, sketching the diagnostic test: 1. The time interval for the field definition is a semiopen interval because the end is not known. 2. We choose random variables: potential angriness A (discrete variable, scale: 0–2, 0 no potential angriness) and C calm (discrete variable, scale: 0–2, 0 lack of calm). 3. The updating and feedback question: What is possible? A change in the values of the variables due to changes in behaviour, environment or errors in our estimates. 4. The sketch of the hodological space of the group is like the one presented above in Fig. 26.1, hence, the distance amounts one. The region to be crossed includes events that may produce angriness or have an influence on the variable C. We should continuously update the variables values.

Table 26.3 The entropy $H(i)$ of the choice of path $\mathbf{c_i}$

$P(i=1)$	$P(i=2)$	$P(i=3)$	H(i)	Measure of uncertainty
0.475	0.475	0.05	0.7801	78.01%

Table 26.4 The entropy of the discrete random variable calm C

$P(C=0)$	$P(C=1)$	$P(C=2)$	H(C)	Measure of uncertainty
0.05	0.15	0.80	0.5579	55.79%

Concerning the social cybernetics, as we are the observer who enters the system, we stipulate our own purpose: to obtain a plausible simulation. We estimated $P(A = 0) = 1$, no potential angriness almost sure, therefore $H(A) = 0$. Tables 26.3 and 26.4 exhibit other entropy values.
We took also a management feedback into account. With some experience we are able to estimate (predict or guess in the past) the position of pedestrians on a known path, if we take account of the uncertainty.

26.6 Conclusions

The mathematical part of path designing is not complicated. We may attempt to predict behaviour if we assume that a person or a society might be seen as a structure of social forces. We share the scepticism of Wiener [20] concerning the appliability of mathematical methods in social sciences, nevertheless we think, as the socio-psychoanalist E. Fromm: 'The knowledge of these forces permits the understanding of the past and to some extent the prediction of the future—prediction not in the sense of events which will necesarilly ocurr, but rather of limited alternatives between which man has to choose' [5]. Path designing might be useful if fast, low cost estimates are required. Applying entropy supplies a measure of uncertainty.

References

1. D. Cox, J. Little, D. O'Shea, *Ideals, Varieties and Algorithms* (Springer, New York, 1992), pp. 20–22
2. F. Dietrich, G. Köster, Gradient navigation model for pedestrian dynamics. Phys. Rev. E **89**, 062801 (2014)
3. H. von Förster, On Self-Organizing Systems and their Environments, *Understanding Understanding*, Essays on Cybernetics and Cognition (Springer, New York, 2003), pp. 1–19

4. H. von Förster, Cybernetics of Cybernetics, *Understanding Understanding*, Essays on Cybernetics and Cognition (Springer, New York, 2003), pp. 283–286
5. E. Fromm, *Beyond the Chains of Illusion* (Continuum, New York, 2009), p. 17
6. D. Helbing, P. Molnar, Social force model for pedestrian dynamics. Phys. Rev. E **51**(5), 4282–4286 (1995)
7. W. Kemloh, U. Armel, M. Chraibi, J. Zhang, G. Lämmel, JuPedSim an open framework for simulating and analyzing the dynamics of pedestrians, in *Proceedings of the 3. Conference of Transportation Research Group of India, Kolkata* (2015)
8. H. Klüepfel, M. Schreckenberger, T. Meyer-König, Models for crowd movement and egress simulation, in *Proceedings of Traffic and Granular Flow '03* (Springer, Heidelberg, 2005), pp. 357–372
9. K. Lewin, *Principles of Topological Psychology* (Mac Graw-Hill, New York, 1936)
10. K. Lewin, *Field Theory in Social Sciences* (Harper & Brothers, New York, 1951)
11. K. Lewin, Defining the field at a given time, in *Symposium on Psychology and Scientific Method Proceedings* (University of Chicago, Chicago, 1941), pp. 292–310
12. J. Méndez Omaña, Field theory and quantitative models for pedestrian dynamics. Transp. Res. Procedia **2**, 442–445 (2014)
13. J. Méndez Omaña, Field theory in practice, in *Proceedings of the 9th International Conference on Pedestrian and Evacuation Dynamics (PED2018)*, Paper No. 111, Lund, Sweden (2018)
14. V.M. Predtechenskii, A.I. Milinski, *Planning for Foot Traffic Flow in Buildings* (Amerind, New Delhi, 1978)
15. Scilab. http://www.scilab.org
16. A. Schadschneider, A. Seyfried, Validation of CA models for pedestrian dynamics with fundamental diagrams. Cybern. Syst. Int. J. **40**(5), 367–389
17. C.E. Shannon, A mathematical theory of communication. Bell Syst. Tech. J. **27**, 379–423 and 623–656 (1948)
18. I. von Sievers, A. Templeton, F. Künzer, G. Köster, J. Drury, A. Philippides, T. Neckel, Bungartz, H: modelling social identification and helping in evacuation simulation. Saf. Sci. **89**, 288–300 (2016)
19. F.L.M. van Wageningen-Kessels, S.P. Hoogedorn, W. Daamen, Extension of Edie's definitions for pedestrian dynamics. Transp. Res. Procedia **2**, 507–512 (2014)
20. N. Wiener, *Cybernetics or Control and Comunication in the Animal and the Machine* (MIT Press, Cambridge, MA, 1965), p. 25

Chapter 27
The Impact of Social Groups on Collective Decision-Making in Evacuations: A Simulation Study

Hisashi Murakami, Claudio Feliciani, Kenichiro Shimura, and Katsuhiro Nishinari

Abstract The study of pedestrian dynamics has attracted the attention in various research fields because of its potential impact on the design of infrastructures and crowd safety. Recent studies have emphasised that the presence of social groups such as families and friends would impact on pedestrian dynamics. However, how social groups affect collective decision-making on the move is largely unknown. Here, we construct a simple simulation model of collective decision-making in evacuation situations, and show that the presence of social groups can result in a robust collective sensing in response to uncertain exit information.

27.1 Introduction

Human crowds show a wide variety of emergent collective behavior [1–3]. Most previous models inspired from "physical particles" where all individuals interact with neighbors in the same way, have provided better understanding of crowd collective motion [4]. Recent studies however emphasized that the presence of social groups such as families and friends have an impact on crowd dynamics [5–8]. In short, homogeneous crowd composition does not always explain the human crowd dynamics. Actually, social groups can be either an advantage or disadvantage in evacuation situations. Most studies stress that it is basically a disadvantage. For example, social groups walking in public spaces generally form a specific structure [5]. They can form a V or U shape like pattern to communicate with each other in their group, and this reduces the flow or the movement efficiency, because of its "non-aerodynamic" shape.

H. Murakami (✉) · C. Feliciani · K. Shimura · K. Nishinari
Research Center for Advanced Science and Technology, The University of Tokyo, 4-6-1 Komaba, Meguro-ku, Tokyo 153-8904, Japan
e-mail: hssh415@gmail.com

K. Nishinari
Department of Aeronautics and Astronautics, School of Engineering, The University of Tokyo, 7-3-1Bunkyo-ku, Hongo 113-8656, Japan

I. Zuriguel et al. (eds.), *Traffic and Granular Flow 2019*, Springer Proceedings in Physics 252,
https://doi.org/10.1007/978-3-030-55973-1_27

In this way, a social group affects crowd dynamics due to its physical structure. But a social group is also affected by its decision making process [6, 8]. The few experiments that have shown that decision making within a group can be advantageous in term of accuracy of decision making. Under certain circumstances, pedestrians need to get information, which would be incomplete and ambiguous (e.g., loudspeaker announcement) [9]. At that time, integrating information from each member, social group can make better decision, leading to effective evacuations.

However, the reasons are still largely unknown especially when they are on the move. We here focus on collective decision making on the move, which is usually studied in animal collective behavior [10]. For example, suppose that there are animal groups such as pigeons moving toward some goals such as foods or migration routes, and that there are two kinds of informed individuals that hold conflicting preferred directions of motion. Then, it is known that the individuals basically tend to maintain as a single group, taking either preferred direction; otherwise, they can take a compromising direction, which depends on the difference between these two preferred directions or number of informed individuals. In this study, we would like to consider how social groups affect collective decision-making on the move, by using a simple simulation model in evacuation situations.

27.2 Methods

We considered a simple evacuation scenario where there are two routes towards the final goal: correct (shorter) path and wrong (longer) path (Fig. 27.1a). Pedestrians in the starting area initially cannot know which route is correct. They may try to get information to escape as soon as possible, but such information would not be always correct, especially in the case of emergency. We assumed that people are given information about which route they should use: each individual is given information of the correct path with probability P (the one relative to the wrong path is $1 - P$). We will call this probability P as "correctness of information". Therefore, there are two kinds of informed individuals that hold conflicting preferred directions of motion. Note that we focused on the behaviour of individuals around the starting area because the issue that matters here is from which path (direction) they leave that area (Fig. 27.1b).

We propose an extended Floor Field cellular automaton (CA) model with individuals selecting one of the Von Neumann neighborhoods composed of its four adjacent sites and its current site. This model is incorporated with social group behavior, which is based on the leader–follower behavior rule that is evident in group dynamics, similar to previous models [11]. In short, individuals intend to move through the shortest path to egress (with their preferred directions), and followers in a group intend to stay close to the leader.

In our model, at each time-step, individuals change the preferred directions by two rules. (1) peer pressure: each individual determines its preferred direction by the majority of directions of its neighbours in the neighbourhood (Fig. 27.1c). Note that we employed the Moore neighbourhood only for this rule, instead of the von

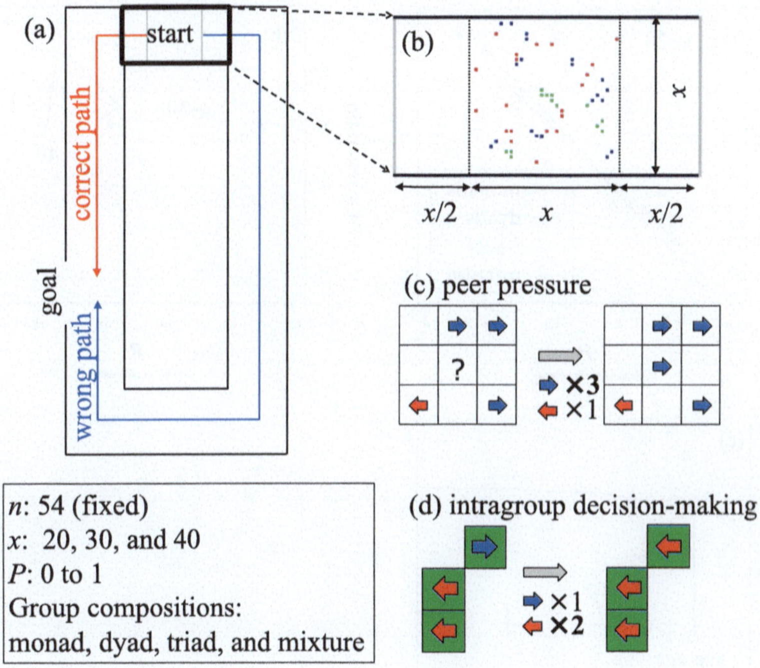

Fig. 27.1 **a**, **b** Considered scenarios. **c**, **d** Model descriptions.

Neumann neighbourhood. (2) intragroup decision-making: in each group, group members unify their preferred directions by the majority of directions of the group (Fig. 27.1d). With the above setting, we calculated the average success ratio defined by the proportion of individuals ($n = 54$) getting out from the starting area with correct path, changing size of the start area ($x = 20$, 30, and 40 cells; x is the side length of the square of the starting area as in Fig. 27.1b), P (from 0 to 1), and group composition (a group of monads (i.e., individuals), a group of dyads (i.e., pairs), a group of triads (i.e., trios), and a mixture of them), with 100 trials.

27.3 Results

As a result, we found that although the success ratio of singles linearly increased as P became larger, that of groups (particularly, triads and mixture) showed sigmoid-like curves against P. This result implies that when there are social groups, the sensitivity to "the probability of giving good correct information" rises compared to when there is no group (Fig. 27.2a).

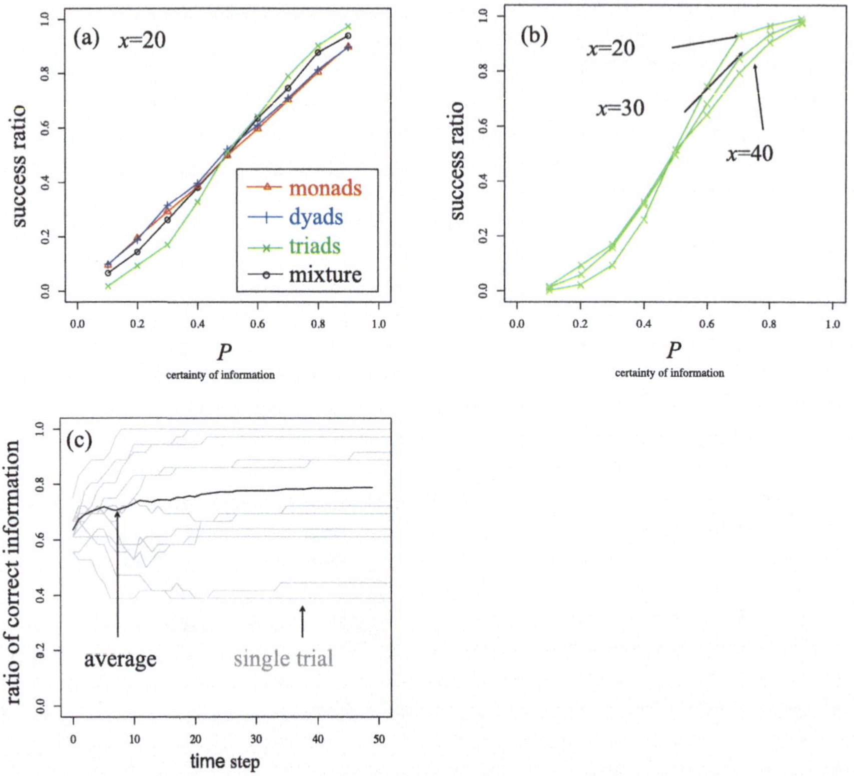

Fig. 27.2 **a** Success ratios plotted against P with different crowd compositions. **b** Success ratios plotted against P with different sizes of the starting area. **c** Time development of the ration of correct information in the triad composition with $x = 20$

When $P > 0.5$, more individuals selected correct path than initially expected, suggesting a positive effect of the presence of social group. On the other hand, when $P < 0.5$, the reverse was true, suggesting a negative effect. These effects became stronger when the density increased (i.e., x decreased) (Fig. 27.2b).

Figure 27.2c presents the time development of the ratio of individuals with correct information, showing that information individuals have can be dynamically changed through their movement dynamics, while the mean of them gradually increases. This means that individuals continued to make decision during movements rather than they kept moving with the decision made at initial state.

27.4 Discussion

Recent study emphasized that the evacuation dynamics is divided into roughly two kind of stages: pre-movement decision making stage and movement stage [e.g., 6]. Pre-movement decision making stage occurs in responding alarm to start evacuation. As noted before, social groups can be advantageous in this stage in the term of accuracy of the decision compared to isolated individuals [9]. In this way, such decision making stage is usually clear-cut from the afterward movement stage. However, this kind of "Decision-then-movement" framework would not be always the case; decision might continue to make during the movement stage especially under uncertain and ambiguous situation. The effect of social groups under "decision-during-movement" framework has been largely unexplored. We here conducted simple simulation under such framework. In this study, we showed that if there are social groups, they can effect both positively and negatively even in decision making process on the move during evacuations. Particularly, we found that individuals continued making decision even after the decision made at initial state. These dynamical process has been ignored through the Decision-then-movement framework. Moreover, we found that if there are social groups, the sensitivity to "the probability of giving good correct information" rises compared to when there is only isolated individuals. Further, we observed that the stronger these effects became the smaller the density is, probably because of higher frequency of individuals interaction with respect to preferred direction (i.e., peer pressure rule). Finally we note that this kind of rise of sensitivity for uncertain information, which occurs not only within-groups but also between-groups, would be similar to the so-called emergent sensing found in animal collective behavior [12].

Overall, our findings show that social roles may contribute in creating unexpected dynamics when incorrect or misleading information are provided and therefore suggest that information provision is an important aspect in crowd management.

Acknowledgements This work was supported by JST-Mirai Program Grant Number JPMJMI17D4, Japan.

References

1. D. Helbing, A. Johansson, Pedestrian, crowd, and evacuation dynamics, in *Encyclopedia of complexity and systems science*. ed. by R.A. Meyers (Springer, New York, NY, 2009), pp. 6476–6495
2. I.D. Couzin, J. Krause, Self-organization and collective behavior in vertebrates. Adv Stud Behav. **32**, 1–75 (2003)
3. H. Murakami, C. Feliciani, K. Nishinari, Lévy walk process in self-organization of pedestrian crowds. J. R. Soc. Interface **16**, 20180939 (2019). https://doi.org/10.1098/rsif.2018.0939
4. D. Helbing, P. Molnar, Social force model for pedestrian dynamics. Phys. Rev. E **51**, 4282–4286 (1995)

5. M. Moussaïd, N. Perozo, S. Garnier, D. Helbing, G. Theraulaz, The walking behaviour of pedestrian social groups and its impact on crowd dynamics. PLoS ONE **5**(4), e10047 (2010). https://doi.org/10.1371/journal.pone.0010047
6. N.W.F. Bode, S. Holl, W. Mehner, A. Seyfried, Disentangling the impact of social groups on response times and movement dynamics in evacuations. PLoS ONE **10**(3), e0121227 (2015). https://doi.org/10.1371/journal.pone.0121227
7. C. Von Krüchten, A. Schadschneider, Empirical study on social groups in pedestrian evacuation dynamics. Phys. A **475**, 129–141 (2017)
8. M. Haghani, M. Sarvi, Simulating dynamics of adaptive exit-choice changing in crowd evacuations: Model implementation and behavioural interpretations. Trans. Res. Part C Emerg. Technol. **103**, 56–82 (2019)
9. R.J.G. Clément, S. Krause, N. von Engelhardt, J.J. Faria, J. Krause, R.H.J.M. Kurvers, Collective cognition in humans: groups outperform their best members in a sentence reconstruction task. PLoS ONE **8**, e77943 (2013). https://doi.org/10.1371/journal.pone.0077943. (PMID: 24147101)
10. I. Pinkoviezky, I.D. Couzin, N.S. Gov, Collective conflict resolution in groups on the move. Phys. Rev. E **97**(3), 032304 (2018)
11. L. Lu, C.-Y. Chan, J. Wang, W. Wang, A study of pedestrian group behaviors in crowd evacuation based on an extended floor field cellular automaton model. Trans. Res. Part C Emerg. Technol. **81**, 317–329 (2017)
12. A. Berdahl, C.J. Torney, C.C. Ioannou, J.J. Faria, I.D. Couzin, Emergent sensing of complex environments by mobile animal groups. Science **339**(6119), 574–576 (2013)

Chapter 28
Set-Up of a Method for People-Counting Using Images from a UAV

Daniel R. Parisi, Juan I. Giribet, Claudio D. Pose, and Ignacio Mas

Abstract We present a new method for obtaining the positions of each person attending an outdoor gathering. From this information it is possible to calculate the density field and the number of people. A dual-camera (visible + infrared (IR)) is mounted on an unmanned aerial vehicle (UAV). In this work, we only use the IR channel and present an initial set-up and calibration of the system along with the characterization of a small group of people.

28.1 Introduction

Counting and monitoring outdoor gatherings, such as street demonstrations, music festivals, parades, religious processions, etc. could be a very important issue for governments, media and, of course, safety managers.

There exist several methods of image processing using visible information from cameras in arbitrary angles [1–3]. In particular, deep neural network approaches are very fulfilling on this task [4–6].

D. R. Parisi (✉)
Dto. Ing. Informática, Instituto Tecnológico de Buenos Aires (ITBA), CONICET, Lavardén 315, C. A. de Buenos Aires, Argentina
e-mail: dparisi@itba.edu.ar

J. I. Giribet · C. D. Pose
Dto. Ing. Electrónica, Facultad de Ingeniería, Universidad de Buenos Aires, Av. Paseo Colón 850, C. A. de Buenos Aires, Argentina

J. I. Giribet
Dto. de Matemática, Facultad de Ingeniería, Universidad de Buenos Aires, Av. Paseo Colón 850, C. A. de Buenos Aires, Argentina

Instituto Argentino de Matemática "Alberto Calderón" CONICET, Av. Paseo Colón 850, C. A. de Buenos Aires, Argentina

I. Mas
Dto. de Matemática, Instituto Tecnológico de Buenos Aires (ITBA), CONICET, Av. Madero 399, C. A. de Buenos Aires, Argentina

I. Zuriguel et al. (eds.), *Traffic and Granular Flow 2019*, Springer Proceedings in Physics 252,
https://doi.org/10.1007/978-3-030-55973-1_28

Our current project consists of adding also the IR information to the analysis which could disambiguate visible information and *vice versa*. However, in this first study, we have only considered the IR channel, which is less sensitive to ambient lighting conditions.

In short, our method is based on the following pillars:—Taking images of the crowd using a unmanned aerial vehicle (UAV, also known as a multirotor drone). It is a key issue that the UAV must be equipped with redundant fail-safe mechanisms to avoid it to fall over the people.—The images comprise both visible and infrared (IR) information from the surface. The IR component allows to discriminate people from non-animated objects and highly improves the correct identification of individuals.—Image processing to obtain the (x, y) positions of each pedestrian.—Post-processing.

Once the positions of pedestrians are determined, we can calculate the density field. By integrating this field over the area of interest, we can count the number of people in the area.

In this work, we present the set-up and calibration of the system composed of the UAV, cameras and other sensors. An initial characterization of a small group of people using only the IR information is also shown.

28.2 Experimental Set-Up

The aerial vehicle was designed and assembled in our laboratory with the following components.

The frame is the Turnigy H.A.L. hexarotor, a strong, lightweight aluminum frame designed for Heavy Aerial Lift, with a distance between rotors of 775 mm. The actuators installed on this frame are Multistar 3508-268 KV brushless motors, with 14 × 5.5″ carbon fiber propellers, driven by 40 A, optocoupled Hobbywing electronic speed controllers (ESC). To power the entire system, a 6S 5000 mAh 20C LiPo battery is used.

The flight controller mounted on the vehicle is the PX4 Pixhawk model, that uses several internal inertial sensors, an N8M GPS plus magnetometer module, and a wireless communication link with a ground station through a 433 MHz, 100 mW telemetry system. The total weight of the vehicle is 3 Kg, and an additional 500 g for the camera with a mounting case and a 3S 2000 mAh battery to power it, allowing for flights of around 10 min.

The camera used is a FLIR DUO PRO R, 640 having the following specifications:—Thermal Imager: Uncooled VOx Microbolometer;—Spectral Band: 7.5–13.5 μm;—Thermal Sensitivity: <50 mK;—Thermal Sensor Resolution: 640 × 512;—Thermal Lens: 19 mm (FOV: 32° × 26°);—Thermal Frame Rate: 9 Hz;—Radiometry Accuracy: ±5 °**C** or 5% of readings in the −25 °**C** to +135 °**C** range:—Visible Sensor Resolution: 4000 × 3000; Visible Camera FOV: 56° × 45°.

A picture of the UAV carrying the dual camera is shown in Fig. 28.1. The final UAV will be enhanced with redundant fault tolerance mechanisms, one of them was developed by Giribet et al. [7, 8].

Fig. 28.1 The unmanned aerial vehicle with the dual camera (visible and infrared)

28.3 Infra-Red Image Processing

Pixels versus cm calibration A first calibration procedure for the IR image consists in obtaining the function that converts pixels into cm depending on the UAV height. A rectangular 50 cm pattern was recorded from different heights and the ratio between cm and number of pixels in the image could be calculated. Results are shown in Fig. 28.2. The heights were measured with the UAV flight controller barometer.

Filtering steps We present here a six-step filtering process that allows to obtain the pedestrian positions in the (x, y) plane from a zenith view. This process only considers the IR channel information. The visible information could have greater variations due to external conditions and it is only considered as reference and illustration of the analyzed scenarios. The six steps are the followings:

Step 1: Raw Infrared Image The IR camera provides images having temperature information that can be calculated from the intensity of each pixel. The first processing is to correct the radial distortion of the lens and generate a transformed matrix with temperature information.

Step 2: Entropy Filter We can filter the thermal information in the spatial domain by using an entropy filter [9]. The idea is to consider a square neighborhood around

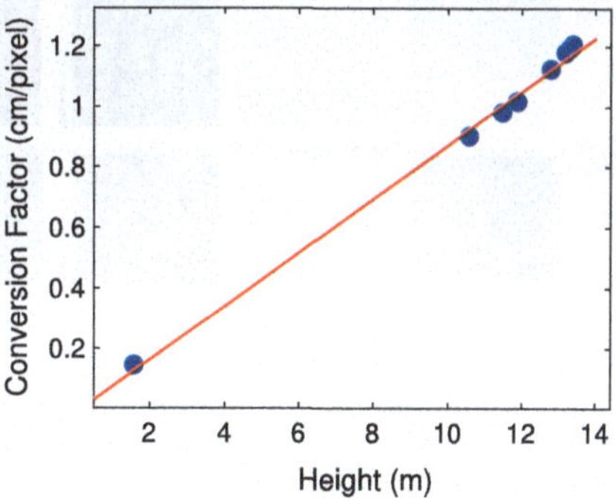

Fig. 28.2 Linear relation between the factor converting pixels to cm and the height at which the image was obtained

each pixel j, then the entropy $S_j = -\Sigma p_i log(p_i)$ is calculated. Where p_i is the probability of a given intensity (in this case, temperature) interval in each pixel i inside the neighborhood.

Step 3: Entropy and Temperature Threshold Only the pixels with values of high entropy (S_j) and human-like temperature (T_j) are kept, i.e., the ones satisfying the condition $S_j > 1 \cap T_j \in [27\,°\mathbf{C}, 38\,°\mathbf{C}]$. The rest of them are set to zero.

Step 4: Smoothing The surviving entropy pixels (S_j) are smoothed with a Gaussian kernel with a standard deviation of $\sigma = 8$, which maximizes the correct identification of people in the current method.

Step 5: Find Peaks Then, the maximum values in a square neighborhood of 40 pixels are selected. This number of pixels corresponds approximately to an area occupied by one person in the image.

Step 6: Final Positions The coordinates of peaks found are approximations of the positions of people in the plane.

28.4 Results Under Different Lighting Conditions

As a case study for the proposed method, we have analyzed a small crowd at an outdoor location (Buenos Aires, summer 2019) at different times. Images were taken at different lighting conditions changing from natural (19 hs.) to artificial illumination (20 hs.).

In Fig. 28.3 the above procedure of six steps (columns) are shown for each of the three lighting conditions (rows). The first column is the visible image displayed only as an illustration, but it is not considered for the six steps analysis. Also the second

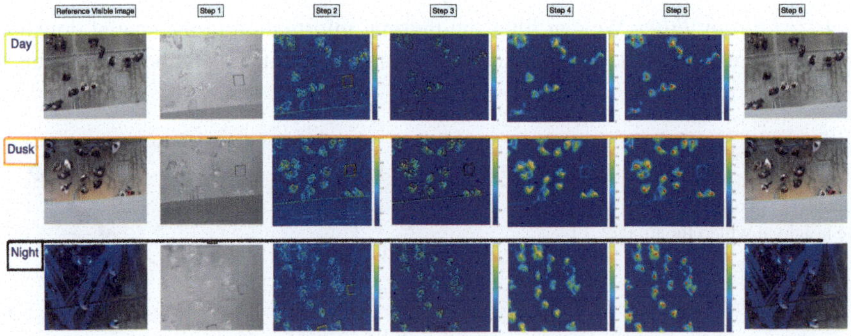

Fig. 28.3 Six steps procedure described in Sect. 28.3 for three different lighting conditions from natural to artificial. The first column is the corresponding visible information shown only as reference. The second column is one possible representation of the temperature. The red dots on image from step 5 and 6 are pedestrian positions

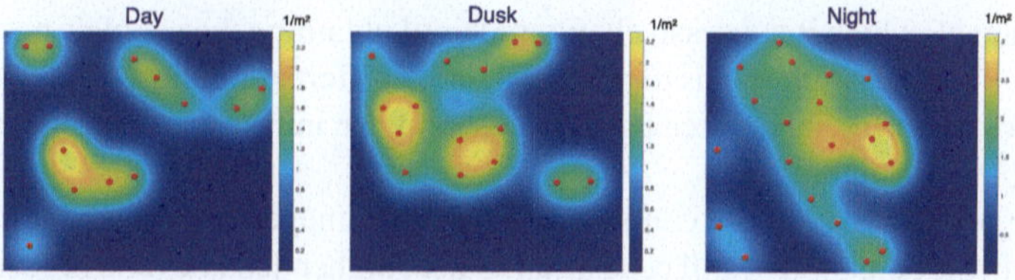

Fig. 28.4 Density fields for the three different lighting conditions indicated

column is just an illustration of the temperature values. The gray scale representation was generated by linearly transforming from maximum and minimum temperatures into the range [1]. Clearly, this representation is arbitrary and there may be other ways of doing it. We must keep in mind, that the following steps are performed over the raw temperature information at each pixel, but not over the arbitrary gray scale values of the artificial images shown in the second column of Fig. 28.3.

As can be seen, the "day" light is the natural light; "dusk" indicates weaker sunlight after the sunset mixed with few artificial light; "night" scene is completely illuminated by artificial and colored light. Precisely, this abrupt color changing of artificial lights that can be present in some gathering as a music festival, inspire us to include the IR information. The temperature information also suffers from ambient contamination, but it changes smoothly over time. In outdoor locations, if the ambient temperature is high, there is a moment of the day in which the floor could have a similar temperature that the human body and at that time our analysis is not valid. But for the rest of the day, in particular, during the night, the proposed method should work properly.

It is important to note that the same method, with the same parameters, performs well for the three different lighting conditions studied.

From the (x, y) positions obtained, the density field can be calculated. There are many ways of doing this, we choose the coarse-graining method in which the pedestrian mass is considered to be spread over the space, we take a Gaussian with its mean value matching the position of the pedestrian and spanning over the whole space, in our case, with a standard deviation $\sigma_g = 0.46$ m. For details on the method see Refs. [10–14]. Figure 28.4 displays the corresponding density fields. One property of the density fields is that integrating it over the whole area provides the number of pedestrians.

28.5 Conclusions and Next Steps

We are developing a solution for people counting and crowd characterizations, based on an fault-tolerant UAV equipped with a visible and an infrared camera.

As a first step, we analyzed a small crowd at an outdoor location at different illumination conditions from natural to artificial light. This first analysis only considers the IR channel, which is successively filtered until obtaining the people's positions in the plane. This approach seems to be robust under different lighting conditions, such as shadows and artificial colored light, representing an improvement over processing only visible information.

Future works will consider combined visible + IR images processing for increased accuracy in the detection of still pedestrians. Also, higher densities and bigger crowds will be studied.

From the people's positions, the density field can be calculated. The convolutional neural network methods mentioned in the introduction, provide as output the density field, taking as input visible images. As a goal for future work, it could be an enhancement for this deep neural network methodology to incorporate infrared information as a four-channel of an extended image (RGB+IR).

Acknowledgements The authors acknowledge financial support via project PID2015-003 (Agencia Nacional de Promoción Científica y Tecnológica, Argentina) and from ITBACyT-2018-42 (Instituto Tecnológico de Buenos Aires).

References

1. V.B. Subburaman, A. Descamps, C. Carincotte, Counting people in the crowd using a generic head detector, in *2012 IEEE Ninth International Conference on Advanced Video and Signal-Based Surveillance* (IEEE, 2012), pp. 470–475
2. A. Bansal, K. Venkatesh, People counting in high density crowds from still images. arXiv:1507.08445 (2015)
3. M. Rodriguez, I. Laptev, J. Sivic, J.Y. Audibert, Density-aware person detection and tracking in crowds, in *2011 International Conference on Computer Vision* (IEEE, 2011), pp. 2423–2430
4. L. Boominathan, S.S. Kruthiventi, R.V. Babu, Crowdnet: a deep convolutional network for dense crowd counting, in *Proceedings of the 24th ACM International Conference on Multimedia* (ACM, 2016), pp. 640–644
5. C. Wang, H. Zhang, L. Yang, S. Liu, X. Cao, Deep people counting in extremely dense crowds, in *Proceedings of the 23rd ACM International Conference on Multimedia* (ACM, 2015), pp. 1299–1302
6. Y. Zhang, D. Zhou, S. Chen, S. Gao, Y. Ma, Single-image crowd counting via multi-column convolutional neural network, in *Proceedings of the IEEE Conference on Computer Vision and Pattern Recognition* (2016), pp. 589–597
7. J.I. Giribet, R.S. Sanchez-Pena, A.S. Ghersin, Analysis and design of a tilted rotor hexacopter for fault tolerance. IEEE Trans. Aerosp. Electron. Syst. **52**(4), 1555–1567 (2016)
8. J.I. Giribet, C.D. Pose, A.S. Ghersin, I. Mas, Experimental validation of a fault tolerant hexacopter with tilted rotors. Int. J. Electr. Electron. Eng. Telecommun. **7**(2), 58–65 (2018). https://doi.org/10.18178/ijeetc.7.2.58-65
9. R.C. Gonzalez, R.E. Woods, S.L. Eddins, *Digital Image Processing using MATLAB*. Pearson Education India (2004)
10. D. Helbing, A. Johansson, H.Z. Al-Abideen, Dynamics of crowd disasters: an empirical study. Phys. Rev. E **75**(4), 046,109 (2007)
11. I. Goldhirsch, Stress, stress asymmetry and couple stress: from discrete particles to continuous fields. Granul. Matter **12**(3), 239–252 (2010)
12. J. Zhang, R.P. Behringer, I. Goldhirsch, Coarse-graining of a physical granular system. Prog. Theor. Phys. Suppl. **184**, 16–30 (2010)

13. T. Weinhart, C. Labra, S. Luding, J.Y. Ooi, Influence of coarse-graining parameters on the analysis of dem simulations of silo flow. Powder Technol. **293**, 138–148 (2016)
14. A. Garcimartín, J.M. Pastor, C. Martín-Gómez, D. Parisi, I. Zuriguel, Pedestrian collective motion in competitive room evacuation. Sci. Rep. **7**(1), 1–9 (2017)

Chapter 29
Modeling of Position Finding in Waiting Processes on Platforms

Tobias Schrödter, Mohcine Chraibi, and Armin Seyfried

Abstract The distribution of passengers waiting for a train is one of the limiting factors when improving the performance of a train station, as it heavily influences the boarding and alighting times of trains. We introduce a probability-based model for the pedestrians' choice of a waiting position. Different factors as the geometry and the positions of other waiting pedestrians are taken into account. To assess the model, simulations on a simplified representation of a platform were used. The results of this simulation show good agreement with observations of previously conducted field studies.

29.1 Introduction

Usually, pedestrian models describe the movement of pedestrians with varying levels of detail. Often complex processes like way-finding in buildings or navigation through a crowd are solved in a simplified way by collision avoidance algorithms or social forces [1, 5, 11]. Moving towards a specific goal is a substantial contribution determining the dynamics of a process, especially in evacuation scenarios. However, in more complex buildings, e.g., platforms or airport gates, the dynamics are heavily influenced by waiting people, who may restrict the space for the movement of others. After reaching their waiting position, pedestrians do not have any need to keep moving unless the event they are waiting for occurs, e.g., a train arrives or board-

T. Schrödter (✉) · M. Chraibi · A. Seyfried
Forschungszentrum Jülich, Institute of Advanced Simulation, Jülich, Germany
e-mail: t.schroedter@fz-juelich.de

M. Chraibi
e-mail: m.chraibi@fz-juelich.de

A. Seyfried
e-mail: a.seyfried@fz-juelich.de

A. Seyfried
Bergische Universität Wupperal, Wuppertal, Germany

I. Zuriguel et al. (eds.), *Traffic and Granular Flow 2019*,
Springer Proceedings in Physics 252,
https://doi.org/10.1007/978-3-030-55973-1_29

ing of the plane starts. In this regard, as opposed to "moving dynamics" resulting from pedestrians evacuating a specific place, we focus on the modelling of "waiting" where pedestrians wait for a certain amount of time by standing or slowly moving without a explicit goal.

Recently, the investigations of inflow processes gained importance in the research of pedestrian dynamics. In [8], different hypotheses of the inflow process are compared with experimental data. Ezaki et al. [3] conducted experiments on the inflow to a confined space and derived a theoretical description of the process. When investigating inflow processes, usually, a confined space with a dedicated entrance/exit is considered. These works point out that the distribution is influenced by the geometry, i.e., the platform's shape and size, and the positioning of entrances/exits. Additionally, the positions of other pedestrians affect the choice of a waiting position.

Multiple studies focusing on the dwell times of trains also report findings on the longitudinal distribution of passengers on platforms of train and metro stations [4, 7, 10, 12]. One of the observation of these studies is the clustering of passengers around the entrances and further platform infrastructure as seats, rain shelters, and vending machines. For more experienced travelers, also the position at the departing station or less crowded coaches influence their position choices. However, no further notions of the distribution of the passengers between the tracks were reported.

Waiting pedestrians were investigated by simulations of train stations in [2, 6], where the influence of standing pedestrians on the flow of passing pedestrians was discussed. In particular it was analyzed how standing pedestrians constrict the flow of passing pedestrians. As waiting positions, arbitrary positions within a designated waiting area were assigned to waiting pedestrians. An approach to model the passengers' distribution on a metro platform with a cost function approach is discussed in [13]. The introduced cost function takes different influences as the distance to a particular waiting area, density, length of the waiting area into account. Contrary to the investigated metro station in Beijing with guiding lines and specific waiting area, we intend to develop a model which is valid in a more general context.

In this paper, we develop a mathematical model to describe the position finding process for waiting pedestrians. We define waiting pedestrians, as pedestrians who enter a specific region, called waiting area (e.g., platform), until the awaited event is triggered, e.g., the arrival of the train. We focus on pedestrians who stand during the waiting process, hence they move towards their waiting position and come to a halt.

29.2 Model

The determination of the waiting position is an optimization problem in which every pedestrian tries to determine a position that is optimal for him or her, taking various factors into account. As the results highly depend on the individual's personal preferences and intentions, we propose a heuristic approach. In our model, space is discretized into small cells of $0.5 \times 0.5\,\text{m}^2$, which can either be empty or occupied by exactly one person. Each of the pedestrians gets an unoccupied cell assigned as waiting position depending on three floor fields. The two fields S and D determine the probability of a cell to be assigned as waiting position. S takes geometrical influences, as the distance to exits, walls, and door areas into account. D considers the distance to a specific position, including detours forced by other pedestrians. R is used as a filter function, to reward a certain distance to other pedestrians. N is a normalization factor such that $\sum_{i,j} P_{i,j} = 1$. The resulting probability for a cell (i, j) to be assigned is given by

$$P_{i,j} = N \cdot \left[\left(S_{i,j} + D_{i,j} \right) \cdot R_{i,j} \right]. \tag{29.1}$$

The floor fields used will be described in detail in the following subsection together with the algorithm for choosing a waiting position.

29.2.1 Static Floor Field S

The static floor field S is defined as

$$S_{i,j} = w_e \cdot E_{i,j} + w_w \cdot W_{i,j} + w_f \cdot F_{i,j}, \tag{29.2}$$

and does not change over time as it is affected by fixed walls, doors and obstacles. It is a combination of multiple probability fields E, W, F, which are scaled by the diameter of the circumcircle of the room w, hence $E_{i,j}, W_{i,j}, F_{i,j} \in [0, 1]$. To model different behavior patterns, the fields are scaled by individual weights $w_e, w_w, w_f \in \mathbb{R}^+$. The influences considered in S are the distance to the designated exits/platform edges, the distance to boundaries, e.g., walls and corners, as well as the distance to an area close to doors where more pedestrians are expected to pass.

In the first two cases, distance to exits and boundaries, passengers will try to minimize their distance to these areas, resulting in a higher probability that pedestrians will choose these areas as their waiting position. The corresponding probabilities are given by $E_{i,j} = 1 - \frac{d_{e_{i,j}}}{w}$ and $W_{i,j} = 1 - \frac{d_{b_{i,j}}}{w}$. Where d_e denotes the distance to the closest point of an exit, and d_b is given $d_b = d_w + 0.5 d_c$ where d_w and d_c denote the distance to the closest point of a wall or corner.

Contrary to the other factors, pedestrians will try to maximize their distance to the area in front of doors where they entered the room, as they will expect more passengers to follow. We modeled the area where more pedestrians are expected to pass and hence are uncomfortable to stand in as an ellipse. The ellipse's center is located in the middle of the door, the semi-minor axis is oriented along the door, whereas the semi-major axis points in the movement direction of the pedestrians. The probability of a cell to be assigned as waiting position is given by $F_{i,j} = \frac{d_{f_{i,j}}}{w}$ where d_f is the distance to the closest point of that ellipse. The resulting static floor field S as in Eq. 29.2 is shown in Fig. 29.1d.

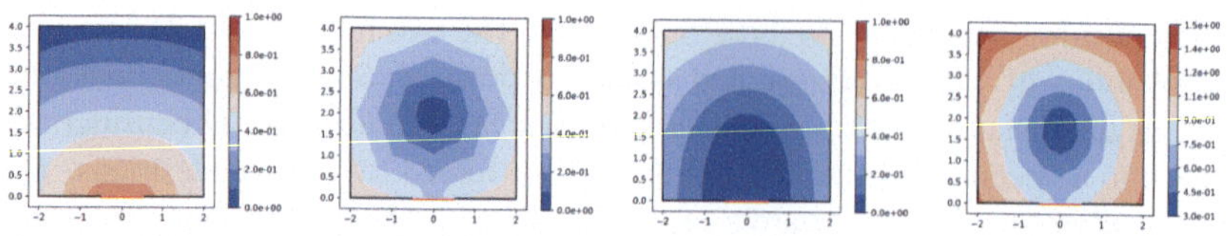

(a) E. (b) W. (c) F. (d) S.

Fig. 29.1 Probability of the static influences. A higher numbers indicates a more preferable waiting position. The door is located in the center at the buttom

29.2.2 Dynamic Floor Fields D and R

The dynamic floor fields D and R take the waiting positions of other pedestrians into account and are changed with every pedestrian entering the room. D is a probability field, and R is as a filter function, ensuring a certain distance between the waiting pedestrians.

Pedestrians tend to minimize the distance they have to walk to reach their waiting position. For the computation of the walking distance d_w, a fast marching approach [9] is used, where the pedestrians are interpreted as circular obstacles with a radius of 0.5 m. The probability for each cell is given by

$$D_{i,j} = w_d \cdot \left(1 - \frac{d_{w_{i,j}}}{w}\right), \tag{29.3}$$

where $w_d \in \mathbb{R}^+$ is the corresponding weight. For unreachable areas the distance is set to $d_w = w$. The resulting floor field is shown in Fig. 29.2a.

Additionally, the pedestrians try to maximize the distance to the closest neighbor d_p to a certain extent. This repulsion gets modeled as

$$R_{i,j} = 1 - \exp\left(-\frac{2 \cdot d_{p_{i,j}}^2}{c}\right). \tag{29.4}$$

where $c \in \mathbb{R}^+$ corresponds to the desired personal space of a pedestrian.

Figure 29.2b shows the resulting floor field. Combining all influence factors as in Eq. 29.1, yields a probability field as in Fig. 29.2c.

The algorithm for determining the waiting position of a pedestrian is given in Algorithm 1. This position can be used as a goal for navigation through crowds in models like [1, 5, 11]. When the waiting position is reached, the pedestrian stands until the awaited event is triggered.

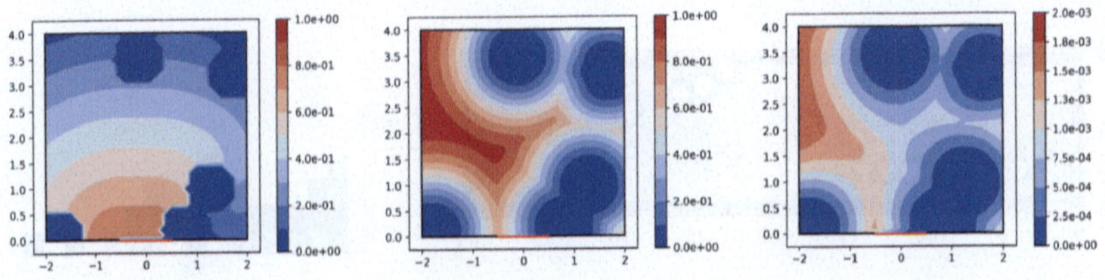

(a) D. (b) R. (c) P.

Fig. 29.2 Probability of the dynamic influences with 5 pedestrians inside the room. A higher numbers indicates a more preferable waiting position

---

Algorithm 1 Choose a waiting position.

$\forall i, \forall j$: compute $S_{i,j}$ according to Eq. 29.2, set $D_{i,j} = 0$ and $R_{i,j} = 1$;
for all pedestrian entering the room/platform **do**
    Compute P according to Eq. 29.1;
    Assign a waiting position to the entering pedestrian based on P;
    Update D and R according to Eq. 29.3 and Eq. 29.4 respectively
end for

29.3 Results

We used the model from Sect. 29.2 to simulate the selection of waiting positions on a simplified platform, as depicted in Fig. 29.3. The platform is 40 m long and 10 m wide with an entrance of 3 m located at the center of the left hand side. Tracks run along the upper and lower edges of the platform. To avoid pedestrian choosing waiting position in dangerous areas close to the tracks, a 0.5 m wide corridor along the top and bottom edges is not used for the computation of the waiting positions. As the distribution tends to be more uniform with higher densities, we restrict our simulation to a maximum of 75 passengers on the platform.

In field observations [4, 7, 10, 12] on different train stations, the positions of entrances were emphasized as the leading factor to the pedestrian longitudinal distri-

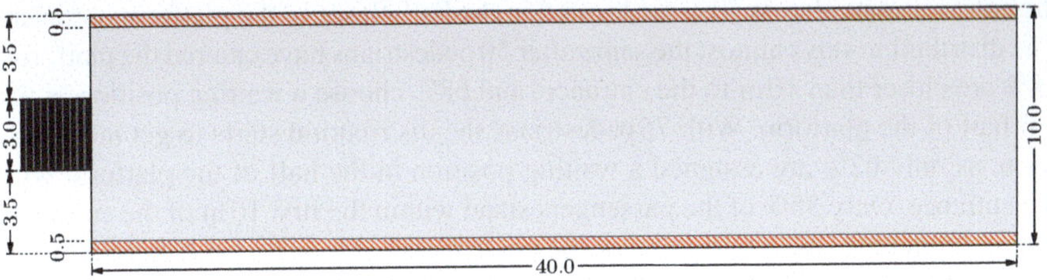

Fig. 29.3 Geometry of the simplified platform. Entrance is located on the left hand side. Tracks are located at the top and bottom. Red areas mark prohibited waiting areas like danger zones close to tracks

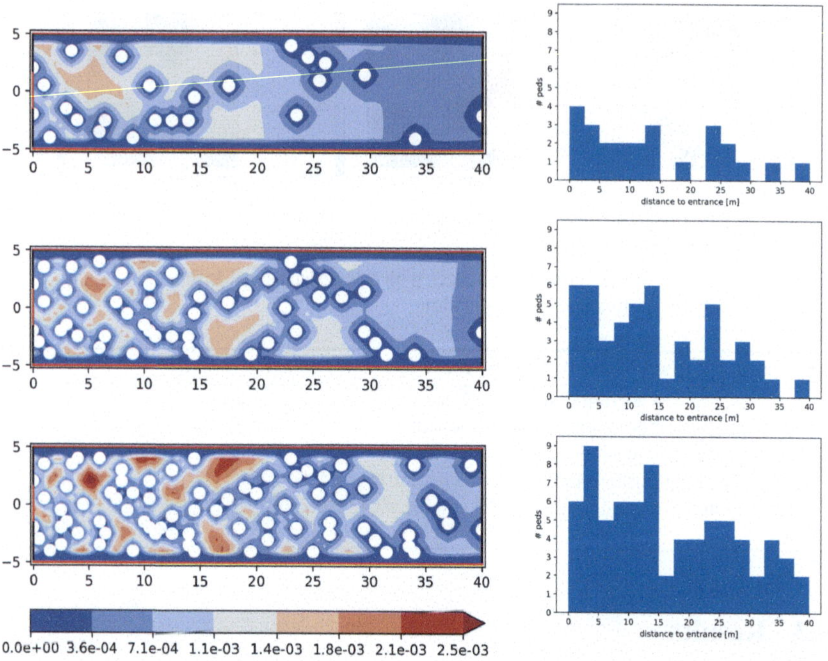

(a) Passenger distribution.

(b) Longitudinal passenger distribution.

Fig. 29.4 Passenger distribution on the platform after 25, 50, and 75 have found a waiting position. Left side shows the distribution and the underlying probability, white circle indicate pedestrian positions. Right side shows the longitudinal distribution of the passengers

butions on platforms. In most cases clustering of pedestrians occurs close to platform entries and exits, leading to a non-uniform distribution and in some cases, leaves parts of the platform empty, which are further away from entrances. This behavior can also be seen in the results from our simulations, as displayed in Fig. 29.4d. Of the first 25 passengers on the platform, 44% will choose a location within 10 m of the entrance, and almost 70% are located in the half closer to the platform entrance. The distribution stays almost the same after 50 pedestrians have entered the platform, 38% are closer than 10 m to the entrance, and 68% choose a waiting position in the left half of the platform. With 75 pedestrians the distribution starts to get more uniform, as only 62% are assigned a waiting position in the half of the platform with the entrance. Only 35% of the passengers stand within the first 10 m of the entrance. In our simulation, the distribution of the passengers tends to become more uniform with higher densities, which qualitatively agrees with field observations.

29.4 Conclusion and Outlook

In this work, we developed an optimization and random based approach to qualitatively model the position finding of pedestrians in a waiting context. The model is based on probability fields derived from different influence factors that are found in experiments as well as field studies. Due to its modularity, extra factors that are needed in a more general context as attraction, repulsion, and danger zones can be easily added to the model.

Due to the strong influence of the entrance on the distribution of the waiting pedestrian, blockage may occur, which would prevent that further pedestrians enter the waiting area. In our simulations, we focused on the process of finding a waiting position, neglecting the interaction between moving pedestrians. In future work, it has to be investigated if the interaction between pedestrians may solve the problem of a blocked entry, as the blocking pedestrian retreat into the room to make space for passing persons. An essential factor that needs to be included in the future is group behavior, as pedestrians would usually stand close to each other during waiting.

Moreover, the model introduced in this paper assumes a global knowledge about layout and occupation of the platform. This has to be replaced by a view-field based knowledge allowing to model the behavior of pedestrians who are unfamiliar with the place and to consider the restricted field of view of passengers just entering the platform. Also, when incorporating the position finding in a movement context, an iterative position finding process is needed, as the pedestrians need to react to other pedestrians taking their preferred spot, and they need to find a new one.

References

1. M. Chraibi, A. Seyfried, A. Schadschneider, Generalized centrifugal-force model for pedestrian dynamics. Phys. Rev. E **82**(4) (2010)
2. M. Davidich, F. Geiss, H.G. Mayer, A. Pfaffinger, C. Royer, Waiting zones for realistic modelling of pedestrian dynamics: a case study using two major German railway stations as examples. Transp. Res. Part C Emerg. Technol. **37**, 210–222 (2013)
3. T. Ezaki, K. Ohtsuka, M. Chraibi, M. Boltes, D. Yanagisawa, A. Seyfried, A. Schadschneider, K. Nishinari, Inflow process of pedestrians to a confined space. Collectiv. Dyn. **1** (2016)
4. W. Heinz, Passenger service times on trains. Technical report, KTH Royal Institute of Technology, Stockholm (2003)
5. D. Helbing, P. Molnár, Social force model for pedestrian dynamics. Phys. Rev. E 51(5), 4282–4286 (1995)
6. F. Johansson, A. Peterson, A. Tapani, Waiting pedestrians in the social force model. Phys. A Stat. Mech. Appl. **419**, 95–107 (2015)
7. N. Krstanoski, Modelling passenger distribution on metro station platform. Int. J. Traffic Transp. Eng. **4**(4), 456–465 (2014)
8. X. Liu, W. Song, L. Fu, W. Lv, Z. Fang, Typical features of pedestrian spatial distribution in the inflow process. Phys. Lett. A **380**(17), 1526–1534 (2016)
9. J.A. Sethian, A fast marching level set method for monotonically advancing fronts. Proc. Natl. Acad. Sci. **93**(4), 1591–1595 (1996)

10. D. Szplett, S.C. Wirasinghe, An investigation of passenger interchange and train standing time at LRT stations: (i) Alighting, boarding and platform distribution of passengers. J. Adv. Transp. **18**(1), 1–12 (1984)
11. A. Tordeux, M. Chraibi, A. Seyfried, Collision-free speed model for pedestrian dynamics, in *Traffic and Granular Flow '15*, ed. by V.L. Knoop, W. Daamen (Springer International Publishing, 2016), pp. 225–232
12. P.B. Wiggenraad, Boarding and Alighting Time in Dutch Rail Stations, Technical report, TRAIL Research School, Delft (2001)
13. X. Yang, X. Yang, Z. Wang, Y. Kang, A cost function approach to the prediction of passenger distribution at the subway platform. J. Adv. Transp., 1–15 (2018)

Chapter 30
Exploring the Effect of Crowd Management Measures on Passengers' Behaviour at Metro Stations

Sebastian Seriani and Taku Fujiyama

Abstract To reduce problems of interaction at the platform train interface (PTI) platform edge doors (PEDs) and markings on the platform are used as door positions indicators. The common methods to study the effect of these measures are based on average values of density using Fruin's Level of Service (LOS), however identification cannot be made of which part of the PTI is more congested. To solve this problem, a new method is proposed. The method included a conceptual model in which the PTI was discretised into 40 cm square cells to identify which part of the platform is more congested. Passengers' behaviour was recorded considering two situations before the train arrives: i) passengers waiting in front of the doors; ii) passengers waiting beside the doors. Observation was done at existing stations at Metro de Santiago and London Underground. Results show that PEDs changed the behaviour of passengers as they were located beside the doors rather than in front of them. In addition, when markings were used on the platform, then this behaviour was reinforced. Therefore, it is recommended to use this method to better design the PTI rather than the LOS which is used to design the whole platform. Further research is needed to study the effect of PEDs on passengers with reduced mobility.

30.1 Introduction

To reduce problems of interaction at the platform train interface (PTI) crowd management measures have been implemented in different urban railway systems worldwide. As an example, platform edge doors (PEDs) are combined with markings on the platform to be used as door positions indicators at the PTI. In developed cities such as

S. Seriani (✉)
Faculty of Engineering and Applied Sciences, Universidad de los Andes, Santiago, Chile
e-mail: sseriani@miuandes.cl

T. Fujiyama
Department of Civil, Environmental and Geomatic Engineering, University College London, London, UK
e-mail: taku.fujiyama@ucl.ac.uk

I. Zuriguel et al. (eds.), *Traffic and Granular Flow 2019*,
Springer Proceedings in Physics 252,
https://doi.org/10.1007/978-3-030-55973-1_30

London, the London Underground installed PEDs in 9 stations of the Jubilee Line, in which a grey line (1.2 m long) is used as marking on the platform in front of the doors. Similarly, in developing cities such as Santiago de Chile, the Metro de Santiago have incorporated PEDs in all stations of the new Line 3 and 6, however not all them have markings on the platform (see Fig. 30.1).

Although PEDs and markings on the platform are used for safety reasons, it is interesting to study the effect of these elements on the behaviour of passengers, which is the main objective of this work. The hypothesis is that crowd management measures used as door position indicators reduce the interaction of passengers, as they know where the doors are. The following specific objectives are presented:

- To define a new method to measure interaction.
- To study the location of passengers and the use of markings on platforms when PEDs are used.

Fig. 30.1 PEDs: **a** Ñuble station (Metro de Santiago) with markings on the platform; **b** Ñuñoa station (Metro de Santiago) without markings on the platform; **c** Westminster station London Underground) with markings on the platform

- To propose some recommendations on how crowd management measures could reduce the interaction between passengers.

30.2 Existing Studies on Crowd Management Measures

With respect to PEDs some authors state that these elements can affect the boarding and alighting time (BAT), and their installation is limited to the number of train doors, number of coaches and design of the platform [1]. However, it is not clear how the authors reached this conclusion and if there is any evidence to support it. Other authors [2] have studied the use of PEDs in evacuation emergencies and they found that these elements improved ventilation and smoke detection in metro tunnels, however, the evacuation time at platforms may increase when using these elements, due to the inconsistency of train stopping at the same position on the platform or by the fragility of their materials.

On the other hand, experiments at University College London's Pedestrian Accessibility Movement Environment Laboratory (PAMELA) showed that PEDs have no relevant impact on the BAT [3, 4]. In the experiments the authors mocked-up a train and the adjacent platform to represent the boarding and alighting process when PEDs are used. The same authors have confirmed these results in existing stations at the London Underground [5].

To study the effect of PEDs, Shen [6] proposed two main areas: circulation and waiting zones. Both areas have their own characteristics and functionality for passengers. When PEDs are installed at the PTI, little demarcation (e.g. markings on the floor) is used on the platform to separate these two areas, and therefore no clear distinction could be identified to measure the interaction between passengers in front of the doors compared to the rest of the platform [7]. Passengers in the waiting areas behave differently from those who are in the circulation zone. For Wu and Ma [7] there are two main types of behaviour of passengers who are waiting: queuing or clustering to the side or in front of the train doors.

Other authors have considered that the platform should be divided into different waiting areas in front of each door for an in-depth analysis of the interaction and behaviour of passengers. For example, Shen [6, 8], states that passengers are not distributed uniformly and waiting areas can be considered as rectangular spaces or as a parabola, while Lu and Dong [9] suggest that this space can be considered as fan or spectrum. Similarly, Seriani and Fernandez [10] proposed that a rectangular area should be used in front of the train doors as a "keep out zone" to prevent passengers boarding from being an obstacle for those who are alighting.

Relevant research has been carried out, showing that PEDs may increase the BAT in emergency situations and when there is inconsistency of train stopping at the same position of train doors. When PEDs are used, waiting areas are fixed values which do not necessarily represent which part of the PTI is more congested.

30.3 Method

The common methods to study the effect of these measures are based on average values of density using Fruin's Level of Service (LOS) [11], however identification cannot be made of which part of the PTI is more congested [12]. To solve this problem a new method is proposed. The method included a conceptual model in which the PTI was discretised into 40 cm square cells to identify which part of the platform is more congested. The discretised PTI enables the identification of where passengers are located at the PTI. Each passenger is represented as a square cell, which is typically used in cellular automata models [13, 14].

Passengers' behaviour was recorded considering two situations before the train arrives: (i) passengers waiting in front of the doors; (ii) passengers waiting beside the doors.

Two stations were compared at London Underground: Westminster station (with PEDs) and Green Park station (without PEDs) of the Jubilee Line. Both stations were observed during one week at the morning and afternoon peak hour (8:15–9:15 am and 5:15–6:15 pm), when trains on that line reach an average frequency of 30 trains/h (approximately 2 min headways). On average Westminster station reached 11 passengers/door boarding, while Green Park station reported 8 passengers/door boarding. The selected door is considered the 'critical door' in which a high demand is reached.

Two stations were compared at Line 6 in Metro de Santiago: Ñuñoa station without markings on the platform; and Ñuble station with markings on the platform. During three days of observation at afternoon peak hour (6:30–7:30 pm), 13 passengers/door were boarding at Ñuñoa station, while at Ñuble station 9 passengers/door were on average. Similarly, in these stations the selected door is considered the 'critical door'.

30.4 Results

Snapshots were taken and displayed for each day of observation. Figure 30.2 represents the average occupation at the PTI. The occupation was registered each time a cell was occupied by a passenger waiting to board the train. The average is obtained considering the 5 days (one week) observed at both stations for the door selected. Manually tracking was used, i.e. the position of each passenger waiting to board the train was assigned manually. The position indicates that a specific cell was used at the PTI. Therefore, as a consequence of the occupation at the PTI, interaction problems can be reached based on the density of passengers and potential risks such as agglomeration, high pressure, "crossing of flows", collision and "confined flow". The green colour represents a low occupation area, whilst the red colour denotes high occupations. Medium occupations are symbolised by an amber colour.

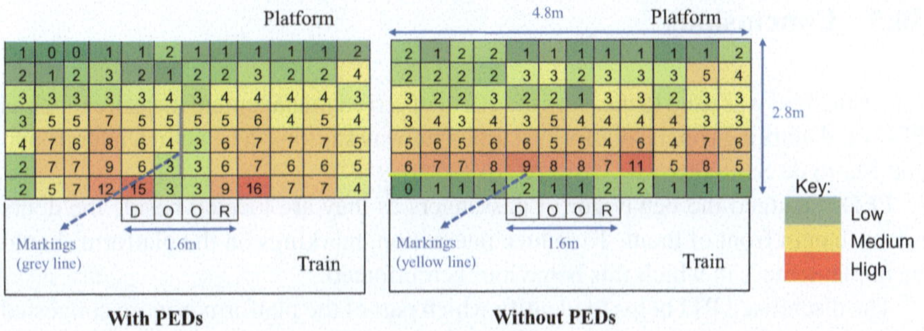

Fig. 30.2 Average occupation at the discretized PTI (40 × 40 cm) just before the doors started to open with (left) and without (right) PEDs at London Underground

In Fig. 30.2 the differences between both stations are clear. In the case of Westminster station, the use of PEDs acting as doors position indications on the platform changed the behaviour of passengers to wait beside the doors rather than in front of them. The cells in front of the doors are less used at Westminster station compared with Green Park station, where no door position indications on platforms are used. Thus, these door indicators help passengers alighting to get off the train with fewer interaction problems. In the case of Green Park (without PEDs) passengers are distributed homogeneously on the PTI and the first row of cells is less used (green colour) due to the yellow line at the edge of the platform which is respected by passengers waiting to board the train.

In the case of PEDs with markings on the platform (Ñuble station), less passengers are located in front of the doors compared to the case without markings on the platform (Ñuñoa station). This is observed in Fig. 30.3, in which the case of Ñuble station presented more green cells in front of the doors compared to Ñuñoa station. The results are presented as an average of the three days observed at both stations for the door selected.

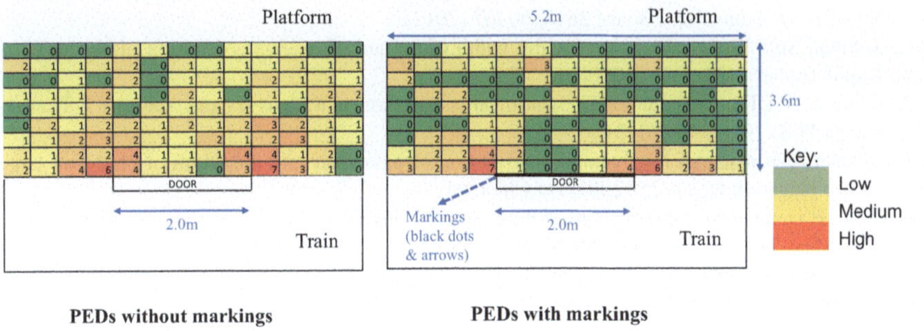

Fig. 30.3 Average occupation at the discretized PTI (40 × 40 cm) just before the doors started to open without (left) and with (right) markings on the platform at Metro de Santiago

30.5 Conclusions

A new method was used to measure the effect of crowd management measures such as PEDs and markings on the platform at urban railway stations in London Underground and Metro de Santiago.

PEDs changed the behaviour of passengers as they are located beside the doors rather than in front of them. To reduce interaction, markings on the platform should be implemented, in which this behaviour is reinforced.

The discretised PTI helps to identify which part of the platform is more congested and therefore where a high interaction occurs. Therefore, it is recommended to use this method to better design the PTI rather than the LOS which is used to design the whole platform.

Further research is needed to explore the use of tracking tools in existing stations. In addition, new experiments are needed to study the effect of PEDs on passengers with reduced mobility.

References

1. S. Coxon, K. Burns, A. De Bono, Design strategies for mitigating passenger door holding behaviour on suburban trains in Paris, in *Paper presented at the 33rd Australasian Transport Research Forum Conference, 29 September–1 October*, Canberra (2010)
2. L. Qu, W.K. Chow, Platform screen doors on emergency evacuation in underground railway stations. Tunn. Undergr. Space Technol. **30**, 1–9 (2012)
3. G. De Ana Rodríguez, S. Seriani, C. Holloway, Impact of platform edge doors on passengers' boarding and alighting time and platform behavior. Transp. Res. Rec. J. Trans. Res. Board **2540**, 102–110 (2016)
4. S. Seriani, T. Fujiyama, C. Holloway, Exploring the pedestrian level of interaction on platform conflict areas at metro stations by real-scale laboratory experiments. Transp. Plan. Technol. **40**(1), 100–118 (2017)
5. S. Seriani, G. de Ana Rodríguez, C. Holloway, The combined effect of platform edge doors and level access on the boarding and alighting process in the London Underground. Transp. Res. Rec. J. Trans. Res. Board **2648**, 60–67 (2017)
6. J. Shen, Simplified calculation for the width of on and off regions of station platform. Urb. Rapid Transit **21**(5), 9–12 (2008) (in Chinese)
7. J. Wu, S. Ma, Division method for waiting areas on island platforms at metro stations. J. Transp. Eng. **139**(4), 339–349 (2013)
8. J. Shen, The research for dynamic distribution of passenger and the width of platform. Urb. Rapid Transit **4**(1), 21–25 (2001) (in Chinese)
9. J. Lu, F. Dong, Statistical analysis of the passenger distribution before getting on subway train. Urb. Mass Transit **13**(7), 53–56 (2010) (in Chinese)
10. S. Seriani, R. Fernandez, Pedestrian traffic management of boarding and alighting in metro stations. Transp. Res. Part C **53**, 76–92 (2015)
11. J.J. Fruin, Designing for pedestrians: a level-of-service concept. Highway Res. Rec. **377**, 1–15 (1971)

12. G.W. Evans, R.E. Wener, Crowding and personal space invasion on the train: please don't make me sit in the middle. J. Environ. Psychol. **27**(1), 90–94 (2007)
13. Q. Zhang, B. Han, D. Li, Modelling and simulation of passenger alighting and boarding movement in Beijing metro stations. Transp. Res. Part C **16**, 635–649 (2008)
14. M. Davidich, F. Geiss, H.G. Mayer, A. Pfaffinger, C. Royer, Waiting zones for realistic modelling of pedestrian dynamics: a case study using two major German railway stations as examples. Transp. Res. Part C **37**, 210–222 (2013)

Chapter 31
Rotation Behaviour of Pedestrians in Bidirectional and Crossing Flows

Martijn Sparnaaij, Dorine C. Duives, and Serge P. Hoogendoorn

Abstract Rotating ones body is a strategy pedestrians commonly use to avoid collisions. Even though this behaviour impacts capacity heavily, this rotation behaviour is seldomly studied. This research aims to increase insight into rotation behaviour of pedestrians in high density bidirectional and crossing flows. Based on data from the CrowdLimits experiments, the effect of density, movement base case, flow ratio and disturbances on the rotation behaviour of pedestrians are studied. The main findings are that all these four factors impact the number of rotations in a flow. Yet, further research is necessary to better identify to what extent and when these factors impact the rotation behaviour most.

31.1 Introduction

One of the core features of any microscopic pedestrian model is the replication of the collision avoidance behaviour of a pedestrian. In these models, the most popular way to represent pedestrians is by using a circle and collision avoidance is performed by changing the speed and/or direction of the pedestrians. However, pedestrians also use body rotation as a strategy to avoid collisions [1] and this cannot be modelled using the circular representation. So, this begs the question; how important is it to be able to represent this body rotation behaviour in our models?

Rotation behaviour has not been studied extensively [1] and this is also the case for both bidirectional and crossing flows. Both [1, 2] studied body rotations in bidirectional flows but neither looked at the relationship with density. For crossing flows, [1] did look at the relationship between body rotation and density. Their two main findings were that body rotations in the range of [0, 20] degrees were observed in a wide range of densities and that large body rotations were only found in the case of (local) densities smaller than 2.5 ped/m^2.

M. Sparnaaij (✉) · D. C. Duives · S. P. Hoogendoorn
Department of Transport and Planning, Delft University of Technology, Stevinweg 1, 2628CN Delft, The Netherlands
e-mail: M.Sparnaaij@tudelft.nl

I. Zuriguel et al. (eds.), *Traffic and Granular Flow 2019*,
Springer Proceedings in Physics 252,
https://doi.org/10.1007/978-3-030-55973-1_31

So, to get more insight into how relevant it is to be able to represent this rotation behaviour in models, this paper aims to provide more insight into the rotation behaviour of pedestrians in bidirectional and crossing flows and how the rotation behaviour relates to the density, the flow ratio and disturbances. Furthermore, the differences between these two movement base cases will also be studied.

The paper is organised as follows. Section 31.2 introduces the CrowdLimits experiment which provides the data used in this study. Section 31.3 accordingly explains how the rotations are extracted from the video data. Section 31.4 discusses the results and lastly Sect. 31.5 provides the conclusions and a discussion of the results.

31.2 CrowdLimits Experiment

The CrowdLimits experiments are large pedestrian experiments performed at the Delft University of Technology in June 2018. The goal of the experiment was to investigate pedestrian behaviour in high density bidirectional and crossings flows. However, given the setup of the experiment, the collected data can also be used to study rotation behaviour of pedestrians in these two flow scenarios. A short summary of the setup of the experiment will be given in the remainder of this section. For a more detailed description the reader is referred to [3].

The experiments consisted of two evenings, whereby on the first evening a bidirectional flow was replicated and on the second evening a two-way crossing flow. Each evening 12 runs were performed whereby two different flow ratios where tested and three different movement assignments. This research uses only 8 of the total of 24 runs and details on these 8 runs can be found in Table 31.1.

Every run took approximately 5 min whereby every minute the inflow was increased in a step-wise manner. The inflow into the infrastructure was controlled using queuing areas with lanes and stop-go lights. All participants were asked to enter the corridor or crossing and exit on the other side and walk as they normally would. After having walked through the infrastructure the participant joined the queue again.

Table 31.1 Overview of the eight runs that are used within this research

Nr.	Scenario	Flow ratio	Assignment	Dimensions
1.	Bidirectional	50-50	None	w = 2.4 m
2.	Bidirectional	80-20	None	w = 2.1 m
3.	Bidirectional	50-50	[C] 10% Fast walk	w = 2.4 m
4.	Bidirectional	80-20	[C] 10% Fast walk	w = 1.8 m
5.	Two-way crossing	50-50	None	w = 1.8 m, l = 1.8 m
6.	Two-way crossing	70-30	None	w = 2.4 m, l = 1.4 m
7.	Two-way crossing	50-50	[C] 10% Fast walk	w = 1.8 m, l = 1.8 m
8.	Two-way crossing	70-30	[C] 10% Fast walk	w = 2.4 m, l = 1.8 m

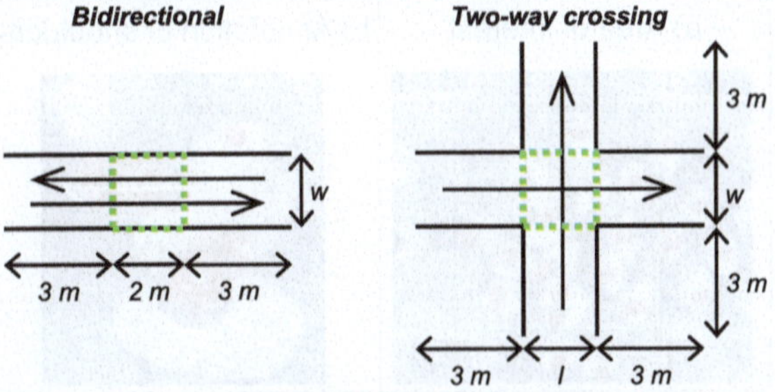

Fig. 31.1 Lay-out of the scenarios including the dimensions, the walking directions and the areas of interest

In the case that assignment C was active, about 10% of the participants who got this assignment were asked to pretend like they were in a hurry and thus walk faster than they normally would. The aim of this assignment was to include a disturbing element to the scenario.

The location of the experiments was a large indoor hall. To construct the corridor and the crossing four l-shaped wooden elements were used whereby in the case of the corridor two additional 2-m wide panel were used. All these elements had a height of 2.4 m to ensure that the participants had the feeling they were in an enclosed space. The exact lay-out and dimensions can be found in Fig. 31.1 and in Table 31.1.

Per evening, approximately 130 people participated, who formed a heterogeneous population consisting of 55% males and 45% was females with age an average of 30.5 (ranging from 18 to 70) and a mean height of 1.76 (ranging from 1.40 to 2.04) meters. 74% of the participants was Dutch and the remainder featured a mixture of different nationalities.

The movement of the participants was captured from above using multiple cameras. To aid the tracking every participant got a red cap and a white t-shirt. To mark the shoulder points blue dots were stuck on the t-shirt to mark the participants' shoulders. See Fig. 31.2 for an example of the camera view.

31.3 Extracting the Rotations from the Video Data

After the experiments the rotation of the participants had to be extracted from the videos. This section details the three-step process that is used to extract the rotations and densities from the videos.

The first step is to determine an area of interest within which the rotations and density will be measured. Figure 31.2a shows an example of the area of interest. Figure 31.1 shows the size and location of the area of interest for both scenarios. The areas are chosen such that we only capture the rotations of the pedestrians related to the movement base cases of interest and not those in the unidirectional flows leading to and from the crossing or the entering and exiting cases at the boundaries of the infrastructure.

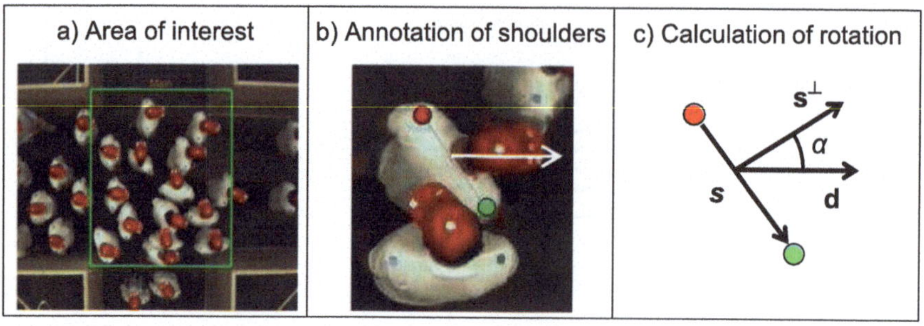

Fig. 31.2 Figure showing the three stages performed to extract the rotations. The green rectangle in **a** shows the area of interest. The red dot in **b** annotates the left shoulder and the green dot the right shoulder and the white arrow indicates the assumed walking direction. **c** shows how the rotation α is determined based on the two shoulder points and the assumed walking direction

The second step is to annotate the shoulders of the participants. This is done for one frame every second for every single person that resides in the area of interest. The choice for using only one frame per second, instead of doing this for every frame, is the fact that the annotation had to be done manually and doing so for every frame would have been too time-consuming. Figure 31.2b shows an example of a pedestrian whose shoulders have been annotated whereby the left shoulder is annotated by the red dot and the right shoulder by the green dot.

The third and final step is to, based on the annotated shoulders and the area of interest, calculate for every frame both the density in the area of interest and the rotations of all participants in the area of interest. The density is the number of pedestrians in the area of interest over the size of the area of interest. Figure 31.2c depicts how the rotation is calculated. The rotation α is calculated by taking the angle between the assumed direction of movement $\mathbf{d}$ and the perpendicular vector $\mathbf{s}^{\perp}$ of the vector connecting the left shoulder point to the right shoulder point $\mathbf{s}$. The assumed directions of movement are displayed in Fig. 31.1. These assumed directions of movement are used instead of the actual directions of movement as trajectory data was not yet available. The influence this assumption has on the results will be discussed in more detail in the last section of the paper.

Using the method described above, the videos of all eight experiments were processed. This resulted in eight datasets which each contained all rotations captured in the video of the respective experiment. The next section discusses what insight these datasets provide.

31.4 Results

Figure 31.3 depicts the results for all eight cases. Graphs 3a and 3b clearly show that, though in all eight cases the distribution shape is similar, there are some differences between the eight cases regarding how the rotations are distributed. The graphs also

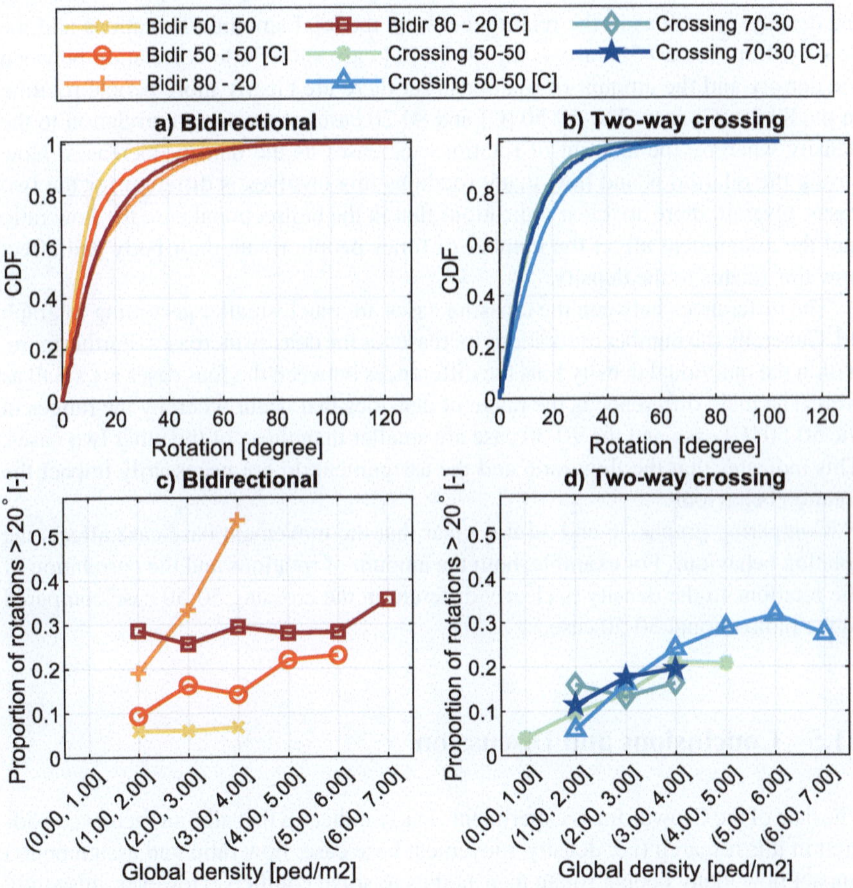

Fig. 31.3 The results whereby graphs **a** and **b** show the cumulative distribution of the rotations per case and graphs **c** and **d** the proportion of rotations per density bin

show that the large majority of the rotations are small (i.e. $\leq 20°$). Previous research [1, 2] suggest that these small rotations are likely not the results of interactions with other pedestrians. To confirm that this is also the case for these datasets, the shoulders of several pedestrians, who were walking in a free flow condition in one of the arms leading to the crossing, were annotated. The results of this analysis are in line with the finding from previous research and confirm that rotations smaller than 20° are likely not the result of interactions with other pedestrians. Hence, only rotations larger than 20° will be used in the remainder of this analysis.

Graphs 3c and 3d show how the rotations larger than 20° relate to the global density ($k = N/A$). In the bidirectional case, graph 3c clearly shows differences between the four cases. These differences show up in the relation between the rotations and the density, as well as in the relation between the total amount of rotations and the density. Both the 50-50 case as the 80-20 [C] case show little correlation between the density and the amount of rotations. Yet there are clearly more people rotating in the 80-20 [C] case. The 50-50 [C] and 80-20 cases show some correlation to the density whereby the amount of rotations increases as the density increases. How strong the relation is and how many rotations this involves is different for the two cases. Overall, there are clear indications that in the bidirectional case the flow ratio and the assignment affect the number of times people rotate their body and affect how this relates to the density.

The differences between the crossing cases are much smaller according to graph 3d. Generally the number of rotations increases as the density increases. Furthermore, within the individual density bins the differences between the four cases are small as well. The main difference is the range of densities that occur whereby the ranges in the 50-50 [C] case and the 70-30 case are smaller than those of the other two cases. This indicates that the flow ratio and the assignment do not necessarily impact the rotation behaviour.

Comparing graphs 3c and 3d it is clear that the movement base case affects the rotation behaviour. For example, both the amount of rotations and the correlation of the rotations to the density is clearly different in the crossing 50-50 case compared to the bidirectional 50-50 case.

31.5 Conclusions and Discussion

The data of the CrowdLimits experiment clearly indicates that all four factors considered in this research (i.e. density, movement base case, flow ratio and assignments) impact how many people rotate their bodies to solve conflicts. However, this study does not have enough power to determine to what degree these factors influence the amount of rotations and in which cases they exactly influence the amount of rotations. Furthermore, this study was also limited by the fact that the assumed walking directions were used to calculate the rotations instead of the actual walking directions. Future research will aim at enhancing the insights by using the actual walking directions, including the other 16 runs, coupling individual rotation behaviour to socio-demographic variables and the assignments and lastly by performing statistical analyses.

References

1. H. Yamamoto, D. Yanagisawa, C. Feliciani, K. Nishinari, Transp. Res. Part B Methodol. **122**, 486 (2019). https://doi.org/10.1016/j.trb.2019.03.008. http://www.sciencedirect.com/science/article/pii/S0191261517300085
2. C.J. Jin, R. Jiang, J.L. Yin, L.Y. Dong, D. Li, Phys. A Stat. Mech. Appl. **482**, 666 (2017). https://doi.org/10.1016/j.physa.2017.04.117. http://www.sciencedirect.com/science/article/pii/S0378437117304338
3. D.C. Duives, M. Sparnaaij, W. Daamen, S.P. Hoogendoorn, How many people can simultaneously move through a pedestrian space? The impact of complex flow situations on the shape of the fundamental diagram (2019). https://arxiv.org/abs/1908.07208. Submitted

Chapter 32
Experimental Study on One-Dimensional Movement with Different Motion Postures

Qiao Wang, Weiguo Song, Jun Zhang, and Siuming Lo

Abstract The motion postures of pedestrians influence the evacuation efficiency in fire and smoke filling buildings. In a building fire, the effective method is to lower the height of mouth and noise to avoid the harm to occupant's health. In this paper, we conduct the one-dimensional pedestrian movement experiment to study the movement characteristics with different motion postures. Firstly, it is observed that the speed of upright walking is the fastest, and the slowest speed is low crawling. However, there is no significant difference in speed between upright walking and walking with a stoop. Through this experiment, the basic speeds of different motion postures can be obtained, which can be used to collect the basic data for evacuation simulation in fire and smoke buildings. These findings can provide insight to evacuation strategy when building fire occurred.

32.1 Introduction

Recently, building fire accidents are frequent. In a building fire accident, the main threats to people's lives and property come from the thermal and non-thermal effects of the combustion process. More than 80% of fatalities were caused by the smoke inhalation, resulting in smoke poisoning or suffocation, according to the analysis of fire deaths from 2003 to 2007 [1]. When an indoor fire occurs, the combustible

Q. Wang · W. Song (✉) · J. Zhang
University of Science and Technology of China, Hefei, Anhui, People's Republic of China
e-mail: wgsong@ustc.edu.cn

Q. Wang
e-mail: qw2014@mail.ustc.edu.cn

J. Zhang
e-mail: junz@ustc.edu.cn

Q. Wang · S. Lo
City University of Hong Kong, Hong Kong, People's Republic of China
e-mail: bcsmli@cityu.edu.hk

I. Zuriguel et al. (eds.), *Traffic and Granular Flow 2019*,
Springer Proceedings in Physics 252,
https://doi.org/10.1007/978-3-030-55973-1_32

materials can produce a great deal of smoke. Once the smoke layer hits the ceiling of the room, the smoke fills the enclosure rapidly [2]. Gradually, the height of the smoke layer drops down. To avoid toxic smoke inhalation and access breathable air, reducing the height of nose and mouse is necessary.

To address the characteristic properties of walkers and crawlers evacuating from the corridor with an exit, Nagai et al. [3] conducted the walking and crawling experiment, and simulated their evacuation process by the lattice gas model. Comparing the speed among normal walking (1.32 m/s), fast walking (2.16 m/s), normal crawling (0.71 m/s) and fast crawling (1.47 m/s), there significant differences among them [4]. Increasing the density of crawlers, the relationship between crawling speed and density conforms to quadratic regression [5]. Furtherly, the walking was divided into upright walking and stoop walking, the crawling included knee and hand crawling and low crawling [6]. For these four motion postures, the average free speeds were 1.9, 1.6, 0.6 and 0.5 m/s, respectively. Additionally, physiological results indicated that crawling needs more physical power, such as higher heart rates and rates of oxygen consumption [7].

Based on existing research, in this paper, we preformed the one-dimensional movement with different motion postures in a narrow channel. First, the average free speeds of upright walking (UW), walking with a stoop (SW), knee and hand crawling (KHC) and low crawling (LC) are obtained through free motion. Next, multiple participants enter the narrow channel, and investigate the impact of density on speed with different motion postures.

32.2 Experiment Setup

In a fire accident, to avoid the damage of toxic smoke, pedestrians have to reduce the height of the mouth and nose. Thus, the motion postures, such as UW, SW, KHC and LC (see Fig. 32.1), should be adopted for safe evacuation. The purpose of this work is to collect the basic data of pedestrians who move in a narrow channel with different motion postures, assuming that they are in a fire accident.

The experiment was conducted in an event square. The experimental area is a narrow channel, in which all participants did one-dimensional movement. There are two parallel narrow channels with a length of 12 m and a width of 0.6 m. The channel is divided into two equal areas: UW area and SW/KHC/LC area, as shown in Fig. 32.1. To increase the authenticity of the fire scene, a fishnet was used to mimic the toxic smoke. In the SW area, the height of the fishnet is set as 1.5 m, where participants can pass this area with walking with a stoop posture. For KHC/LC area, the height of fishnet is 0.7 m, and the foam mats are paved on the ground for KHC and LC. All participants were required to pass the narrow channel quickly, thus the running is allowed. To record the whole experiment, one video camera (with a resolution of 1440 $\times$ 1080 pixels and a framerate of 25 fps) was mounted on the roof of a four-story building. Besides, all participants were asked to wear a colored hat for facilitating the extraction of trajectories.

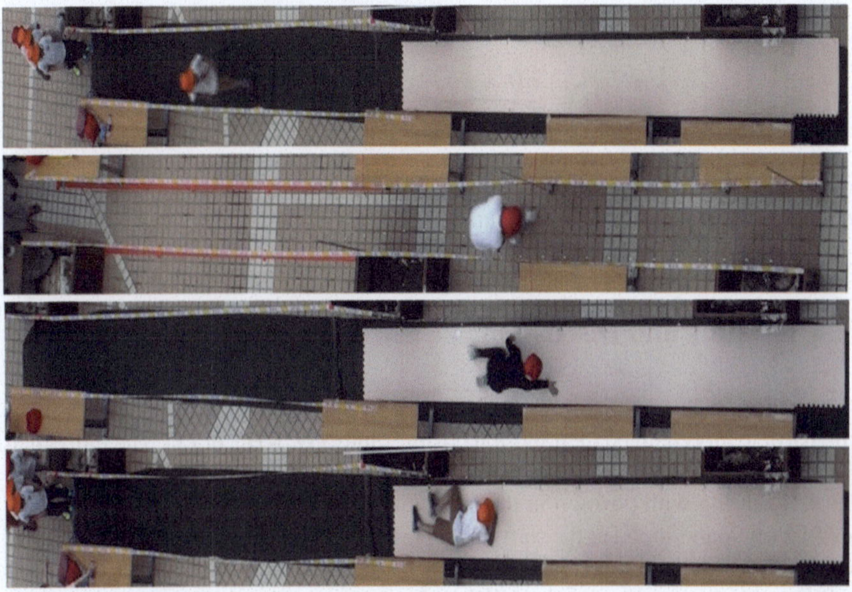

Fig. 32.1 Participants passed the narrow channel with different motion postures. From top to bottom: UW, SW, KHC, LC

Table 32.1 Number of participants attending the free motion

Motion posture	UW-SW	UW-KHC	UW-LC
Number of participants	52	75	26

In this experiment, the main participants are elementary and middle school students, and little children with their parents. Before they entered the narrow channel, all participants stood at the entrance, and arranged in one line. To obtain the free speed of UW, SW, KHC and LC, only one participant passed through the narrow channel (free motion). The number of participants who participated in the free motion is displayed in Table 32.1. After a while, multiple participants entered the narrow channel continuously, and did single-file movement. Each person participated only once in the experiment.

32.3 Measurements and Methods

To obtain the trajectories of participants, the software *PeTrack* [8] was used to extract the plane coordinates automatically. In this experiment, we only focus on one-dimensional characteristics of participants. Thus, the coordinate x is used to calculate the instantaneous speed of participants, whereas, the coordinate y is ignored because of less lateral oscillation of participant movement. For participant i at time

t, the individual instantaneous speed $v_i(t)$ is calculated as.

$$v_i(t) = \frac{x_i(t + \Delta t/2) - x_i(t - \Delta t/2)}{\Delta t} \quad (32.1)$$

where $x_i(t)$ is the coordinate of participant i in x-axis at time t. Δt is the time interval and the value of Δt is 0.4 s in this paper.

32.4 Results and Discussion

In each narrow channel, participants moved with upright walking in first 6 m. After 6 m, participants changed their motion postures into SW, KHC or LC. The individual instantaneous speeds of three participants who moved with UW-SW, UW-KHC and UW-LC varying with position in the free motion are plotted in Fig. 32.2. However, not all participants stood at the start line of entrance of the channel (zero of position) before they started moving. Notably, in Fig. 32.2, the individual instantaneous speed of SW (without considering the UW) is the fastest, and the lowest instantaneous speed of motion posture is LC.

In the process of free motion, the average free speeds of UW, SW, KHC and LC are calculated and shown in Fig. 32.3. The average free speed excludes the effect of acceleration, deceleration and the transition of posture. Here, the value of speed calculated the average free speed is in the stable phase that is selected manually. The average pair speeds for UW-SW, UW-KHC and UW-LC are 2.31–2.43, 2.07–1.04 and 1.80–0.68 m/s, respectively. The average free speed has no significant difference between UW and SW evaluated by T-test (T-test: $T = -0.88$, $P > 0.05$). However,

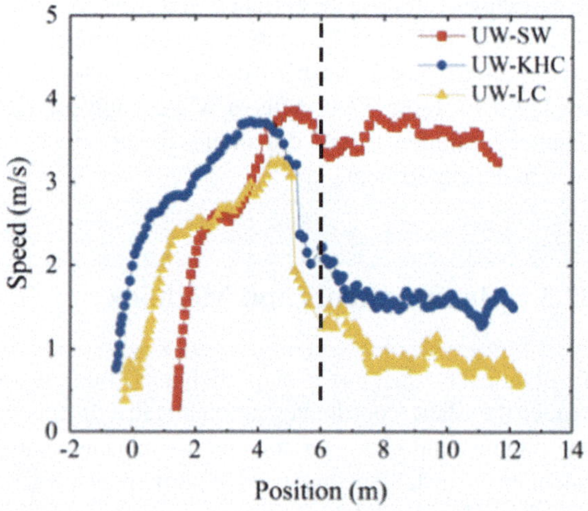

Fig. 32.2 Individual instantaneous speed varying with position for three randomly selected participants in the free motion

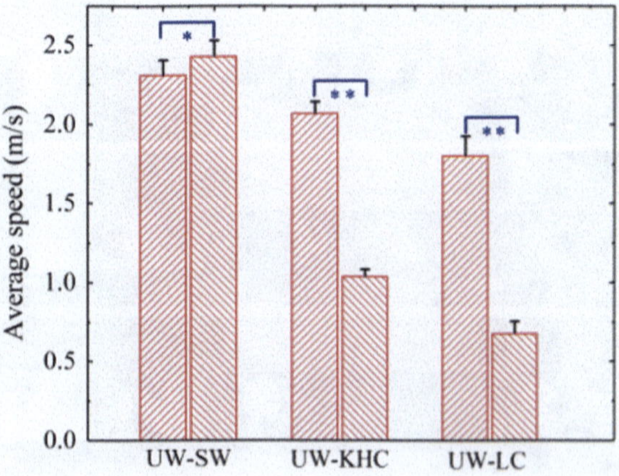

Fig. 32.3 Average speed of UW, SW, KHC and LC. * means there is no significant difference, and ** represents that there is significant difference

there is significantly difference for the average free speed of UW-KHC (T-test: T = 12.21, P < 0.05) and UW-LC (T-test: T = 7.78, P < 0.05). The average free speed of KHC and LC reduce 50% and 62% compared with the corresponding UW speed, respectively. In previous study, it was demonstrated that crawling requires more physical strength than walking [4]. As can be seen in Fig. 32.3, the average free speed of KHC and LC is lower than UW and SW notably. Additionally, it observed that for different motion postures in SW/KHC/LC area, the average free speed of UW is different. Observably, larger SW speed corresponds to higher UW speed. In this experiment, the UW and SW/KHC/LC areas are 6 m, and all participants knew where to change their motion posture. To adapt the speed in KHC/LC area, participants may decrease their speed in UW area consciously. In emergency conditions, to avoid the harm of toxic smoke, the best way is adopting the UW to escape as quickly as possible. For lower layer of toxic smoke, the postures of KHC and LC have to be adopted, however, these postures reduce the movement efficiency greatly.

After free motion, multiple participants entered the narrow channel successively, as displayed in Fig. 32.4. It can be seen that in UW area, it can accommodate more participants. Contrary, participants moving with LC need longer distance in front to move forward. Additionally, before entered the SW/KHC/LC area, participants waited at the entrance until the available space in front is large enough.

Figure 32.5 shows the trajectories and the individual instantaneous speed of successive participants passing through the narrow channel. In the case of that participants moved with UW first, and then changed into SW and KHC postures, the speed of UW is lower than the speed of SW and KHC. For the successive participants, the participants who entered the SW/KHC area passed through this area quickly, whereas, the participants in UW area waited for enough space to change their postures, as presenting in Fig. 32.5a, b. However, the speed of UW is higher than LC in UW-LC experiment (seeing Fig. 32.5c). The main reasons are the posture of LC needs more physical strength, and this posture is unskilled for participants compared with UW.

Fig. 32.4 Snapshots from the raw footage of UW-SW (up), UW-KHC (middle) and UW-LC (bottom) experimental runs

Consequently, in emergency conditions, the evacuation efficiency of motion posture LC is lowest.

Summarily, in emergency conditions, such as building fire, occupants should evacuate the building with UW or SW motion posture quickly when the toxic smoke layer is not very low. This can improve the evacuation efficiency. For the lower smoke layer, occupants have to reduce the height of nose and mouth to avoid inhaling toxic smoke. In this case, the motion postures of KHC and LC are adopted. However, KHC and LC greatly reduce the evacuation efficiency.

32.5 Conclusions

The experiment investigating the characteristics of different motion postures is conducted in the narrow channel. First, the free motion was conducted to obtain the average free speed. It is found that the speed of crawling is significantly lower than the walking speed. The average free speed of KHC and LC reduce 50 and 62% compared with the corresponding UW speed. When multiple participants entered the narrow channel continuously, following participants would wait at the entrance of SW/KHC/LC area, until the space in front is large enough to changing their motion posture. Thus, for UW-SW and UW-KHC, the speed of UW is smaller than that of SW and KHC. To evacuation more quickly, occupants should evacuate from the building with UW and SW. Once the smoke layer is too low, it is time to change UW/SW into KHC or LC.

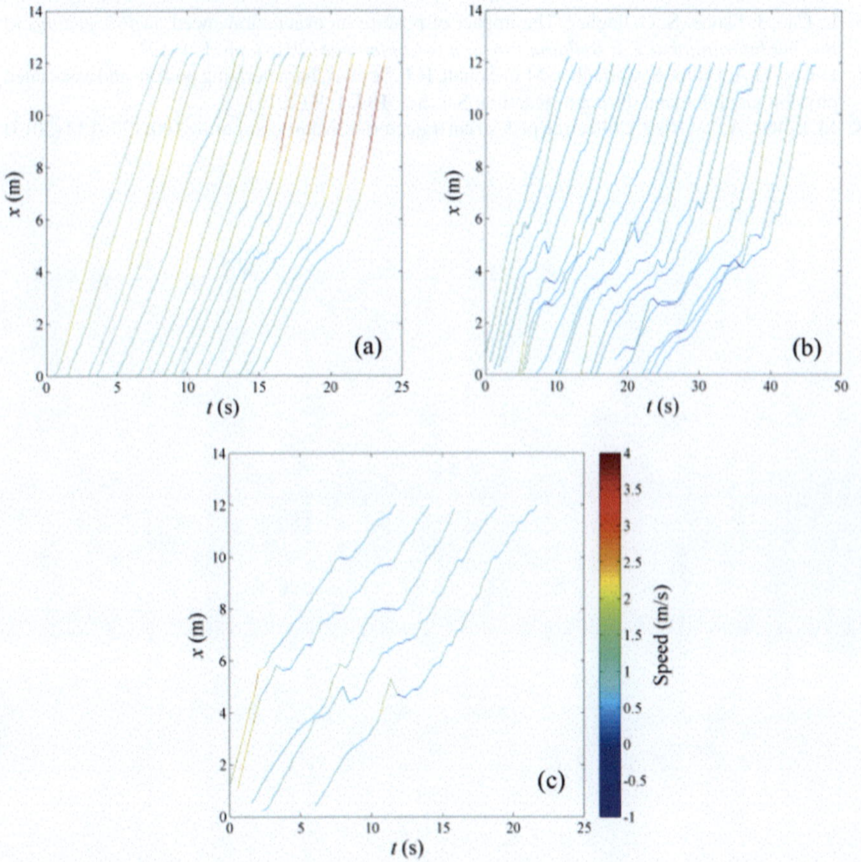

Fig. 32.5 Trajectories of successive participants passing the narrow channel with different motion postures. **a** UW-SW; **b** UW-KHC; **c** UW-LC

References

1. D.M. Gordon, *Ant Encounters: Interaction Networks and Colony Behavior* (Princeton University Press, 2010)
2. Z. Shahhoseini, M. Sarvi, Pedestrian crowd flows in shared spaces: Investigating the impact of geometry based on micro and macro scale measures. Trans. Res. Part B Methodol **122**, 57–87 (2019)
3. R. Nagai, M. Fukamachi, T. Nagatani, Evacuation of crawlers and walkers from corridor through an exit. Phys. A Stat. Mech. Appl. **367**, 449-460 (2006)
4. R. Muhdi, J. Davis, T. Blackburn, Improving occupant characteristics in performance-based evacuation modeling, in *Proceedings of the Human Factors and Ergonomics Society Annual Meeting* (Sage Publications, Los Angeles, CA, 2006), pp. 1199–1203.
5. R.A. Kady, The development of a movement–density relationship for people going on four in evacuation. Saf. Sci. **50**, 253–258 (2012)

6. L. Cao, J. Davis, S. Gallagher, The impact of posture on evacuation speed, in *Proceedings of the 2nd International Tall Building Fire Safety Conference* (2014), pp. 111–121.
7. L. Cao, G.A. Davis, S. Gallagher, M.C. Schall, R.F. Sesek, Characterizing posture and associated physiological demand during evacuation. Saf. Sci. **104**, 1–9 (2018)
8. M. Boltes, A. Seyfried, Collecting pedestrian trajectories. Neurocomputing **100**, 127–133 (2013)

Chapter 33
A Decision Model for Pre-evacuation Time Prediction Based on Fuzzy Logic Theory

Ke Wang, Shunzhi Qian, and Zhijian Fu

Abstract Efficient evacuation is crucial for reducing deaths and injuries caused by disastrous events such as earthquakes. Notably, pre-evacuation time constitutes a large proportion of the total evacuation time; whether and when to initiate the evacuation largely determines the outcome of the evacuation in an emergency. Despite considerable efforts made to elaborate the pre-evacuation process, the evacuees' vague and imprecise cognitive evaluation on the environment in pre-evacuation decision-making process is underrepresented in these studies. This study aims to enrich behavioral knowledge in the evacuation process during earthquakes and to explore modeling methods for characterization of the pre-evacuation process. As such, we conducted detailed analysis of real earthquake evacuation records to gain some insight into evacuees' behavioral features. The extracted information from the records, together with the empirical knowledge formed the basis of building a fuzzy logic based decision-making model. The proposed model allowed the prediction of investigating/evacuating decision time with the consideration of individual heterogeneity and changes of cues. The validity of this model was validated against real-case data with reasonable agreement in average pre-evacuation time. A further parametric study was conducted to investigate the influence of features of physical signals and those of instructions on the investigating/evacuating decisions.

K. Wang · S. Qian (✉)
School of Civil and Environmental Engineering, Nanyang Technological University, Singapore 639798, Singapore
e-mail: szqian@ntu.edu.sg

Z. Fu
School of Transportation and Logistics, Southwest Jiaotong University, Chengdu 610031, PR China

I. Zuriguel et al. (eds.), *Traffic and Granular Flow 2019*, Springer Proceedings in Physics 252,
https://doi.org/10.1007/978-3-030-55973-1_33

33.1 Introduction

Disasters in indoor spaces, such as arenas, apartments, and schools, bring about destructive consequences, demonstrating the dire need to develop better evacuation management strategies. Upon the onset of a disaster, the evacuee goes through an evacuation process, which comprises a pre-evacuation process and a movement process [1]. The duration of the pre-evacuation process, i.e. pre-evacuation time, is defined as the duration from the occurrence of the disaster to one's initiation of movement to a safe site [2]. It has been recognized that pre-evacuation time constitutes an unneglectable portion in the total evacuation time [3]. Nevertheless, the understanding of pre-evacuation behaviors is not sufficient due to the scarcity of real-life evacuation data [4, 5].

Recently, a few studies emerged on the pre-evacuation process. Some behavioral rules have been summarized based on extensive reviews [6], post-disaster questionnaire [7] and experiment [2]. Built upon these behavioral knowledge, pre-evacuation models have been developed. Evacuation Decision Model was firstly proposed, calculating evacuee's perceived risk based on constant cue, non-constant cue, prior knowledge cue, and social influence cue [8]. This model was further improved by incorporating behavioral uncertainty [9]. Another probability-based method assesses the level of perceived risk by Bayesian Belief Network [10]. Recently, an econometric hazard model was applied to study the effectiveness of waiting strategy in pre-evacuation process [11].

Given the aforementioned efforts, there is a scarcity of studies on pre-evacuation behaviors under earthquake [12, 13], especially in high school with leadership. Moreover, it is increasingly recognized that human's cognitive process is imprecise [14, 15], which is not captured by current models such as discrete models. During an evacuation, evacuee's perception and decisions are usually presented by natural language. Fuzzy inference system is capable of describing the cognitive process with a set of verbal terms [14]. This process is easy to implement in practice [16, 17], which could possibly dispel the concern of other models on computational complexity issue [9]. To fill in both gaps, this study aims to enrich behavioral knowledge with regards to earthquake scenarios and explore the application of fuzzy logic-based modeling methods for pre-evacuation time prediction.

33.2 Video Analysis

Two videos gathered from the Internet were used for pre-evacuation process study, as both happened in high schools under earthquakes (Fig. 33.1). In this study, three behavioral state concept was adopted, where evacuees' behavioral state cross the whole evacuation process can be categorized as normal, investigating, and evacuating state [8]. An evacuee is in his/her normal state if he/she continues his/her pre-evacuation activities, and in investigating state if he/she has perceived the abnormity

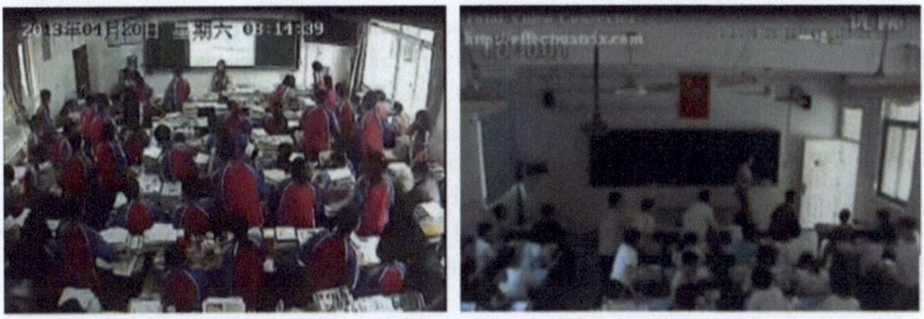

Fig. 33.1 Snapshot of evacuation process under evacuation situation in (left) Mingshan High School [18] and (right) Wenchuan High School [19]

and begins to search for more information. If he/she starts to flee from the current location or take some protective actions, he/she is considered in the evacuating state. Here, evacuee's state is analyzed and identified frame by frame with a time interval of 1 s. The result, as well as the background information is shown in Table 33.1.

Table 33.1 Evacuees' decision-making time and the observed cues during the pre-evacuation process (Time was set as zero at the beginning of the recorded videos)

	Mingshan High School	Wenchuan High School
1. Date of earthquake	April 20, 2013	May 12, 2008
2. Location of schools	Lu shan, Ya'an, Sichuan [18]	Chengdu, Sichuan [19]
3. Surface-wave magnitude scale	Ms 7.0[a]	Ms 8.0[b]
4. No of students (teachers)[c]	49 (1)	27 (1)
5. Prior-earthquake evacuation training	Yes [18]	Not clear
6. Investigating decision time[d] (s)	3.02 (±0.65)	11.85(±2.07)
7. Evacuating decision time[e] (s)	5.41 (±1.86)	45.37 (±0.91)
8. Observed cues[f]	0:00:00 s, slight shake, increased rapidly 0:00:04 s, the teacher instructed students to evacuate	0:00:00 s, slight shake, increased slowly 0:00:42 s, evacuation message broadcast

Note [a]https://en.wikipedia.org/wiki/2013_Lushan_earthquake
[b]https://en.wikipedia.org/wiki/2008_Sichuan_earthquake
[c]"No. of students(teachers)" refers to the evacuees visible in the video records
[d]"Investigating decision time" of an evacuee refers to duration from the beginning of the video to the time when an evacuee changes to investigating state
[e]"Evacuating decision time" refers to duration from the beginning of the video to the time when an evacuee changes to evacuating state
[f]"Observed cues" refers to the signals that may influence evacuee's perception of risk and thus their pre-evacuation decisions

It is obvious that the pre-evacuation time cannot be neglected as it is actually a long duration (Table 33.1), especially in Wenchuan High School scenario (45.37 s). This is likely because that the intensity of the shake was low and this abnormity could not be easily perceived. The possible lack of prior-disaster training might be another reason for such a large delay. In both cases, evacuees did not respond to the disasters immediately. Notably, the average time taken by the evacuees to make the investigating/evacuation decision varied significantly between the two scenarios, which supports the argument that the evacuation process is highly context-dependent [20].

In each of scenario, the two decision times are analyzed from the aspect of temporal distribution and spatial distribution. The large variance in the two decision times was observed among evacuees within each scenario (Fig. 33.2(a.1, b.1) and 33.2(a.3, b.3)). This suggests the existence of individual heterogeneity: evacuee may have different perception and reaction under the same environment. Compared with investigating decision time, evacuating decision time was more centralized. This testifies the significant effect of instructions on one's evacuating decision: in Mingshan High School scenario (Fig. 33.2(1.3)), the teacher instructed students to evacuate at 0:00:04 s; in Wenchuan High School scenario (Fig. 33.2(b.3)), the broadcast system gave the instruction at 0:00:42 s, The sharp increase in the number of pedestrians starting to evacuate indicates that they followed these instructions immediately.

In terms of the spatial distribution, concentric circles were observed in both investigating decision time (Fig. 33.2(a.2 and b.2)) and evacuating decision time (Fig. 33.2(a.4 and b.4)). This indicates that evacuees tended to be influenced by others' behaviors especially surrounding ones', which can be perceived as a confirmation of social proof theory [21]. In this theory, people tend to follow the actions of others in a given situation.

Based on the analyses above, it is summarized that evacuee's behavioral state could be influenced by physical signal, others' behaviors, and instruction. In addition, large

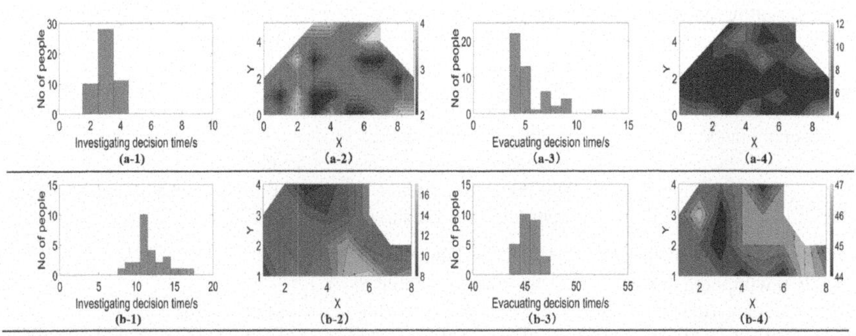

Fig. 33.2 Distribution of investigating decision time and evacuating decision time for **a** Mingshan High School and **b** Wenchuan High School scenario. In each of the scenario, four figures are provided: (x.1) temporal and (x.2) spatial distribution of investigating decision time, (x.3) temporal and (x.4) spatial distribution of evacuating decision time, where x represents **a/b**. Some areas are left blank due to the lack of data

behavioral heterogeneity exists in the perception of environment and reaction during pre-evacuation process.

33.3 Model Description and Validation

33.3.1 Model Description

Enlightened by Evacuation Decision Model [8], we propose a fuzzy-logic embedded decision-making model to predict evacuee's pre-evacuation time (Fig. 33.3). Some external cues can affect one's risk perception, such as physical signals, instructions, and others' behaviors as identified in Sect. 33.2. The level of perceived risk is calculated through Fuzzy Inference System (FIS) and then compared with one's decision threshold. When the perceived risk reaches the threshold for investigating/evacuating action, the evacuee will change to investigating/evacuating state. Notably, IBU and PPBU are considered due to aforementioned individual heterogeneity.

Fuzzy inference is a process of mapping from inputs to outputs through fuzzy logic. In Fuzzy Logic Theory, linguistic terms are used to describe one's perception and if–then rules are applied to the reasoning process. Each variable belongs to a class (the linguistic description of an input or output, such as "high" and "low") to a partial degree, where Membership Functions (MFs) are used to transform crisp value into likelihood degree. If–then rules are applied to infer the degree of output variables.

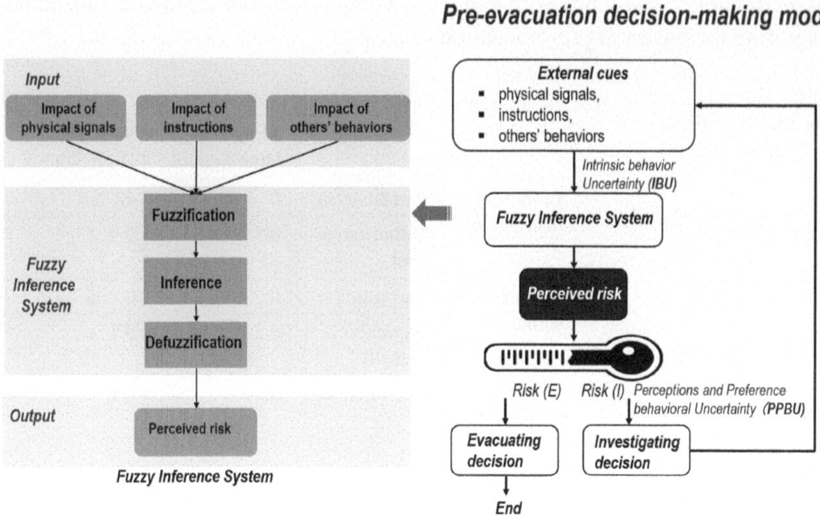

Fig. 33.3 Flowchart of pre-evacuation decision-making model

In our proposed model, cues are differentiated into physical signals, instructions, and other's behaviors. Physical signal is characterized by its original intensity, rate of change, and duration; instruction is represented by its authoritative level and duration; cue from others' evacuation behavior can be described by the number of people in normal/investigating/evacuating state. The impact of cues in each type will be aggregated following sub-FISs similar to main FIS as shown in Fig. 33.3. Output from these sub-FISs—impact of physical signals, impact of instructions, and impact of others' behaviors—are subsequently used as the input for the main FIS.

33.3.2 Model Calibration and Validation

Mingshan High School evacuation was used to obtain the threshold values for investigating/evacuating decisions ($Risk_{I,} Risk_E$). Loss function Eq. (33.1) was minimized based on the model in Fig. 33.3 and parameters specified in Table 33.2. Final value of $Risk_I = 0.29$, $Risk_E = 0.49$ was determined after extensive simulation.

$$\mathrm{L}(T_{I,real}, T_{I,real}, f_{Main}) = \min((f_{Main,I} - T_{I,real})^2 + (f_{Main,E} - T_{E,real})^2) \quad (33.1)$$

where: $T_{I,real} = 3.02$s, investigating time for Mingshan High School scenario;
$T_{E,real} = 5.41$s, evacuating time for Mingshan High School scenario.

Wenchuan High School scenario was then used to validate the performance of the model. The prediction variances of the average investigating and evacuating decision time are 0.71 s, 1.78 s respectively. Both the overall mean and standard derivation show reasonable consistency with real data, which proves the validity of this model in capturing the basics of pre-evacuation process.

Table 33.2 Parameter specification in the two scenarios

		Mingshan High School	Wenchuan High School
Feature of instruction	Start time (s)	0	42
	Authoritative level	0.5	0.9
Feature of signal	Start time (s)	0	0
	Original intensity	0.2	0.1
	Rate of change	0.4	0.1

33.4 Parametric Study

To test how evacuees would react under various physical signals and instructions, a series of simulations were conducted. All the simulations are based on the Mingshan High School evacuation scenario with some adjustments. It was assumed that there are 49 students evacuating from an earthquake. A physical signal and instruction was given at 0:00:00 s and 0:00:15 s respectively. The threshold value of perceived risk for the investigating and evacuating decision was 0.29, 0.49 respectively.

33.4.1 Influence of Physical Signals on Investigating/evacuating Decision

Overall decreasing trend of the investigating decision time was observed with the increase in the original intensity of the physical signals (Fig. 33.4a). When the original intensity was low (< 0.5), the downward trend was very pronounced. Under high original intensity (>0.5), the downward trend was stabilized. This was probably because evacuees were more sensitive to the occurrence of abnormity than the magnitude of abnormity. The moment a large shake or huge sound comes about, one would soon be alert and prepared to take actions in response.

In term of the influence of rate of change of physical signals, increasing ones and decreasing ones showed distinct effect. The increasing intensity of the physical signals convinced evacuees of their first thought, which leaded to the consistent shortening of investigating decision time (Rate of change = 0.2/0.4); Conversely, evacuees may doubt whether a "real" emergency event is happening when the physical signal is decreasing (Rate of change = –0.2/–0.4). In this sense, evacuees were more sensitive to gradual decrease than increase in the signal intensity.

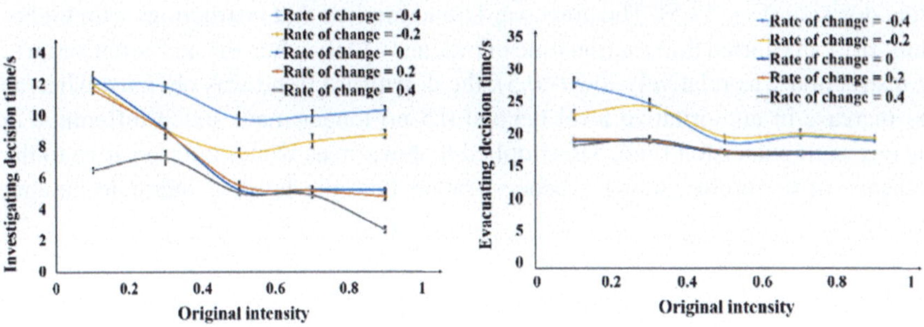

(a) (b)

Fig. 33.4 Plot of (left) investigating decision time and (right) evacuating decision time against the original intensity of the signal under different rate of change of intensity. The error bars indicate standard error for the mean value of 5 runs. Authoritative level of instruction = 0.5

Fig. 33.5 Plot of evacuating decision time against authoritative levels of instruction under different original intensities of the physical signal. The error bars indicate standard error for the mean value of 5 runs. Rate of change of intensity = 0

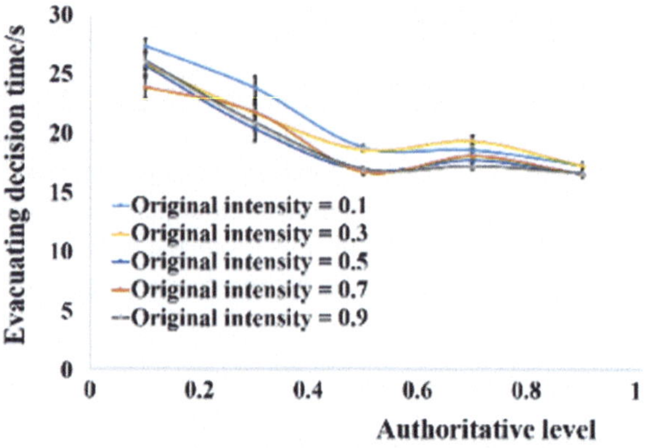

Similarly, negative correlation was found between original intensity and evacuating time albeit smaller slope (Fig. 33.4b). Conforming with common sense, increasing signals (positive rate of change) resulted in shorter evacuating time than decreasing signals (negative rate of change). At higher rate of change (0/0.2/0.4), the overlapping and almost constant curves suggested that the evacuating decision time was neither sensitive to the original intensity nor to the rate of change of physical signals. Nevertheless, the impact of negative rate of change (decreasing signal) was apparent, which indicated that decreasing signal tended to delay the evacuating decision. Accordingly, for the evacuation management, we suggest that disaster monitoring tools should be able to capture small signals and emit amplified signals.

33.4.2 Influence of Instructions on Evacuating Decision

The authoritative level of the instruction had a significant impact on the evacuating decision (Fig. 33.5). The downward trend implied that instructions with higher authority can shorten the reaction time of evacuees. Only when the authoritative level of instruction was relatively low (<0.5), the downward trend was obvious, whereas the increase in authoritative level beyond 0.5 no longer made much difference in the evacuating decision time. This implied that evacuees were more sensitive to the presence of hierarchy among evacuees, rather than the level of hierarchy among evacuees.

33.5 Conclusion and Future Work

In this study, a pre-evacuation decision-making model is proposed based on observed facts from real earthquake evacuation, and fuzzy logic theory addressing the vagueness of evacuees' perception of the event and thinking process. The consistency between the simulation results and the real-case data validates the effectiveness of this model. The main conclusions from video analysis and parametric studies are:

(1) Influencing factors of pre-evacuation process was summarized from qualitative analysis of two real-world evacuation videos: physical signals such as shake, instructions, and other's behaviors. Individuals showed behavioral heterogeneity in terms of the alertness to abnormity, the trust in instructions, and the conformity with group behaviors etc.
(2) Investigating decision time was mainly determined by physical signals. Evacuees were more sensitive to the onset of physical abnormity than the magnitude of abnormity, i.e. evacuees would more readily get prepared when something unusual happens. Moreover, an increase in the intensity of the physical signals could give support to one's judgment of emergency event whereas a decrease could lead to one's hesitation in the action.
(3) Evacuating decision was influenced by both physical signals and instructions. Instruction of higher authoritative level of can lead to faster pre-evacuation process, whereas the marginal effect of the authoritative level of instructions was significant only when the level was low (authoritative level <0.5).

However, this study applied a qualitative video-analysis method, to which a superior choice would be a quantitative based one. This limitation is caused by the poor quality of the videos [18] and the shaking of the cameras [18, 19]. With more high-quality videos added in, post-disaster questionnaires, and scaling methods such as Likert scale, Stapel scale etc. could be combined to give more rigorous video analysis. Besides, the calibration of fuzzy inference system could be further conducted following the method in [22]. Future work will focus on the gathering of real-case data and calibration of the parameters in the fuzzy inference system.

References

1. E.D. Kuligowski, R.D. Peacock, *A Review of Building Evacuation Models* (US Department of Commerce, National Institute of Standards and Technology Gaithersburg, MD, 2005)
2. R. Lovreglio, E. Ronchi, D. Nilsson, A model of the decision-making process during pre-evacuation. Fire Saf. J. **78**, 168–179 (2015). https://doi.org/10.1016/j.firesaf.2015.07.001
3. M. Kobes, I. Helsloot, B. De Vries, J. Post, Exit choice, (pre-)movement time and (pre-)evacuation behaviour in hotel fire evacuation—behavioural analysis and validation of the use of serious gaming in experimental research. Procedia Eng, pp. 37–51 (2010). https://doi.org/10.1016/j.proeng.2010.07.006.
4. M. Kobes, I. Helsloot, B. de Vries, J.G. Post, N. Oberijé, K. Groenewegen, Way finding during fire evacuation: an analysis of unannounced fire drills in a hotel at night. Build. Environ. **45**, 537–548 (2010). https://doi.org/10.1016/j.buildenv.2009.07.004
5. R.D. Peacock, B.L. Hoskins, E.D. Kuligowski, Overall and local movement speeds during fire drill evacuations in buildings up to 31 stories. Saf. Sci. **50**, 1655–1664 (2012). https://doi.org/10.1016/j.ssci.2012.01.003
6. H. Meltzer, The Process of Human Behavior. J. Educ. Psychol. **20**, 717–718 (1929). https://doi.org/10.1037/h0066738

7. E.D. Kuligowski, D.S. Mileti, Modeling pre-evacuation delay by occupants in World Trade Center Towers 1 and 2 on September 11, Fire Saf. J. **44**(2009), 487–496 (2001). https://doi.org/10.1016/j.firesaf.2008.10.001
8. P. Reneke, Evacuation decision model, US Department of Commerce, National Institute of Standards and Technology, Gaithersburg, MD (2013). https://doi.org/10.6028/NIST.IR.7914.
9. R. Lovreglio, E. Ronchi, D. Nilsson, An evacuation decision model based on perceived risk, social influence and behavioural uncertainty. Simul. Model. Pract. Theory. **66**, 226–242 (2016). https://doi.org/10.1016/j.simpat.2016.03.006
10. S. Li, J. Zhuang, S. Shen, A three-stage evacuation decision-making and behavior model for the onset of an attack. Transp. Res. Part C Emerg. Technol. **79**, 119–135 (2017). https://doi.org/10.1016/j.trc.2017.03.008
11. M. Haghani, M. Sarvi, L. Scanlon, Simulating pre-evacuation times using hazard-based duration models: Is waiting strategy more efficient than instant response? Saf. Sci. **117**, 339–351 (2019). https://doi.org/10.1016/j.ssci.2019.04.035
12. G. Bernardini, E. Quagliarini, M. D'Orazio, Towards creating a combined database for earthquake pedestrians' evacuation models. Saf. Sci. **82**, 77–94 (2016). https://doi.org/10.1016/j.ssci.2015.09.001
13. X. Yang, Z. Wu, Civilian monitoring video records for earthquake intensity: a potentially unbiased online information source of macro-seismology. Nat. Hazards. **65**, 1765–1781 (2013). https://doi.org/10.1007/s11069-012-0447-3
14. Z. Xue, Q. Dong, X. Fan, Q. Jin, H. Jian, J. Liu, Fuzzy logic-based model that incorporates personality traits for heterogeneous pedestrians. Symmetry (Basel). **9**, 239 (2017). https://doi.org/10.3390/sym9100239
15. H. Fu, N. Liu, J. Liang, A.J. Pel, S.P. Hoogendoorn, modeling and simulation of evacuation route choice behavior using fuzzy set theory, in *IEEE Conference Intelligent Transporation System Proceedings, ITSC*, October 2015, pp. 1327–1332. https://doi.org/10.1109/ITSC.2015.218
16. L.A. Zadeh, Fuzzy sets. Inf. Control. **8**, 338–353 (1965). https://doi.org/10.1016/S0019-9958(65)90241-X
17. I. Gerakakis, P. Gavriilidis, N.I. Dourvas, I.G. Georgoudas, G.A. Trunfio, G.C. Sirakoulis, Accelerating fuzzy cellular automata for modeling crowd dynamics. J. Comput. Sci. **32**, 125–140 (2019). https://doi.org/10.1016/j.jocs.2018.10.007
18. M. Li, Y. Zhao, L. He, W. Chen, X. Xu, The parameter calibration and optimization of social force model for the real-life 2013 Ya'an earthquake evacuation in China, Saf. Sci. **79**, 243–253 (2015). https://doi.org/10.1016/j.ssci.2015.06.018
19. X. Yang, Z. Wu, Y. Li, Difference between real-life escape panic and mimic exercises in simulated situation with implications to the statistical physics models of emergency evacuation: the 2008 Wenchuan earthquake, Phys. A Stat. Mech. Appl. **390**, 2375–2380 (2011). https://doi.org/10.1016/j.physa.2010.10.019
20. M. Haghani, M. Sarvi, Crowd behaviour and motion: empirical methods. Transp. Res. Part B Methodol. **107**, 253–294 (2018). https://doi.org/10.1016/j.trb.2017.06.017
21. B. Latane, J.M. Darley, Group inhibition of bystander intervention in emergencies. J. Pers. Soc. Psychol. **10**, 215–221 (1968). https://doi.org/10.1037/h0026570
22. C. Chai, Y.D. Wong, Fuzzy cellular automata model for signalized intersections. Comput. Civ. Infrastruct. Eng. **30**, 951–964 (2015). https://doi.org/10.1111/mice.12181

Chapter 34
Clogging in Velocity-Based Models for Pedestrian Dynamics

Qiancheng Xu, Mohcine Chraibi, and Armin Seyfried

Abstract In the simulation of pedestrian bottleneck flow with velocity-based models stable clogging could be observed, which is not in line with the movement of pedestrians in the real world. Four velocity-based models with different structures are selected in this paper to explore the cause of the phenomenon by identifying and analysing stable clogging. A series of simulations of pedestrians moving through a bottleneck are designed to investigate the decisive factors behind the unrealistic stable clogging, which includes the width of the exit, the flow of the pedestrian at the entrance and the size of the simulation time-step. The result of simulations with different models are compared and analyzed.

34.1 Introduction

Clogging is a phenomenon that can be described as a jamming arch formed in spatial structure by inert particles [1], animals [2] or pedestrians [3], which decreases or even blocks the flow of the exit [4].

The clogging of inert particles in a system without external input is usually stable [5, 6] and can not be solved spontaneously, whereas in living systems with animals or humans its duration is limited and relatively short due to the flexibility and elasticity of the body together with complex steering mechanisms that include cognitive processes, communication, cooperation and control of the body.

Q. Xu (✉) · M. Chraibi · A. Seyfried
Institute for Advanced Simulation, Forschungszentrum Jülich GmbH, 52425 Jülich, Germany
e-mail: q.xu@fz-juelich.de

M. Chraibi
e-mail: m.chraibi@fz-juelich.de

A. Seyfried
e-mail: a.seyfried@fz-juelich.de

A. Seyfried
School of Architecture and Civil Engineering, University of Wuppertal, 42119 Wuppertal, Germany

I. Zuriguel et al. (eds.), *Traffic and Granular Flow 2019*,
Springer Proceedings in Physics 252,
https://doi.org/10.1007/978-3-030-55973-1_34

However, microscopic models for pedestrian dynamics based on physical principles do not take communication and cooperation sufficiently into account, which could lead to the stable clogging similar to that in the system with inert particles. Several approaches have been proposed to solve this issue for different types of microscopic models. A friction parameter is introduced as the conditional probability in a cellular automation model [7] to solve a situation similar to the stable clogging, which is caused by two or more pedestrians want to occupy one cell, and [8] adopts a friction function instead of the friction parameter. Besides, game-theory is introduced in CA to solve the issue [9]. For social force models, introducing a random behavioral variation is widely used to dissipate stable clogging [10, 11].

In this paper, we focus on the stable clogging that occurs in several velocity-based models with increasing complexity. The effect of three decisive factors, including the width of the exit, the flow of the pedestrian at the entrance and the size of the simulation time-step are explored. The simulation results are used to investigate the relationship between the occurrence of stable clogging and the factors mentioned above.

34.2 Velocity-Based Models

The velocity-based models used in this work describe pedestrians as a circle of diameter ℓ. These models are all composed of a direction sub-model and a speed sub-model. In general we can write

$$\dot{\mathbf{x}}_i = V_i(\mathbf{x}_i, \mathbf{x}_j, \ldots) \cdot \mathbf{e}_i(\mathbf{x}_i, \mathbf{x}_j, \ldots), \tag{34.1}$$

where $\mathbf{x}_i$ and $\mathbf{e}_i$ represent the position and the moving direction of pedestrian i, respectively. V_i is a speed function.

For a given pedestrian i, the moving direction is firstly calculated with the direction sub-model. Two direction sub-model with different structures are used. The first one can be described as

$$\mathbf{e}_i(\mathbf{x}_i, \mathbf{x}_j, \ldots) = \mathbf{d}_i(\mathbf{x}_i), \tag{34.2}$$

where $\mathbf{d}_i$ is the unit vector representing the desired moving direction of the pedestrian depending only on its position. The second one was introduced in [12] and takes into account the position of the neighbors to model the avoidance of collisions. It can be expressed as

$$\mathbf{e}_i(\mathbf{x}_i, \mathbf{x}_j, \ldots) = u \cdot \Big(\mathbf{d}_i(\mathbf{x}_i) + \sum_{j \in N_i} R(s_{i,j}) \cdot \mathbf{e}_{i,j}\Big), \tag{34.3}$$

where u is a normalization constant such that $||\mathbf{e_i}|| = 1$, N_i is the set of all neighbours, $s_{i,j} = ||\mathbf{x}_i - \mathbf{x}_j||$, $\mathbf{e}_{i,j} = (\mathbf{x}_i - \mathbf{x}_j)/s_{i,j}$ and $R(s_{i,j}) = a \cdot \exp\big((\ell - s_{i,j})/D\big)$ where $a > 0$ and $D > 0$ are coefficients for calibration.

Table 34.1 The definition of the velocity-based models in this work

Model	Direction sub-model	Speed sub-model
1	Eq. (34.2)	Eq. (34.5)
2	Eq. (34.3)	Eq. (34.5)
3	Eq. (34.2)	Eq. (34.6)
4	Eq. (34.3)	Eq. (34.6)

The speed on the moving direction is calculated with the speed sub-model according to the minimum distance in front of pedestrian i, which is defined as

$$s_i(\mathbf{x}_i, \mathbf{x}_j) = \min_{j \in J_i} s_{i,j}, \tag{34.4}$$

where $J_i = \left\{j, \mathbf{e}_i \cdot \mathbf{e}_{i,j} \leq 0 \text{ and } \left|\mathbf{e}_i^{\perp} \cdot \mathbf{e}_{i,j}\right| \leq \ell/s_{i,j}\right\}$ is the set of the pedestrian in front. In this work we investigate the influence of two types of velocity sub-models. In the first, a pedestrian has only two states: moving with the free flow speed v_i or standing still. It is described by

$$V_i\Big(s_i(\mathbf{x}_i, \mathbf{x}_j)\Big) = \begin{cases} v_i, & s_i > \ell \\ 0, & s_i \leq \ell. \end{cases} \tag{34.5}$$

The second one considers a simplified fundamental diagram by introducing a coefficient $T > 0$ to adjust the speed according to the gap between two pedestrians [12], which is

$$V_i\Big(s_i(\mathbf{x}_i, \mathbf{x}_j)\Big) = \min\left\{v_i, \max\left\{0, \frac{s_i - \ell}{T}\right\}\right\}. \tag{34.6}$$

Based of the above-mentioned sub-models we investigate four different model variations (see Table 34.1).

34.3 How to Identify a Stable Clogging

Clogging in velocity-based models can be easily observed in a bottleneck scenario with a narrow exit. As shown in Fig. 34.1a, we can observe a stable arch formed by three blue agents in front of the narrow exit. The direct cause of the clogging can be regarded as two pedestrians blocking each other's movement as shown in Fig. 34.1b. We express this state as

$$s_{i,j} \leq \ell \text{ and } \mathbf{e}_{i,j} \cdot \mathbf{e}_i < 0 \text{ and } \mathbf{e}_{i,j} \cdot \mathbf{e}_j > 0. \tag{34.7}$$

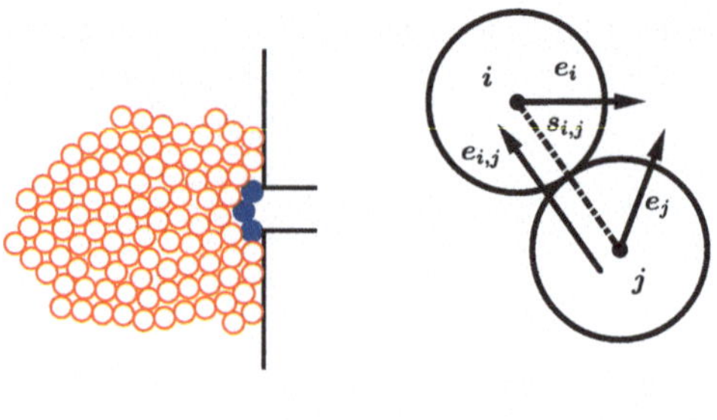

(a) (b)

Fig. 34.1 Clogging in velocity-based models. Left: Clogging in the bottleneck. Right: Direct cause of the clogging

Pedestrian i and j are both on each other's moving path, which means $i \in J_j$ and $j \in J_i$. Besides, since they are already in contact, the distance between their centers is equal to ℓ. Therefore, according to the definitions in Eqs. (34.5) and (34.6) their speed is simultaneously zero.

Equation (34.7) can be seen as the necessary and sufficient condition of the clogging in the hereby used models. Pedestrians in the clogging state can only regain a nonzero speed by changing their moving direction. If, however, the clogging can not be solved swiftly, more pedestrians will gather around, which significantly reduces the available manoeuvring space of the two blocking pedestrians leading to the development of a stable clogging. Therefore, two additional conditions are introduced to identify a stable clogging with respect to time and space. The first one is that the clogging should last longer than 1 s. The second condition is only for the model using Eq. (34.3) as the direction sub-model (models 2 and 4). Here the further condition refers to the location where the clogging occurs. The position should be within one meter distance to the center of the exit. Since agents in the simulation can not solve the clogging by themselves when using model 1 and model 3, which means all the clogging in models 1 and 3 are stable, the second condition is only for models 2 and 4. Only cloggings meeting Eq. (34.7) and the two further conditions will be investigated as "stable clogging".

34.4 Simulations

A bottleneck scenario as shown in Fig. 34.2a is simulated. In each simulation, 400 agents are generated in the "Source Area" with a constant rate. Then these pedestrians move through the "Moving Area" to leave the area from the middle "Exit". The size of the "Source Area" is 8×8 m^2, the size of the "Moving Area" is 10×8 m^2 and the width of the "Exit" is variable (w). During the evacuation process, we record the position and the moment of each clogging that satisfies the condition defined in the previous section, and randomly remove one of the agents involved in the clogging to avoid standstill of the outflow.

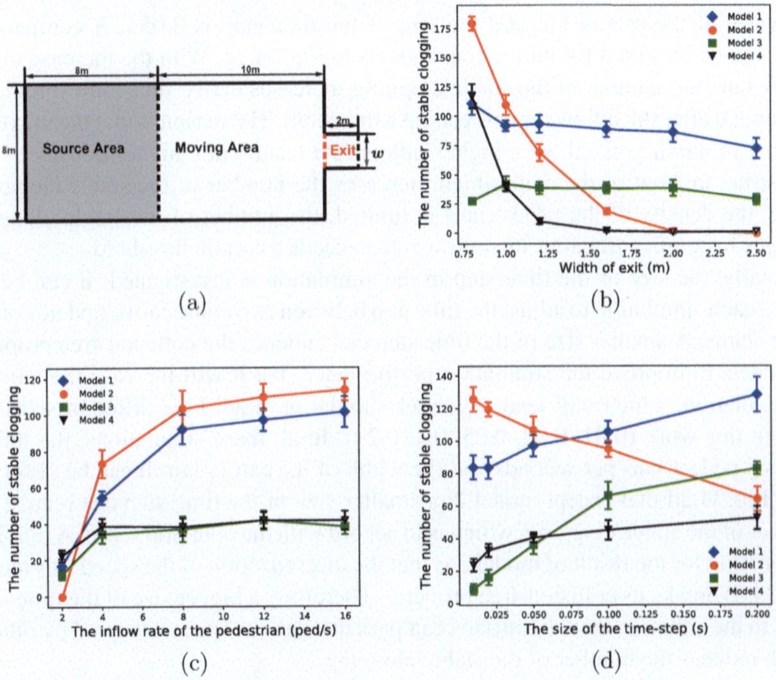

Fig. 34.2 The simulation scenario and result. Upper left: The bottleneck scenario for the simulation. Upper right: The trend of the number of the stable clogging over the width of the exit. Lower left: The trend of the number of the stable clogging over the inflow rate of the "Moving area". Lower right: The trend of the number of the stable clogging over the size of the time-step in the simulation

Firstly, the effect of the exit width on the occurrence of the stable clogging is analysed. Six different widths are chosen for the simulation ($w = 0.8, 1.0, 1.2, 1.6, 2.0, 2.5$ m). In all these simulations, 8 agents per second are generated in the "Source Area". The time-step is 0.05 s.

Figure 34.2b shows the change of the number of the stable clogging with respect to the width of the exit for four different models. When the direction sub-model is Eq. (34.3) (models 2 and 4), the number of the stable clogging decreases as the width of the exit increases. The reason is that a wider exit can provide more space to the pedestrians to avoid the clogging by changing their moving direction, thereby reducing the number of the stable clogging. However, when the direction sub-model Eq. (34.2) is used (models 1 and 3), pedestrians do not change their moving direction according to their neighbours, which explains why the width of the exit has little influence on the number of the stable clogging.

Furthermore, the influence of the inflow rate of the "Moving area" on the occurrence of the stable clogging is investigated. Five different inflow rates are selected for the simulation (2, 4, 8, 12, 16 pedestrians per seconds), which is controlled by the generation rate of the pedestrians in the "Source Area". In all these simulations, the width of the exit is 1 m and the size of the time-step is 0.05 s. A comparable trend can be observed for all the four models in Fig. 34.2c. With the increase of the inflow rate, the number of the stable clogging increases firstly, then almost remains unchanged after the inflow rate exceeding a threshold. The reason is that the maximal outflow of a exit is fixed, so a higher inflow rate results in a higher density of the pedestrian in front of the exit, which increases the number of the stable clogging. Since, the density of the pedestrians is limited, the number of the stable clogging changes insignificantly after the inflow rate exceeds a certain threshold.

Finally, the size of the time-step in the simulation is investigated. It can be set before each simulation to adjust the time gap between two consecutive updates of the agent status. A smaller size of the time step can enhance the collision-free property of models to improve the simulation performance, but it will increase the amount of calculation, which will lead to longer simulation time. Five different sizes are used in this work (0.01, 0.02, 0.05, 0.1, 0.2 s). In all these simulations, the inflow rate is 8 pedestrians per seconds and the width of the exit is 1 m. It can be observed from Fig. 34.2d that except model 2, a smaller size of the time-step can reduce the number of the stable clogging, which is in accord with the common sense. A possible explanation for the result of model 2 is that the discretization of the speed sub-model Eq. (34.5) breaks its collision-free property. Therefore, a bigger size of the time-step leads to the situation that pedestrians can pass through the space occupied by others, which reduces the number of the stable clogging.

34.5 Conclusion

In this work, stable clogging in different variations of the collision-free velocity model [12] is investigated numerically. Model variants considering the position of neighbour in the direction sub-model, leads to lower number of the stable clogging as the width of the exit increases, otherwise the width of the exit has little influence. Besides, with the increase of the inflow rate, the number of the stable clogging increases firstly, then changes little after the inflow rate exceeding a threshold. A lowering of the time step decreases the number of stable clogging in models considering a non-trivial fundamental diagram where the speed decreases with decreasing distance. Moreover, when the direction sub-models are the same, using Eq. (34.6) as the speed sub-model leads to fewer stable clogging than Eq. (34.5), which can be observed in all simulations in this work.

A threshold of the duration time will be obtained with further simulations to distinguish the stable and the temporary clogging more accurate, and a new condition about the space will be introduced to ignore the fake stable clogging in simulations which is caused by other stable clogging arisen earlier. In addition, the effect of more factors is going to be researched in the future work.

References

1. R. Arévalo, I. Zuriguel, Soft Matter **12**(1), 123 (2016)
2. A. Garcimartín, J. Pastor, L. Ferrer, J. Ramos, C. Martín-Gómez, I. Zuriguel, Phys. Rev. E **91**(2), 022808 (2015)
3. A. Garcimartín, D.R. Parisi, J.M. Pastor, C. Martín-Gómez, I. Zuriguel, J. Stat. Mech.: Theory Exp. **2016**(4), 043402 (2016)
4. J. Adrian, M. Amos, M. Baratchi, M. Beermann, N. Bode, M. Boltes, A. Corbetta, G. Dezecache, J. Drury, Z. Fu et al., Collect. Dyn. **4**(A19), 1 (2019)
5. I. Zuriguel, Pap. Phys. **6**, 060014 (2014)
6. C. Lozano, G. Lumay, I. Zuriguel, R. Hidalgo, A. Garcimartín, Phys. Rev. Lett. **109**(6), 068001 (2012)
7. A. Kirchner, K. Nishinari, A. Schadschneider, Phys. Rev. E **67**(5), 056122 (2003)
8. D. Yanagisawa, A. Kimura, A. Tomoeda, R. Nishi, Y. Suma, K. Ohtsuka, K. Nishinari, Phys. Rev. E **80**(3), 036110 (2009)
9. A. von Schantz, H. Ehtamo, Phys. Rev. E **92**(5), 052805 (2015)
10. D. Helbing, L. Buzna, A. Johansson, T. Werner, Transp. Sci. **39**(1), 1 (2005)
11. R.C. Hidalgo, D.R. Parisi, I. Zuriguel, Phys. Rev. E **95**(4), 042319 (2017)
12. A. Tordeux, M. Chraibi, A. Seyfried, in *Traffic and Granular Flow'15* (Springer, 2016), pp. 225–232

Chapter 35
Exit-Choice Behavior in Evacuation Through an L-Shaped Corridor

Daichi Yanagisawa, Milad Haghani, and Majid Sarvi

Abstract We studied exit-choice behavior in an L-shaped corridor, which included two exits. In the experiments, it was observed that more than half of the evacuees chose the farther exit. The fraction of evacuees choosing the farther exit was larger in the running scenarios than in the walking scenarios. To investigate the significant factors of the exit-choice behavior, we conducted simulations by the floor-field cellular-automata model, which included the effects of herding, avoidance of exit crowding, and inertia. Only the inertia effect succeeded to reproduce the experimental results, i.e., when the inertia effect increased, the fraction of choosing the farther exit increased and finally exceeded 50%. The results of the experiments and simulations correspond well with each other if we consider the relation between the speed and inertia. Hence, our results indicate that the inertia effect is one of the important factors in the exit-choice behavior in the L-shaped corridor.

D. Yanagisawa (✉)
Research Center for Advanced Science and Technology, The University of Tokyo,
4-6-1, Komaba, Meguro-ku, Tokyo 153-8904, Japan
e-mail: tDaichi@mail.ecc.u-tokyo.ac.jp
URL: https://sites.google.com/g.ecc.u-tokyo.ac.jp/yana

Department of Aeronautics and Astronautics, School of Engineering, The University of Tokyo,
7-3-1, Hongo, Bunkyo-ku, Tokyo 113-8656, Japan

M. Haghani
Institute of Transport and Logistics Studies—Business School, The University of Sydney,
Darlington, NSW 2008, Australia

M. Sarvi
Department of Infrastructure Engineering, The University of Melbourne,
Parkville, VIC 3010, Australia

I. Zuriguel et al. (eds.), *Traffic and Granular Flow 2019*,
Springer Proceedings in Physics 252,
https://doi.org/10.1007/978-3-030-55973-1_35

35.1 Introduction

Exit-choice behavior in evacuation has been studied vigorously in recent years [1–4]. It is not only interesting from the academic point of view, but also important for developing effective guiding methods and designing buildings where pedestrians can evacuate easily.

To understand the exit-choice behavior of evacuees and pedestrians, we need to know what factors affect the exit-choice behavior. Therefore, significance of various factors in the exit-choice behavior was investigated in the previous researches. Distance to the exits, herding behavior, size of exit crowding (queue length), exit width, exit familiarity, and exit visibility are the examples of the important factors, which strongly influence on the exit-choice behavior. In many studies, statistical results elucidated that not only the significance of each factor but also quantitative strength of influence by comparing several factors.

Evacuations considered in these previous studies were conducted in either simple rectangular rooms or realistic complex venues. In other words, to the best of our knowledge, it has been not investigated so far how the important factors changes due to the geometrical setting. Therefore, we studied the exit-choice behavior in an L-shaped corridor and clarified the important factors when the evacuees turn at a corner.

35.2 Experiment

We analyzed the experimental data in [3]. A part of the original experimental area in [3] as in Fig. 35.1a was focused in this study. Depending on the scenario, approximately forty participants entered into the corridor from the entrance at the bottom left and exited through either Exit A or Exit B at the top right. A total of ten scenarios were selected for performing the analysis. The evacuees walked in four of those scenarios and ran in the remaining six scenarios.

Figure 35.1b depicts the fraction of Exit B utilization (F_B) and its 95% confidence interval in the walking and running scenarios. We see that $F_B > 0.5$ in both the walking and running scenarios. As can be observed from Fig. 35.1a, Exit B was farther than Exit A. Distance is considered to be one of the fundamental factors influencing the exit-choice behavior [2]; however, it seems that more significant factors determined the exit choices in the L-shaped corridor.

The results of the binomial-proportion tests (Clopper-Pearson) for the null hypothesis $F_B = 0.5$ for the walking and running scenarios were $p = 0.28$ and 3.9×10^{-9}, respectively; hence, the null hypothesis was rejected only in the running scenarios. Moreover, since F_B(Walking) $<$ F_B(Running), we consider that moving speed strongly affected on the exit-choice behavior in the L-shaped corridor.

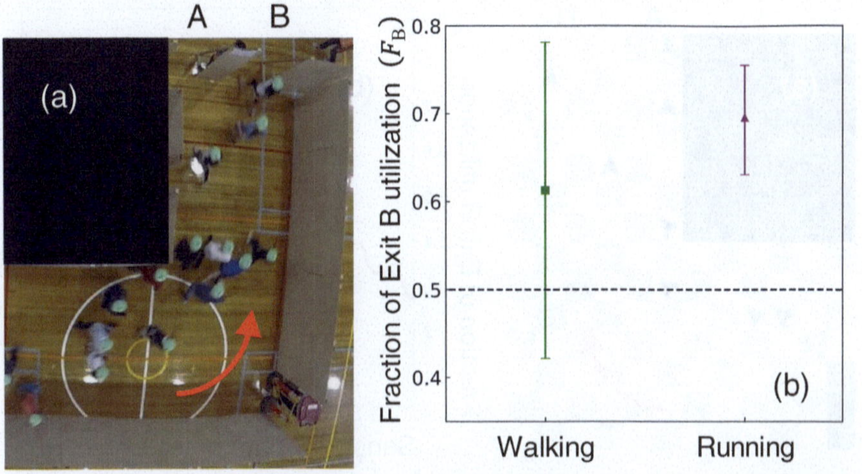

Fig. 35.1 **a** Schematic of the experiment. The evacuees entered into the L-shaped corridor from the entrance in the lower-left part of the figure and evacuated from the exits (Exit A and B) at the upper-right part of the figure. The lateral and longitudinal lengths of the corridor were 6.0 m and 7.15 m, respectively. The lateral and longitudinal lengths of the obstacle (black part) were 2.75 m and 3.6 m, respectively. The widths of Exit A and B were both 0.9 m. Exit A was in contact with the obstacle. The distance between the centers of Exit A and B was 1.5 m. **b** Experimental results of the fraction of Exit B utilization (F_B) for the walking and running scenarios. The error bars represent the 95% confidence intervals

35.3 Floor Field Model and Simulation

To investigate the significant factors influencing the exit-choice behavior in the L-shaped corridor in the running scenarios, we conducted an evacuation simulation using the floor-field cellular-automata model (FFCAM) [5, 6]. Figure 35.2a shows the schematic of the FFCAM for evacuation from the L-shaped corridor. The corridor was divided into cells whose sizes are 50 cm × 50 cm. Each cell can include one evacuee at most. The time is discrete, and the positions of the evacuees are updated in parallel. The evacuees stochastically move to the Moore neighboring cells (eight directions) or stay in their present cell in one time step. The moving direction of each evacuee in every time step is determined by the following four factors: the distances to the exits; herding behavior; exit crowding; and inertia effect, which are explained in the following subsections.

In addition to the floor cells (white and gray), which evacuees can move, there are obstacle (black), entrance (blue), and exit (green and pink) cells as in Fig. 35.2a. (For interpretation of the references to color of the cells in Fig. 35.2a, the reader is referred to the web version of this article.) Obstacle cells cannot be entered by evacuees. From the entrance cells, evacuees stochastically enter into the corridor. The number of entering evacuees in one time step obeys Poisson distribution with the parameter λ. Note that the maximum number of entering evacuees in one time step

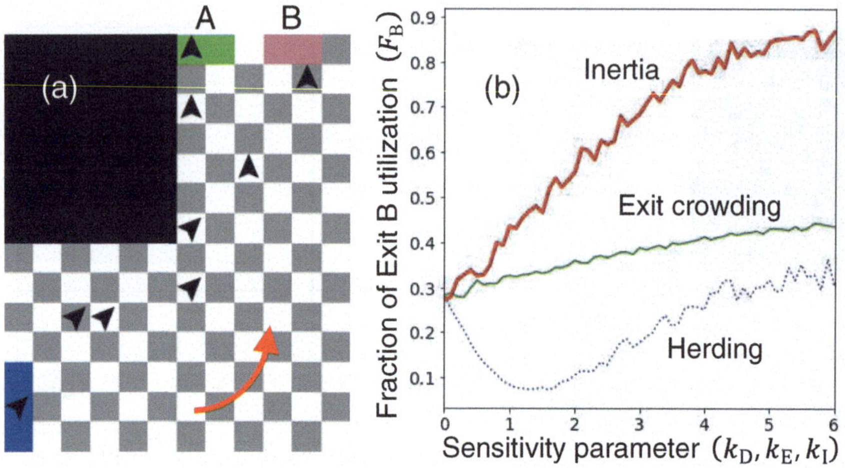

Fig. 35.2 **a** Schematic of the simulation using the floor-field model in the L-shaped corridor. **b** Simulation results of the fraction of Exit B utilization (F_B) for the walking and running scenarios. The parameters were $\lambda = 0.5$, $k_S = 3.0$, $\alpha = 0.2$, $\delta = 0.2$, $d_{ea} = 1$, and $\mu = 0.2$

is restricted by the number of the entrance cells. At the exit cell, evacuees evacuate from the room with the probability 1.

35.3.1 Distance to Exits (Static Floor Field)

Effect of the distance to the exits was implemented by the static floor field (SFF) S. Each cell has SFF, which represents the Euclidean distance from the exit cells. When there are multiple exit cells, the minimum value, i.e., the Euclidean distance from the closest exit cell, is set as SFF. The evacuees move to the cells which have small SFF with high probabilities; therefore, the evacuees can move to the exits and tend to choose the nearer exits.

35.3.2 Herding Behavior (Dynamic Floor Field)

Herding behavior was implemented by the dynamic floor field (DFF) D, which are footprints of the evacuees. Each evacuee leaves a footprint at the previous cell when they move. Evacuees refer to the sum of the footprints (DFF) at the Moore neighboring and present cells and move to the cells which have large DFF with high probabilities. The DFF of a cell c diffuse and decay as,

$$D_c(t+1) = (1-\delta)\left[(1-\alpha)D_c(t) + \frac{1}{8}\alpha \sum_{\tilde{c}\in\mathbb{M}} D_{\tilde{c}}(t)\right], \quad (35.1)$$

where t represents the time step, and α and δ are the diffusion and decay parameters, respectively. $\mathbb{M}$ is the set of the Moore neighboring cells of the cell c. Note that own footprint left just one time step before is not included in his/her DFF, otherwise, the evacuees are trapped with own DFF and keep moving back and forth between the same two cells forever.

35.3.3 Exit Crowding

Exit crowding was considered in several previous studies. We applied the method in [5] and defined the exit area, which is the cells whose euclidean distance from the nearest exit X are smaller than or equal to the threshold d_{ea}. If a cell satisfies the condition of the exit area for multiple exits, it is removed from all the exit area.

The density in the exit X area is computed as ρ_{X}. The transition probability to the neighboring cell which is the closest to the exit X is modified depending on ρ_{X}, thus, the evacuees try to avoid crowded exits and move to the vacant exits.

35.3.4 Inertia

Inertia effect for FFCAM was considered in [6]. When an evacuee moved to upper neighboring cell at t, the transition probabilities to the upper, upper-right, and upper-left cells are increased at $t+1$. By this simple rule, evacuees tend to move in the same direction as in the previous time step, hence, the inertia effect is implemented.

35.3.5 Transition Probability

By considering the four aforementioned factors, the transition probability to the neighboring cell c from the present cell is described as

$$p_c = Z^{-1}\exp\left(-k_{\mathrm{S}}S_c + k_{\mathrm{D}}D_c + k_{\mathrm{E}}\sum_{\mathrm{X}\in\mathbb{E}} E_c(\mathrm{X}) + k_{\mathrm{I}}I_c\right)\xi_c, \quad (35.2)$$

where Z is the normalization, and k_{S}, k_{D}, k_{E}, and k_{I} are the sensitivity parameters for SFF, DFF, exit crowding, and inertia effect, respectively. $\mathbb{E}$ is the set of exits. $E_c(\mathrm{X})$ represents the effect of exit crowding ($1-\rho_{\mathrm{X}}$ if the cell c is the closest to the exit X of the Moore neighboring cells and 0 otherwise). I_c represents the effect of inertia

(1 if the moving direction to the cell c is the same as in the previous time step, 1/2 if the difference of the moving direction to the cell c and that in the previous time step is 45°, 0 in other cases). ξ_c represents the effect of obstacle (0 for obstacle cells and 1 otherwise).

Evacuees stochastically choose their target cells according to the transition probability (35.2) and move. Note that occupied cells are included in the options of moving directions; however, evacuees fail to move and stay at their present cells when they choose the occupied cells.

If more than one evacuee select the same cell and a conflict occurs, one of the evacuees is randomly selected and move to his/her target cell, and others stay at their present cells with the probability $1 - \mu$. With the probability μ, the conflict is not solved, and all the evacuees stay at their present cells.

35.3.6 Results

Figure 35.2b depicts the results of the simulation (F_{B}) as a function of the sensitivity parameters. We fixed the sensitivity parameter of the distance effect and controlled one among the remaining three parameters (two parameters were set to zero) in each plot. Neither herding behavior nor avoidance of exit crowding overcomes the distance effect and achieves $F_B > 0.5$. In contrast, when the inertia effect increases, F_{B} increases, exceeds 0.5 and becomes 0.7, which can be observed in the running scenarios in real-time experiments. The inertia effect depends on the speed of the evacuees; hence, we can consider that the inertia effect was larger in the running scenarios than in the walking scenarios in the experiments. Therefore, the results of the experiment and simulation are in good agreement, i.e., F_{B} increase as the inertia effect increases. Thus, our results indicate that the inertia effect was the main cause of the large F_B in the L-shaped corridor.

35.4 Conclusion

In this paper, we studied the exit-choice behavior of evacuees in a L-shaped corridor. It was observed that more evacuees chose to evacuate from the farther exit from the entrance in the experiments. The fraction of choosing the farther exit increased when evacuees increased their moving speeds.

To investigate the cause of the experimental phenomena, we performed simulations using the floor-field cellular-automata model for evacuation, which includes the effect of distance to the exits, herding behavior, avoidance of exit crowding, and inertia. The simulation results indicated that only the inertia effect could reproduce the large fraction of choosing the farther exit observed in the experiments. Furthermore, the fraction increased as the inertia effect increased. We consider that this result also corresponds the experimental result because the inertia effect depends on the moving speed. Therefore, our results indicate that the inertia effect is one of the significant factors of the exit-choice behavior when evacuees need to turn just before the exits.

Acknowledgements The experimental videos were provided by the crowd dynamics group at The University of Melbourne. This work was partially supported by JSPS KAKENHI Grant Number JP15K17583, The 37th LNest Grant JR East Prize, JST-Mirai Program Grant Number JPMJMI17D4, Japan, and MEXT as "Priority Issues and Exploratory Challenges on post-K (Supercomputer Fugaku)" (Project ID: hp190163).

References

1. N.W.F. Bode, A.U. Kemloh Wagoum, E.A. Codling, Information use by humans during dynamic route choice in virtual crowd evacuations. R. Soc. Open Sci. **2**(1), 140,410 (2015). https://doi.org/10.1098/rsos.140410
2. Z. Fang, W. Song, J. Zhang, H. Wu, Experiment and modeling of exit-selecting behaviors during a building evacuation. Phys. A Stat. Mech. Appl. **389**(4), 815–824 (2010). https://doi.org/10.1016/j.physa.2009.10.019
3. M. Haghani, M. Sarvi, Hypothetical bias and decision-rule effect in modelling discrete directional choices. Transp. Res. Part A Polic. Pract. **116**(June), 361–388 (2018). https://doi.org/10.1016/j.tra.2018.06.012
4. M. Kinateder, B. Comunale, W.H. Warren, Exit choice in an emergency evacuation scenario is influenced by exit familiarity and neighbor behavior. Saf. Sci. **106**(March), 170–175 (2018). https://doi.org/10.1016/j.ssci.2018.03.015
5. S. Liu, L. Yang, T. Fang, J. Li, Evacuation from a classroom considering the occupant density around exits. Phys. A Stat. Mech. Appl. **388**(9), 1921–1928 (2009). https://doi.org/10.1016/j.physa.2009.01.008
6. K. Nishinari, A. Kirchner, A. Namazi, A. Schadschneider, Extended floor field CA model for evacuation dynamics. IEICE Trans. Inform. Syst. **E87-D**(3), 726–732 (2004)

Chapter 36
Bidirectional Flow on Stairs at Different Flow Ratios

Rui Ye, Jun Zhang, Mohcine Chraibi, and Weiguo Song

Abstract Bidirectional flow happens not only on horizontal planes, but also on stairs. So far, bidirectional horizontal movements have been investigated through various field studies and controlled experiments, but such movements on stairs are rarely paid attention to, especially those based on trajectories. In this paper we investigate bidirectional flow on stairs at different flow ratios ranging from 0.1 to 0.9. Preliminary results such as lane formation, crossing time and fundamental diagrams in three movement scenarios are presented, and the characteristics of the descending and ascending pedestrians are compared. Our results can be applied for the validation and calibration of simulation models on stairs.

36.1 Introduction

Over the last few decades, city population has been increasing dramatically, especially in the developing countries. Nowadays dense crowds are frequently observed in various public facilities, such as subway stations, airports, shopping malls and stadiums. To improve the traffic efficiency in these facilities as well as maintain safety and comfort, the characteristics of pedestrian movement under different geometries should be carefully understood and considered. As a typical pedestrian flow that one can experience in daily life, the bidirectional stream has attracted many researchers'

R. Ye · J. Zhang · W. Song (✉)
State Key Laboratory of Fire Science, University of Science and Technology of China, Hefei 230027, People's Republic of China
e-mail: wgsong@ustc.edu.cn

R. Ye
e-mail: yerui@mail.ustc.edu.cn

J. Zhang
e-mail: junz@ustc.edu.cn

M. Chraibi
Institute for Advanced Simulation, Forschungszentrum Jülich, 52425 Jülich, Germany
e-mail: m.chraibi@fz-juelich.de

I. Zuriguel et al. (eds.), *Traffic and Granular Flow 2019*,
Springer Proceedings in Physics 252,
https://doi.org/10.1007/978-3-030-55973-1_36

Fig. 36.1 Typical bidirectional movement on stairs observed in Hefei, China

interests. With laboratory controlled experiments, the differences of fundamental diagrams between uni- and bidirectional flows [1, 2], with and without walls [3], and at varying flow ratios [3, 4] have been investigated in normal straight corridors. Apart from these, bidirectional experiments have also been conducted in other geometries. For example, the oscillation flow has been experimentally verified when pedestrians from two directions pass through the bottleneck at the same time [5]. Besides, the velocity during the bidirectional turning is quantified in a corner experiment at different densities [6]. However, when looking back on all the related works, it's found that they only focus on planar motion. A question therefore arises: whether the bidirectional movement only occurs on the horizontal plane? Of course the answer is no. Actually the bidirectional motion can also be found on stairs. Here in Fig. 36.1 we show two typical scenes found in Hefei, China, where such movements are observed.

As far as the authors are concerned, most studies about stairs so far focus on the unidirectional descending process when pedestrians evacuate downstairs, and the bidirectional movement is barely paid attention to except for a model proposed in [7]. Following this work, in this paper we focus on the experimental study of bidirectional flow on stairs at different flow ratios. The experimental setup will be explained in Sect. 36.2. The preliminary results will be given in Sect. 36.3. Finally, the conclusion and outlook will be presented in Sect. 36.4.

36.2 Experiment Setup

A sketch and snapshot of the experiment setup are shown in Fig. 36.2. The experiment is conducted on an open staircase with totally 17 steps. Each step has almost the same size of step riser and step tread and they're 15 cm and 30 cm, respectively. The corridor is built using desks and its width is 2 m. 100 university students with the average height of 167.9 cm are recruited as participants. The flow ratio r is the proportion between the descending flow and total flow. Detailed illustrations about each run during the experiment can be found in Table 36.1. Before each run, the numbers of participants in the waiting areas will be adjusted according to the flow ratio at that

Fig. 36.2 Sketch and snapshot of the experiment setup

Table 36.1 Illustration about each run during the experiment

Chronological order	1	2	3	4	5	6	7	8	9
r (rounding)	0.1	0.9	0.2	0.8	0.3	0.7	0.4	0.6	0.5
$N_{descending}$	10	90	21	79	30	71	40	60	50
N_{total}	100	100	100	100	100	101	100	100	100

run. The whole walking process is recorded by a camera, and pedestrians' trajectories are further extracted with *PeTrack* software, with the maximum error of 5 cm [8]. Besides, the coordinate system is also displayed in the snapshot of Fig. 36.2.

36.3 Preliminary Results

Before analyzing the results, we project the trajectories onto the horizontal plane. Typical trajectories at four flow ratios are shown in Fig. 36.3, with the red and blue lines indicating the descending and ascending movement, respectively. Similar to planar motion, the lane formation phenomenon can be clearly observed in all runs.

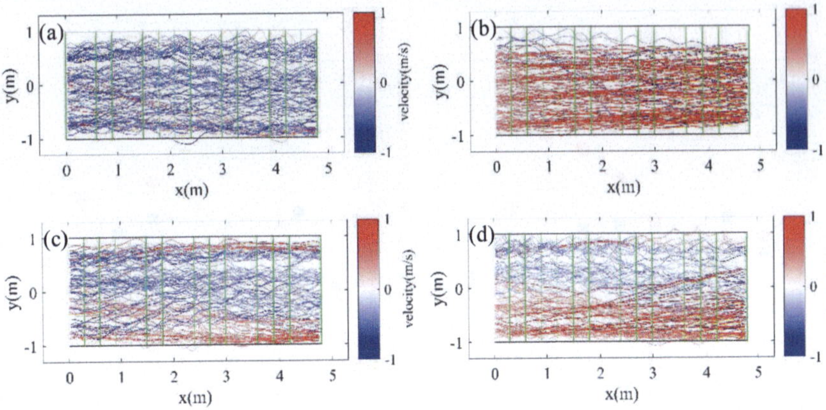

Fig. 36.3 Typical trajectories at four runs. **a** $r = 0.1$. **b** $r = 0.9$. **c** $r = 0.4$. **d** $r = 0.6$

Through watching the video recording of the experiment process, two major modes of collision avoidance are found. For the runs when $r < 0.3$ or $r > 0.7$, the minor flow tends to be more active and flexible in avoiding the opposite pedestrians by having strong right walking preference, leading to a two-lane case. However, when there are at least 30 pedestrians at both sides, the descending flow will be more likely to play the dominant role in collision avoidance by splitting into two lanes and a three-lane case therefore emerges. Besides, with the bidirectional flow becoming more balanced, a significant decrease of velocity can be observed according to the intensity of the color in the trajectories in each plot. Also, the ascending pedestrians seem to have larger head sway than the descending ones, which may be due to the lower walking velocity. What's more, because of the unequal numbers of pedestrians in two waiting areas, unidirectional movement either along the descending or ascending direction will occur when pedestrians from one side have run out. That is to say, we can obtain data from both uni- and bidirectional flows in one single run.

The crossing time of an individual is the duration between the first and last time when he/she appears in the corridor. For one run, we treat pedestrians in opposite moving directions separately, and values of the mean and standard deviation are displayed in Fig. 36.4. Generally speaking, the mean crossing time in the descending flow is always smaller than that in the ascending flow apart from $r = 0.2$. Although pedestrians in the descending flow are thought to walk faster, their number is smaller compared with the ascending flow when $r < 0.5$, which is the reason why the difference in mean value is not significant in these runs. After the descending flow becomes the major one, such difference is enlarged, and the maximal value can be up to around 10 s when the flow ratio is 0.7. As for the standard deviation, we find the value in the minor flow is larger. No matter for the major or minor flow, the first few pedestrians always walk fast, but the minor flow will soon be affected by the major flow as there are more pedestrians, thus lengthening the crossing time and making the standard deviation become larger in the minor flow.

We use Voronoi cells to collect data in the fundamental diagrams [9]. The generated Voronoi diagram at a certain frame when the flow ratio equals to 0.5 is presented

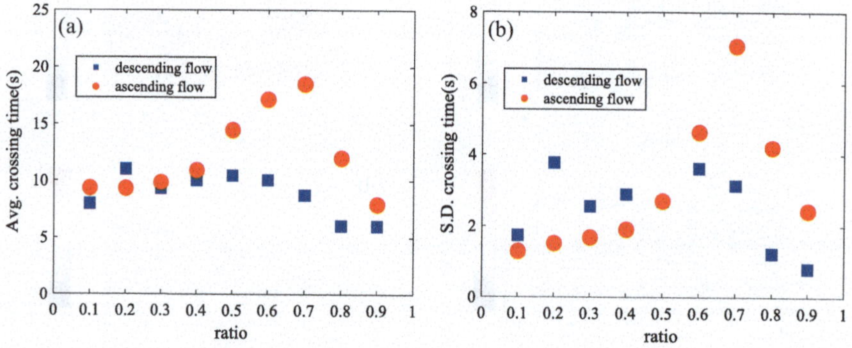

Fig. 36.4 Statistical results of the crossing time. **a** Mean. **b** Standard deviation

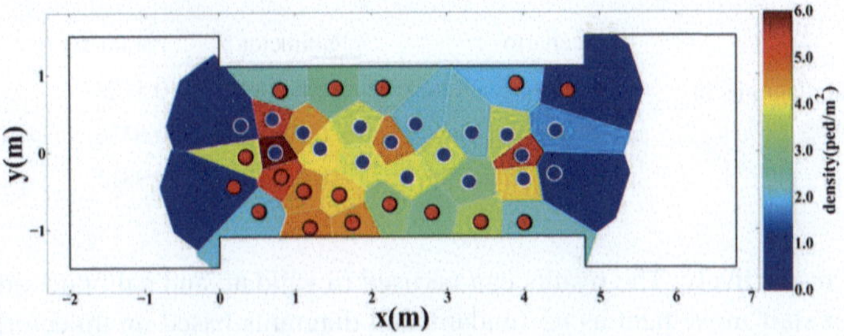

Fig. 36.5 The generated Voronoi cells according to pedestrians' positions at a certain frame

in Fig. 36.5 with colors indicating different values of density. The measurement area is set as $x \in [0.5\text{ m}, 4.3\text{ m}]$, and $y \in [-1.0\text{ m}, 1.0\text{ m}]$ to eliminate the boundary effect.

As mentioned earlier, we can obtain data from both uni- and bidirectional scenarios at one single run. Those data from the Voronoi cells are further classified to form fundamental diagrams in three scenarios, namely the bidirectional flow, ascending flow and descending flow. The raw scatter data are plotted in Fig. 36.6a. From the scatter plot, it looks that even at the same density, the difference in velocity is rather large no matter in uni- or bidirectional scenarios, and it can also be roughly found that pedestrians in the descending flow can walk the fastest. To better compare the fundamental diagrams in the three scenarios, we average the data with a density interval of 0.1 ped/m^2, and the results are shown in Fig. 36.6b, where the lines and transparent areas represent the mean values and standard deviations. Through the average lines, the velocity almost decreases linearly with the increase of density. We further fit these lines with the following form: $v = a \cdot \rho + b$ and the fitting parameters are listed in Table 36.2. From parameter a, it's found that the velocity decreases the fastest in the descending flow. Besides, from parameter b, we can know that the horizontal free velocities in three scenarios are around 0.52 m/s, 0.64 m/s and

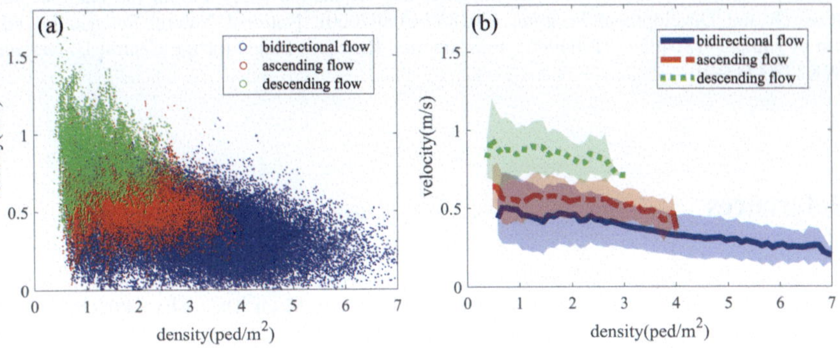

Fig. 36.6 The fundamental diagrams in three scenarios. **a** Raw scatter data. **b** Mean ± standard deviation

Table 36.2 Fitting parameters for the fundamental diagrams in three scenarios

Scenario	Parameter a	Parameter b	R^2
Bidirectional flow	−0.0424	0.5158	0.96
Ascending flow	−0.0444	0.6356	0.69
Descending flow	−0.0552	0.9312	0.62

0.93 m/s, respectively. The results can be used to validate and calibrate simulation models for stair movement as no fundamental diagrams based on trajectories have been obtained for both uni- and bidirectional scenarios on stairs.

36.4 Conclusion and Outlook

In this paper, we investigate bidirectional flow on stairs at different flow ratios.

We find that for $r < 0.3$ or $r > 0.7$, namely the unbalanced runs, the minor flow is more active in collision avoidance, while in more balanced runs, the descending flow will play the dominant role. Besides, on average descending pedestrians tend to walk faster than those in the ascending flow. But the standard deviation of the crossing time is always larger in the minor flow. Also, we obtain fundamental diagrams in three scenarios, namely the bidirectional flow, ascending flow and descending flow, due to the unequal numbers of pedestrians in two waiting areas, and these fundamental diagrams are further fitted and compared.

In our future work, the collision avoidance of pedestrians on stairs will be further analyzed based on the precise trajectories. Besides, it will also be interesting to quantify the head sways and compare those from the descending and ascending movements. In addition, the stepping behavior of pedestrians on stairs is also an important aspect that needs to be paid attention to.

Acknowledgements The authors acknowledge the foundation support from the National Key Research and Development Program (2018YFC0807000), National Natural Science Foundation of China (71804026, 71904006), Fundamental Research Funds for the Central Universities (WK2320000040).

References

1. J. Zhang, W. Klingsch, A. Schadschneider, A. Seyfried, Ordering in bidirectional pedestrian flows and its influence on the fundamental diagram. J. Stat. Mech. Theory Exp. **P02002**, P02002 (2011)
2. C.J. Jin, J. Rui, S.C. Wong, D. Li, G. Ning, W. Wei, Large-scale pedestrian flow experiments under high-density conditions (2017)
3. J. Zhang, Q. Wang, Y. Hu, S. Cao, L. Xia, W. Song, The effect of a directional split flow ratio on bidirectional pedestrian streams at signalized crosswalks. J. Stat. Mech. Theory Exp. **2018**, 073408 (2018)

4. C. Feliciani, K. Nishinari, Empirical analysis of the lane formation process in bidirectional pedestrian flow. Phys. Rev. **94**, 032304 (2016)
5. X.-D. Liu, W.-G. Song, W. Lv, Empirical data for pedestrian counterflow through bottlenecks in the channel. Transp. Res. Procedia **2**, 34–42 (2014)
6. R. Ye, M. Chraibi, C. Liu, L. Lian, Y. Zeng, J. Zhang, W. Song, Experimental study of pedestrian flow through right-angled corridor: uni- and bidirectional scenarios. J. Stat. Mech. Theory Exp. **2019**, 043401 (2019)
7. M. Liu, S. Wang, Y. Oeda, T. Sumi, Simulating uni-and bi-directional pedestrian movement on stairs by considering specifications of personal space. Accid. Anal. Prev. **122**, 350–364 (2019)
8. M. Boltes, A. Seyfried, Collecting pedestrian trajectories. Neurocomputing **100**, 127–133 (2013)
9. B. Steffen, A. Seyfried, Methods for measuring pedestrian density, flow, speed and direction with minimal scatter. Phys. A **389**, 1902–1910 (2010)

Chapter 37
Gender Profiling of Pedestrian Dyads

Zeynep Yücel, Francesco Zanlungo, and Takayuki Kanda

Abstract In traffic safety community, behavioral differences between genders have been attracting considerable attention in recent decades. Various empirical studies have proven that gender has a significant relation to drivers', cyclists' or pedestrians' decision making, route choice, rule compliance, as well as risk taking/perception. However, most studies examine behavior of individuals, and only very few consider (pedestrian) groups with different gender profiles. Therefore, this study investigates effect of gender composition of pedestrian dyads on the tangible dynamics, which may potentially help in automatically understanding and interpreting higher level behaviors such as decision making. We first propose a set of variables to represent dyads's physical/dynamical state. Observing empirical distributions, we comment on the effect of gender interplay on locomotion preferences. In order to verify our inferences quantitatively, we propose a gender profile recognition algorithm. Removing one variable at a time, contribution of each variable to recognition is evaluated. Our findings indicate that height related variables have a more strict relation to gender, followed by group velocity and inter-personal distance. Moreover, the "male" effect on dyad motion is found to somehow diminish when the male is paired with a female.

This work was supported by JSPS KAKENHI Grant Number JP18K18168.

Z. Yücel (✉)
Okayama University, Okayama, Japan
e-mail: zeynep@okayama-u.ac.jp

Z. Yücel · F. Zanlungo · T. Kanda
ATR, Kyoto, Japan
e-mail: zanlungo@atr.jp

T. Kanda
Kyoto University, Kyoto, Japan
e-mail: kanda@i.kyoto-u.ac.jp

I. Zuriguel et al. (eds.), *Traffic and Granular Flow 2019*,
Springer Proceedings in Physics 252,
https://doi.org/10.1007/978-3-030-55973-1_37

37.1 Introduction and Related Work

A deep understanding of pedestrians' movement patterns is crucial for various applications including designing of urban spaces [1], planning of transportation facilities, building more realistic crowd simulation systems [2], improving traffic safety [3] and integrating artificial intelligence in smart vehicles etc. [4, 5]. To achieve this understanding, it is important to analyze the significant elements that act on pedestrian locomotion, which can be categorized roughly into two as extrinsic and intrinsic factors.

Extrinsic factors involve location, built environment, time of the day and like [6], whereas intrinsic factors appertain to age, gender, level of mobility [7] etc. Obviously, an interplay of the extrinsic and intrinsic factors shape the behavior of all constituents of traffic (e.g. drivers or cyclists), but here we contain ourselves to only one of the intrinsic factors, gender, and only of the constituents of traffic, pedestrians [8, 9]. Therefore, this study focuses particularly on the effect of gender on pedestrian motion, which is shown to be an important element in decision making, rule compliance and risk perception/taking [10–12]. Although the effect of gender on individual's motion is investigated in many studies, gender composition of pedestrian groups is not yet treated in detail apart from very few works [1, 13]. Therefore, this study aims to fill the void in literature by addressing diversity of locomotion patterns of pedestrian groups with various gender profiles (i.e. gender composition).

To that end, we focus initially on dyads (i.e. groups of two people), which are straightforward to categorize in terms of gender composition, as all-female, all-male or mixed gender. We propose a few variables derived from trajectory and physical features, which are first evaluated qualitatively concerning their relation to gender profile. Subsequently, we propose a method to automatically recognize gender composition based on the approach of [14]. In addition, by removing one variable at a time, we point out to the contribution of each variable to recognition of gender composition.

37.2 Data Set and Variables

We used the data set introduced by [15], which is collected in the atrium of a business center in Osaka, Japan. The atrium covers roughly 900 m^2; connects the business center to a ferry terminal, a train station and a shopping mall; and thus is visited by a diverse demographic profile of pedestrians. Specifically, the data is recorded on two week days and one weekend over a one year time period. The recordings involve readings of 3D range sensors and video footage. The range sensors are densely populated so as to track pedestrians over their entire trajectory [16]. In addition to the trajectory, thanks to their overhead configuration, range sensors provide estimations of pedestrian height as well. By watching the video footage, a set of coders annotate groups (i.e. pedestrians constituting a social group), and gender composition of those groups (i.e. all-female F, all-male M or mixed gender X). In this manner, the distribution of dyad gender profiles in our data set is found to be as follows: 227 F, 424 M, 311 X.[1]

It is known that individuals have significantly different velocities depending on their gender and age [17–19]. In addition, studies on pedestrian dyads and triads show that all-male, all-female and mixed gender dyads present different velocity patterns [1, 13]. Using the data set and the associated ground truth, we examine such motion attributes, in addition to a set of features relating the height of peers.

Inter-personal distance δ relating various gender profiles is found as in Fig. 37.1a, which ascertains that M dyads have a significantly larger inter-personal distance, whereas F and X dyads stay in closer proximity. Nevertheless, the tails of δ distributions indicate that X dyads may have a heavier tail beyond 1 m than F dyads.

Relating velocity, we consider two variables, namely, group velocity $\bar{v}$ and velocity difference ω between peers. Group velocity $\bar{v}$ is described as the average velocity of the peers and is found as in Fig. 37.1b. It is clear that the typical attribute of—individual—males to have a higher velocity than females is sustained, as long as they move as part of an all-male dyad, whereas it diminishes as they move as part of a mixed gender dyad. In that case, the dyad adapts to the velocity of the slower peer, which is the female. Therefore, the velocity distribution of F and X dyads do not present any significant difference, whereas the velocity distribution of M dyads is in line with the velocity pattern of individual males. Velocity difference ω is defined as the norm of the difference vector between peers' velocities. It can be seen in Fig. 37.1c that M dyads have a slightly higher velocity difference, whereas F and X dyads have virtually the same ω distribution.

Average height of peers denoted with $\bar{\eta}$ is found as in Fig. 37.2a.[2] As expected M dyads are significantly taller than F and X, where F dyads have the shortest average height. Although locations of the peaks may depend on the geographical location (thus, indirectly on genetic factors), we consider these findings to be persistent among countries. Despite the dependency of height on demographic factors, the difference of height, denoted by $\Delta\eta$, between males and females in different countries is found to be rather similar [20, 21], which is supported by Fig. 37.2b. Namely, height difference between two males and two females follow very similar distributions around 0, whereas $\Delta\eta$ concerning X has a distribution with a much larger deviation and a heavier tail.

[1] A detailed analysis on inter-rater agreement confirms that the coders have significant agreement rates [13].

[2] Since height depends to a great extent on age particularly until the end of adolescence, we consider pedestrians whose apparent age are labeled to be more than 20 years old.

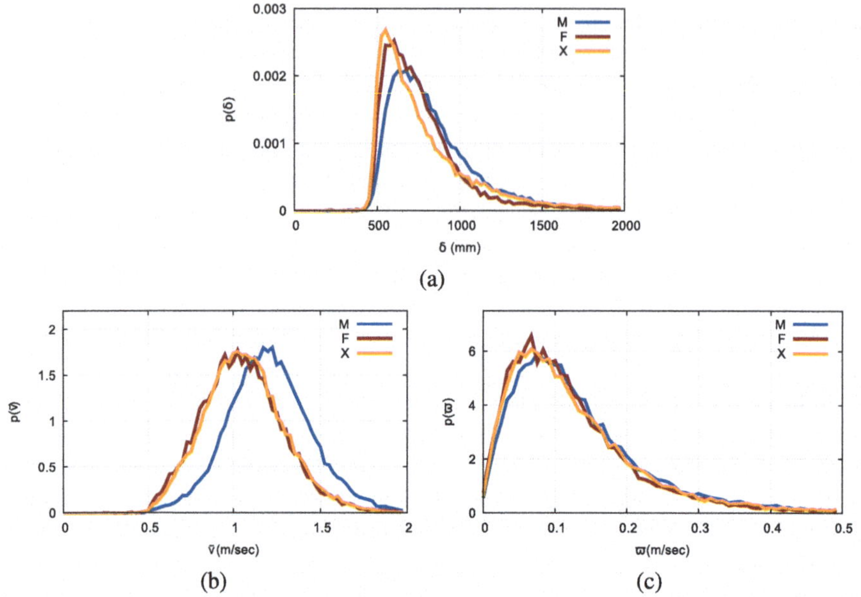

Fig. 37.1 **a** Interpersonal distance, **b** group velocity and **c** velocity difference

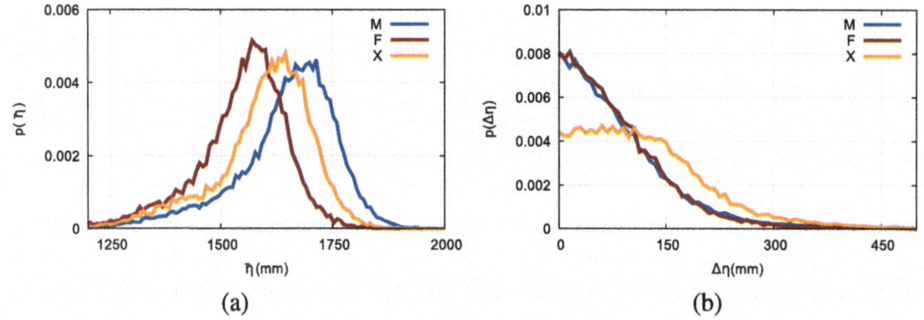

Fig. 37.2 **a** Average height and **b** height difference for dyads of various gender composition

37.3 Automatic Recognition of Gender Profiles

Inspired by the Bayesian approach of [14], we propose a method to assess the probability at each time instant that a dyad belongs to one of the three gender compositions M, F, or X. Suppose that at each time instant we collect an observation set Σ composed of the five variables listed in Sect. 37.2, $\Sigma = \{\delta,\ \bar{v},\ \omega,\ \bar{\eta},\ \Delta\eta\}$. Let $g \in \{M,\ F,\ X\}$ denote the gender composition. Given Σ, the (posterior) probability at time t that a dyad belongs to a gender composition g, is updated in the Bayesian sense,

$$P_t(g|\Sigma) = \frac{P_t(\Sigma|g)P_t(g)}{P_t(\Sigma)}. \tag{37.1}$$

Although probability density function (pdf) of Σ can be computed in a five dimensional space, it is very hard to overcome sparsity. Thus, by verifying that any two pair of variables can be considered to be independent [14], we break down $P_t(\Sigma|g)$ into a product of individual pdfs.

In order to avoid any bias, we consider equal probabilities for the priors, $P_0(g)$. Furthermore, so as to avoid computing the exact value of the marginal likelihood $P_t(\Sigma)$, we make use of the fact that the sum of $P_t(g)$ for all g must be 1. In order to propagate the information from past observations, while also preventing any drift due to an accumulation of error; we update the prior at every instant as a weighted average of the first value of the prior $P_0(g)$ and the previous value of the posterior $P_{t-1}(g)$. In this manner, we obtain a probability of belonging to any one of the gender combinations at every instant.

Subsequently, probability of belonging to a particular gender profile is computed as the average of concerning instantaneous probabilities. Suppose that the "ground truth" gender profile assigned to a dyad, d, is represented with r_d^{GT}. Probability of belonging to g can be defined as the average of all posterior probabilities along its trajectory,

$$\overline{P_d}(r) = \frac{\sum_t P_t(r|\Sigma)}{n_d}, \tag{37.2}$$

where n_d is the number of time instants (i.e. number of trajectory data points). We then define the estimated gender composition as the one with the maximum probability[3]

$$r_d^p = \underset{r}{argmax}(\overline{P}_d(r)). \tag{37.3}$$

In this sense, a single estimation result of gender composition is yielded for each dyad, which implies that the length of their trajectory is not paid regard.

37.4 Results and Discussion

First, we run the method proposed in Sect. 37.3 using all five variables depicted in Figs. 37.1 and 37.2. Subsequently, we remove one variable at a time and repeat the same procedure for judging the contribution of each variable.[4] The performance rates are found as in Table 37.1. Using the set of all five variables yields the best performance. Moreover, removal of δ degrades particularly the detection of X dyads. Although the location of the peak regarding δ distribution of X dyads is positioned not far from those of M and F dyads, its lighter tail and smaller variation help distinguishing of X from other gender profiles, in particular from M dyads. Removal of $\bar{v}$ has roughly the same effect on M and X dyads, whereas it slightly works in

[3]If the maximum occurs multiple times, one of the relating gender profiles is picked randomly.

[4]We consider 30% of the data set as training set and the remaining 70% as test set; and repeat this procedure 20 times.

favor of F dyads. From Fig. 37.1b $\bar{v}$ of M dyads distinguishes from that of F and X. However, since M dyads have other prominent characteristics, removal of $\bar{v}$ results in a limited degradation in identification of M dyads. Additionally, removal of ω has the least effect on overall performance.

On the other hand, variables relating height have obviously a larger impact on gender profiling. Among the five variables, average height $\bar{\eta}$ emerges as the most influential one. Although height (and in turn the overall body volume) has a direct impact on how large is the step size—a plain rationale of velocity—and how far people position themselves from others (i.e. δ), it is certainly much more self-evident to observe directly the height of individuals than its indirect implications on δ and $\bar{v}$. Considering also that the behavior of X dyads converges to that of F dyads in terms of particularly $\bar{v}$ and ω but also δ, we can say that "male" effect on velocity and inter-personal distance is somehow diminished when a male is paired with a female. Interestingly, this does not reflect as a degradation in the identification of M dyads but F dyads, which is probably due to the fact there are almost twice as many M samples in our dataset (see Sect. 37.2) as F dyads. In addition, height difference $\Delta\eta$ is another influential variable, yet not as much as $\bar{\eta}$. As expected, removal of $\Delta\eta$ lowers the detection rates of particularly X dyads to a large extent. In addition, in Table 37.1, diagonals have always the largest entry, i.e. we most often correctly identify the gender composition as the true class.

Table 37.1 Estimation results based on **a** Σ, **b** $\Sigma\backslash\delta$, **c** $\Sigma\backslash\bar{v}$, **d** $\Sigma\backslash\omega$, **e** $\Sigma\backslash\bar{\eta}$ and **f** $\Sigma\backslash\Delta\eta$

(a)	M	F	X	(b)	M	F	X	(c)	M	F	X
M	83.50	8.64	7.86	M	82.38	9.87	7.76	M	79.14	11.69	9.17
F	10.97	76.01	13.02	F	11.78	73.10	15.12	F	9.97	78.19	11.84
X	23.51	28.58	47.91	X	28.61	31.61	39.78	X	26.19	32.03	41.78
Tot		70.22		Tot		66.19		Tot		66.64	

(d)	M	F	X	(e)	M	F	X	(f)	M	F	X
M	82.84	8.62	8.54	M	74.40	15.03	10.56	M	82.94	8.16	8.90
F	12.73	73.07	14.20	F	41.63	36.66	21.72	F	15.74	70.15	14.11
X	26.48	28.41	45.11	X	29.87	25.26	44.87	X	32.20	33.61	34.19
Tot		68.14		Tot		55.80		Tot		63.91	

37.5 Conclusion

This study makes an effort to demonstrate the effect gender at a granular scale by presenting the locomotion distinctions regarding 5 variables, which reflect tangible and quantified dynamics/state of dyads. Moreover, we propose a method to rate the impact of gender profile on each of those, through an automatic recognition method for gender profile, which to the best of our knowledge is the first method for automatic recognition of gender composition of social groups. By feeding the recognition algorithm a different set of inputs—a set which lacks exactly one variable out of the five—we demonstrate that average height is the feature that reflects the effect of gender profile the most, which is followed by height difference. Interpersonal distance and group velocity have yet a limited effect, though they contribute recognition of different gender compositions.

References

1. A. Willis, N. Gjersoe, C. Havard, J. Kerridge, R. Kukla, Environ. Plan. B **31**(6), 805 (2004)
2. C. Holland, R. Hill, Accid. Anal. Prev. **42**(4), 1097 (2010)
3. M.A. Granié, Accid. Anal. Prev. **42**(2), 726 (2010)
4. I.Y. Noy, D. Shinar, W.J. Horrey, Saf. Sci. **102**, 68 (2018)
5. J. Son, M. Park, B.B. Park, Transp. Res. Part F Traffic Psychol. Behav. **31**, 12 (2015)
6. P. Stoker, A. Garfinkel-Castro, M. Khayesi, W. Odero, M.N. Mwangi, M. Peden, R. Ewing, J. Plann. Literat. **30**(4), 377 (2015)
7. E. Papadimitriou, S. Lassarre, G. Yannis, Transp. Res. Part F **36**, 69 (2016)
8. K.J. Anstey, R. Eramudugolla, L.A. Ross, N.T. Lautenschlager, J. Wood, Int. Psychogeriatrics **28**(3), 349 (2016)
9. I.M. Bernhoft, G. Carstensen, Transp. Res. Part F Traffic Psychol. Behav. **11**(2), 83 (2008)
10. M.A. Granié, Transp. Res. Part F Traffic Psychol. Behav. **10**(5), 371 (2007)
11. M.A. Granié, Saf. Sci. **47**(9), 1277 (2009)
12. A. Tom, M.A. Granié, Accid. Anal. Prev. **43**(5), 1794 (2011)
13. F. Zanlungo, Z. Yücel, D. Bršćić, T. Kanda, N. Hagita, PLoS One **12**(11), e0187253 (2017)
14. Z. Yucel, F. Zanlungo, C. Feliciani, A. Gregorj, T. Kanda, *Pedestrian and Evacuation Dynamics* (2018) (to appear)
15. F. Zanlungo, D. Bršćić, T. Kanda, Phys. Rev. E **91**(6), 062810 (2015)
16. D. Bršcic, T. Kanda, T. Ikeda, T. Miyashita, I.E.E.E. Tran, Hum.-Mach. Syst. **43**(6), 522 (2013)
17. R. Lobjois, N. Benguigui, V. Cavallo, Accid. Anal. Prev. **53**, 166 (2013)
18. A. Dommes, V. Cavallo, J.B. Dubuisson, I. Tournier, F. Vienne, J. Saf. Res. **50**, 27 (2014)
19. N.N. Ferenchak, J. Traffic Transp. Eng. **3**(4), 345 (2016)
20. MEXT. Official Health Statistics Survey, http://www.mext.go.jp/component/b_menu/other/__icsFiles/afieldfile/2014/03/28/1345147_1.pdf. Accessed Jan 2019
21. OECD iLibrary. OECD social indicators, https://www.oecd-ilibrary.org/social-issues-migration-health/society-at-a-glance-2009/height_soc_glance-2008-26-en. Accessed Jan 2019

Chapter 38
The Effect of Social Groups on the Dynamics of Bi-Directional Pedestrian Flow: A Numerical Study

Francesco Zanlungo, Luca Crociani, Zeynep Yücel, and Takayuki Kanda

Abstract We investigate the effect of groups on a bi-directional flow, focusing on self-organisation phenomena, and more specifically on the time needed for the occurrence of pedestrian lanes, their stability and their effect on the velocity-density relation, and the amount of physical contact in the crowd. We use a novel collision avoidance model considering the asymmetrical shape of the human body, and combine it to a mathematical model of group behaviour. The presence of groups results to have a significant effect on velocity and lane organisation, and a dramatic one on collision. Despite the limitations of our approach, we believe that our results show the great theoretical and practical relevance of group behaviour in pedestrian models, and suggest that realistic results may hardly be achieved simply by adding together modular models.

38.1 Introduction

Pedestrian physical crowds, i.e. not sharing a social identity [1], are nevertheless characterised by the presence of a large number of social groups [2]. It has to be expected that these pedestrians, that move together with peculiar velocity patterns and spatial formations [3, 4], may strongly influence the dynamics of the crowd [5]. Although a few microscopic models of group have been recently introduced [6–16], a quantitative assessing of the effect of groups on crowd dynamics has not been

F. Zanlungo (✉)
ATR International, Kyoto, Japan
e-mail: zanlungo@atr.jp

L. Crociani
Bicocca University, Milan, Italy

Z. Yücel
Okayama University, Okayama, Japan

T. Kanda
Kyoto University, Kyoto, Japan

I. Zuriguel et al. (eds.), *Traffic and Granular Flow 2019*,
Springer Proceedings in Physics 252,
https://doi.org/10.1007/978-3-030-55973-1_38

achieved. This is also due to the fact that, related to the absence of quantitative empirical data, we are still lacking realistic models of how groups behave *at different density ranges* (not to mention how they behave under different conditions e.g. commuting vs shopping vs evacuation; [17] shows that purpose, relation, gender, age and height have a strong impact on group dynamics). To investigate the relevance of these issues, we are going to: (i) use a realistic mathematical model for the behaviour of pedestrian groups in low to moderate density settings, (ii) combine it to a realistic and efficient collision avoidance model, and (iii) use the resulting model to investigate the effect of the presence of groups on self-organisation properties of a bi-directional flow.

38.2 Computational Model

In [3, 18], we introduced a pair-wise interaction potential describing the dynamics of socially interacting pedestrian groups.[1] This model is based on observations of pedestrian behaviour at low to medium density settings, and it is able to reproduce the behaviour of pedestrians in such situations. In this work, we are using the strong (and probably wrong) assumption that the model may be used under general density conditions, and study its consequences.

Reference [21] shows that the group model of [3, 18] may be effectively combined with the collision avoidance module of [19], which is a second order (force) model that implicitly introduces a velocity dependence by using repulsive forces based on, instead of current distances, future distances at the time of maximum approach to an obstacle or pedestrian (computed using a linear approximation).[1]

The model of [19] was designed for moderate densities and did not take in account the shape of the human body. In order to describe the motion of pedestrians at high density is necessary to consider at least the fact that the 2D projection of the human body is not symmetrical (this asymmetry may be first approximated by using ellipses instead of circles). When such an asymmetry is introduced, even if we still limit ourselves to the motion of pedestrians on a 2D plane, we need to introduce a new degree of freedom, body orientation angle θ. Assuming body orientation to be equal to velocity orientation would be a too strong limitation, since it would not allow pedestrians to rotate their torso while avoiding a collision without changing considerably their motion direction [22]. We thus introduce a model that approximates human bodies as ellipses with axes (A, B) (45 and 20 cms), and use a system of coupled second order differential equations for the pedestrian linear and angular acceleration $\ddot{\mathbf{r}}$ and $\ddot{\theta}$, as functions of $\mathbf{r}$, $\dot{\mathbf{r}}$, θ and $\dot{\theta}$. For reasons of space, we direct the reader interested in the details of the model to the technical document [20], where it is explained how acceleration and torque are computed on the basis of predicted collisions between ellipses and obstacles (including other ellipses, i.e., other pedestrians), which are

[1]Details on cited models [3, 18, 19] may be found in the original works, while a short introduction concerning their use in this work may be found at [20].

estimated using a linear approximation on velocity and angular velocity, and an event driven algorithm [23]. As explained in detail in [20], the proposed collision avoidance model is actually a linear combination between the model of [19], that does not use information concerning the shape of the human body, and the proposed module using the elliptical approximation. The combination is performed in such a way that information concerning the shape of the human body is used only for collisions that will happen in short time.

As explained in detail in [20], the parameters of the collision model are optimised in such a way that pedestrians try to minimise collisions while walking as straight as possible watching towards their goal, at their preferred velocity and while keeping groups together. The important point to keep in mind is that the collision avoidance module is optimised to be effective at relative high densities (optimisation is performed in an environment with an average density of 1 ped/m^2) in presence of groups, while the group model uses the original parameters calibrated on human behaviour *at low densities* in [3].

38.3 Experiments

38.3.1 Experimental Setting

In our simulations we use a corridor of width 3 m and length 20 m, and periodic boundary conditions. The width of the corridor is chosen is such a way that two 3 people groups walking in opposite directions may not cross without changing their formation, based on the group spatial size as reported in [3]. We use four different density conditions, $\rho = 1$ ped/m^2, i.e. 60 pedestrians, $\rho = 2$ ped/m^2, (120 pedestrians), $\rho = 3$ ped/m^2, (180 pedestrians) and $\rho = 4$ ped/m^2 (240 pedestrians). We also use two different group rate r_g conditions, namely without groups $r_g = 0$ and with half of the pedestrians in groups, $r_g = 0.5$. When groups are present, 20% of pedestrians are in triads, and 30% of them in dyads. We thus have 8 conditions depending on the values of ρ and r_g. For example, when $\rho = 4$ ped/m^2 and $r_g = 0.5$ we have 120 individual pedestrians, 36 dyads and 16 triads. Initial conditions are determined by dividing the corridor in cells of equal size, and placing pedestrians in randomly chosen cells, regardless of their flow direction.[2] The pedestrians' preferred velocities are chosen from a normal distribution with $\mu = 1.2$m/s and $\sigma = 0.2$ m/s.[3] Each group or pedestrian has a probability 0.5 of belonging to each one of the flows.[4] For each condition, 10 different simulations with independent initial conditions are used and the observables defined below are averaged over these initial conditions.

[2]Pedestrians in the same group are nevertheless located in neighbouring cells.

[3]Pedestrians in the same group have the same preferred velocity.

[4]Thus flows have the same weight only in average.

38.3.2 Observables

For each experimental condition, we define the following observables:

1. rate of velocity over preferred velocity, $\nu = v/v_p$,
2. energy exchanged in collisions E,
3. number of lanes n_l,
4. rate of pedestrians in lanes, $r_l = (\sum_{i \in n_l} s_i)/N$, where s_i is the number of pedestrians in lane i and N is the overall number of pedestrians.

Lane recognition is performed using the algorithm introduced in [24], whose relation to the current work is explained in detail in [20]. For each observable we first compute the average over pedestrians and time for each independent initial condition, and then compute the average, standard deviation and standard error over the different independent conditions. The latter are the values shown in the figures. In order to show also time dependence, time averages are computed over 10 slots of length $T_i = 20$ s.

38.4 Results

Figure 38.1 shows the time evolution of the ν observable in the $\rho = 1$ and $\rho = 4$ ped/m^2 conditions. We may observe that the effect of groups on velocity is very strong in the medium density range, in which flows without groups present a clearly higher velocity. The velocity difference is higher than the one between groups and individuals in moderate density settings [4]. On the other hand, the effect of the presence of groups on velocity is not present at very high densities. Figure 38.2 shows the time evolution of the E observable in the $\rho = 1$ ped/m^2 and $\rho = 4$ ped/m^2 conditions. We may observe that the amount of collision is strongly increased in presence of groups for any value of ρ. Figure 38.3 on the left shows the number of lanes in the $\rho = 3$ ped/m^2 condition in absence and presence of groups. We may see that in presence of groups, the system needs more time to converge to an asymptotic state, furthermore the asymptotic number of lanes is different (3 in absence of groups, and 2 in presence of groups; the presence of groups occupying a wider space makes the formation of three lanes more difficult [20]). Finally, Figure 38.3 on the right shows the rate of pedestrians in lanes in the $\rho = 2$ condition in absence and presence of groups. As it happens in general, a larger number of pedestrians is organised in lanes when groups are absent.

38.5 Conclusions

We have verified that by combining a realistic model for group behaviour at moderate crowd density with an efficient collision avoidance model, the presence of groups

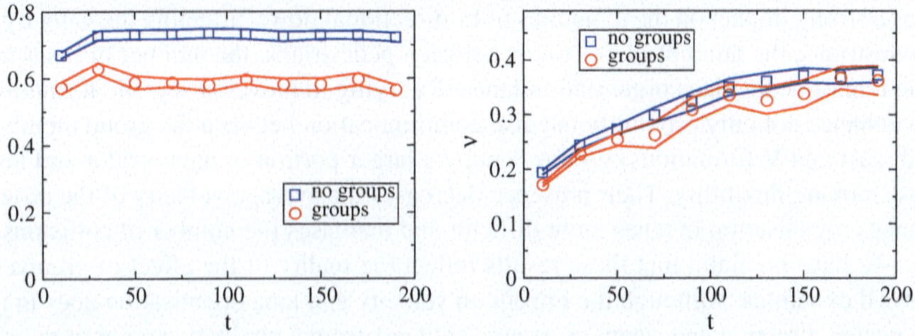

Fig. 38.1 Left: $v(t)$ for $\rho = 1$ ped/m^2 (blue squares: $r_g = 0$; red circles: $r_g = 0.5$; continuous lines: standard error intervals). Right: corresponding result for $\rho = 4$ ped/m^2

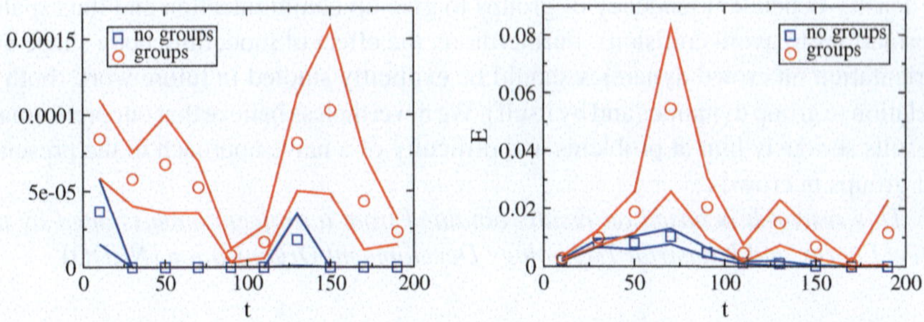

Fig. 38.2 Left: $E(t)$ for $\rho = 1$ ped/m^2 (blue squares: $r_g = 0$; red circles: $r_g = 0.5$; continuous lines: standard error intervals). E measured as unit mass energy exchanged in average by a pedestrian each second. Right: corresponding result for $\rho = 4$ ped/m^2

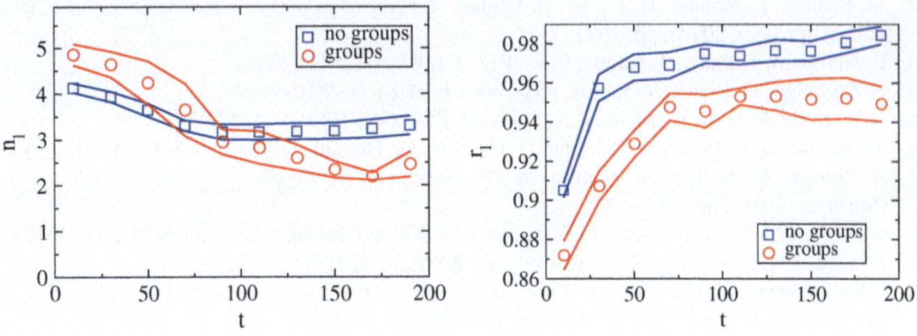

Fig. 38.3 Left: $n_l(t)$ for $\rho = 3$ ped/m^2. Right: $r_l(t)$ for $\rho = 2$ ped/m^2 (blue squares: $r_g = 0$; red circles: $r_g = 0.5$; continuous lines: standard error intervals)

has a strong impact on the dynamics of bi-directional flows, affecting the velocity of pedestrians, the amount of collisions between pedestrians, the number of lanes and the rate of pedestrians organised in lanes. By trying to move on specific formations to enhance not only proximity but also communication between the group members (abreast and V formations) groups occupy a larger portion of the corridor and have less moving flexibility. Their presence decreases the average velocity of the crowd, makes organisation in lanes more difficult and increases the number of collisions.

We have no claim that these results reflect the reality of the effect of groups on crowd dynamics. Although the impact on velocity and lane organisation goes in the expected direction and seems reasonable at least from a qualitative point of view, it is questionable that crowds with groups present such a larger amount of collisions. Just combining a collision avoidance and a group behaviour model could overlook important aspects such as a specific behaviour of pedestrians *towards* groups, and a density-dependent tendency of groups to give up communication and thus spatial formations to avoid collisions. Furthermore, the effect of modelling body shape and orientation on crowd dynamics should be explicitly studied in future work (both in relation to group dynamics and by itself). We nevertheless believe that our preliminary results seriously hint at problems and difficulty of a naive approach to the presence of groups in crowds.

This research is based on results obtained from a project commissioned by the New Energy and Industrial Technology Development Organization (NEDO).

References

1. S.D. Reicher, *The Psychology of Crowd Dynamics*, vol. 44 (Wiley Online Library, 2001)
2. M. Schultz, L. Rößger, H. Fricke, B. Schlag, in *Pedestrian and Evacuation Dynamics 2012*, vol. 2 (Springer, 2014), pp. 1097–1111
3. F. Zanlungo, T. Ikeda, T. Kanda, Phys. Rev. E **89**(1), 012811 (2014)
4. F. Zanlungo, D. Brščić, T. Kanda, Phys. Rev. E **91**(6), 062810 (2015)
5. N.W. Bode, S. Holl, W. Mehner, A. Seyfried, PloS One **10**(3), e0121227 (2015)
6. M. Moussaïd, N. Perozo, S. Garnier, D. Helbing, G. Theraulaz, PloS One **5**(4), e10047 (2010)
7. A. Gorrini, G. Vizzari, S. Bandini, in *Proceedings of the Traffic and Granular Flow 2015* (Springer, 2016), pp. 273–280
8. G. Köster, M. Seitz, F. Treml, D. Hartmann, W. Klein, Contemp. Soc. Sci. **6**(3), 397 (2011)
9. X. Wei, W. Lv, W. Song, X. Li, Complexity **20**(6), 87 (2015)
10. I. Karamouzas, M. Overmars, in *Proceedings of the ACM Symposium on Virtual Reality Software and Technology* (ACM, 2010), pp. 183–190
11. Y. Zhang, J. Pettre, X. Qin, S. Donikian, Q. Peng, in *Proceedings of the International Conference on Computer-Aided Design and Computer Graphics* (IEEE, 2011), pp. 275–281
12. A. Gorrini, L. Crociani, C. Feliciani, P. Zhao, K. Nishinari, S. Bandini, arXiv preprint arXiv:1610.08325 (2016)
13. P. Zhao, L. Sun, L. Cui, W. Luo, Y. Ding, in *Proceedings of the 2016 Pedestrian and Evacuation Dynamics Conference* (2016)
14. C. von Krüchten, F. Müller, A. Svachiy, O. Wohak, A. Schadschneider, in *Proceedings of Traffic and Granular Flow* (Springer, 2016), pp. 65–72
15. W. Wang, S. Lo, S. Liu, J. Ma, in *Proceedings of the International Conference on Pedestrian and Evacuation Dynamics* (2016)

16. A. Templeton, J. Drury, A. Philippides, Rev. Gen. Psychol. **19**(3), 215 (2015)
17. F. Zanlungo, Z. Yücel, D. Bršćić, T. Kanda, N. Hagita, PLoS One **12**(11), e0187253 (2017)
18. F. Zanlungo, T. Kanda, EPL Europhys. Lett. **111**(3), 38007 (2015)
19. F. Zanlungo, T. Ikeda, T. Kanda, EPL Europhys. Lett. **93**(6), 68005 (2011)
20. F. Zanlungo, L. Crociani, Z. Yücel, K. Takayuki, arXiv preprint arXiv:1910.04337 (2019)
21. F. Zanlungo, Z. Yücel, F. Ferreri, J. Even, L.Y.M. Saiki, T. Kanda, in *International Conference on Social Robotics* (Springer, 2017), pp. 474–484
22. H. Yamamoto, D. Yanagisawa, C. Feliciani, K. Nishinari, Transp. Res. part B Methodol. **122**, 486 (2019)
23. M.E. Foulaadvand, M. Yarifard, Phys. Rev. E **88**(5), 052504 (2013)
24. L. Crociani, G. Vizzari, A. Gorrini, S. Bandini, in *International Conference of the Italian Association for Artificial Intelligence* (Springer, 2018), pp. 71–82

Chapter 39
Experimental Study on Pedestrian Flow Under Different Age Groups and Movement Motivations

Jun Zhang, Xiangxia Ren, Hongliu Li, and Weiguo Song

Abstract In this work, we performed series of laboratory experiments with preschool students, university students and the elderly older than 60 years old. From these experiments, different movement behaviors among these groups are observed. For example, the preschool children are the most active ones and like competition but easy to fall, while the elderly move always very carefully. High movement motivation leads to competition among pedestrians, which results in different spatial-temporal distribution of the density and speed compared to normal movement of adults. Based on the Voronoi method, the spatial-temporal distribution of density is observed and compared for different types of age groups with movement motivations. It is found that the movement motivation has strong influence on the density distribution for semicircle and teardrop shaped high-density regions are observed from the experiments. These findings can be useful for the improvement of pedestrian modelling and design of pedestrian facilities.

J. Zhang (✉) · X. Ren · H. Li · W. Song
State Key Laboratory of Fire Science, University of Science and Technology of China, Hefei, China
e-mail: junz@ustc.edu.cn

X. Ren
e-mail: renxia@mail.ustc.edu.cn

H. Li
e-mail: hongliul@mail.ustc.edu.cn

W. Song
e-mail: wgsong@ustc.edu.cn

H. Li
Department of Architecture and Civil Engineering, City University of Hong Kong, Hong Kong, China

I. Zuriguel et al. (eds.), *Traffic and Granular Flow 2019*,
Springer Proceedings in Physics 252,
https://doi.org/10.1007/978-3-030-55973-1_39

39.1 Introduction

The worldwide population is aging and countries are facing ongoing challenges in meeting the transportation demand of the elderly. Meanwhile, children under 9 years represents about 17.9% of the world's population [1]. Among them the ones between 3 and 5 years are a special group, since most of them stay in kindergarten for a long time according to National Center for Education Statistics [2] but with weaker self-care ability compared to the pupils. For these vulnerable groups in the society, guaranteeing their safety under emergency and improving their comfort during daily life are necessary. The knowledge on the movement characteristics of these groups are the key to achieving the aim. However, there are fewer studies related to these groups up to now.

Numerous experimental studies have been performed in the last few decades to study the movement of pedestrians in different kinds of facilities including straight corridor [3, 4], corner [5], T-junction [6], and crossing [7] and so on. Several factors like cultural difference [8], movement motivation [9–11], facility geometries [12–15] on the pedestrian dynamics have been investigated. Taking the bottleneck as an example, arc-shaped [16] and tear-shaped [17] jam regions are observed previously. A large number of studies [8, 12, 18] focused on the relation between the specific flow and bottleneck width and linear functions are obtained. Besides, the influence of the bottleneck length [12, 13], the ratio between the bottleneck width and corridor [19, 20], the number of total evacuees [21] on the bottleneck flow are also well studied. However, the participants in most of these experiments are university students. The age composition and movement ability are less considered in controlled laboratory experiments up to now. How will the human age effect their movement characteristics in micro and macro levels? The similarity and connections among pedestrian flow of different types of populations are still less studied.

Based on these considerations, in this work we study bottleneck pedestrian flow via laboratory experiments with preschool students, university students and the elderly older than 60 years old. The spatial-temporal distributions of the density are compared. The remainder of the paper is as following. In Sect. 39.2, the experiment setup is briefly described. Spatial-temporal distributions of density are shown in Sect. 39.3. Section 39.4 summarizes this work.

39.2 Experimental Setup

In this section, we will give an overview of the experiment performed. As mentioned above, there are plenty of possible factors on the characteristics of pedestrian crowd movement. To reduce uncontrolled influences as much as possible, we performed the experiment with pedestrians with different ages separately. With this kind of setup, the results from special conditions are not suited for facility design recommendation. However, we could have the possible to compare the pedestrian movement for

Table 39.1 Table captions should be placed above the tables

Participants	Runs	Motivation	Age	Total participants	Bottleneck width
Kids	8	High	3–5	54	0.4–1.1 m
University students	19	Low/high	20–23	86	0.6–1.65 m (high)/0.5–1.8 m (low)
Elders	9	Low	52–81	63	0.5–1.8 m

pedestrians with different age range easily and gain deeper understanding on their movement characteristics as a first step.

The experiments were performed in 2017 in China. Kids (3–5 year-old) from a kindergarten, university students (20–23 year-old) and elders (52–81 year-old) from a senior center were recruited as participants of the experiments. Different runs were performed by changing the length and width of a bottleneck connecting a 3.5 m wide corridor whose length is longer than 5 m. The bottleneck was always set in the middle of corridor. The detailed information on the experiments can be found in Table 39.1. Before the experiment, all the participants were arranged in a holding area outside the corridor. When the experiment started, they walked or ran to the bottleneck according to different instructions from the organizer. Especially, in the experiment the kids were asked to run out of the experimental scenario as fast as possible (High motivation) due to their limited understanding on the instructions. The elders were asked to go outside the scenario in normal speed (Low motivation) by considering their body condition and safety issues. As for the students' group, we performed experiment in both of the two conditions to investigate the effect of movement motivation on the pedestrian dynamics. Figure 39.1 shows three snapshots during the experiments for different types of participants.

The whole process of the experiments was recorded by two HDR-SR11 cameras (resolution: 1920 × 1080 pixels, frame rate: 25 fps. Except for high motivated students' experiment, frame rate: 50 fps.) mounted at places more than 10 m high from the ground. All the participants were asked to wear pure color hats for easily recognition afterwards. The high precision trajectories of each participants (as shown in Fig. 39.2) were extracted from the video recording automatically and calibrated manually again for the following analysis.

Fig. 39.1 Snapshots during the experiments. **a** Running of kids. **b** Walking of university students. **c** Walking of elders

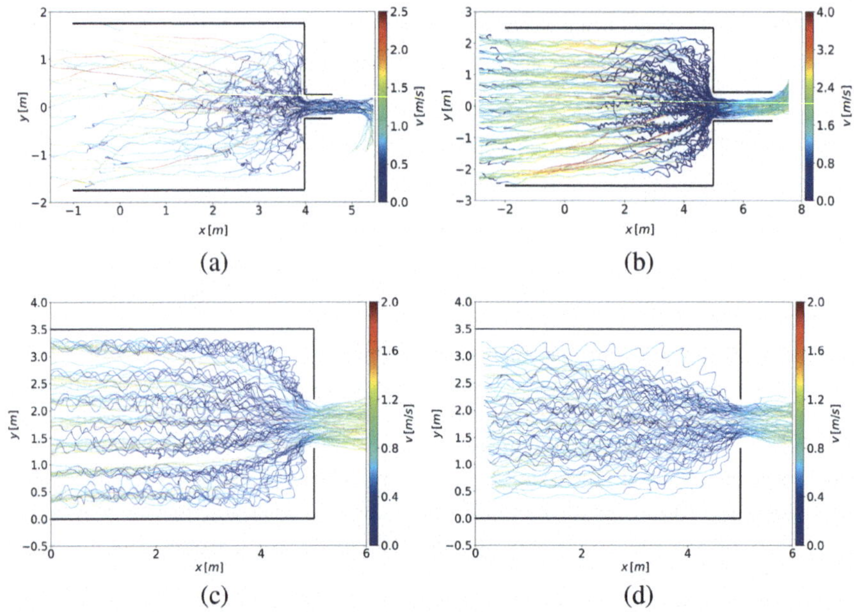

Fig. 39.2 Trajectories of the pedestrians in the experiment and the instantaneous speeds of each one are indicated by the color. **a** Running of kids. **b** Running of university students. **c** Walking of university students. **d** Walking of elders

39.3 Spatial-Temporal Distribution of Density

In this section, we focus on the spatial-temporal distribution of density in front of the bottleneck, which provides insights into the dynamics of the motion and the sensitivity of the integrated quantities to influencing factors such as boundary conditions. We aim to see the influence of age and movement motivation on the shape of jams in each scenario, on which it does not reach a consensus yet. With the Voronoi method proposed in [22] we calculate the density integrated over small regions (0.1 m $\times$ 0.1 m) for each scenario and the Gaussian filtering procedure is used to filter the data, as shown in Fig. 39.3. For the high motivation cases, the pedestrians distribute in a semicircle area in front of the bottleneck and the high density region presents a circle for both kids and university students. Due to the different body size, the highest density for kids is beyond 12 ped/m^2 and it is about 8 ped/m^2 for university students. As for the low motivation situation, the pedestrians locate in a teardrop-shape area for university students and the elders. However, the shapes of the high density in front of the bottleneck for these two groups are different obviously. For the elders, it is slender than that of the young students. The pick-density regions of the elderly display an ellipse shape with a longer major axis.

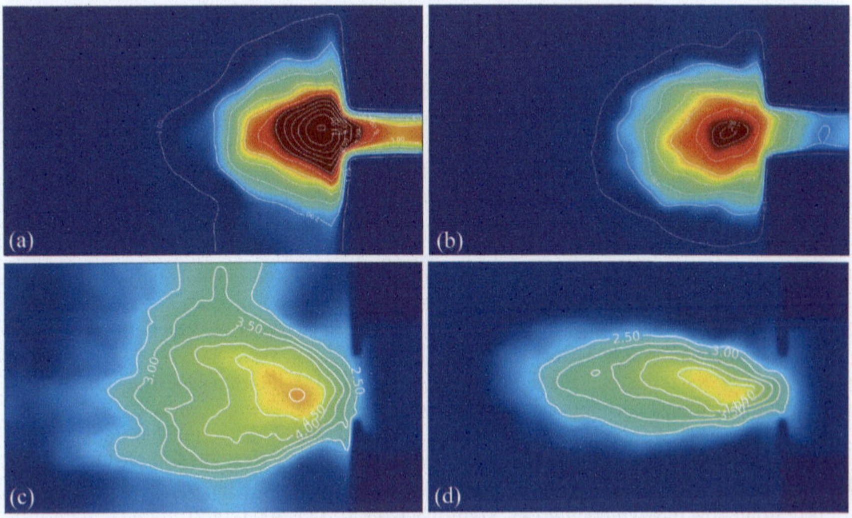

Fig. 39.3 Density profiles for bottleneck flow. The shape of the pedestrian density distribution in front of the bottleneck is arch-like and teardrop shaped for running and walking cases respectively. **a** Kids running for bottleneck width of 0.5 m. **b** University students running for bottleneck width of 0.9 m. **c** University students walking. **d** Elders walking

39.4 Summary

In this study, we investigate the influence of age and movement motivation on pedestrian dynamics through bottlenecks. A series of laboratory experiments with the participants of kids, university students and elders were performed. From these experiments, pedestrian movement behaviors for different groups are observed. For example, the kids in kindergarten are the most active ones and like competition but easy to fall, while the elderly move always very carefully. At the same densities, the speeds of the elderly are always lower than those of the young pedestrians in the observed density range. Besides, high movement motivation leads to competition among pedestrians, which results in different spatial-temporal distribution of the density compared to normal movement. Based on Voronoi method, semicircle high-density regions are observed for the movement in high motivation, whereas it is teardrop-shaped for the adults in low motivation. These findings can be useful for the improvement of pedestrian modelling and design of pedestrian facilities.

Acknowledgements The authors acknowledge the foundation support from the National Key Research and Development Program of China (Grant No. 2018YFC0808600), the National Natural Science Foundation of China (Grant No. 71704168, U1933105), from Anhui Provincial Natural Science Foundation (Grant No. 1808085MG217) and the Fundamental Research Funds for the Central Universities (Grant No. WK2320000040).

References

1. United Nations, *World Population Prospects 2017* (2017, December). Available https://esa.un.org/unpd/wpp/Download/Standard/Population/
2. NCES, *The Condition of Education: Preschool and Kindergarten Enrollment* (2017). Available https://nces.ed.gov/programs/coe/indicator_cfa.asp
3. X. Ren, J. Zhang, W. Song, C. Shuchao, The fundamental diagrams of elderly pedestrian flow in straight corridors under different densities. J. Stat. Mech. Theory Exp. **2019**(2), 023403–023403 (2019)
4. W. Tian, W. Song, J. Ma, Z. Fang, A. Seyfried, J. Liddle, Experimental study of pedestrian behaviors in a corridor based on digital image processing. Fire Saf. J. **47**, 8–15 (2012)
5. J.M. Chen, P. Lin, F.Y. Wu, D.L. Gao, G.Y. Wang, Revisit the faster-is-slower effect for an exit at a corner. J. Stat. Mech. Theory Exp. **2**, 2018 (2018)
6. J. Zhang, W. Klingsch, A. Schadschneider, A. Seyfried, Transitions in pedestrian fundamental diagrams of straight corridors and T-junctions. J. Stat. Mech. Theory Exp. **6**, 2011 (2011)
7. L. Lian, X. Mai, W. Song, Y.K. Kit Richard, X. Wei, J. Ma, An experimental study on four-directional intersecting pedestrian flows. J. Stat. Mech. Theory Exp. **2015**(8) (2015)
8. A. Seyfried, O. Passon, B. Steffen, M. Boltes, T. Rupprecht, W. Klingsch, New insights into pedestrian flow through bottlenecks. Transp. Sci. **43**(3), 395–406 (2009)
9. A. Garcimartin, J.M. Pastor, C. Martin-Gomez, D. Parisi, I. Zuriguel, Pedestrian collective motion in competitive room evacuation. Sci. Rep. **7**(1), 10792 (2017)
10. A. Garcimartín, D.R. Parisi, J.M. Pastor, C. Martín-Gómez, I. Zuriguel, Flow of pedestrians through narrow doors with different competitiveness. J. Stat. Mech. Theory Exp. **2016**(4), 043402 (2016)
11. H.C. Muir, D.M. Bottomley, C. Marrison, Effects of motivation and cabin configuration on emergency aircraft evacuation behavior and rates of egress. Int. J. Aviat. Psychol. **6**(1), 57–77 (1996)
12. J. Liddle, A. Seyfried, W. Klingsch, T. Rupprecht, A. Schadschneider, A. Winkens, An experimental study of pedestrian congestions: influence of bottleneck width and length. arXiv preprint arXiv:0911.4350 (2009)
13. D.R. Parisi, G.A. Patterson, Influence of bottleneck lengths and position on simulated pedestrian egress. Pap. Phys. **9**(0) (2017)
14. T. Rupprecht, W. Klingsch, A. Seyfried, Influence of geometry parameters on pedestrian flow through Bottleneck (2011), pp. 71–80
15. L. Sun, W. Luo, L. Yao, S. Qiu, J. Rong, A comparative study of funnel shape bottlenecks in subway stations (in English). Transp. Res. Part A-Policy Pract. **98**, 14–27 (2017)
16. D. Helbing, I. Farkas, T. Vicsek, Simulating dynamical features of escape panic. Nature **407**(6803), 487–490 (2000)
17. J. Liddle, A. Seyfried, B. Steffen, *Analysis of Bottleneck Motion Using Voronoi Diagrams* (2011), pp. 833–836
18. T. Kretz, A. Grünebohm, M. Schreckenberg, Experimental study of pedestrian flow through a bottleneck. J. Stat. Mech. Theory Exp. **2006**(10), P10014 (2006)
19. J. Zhang, A. Seyfried, Quantification of bottleneck effects for different types of facilities. Transp. Res. Procedia **2**, 51–59 (2014)
20. W. Liao, A. Seyfried, J. Zhang, M. Boltes, X. Zheng, Y. Zhao, Experimental study on pedestrian flow through wide bottleneck. Transp. Res. Procedia **2**, 26–33 (2014)
21. D.A. Purser, M. Bensilum, Quantification of behaviour for engineering design standards and escape time calculations. Saf. Sci. **38** (2001)
22. B. Steffen, A. Seyfried, Methods for measuring pedestrian density, flow, speed and direction with minimal scatter (in English). Physica A-Stat. Mech. Appl. **389**(9), 1902–1910 (2010)

Chapter 40
Experimental Analysis of the Restriction Mechanisms of Queuing on Pedestrian Flow at Bottleneck

Yifan Zhuang, Zhigang Liu, Andreas Schadschneider, Zhijian Fu, and Lizhong Yang

Abstract In this study we analyze the underlying restriction mechanisms when pedestrians queue through a bottleneck, and also compare them with that of setting up corridors in front of a bottleneck. A controlled laboratory experiment was conducted, by using Ultra-wideband (UWB) technology extracting the real-time pedestrians' trajectories. Different sides when pedestrians enter into the experiment room could have great impacts on the their effective speeds in the constriction process. Further, how the queuing affects the time variations of the inflow rate and the outflow rate, as well the fundamental diagrams near the exit are disclosed. The experimental results show that queuing before a bottleneck ensures orderliness and justness even if longer passing time is required, which could be favorable for non-urgent supervision.

Y. Zhuang (✉) · Z. Liu
College of Urban Railway Transportation, Shanghai University of Engineering Science, Shanghai 201620, People's Republic of China
e-mail: zhuangyf@sues.edu.cn

Z. Liu
e-mail: liuzhigang@sues.edu.cn

A. Schadschneider
Institut Fuer Theoretische Physik, Universitaet Zu Koeln, 50937 Koeln, Germany
e-mail: as@thp.uni-koeln.de

Z. Fu
Southwest Jiaotong University, Chengdu 610031, People's Republic of China
e-mail: zjfu@home.swjtu.edu.cn

L. Yang
University of Science and Technology of China, Hefei 230026, People's Republic of China
e-mail: yanglz@ustc.edu.cn

I. Zuriguel et al. (eds.), *Traffic and Granular Flow 2019*,
Springer Proceedings in Physics 252,
https://doi.org/10.1007/978-3-030-55973-1_40

40.1 Introduction

In recent years researchers have paid more and more attentions to controlling the large passenger flow, which has been a serious social problem in Chinese big cities. If no restriction measures are taken for the dense crowd, it is likely to cause casualties such as crowded trampling or accidents [1]. According to our known to the flow restrictions at bottleneck, there are two common physical measures are widely studied: one is to set an obstacle at a proper position near an outlet or an exit [2–5], which can increase the flow rate and reduce the clogging probability. The other is to set corridor-type restrictions before an entrance or an egress [1, 6–9], which can force pedestrians to form loose lanes and increase their well-being and justness. In Sieben's [7] and Adrian's [8] researches, the corridor setup forms an ordering structure, so pedestrians show a queuing behavior rather than a pushing behavior. In Zhuang's [1, 9] study, the longer of the entry railing the increase of the motion consistency. Both the corridor-type restrictions and the queuing behavior can narrow down the area pedestrians collected in front of the bottleneck. If only queuing can achieve the goal to control the outflow and ensure the fairness, it will be convenient to crowd management without needing to set any barriers. Thus, it is important to disclose the underlying restriction effects of queuing before a bottleneck, and compare with the effects of such physical measures.

In this study, pedestrians from different entry directions were arranged to pass through a room with one exit. The real-time pedestrians' trajectories are extracted through a high accuracy indoor positioning technology—Ultra-wideband (UWB). Macroscopic pedestrian moving characteristics have been analyzed by the trajectories. The results of the queuing experiment are consistent with the corridor-type restrictions at the bottleneck under low-motivation situations, but are different under high-motivation reality. This study represents useful empirical data for understanding the restriction mechanisms with or without physical measures. Besides, the automated positioning technique for data collection has been well employed and developed.

40.2 Experiment

40.2.1 Settings

Figure 40.1 shows the scene in our experiment. The length and the width of the room are 8.2 m and 7.6 m, respectively. The walls are structured by plastic boards with metal bases, whose size are 2.0 m height and 1.2 m width. The width of the exit is 1.0 m and set in the center of the room's length. Three 2.4 m-width entrances allowing for participants to enter the room are set: one is in the center of the room length facing the exit (hereinafter referred to as "opposite entrance (OE)"), and the other two are in the edges of the room widths (hereinafter referred to as "right entrance (RE)" and "left entrance (LE)").

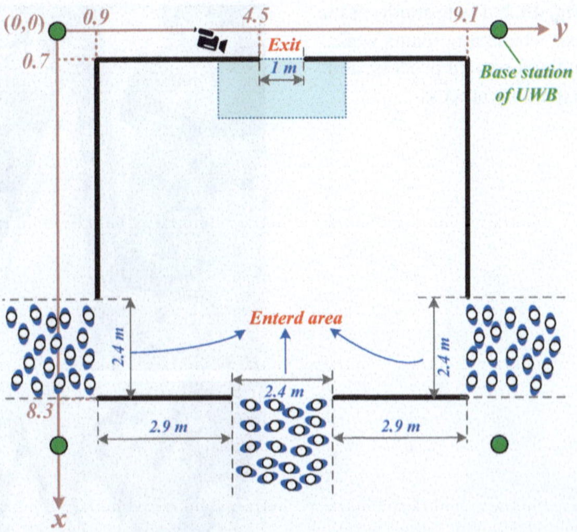

Fig. 40.1 The sketch of the experiment site

The experiment was conducted in a students' activity center of a university in Hefei, China, in January 2018. 107 college students participated the experiment, who were 18–20 years old and 158–178 m tall. They were asked to stand closely before the entrance and wait for 'GO!' announcement to walk towards the exit. In this experiment, five different conditions were conducted as below and each was repeated two times:

(1) Normal process with all participants entering from the RE;
(2) Queuing process with all participants entering from the RE;
(3) Queuing process with all participants entering from the OE;
(4) Queuing process with half entering from the RE and the other entering from the LE;
(5) Queuing egress with half entering from the RE and the other entering from the OE;

Note here that, the width of each entrance was set to 1.2 m in condition (4) and (5) to keep the inflow rate at a same level. NORMAL and QUEUING process are participants walk randomly and under the 'Please line up!' instruction, respectively, as seen in Fig. 40.2.

40.2.2 Data Processing

We employed an indoor positioning technology—Ultra-wideband (UWB) to automatically record and display the real movement trajectories. Four base stations were fixed atop 4 supports in 3.0 m high at four outside corners in Fig. 40.1. An electronic tag was affixed to the headgear of each participant (Fig. 40.2), so each trajectory

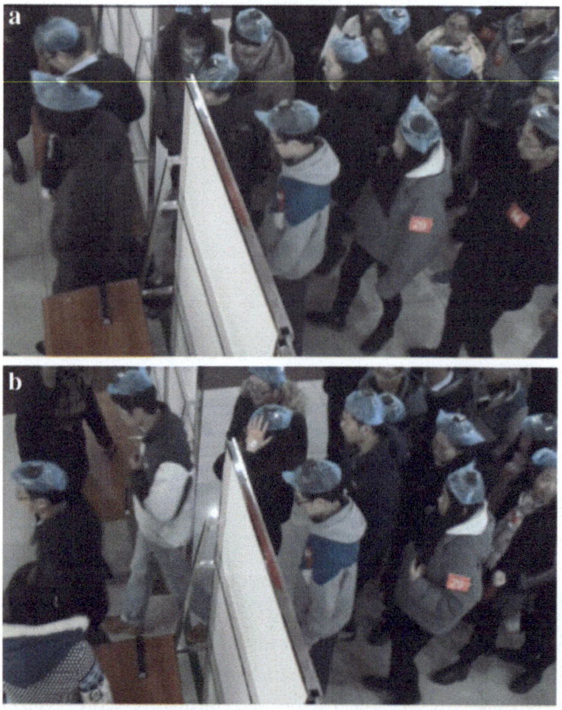

Fig. 40.2 Photographs at the exit when participants were in the **a** normal process, and **b** queuing process

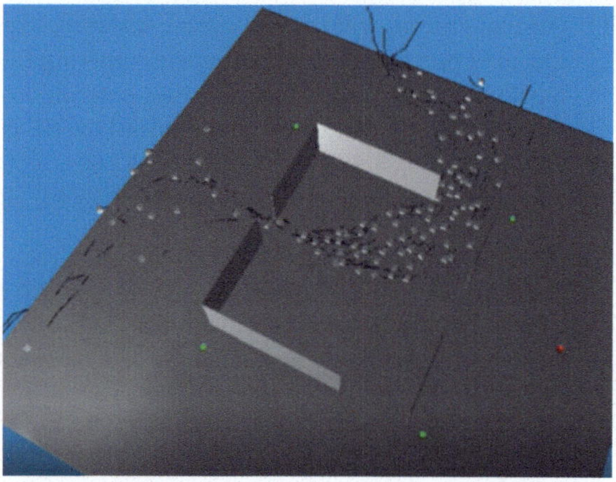

Fig. 40.3 Screenshot of the trajectory animation during the queuing process

can be distinguished and connected to each person's number. Besides, the trajectory animation was displayed in real time, as seen in Fig. 40.3. The time and the spatial accuracy of this method are up to 0.1 s, 0.01 m, respectively.

40.3 Results and Discussion

40.3.1 Trajectory Analysis

Figure 40.4 shows the participants' trajectories throughout five types of runs, as introduced in Sect. 2.1. Start with Fig. 40.4a, mauve and cyan lines represent participants enter from the right entrance (RE) with and without the instruction of 'Please line up!', respectively. When participants gather with no instruction, they tend to walk in the shortest path. The trajectories from the left of RE are obviously denser than the right, because pedestrians tend to detour to avoid crowding if there are enough space to use. But for queuing process, participants tend to assemble to the central axis first and then move straightforwards the exit, especially for whom entered from the left of RE. For the other queuing processes in Fig. 40.4b–d, trajectories are frequently observed perpendicular to the exit/entrance, and detours sometimes occur due to the congestion avoidance mentality.

40.3.2 Relation Between Time and Distance to Exit

The relations between the time to the exit and the distance to exit are distinguished in Fig. 40.5. The reciprocal of the line slope represents the effective speed. The sharper the slope, the lower the effective speed, and the longer waiting time participants have to cost from the entrance to the exit. Thus, participants can obtain higher time efficiency by queuing up (comparing Fig. 40.5a to Fig. 40.5b–e), entering from the direction facing the exit rather than from the side direction (comparing Fig. 40.5c to Fig. 40.5a, b, d, e).

Besides, the more uniform the effective speeds, the fairer the egress process. Thus, participants could feel juster by queuing up (comparing Fig. 40.5a to Fig. 40.5b–e), entering from one direction than from two different directions (comparing Fig. 40.5b to Fig. 40.5d and e), and entering from symmetry directions than from unsymmetrical directions (comparing Fig. 40.5d to Fig. 40.5 e). Participants near the group edges made obvious detours before the confluence (Fig. 40.4d), leading to a hump of the curves in Fig. 40.5e. Detours were commonly observed in the normal process and the

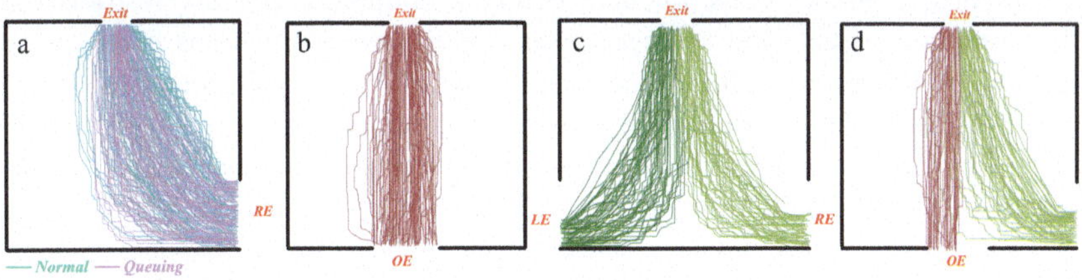

Fig. 40.4 Trajectories of participants in different conditions

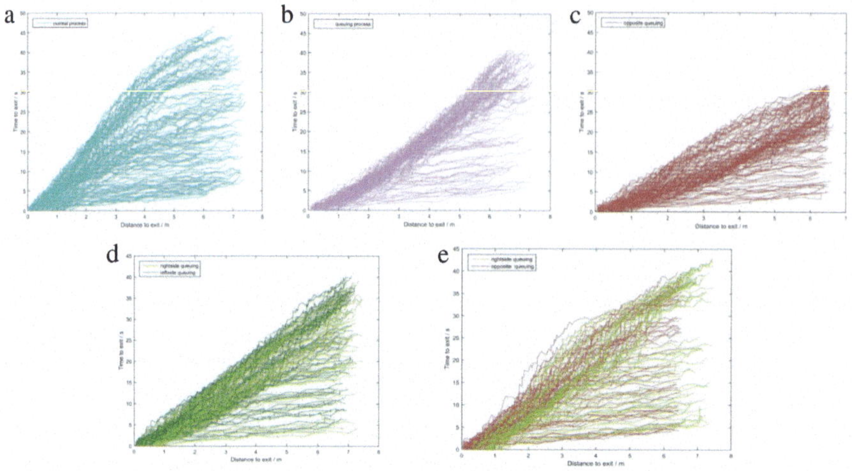

Fig. 40.5 The time to the exit versus the linear distance to the exit

queuing process from unsymmetrical directions (seen in Fig. 40.4a and d, Fig. 40.5a and e), which could make the whole egress process unfairer.

The results are worth further comparing with setting a corridor restriction before a bottleneck. When participants walk in a corridor setup, their effective speeds can be higher than the normal semicircle setup [7]. The narrower the corridor width, the more uniform the effective speeds [8]. So, in order to ensure the time efficiency and justness, queuing before a bottleneck could be an alternative to setting corridor-type restrictions.

40.3.3 Effective Speed and Flow Rate Time-Series

The effective speed, calculated by the mean inverse of the slopes of the curves in Fig. 40.5, shows clearly difference between random walking and queuing processes. According to Fig. 40.6a, participants can obtain a nearly 0.1 m/s higher effective speed by queuing than random walking to the exit, but the entire time becomes 10–15 s longer. Relatively lower but stabler outflow rates occur throughout the queuing processes in Fig. 40.6c. Thus, queuing before a bottleneck actually restricts the outflow rate and reduces the average speed of the group, which has been barely mentioned by other researchers. The inflow rate and the total entered time are consistent with the outflow rate and the exited time, according to Fig. 40.6b and c.

Apart from the gathering pattern itself (normal/queuing), other uncertainties could also cause different variations in the effective speed time-series in Fig. 40.6a. One is the entry directions before queuing. The effective speed of the participants queuing from the opposite entrance (OE) is slightly higher than other processes, so moving straightforwards to the exit can be time-saving than from other directions. But since

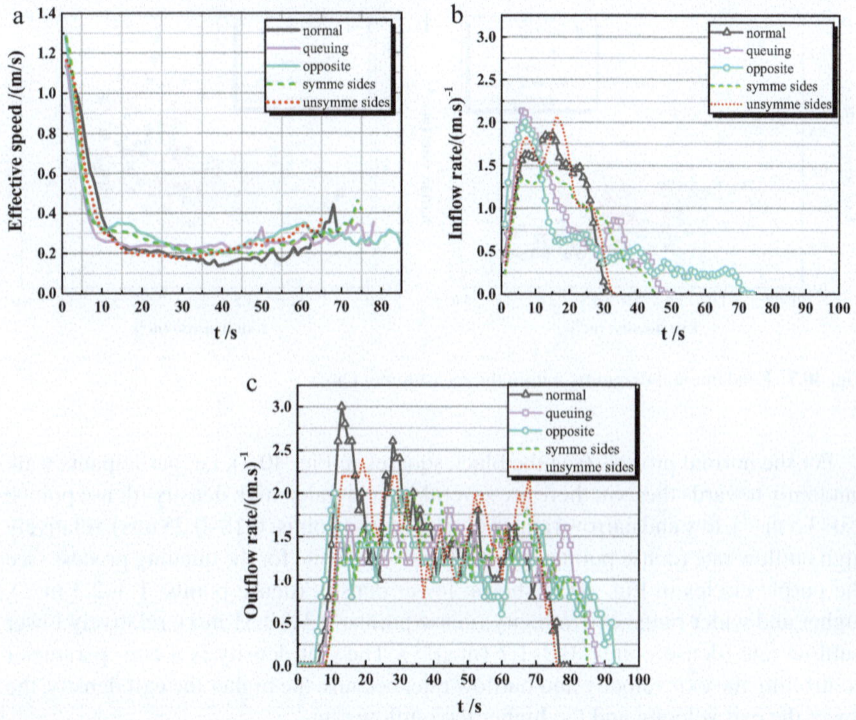

Fig. 40.6 Time-series of **a** average effective speed, **b** inflow rate and **c** outflow rate

participants have to spend extra time on waiting at the entrance, the inflow rate decreases and entering time increases in Fig. 40.6b. Besides, the symmetry of the entry directions. When the pedestrian flow from the OE mixes with that from the RE—an unsymmetrical entering process, the average effective speed decreases firstly and then increases. It may due to the occurrence of more waiting and competing for the same path in the symmetrical process. So the entire time of the unsymmetrical entering process could be 10 s shorter than the symmetrical process.

The outflow rate of the queuing processes is more stable than the non-queuing process in Fig. 40.6c. Thus queuing before a bottleneck can greatly enhance the flow stability.

40.3.4 Fundamental Diagrams

The fundamental diagrams of the average velocity, density and the outflow rate of the participants within the measurement area are shown in Fig. 40.7. The size of the measurement area is 3.2 m by 1.6 m, as seen the light blue rectangular box in Fig. 40.1.

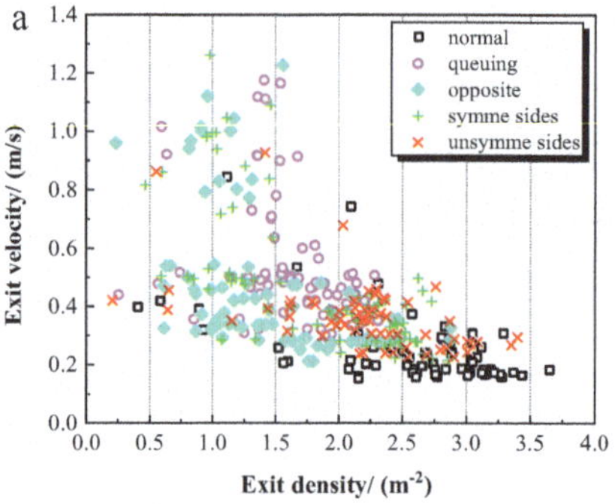

Fig. 40.7 Fundamental diagrams within the measurement area

For the normal process (see the black squares in Fig. 40.7), i.e. participants walk randomly towards the exit, there are several typical traits: high density (dense points: 2.0–3.5 m^{-2}), low and narrow range velocity (dense points: 0.18–0.25 m/s), relatively high outflow rate (dense points: 1.2–2.0 $(m.s)^{-1}$). While for the queuing process (see the purple circles in Fig. 40.7) shows: lower density (dense points: 1.3–2.3 m^{-2}), higher and wider range exit velocity (dense points: 0.31–0.51 m/s), relatively lower outflow rate (dense points: 0.8–1.7 $(m.s)^{-1}$). The exit density is a core parameter controlling the exit velocity and outflow rate, because the higher the exit density, the lower the exit velocity and the higher the outflow rate.

In this study, participants walking under low motivation (they were not told to move quickly as in Adrian's study [8]) tend to accumulate around the exit in a fan-shaped region, but in a right trapezoid region when they queuing. So there are hardly any pushing behaviors during the queuing process. This result is consistent with the walking process under the corridor-type restrictions [7, 8]. However, it could be very different under high motivation situations. In Adrian's laboratory experiment [8], the measured densities are nearly the same in a narrow (2.3 m) and a wider (5.6 m) corridor setup when participants had strong motivation moving out. In Zhuang's observational experiment [1], the measured density in medium length fences (2.5 m) is higher than in the long fences (3.5 m) when pedestrians rush to take the escalator. Because more pushing or squeezing behaviors are permitted under the physical restrictions of corridors/fences, even narrow or short corridor setup could accommodate denser pedestrians.

40.4 Conclusions

We investigated the effects of the queuing behavior and the entry directions on pedestrians' trajectories, effective speed, flow rate time-series and fundamental diagrams, by applying UWB location system into a laboratory experiment. The results could help develop the restriction mechanism of the bottleneck for better behavior regulation in crowd management.

By queuing up to the exit, participants can cost less waiting time and feel juster during the whole process. Also, participant's perception of justness can be enhanced by entering from one direction facing the exit, or from two symmetry directions. Detours observed in the non-queuing and the unsymmetrical queuing process reduce the fairness but accelerate the egress process. Because the inflow rate and outflow rate have been restricted by queuing, longer passing time participants have to spend both on the entrance and exit.

Comparing to corridor-type restrictions, queuing could be a good alternative in ensuring the flow stability and group orderliness, reducing the crowding near the exit under non-urgent or low motivation situations. However, the corridor-type restrictions could be superior when pedestrians have strong motivations. Since the physical restrictions can accommodate more people, both high pedestrian flow and orderliness can be maintained.

Based on the presented results, the area size where pedestrians accumulate is quite important for the fundamental diagrams. That's how the queuing and corridor-type restrictions work on regulating the bottleneck flow. Therefore, we suggest to investigate the queuing behavior within limited space of different shapes before a bottleneck and consider the urgent situations in future experiments.

References

1. Y. Zhuang, X. Zhou, Y. Ni, L. Yang, KSCE J. Civil Eng. **21**, 2392–2402 (2017)
2. F. Alonso-Marroquin, S.I. Azeezullah, S.A. Galindo-Torres, L.M. Olsen-Kettle, Phys. Rev. E **85**, 020301 (2012)
3. I. Zuriguel, A. et al. *Phys. Rev. Lett.* **107**, 278001 (2011)
4. A. Kirchner, K. Nishinari, A. Schadschneider, Phys. Rev. E **67**, 056122 (2003)
5. G.A. Frank, C.O. Dorso, Phys. A **390**, 2135–2145 (2011)
6. S. Seer, D. Bauer, N. Brandle, M. Ray, *11th Inter. IEEE Conf. Intel. Trans. Sys.* 742–747 (2008)
7. A. Sieben, J. Schumann, A. Seyfried, PLoS ONE **12**, e0177328 (2017)
8. J. Adrian, M. Boltes, S. Holl, A. Sieben, A. Seyfried. *arXiv preprint arXiv* 1810.07424 (2018)
9. Y. Zhuang, A. Schadschneider, H. Cheng, L. Yang. *KSCE J. Civil Eng.* **22**, 5203–5214. (2018)

Chapter 41
Vadere—A Simulation Framework to Compare Locomotion Models

Benedikt Zönnchen, Benedikt Kleinmeier, and Gerta Köster

Abstract Unlike many dynamical systems in physics, there is no universally accepted locomotion model for crowd dynamics. On the contrary, many different approaches compete, of which some are more suitable for a specific scenario, like evacuations, than others. We showcase how to compare microscopic models based on a real-world experiment using the open-source simulation framework Vadere and two models: the Optimal Steps Model and the Behavioral Heuristics Model. Aside from quantitative aspects, we discuss visual results. Both models are able to reproduce the density-speed relation to a reasonable degree. We also identify model characteristics that led to deviations, thus enhancing our understanding of both models and facilitating the decision which model to choose to investigate a particular real-world situation.

41.1 Introduction

Nowadays, simulations of pedestrian streams are a tool for researchers from different disciplines and practitioners like crowd managers. They help to deepen the understanding of pedestrian models as well as real-world crowd behavior. In practice, they support crowd managers to identify risks early in the planning phase of an event. A list of current pedestrian simulators, both commercial and free, can be found in [1].

B. Zönnchen (✉) · B. Kleinmeier · G. Köster
Department of Computer Science and Mathematics, Munich University of Applied Sciences, 80335 Munich, Germany
e-mail: zoennchen.benedikt@hm.edu

B. Kleinmeier
e-mail: benedikt.kleinmeier@hm.edu

G. Köster
e-mail: gerta.koester@hm.edu

B. Zönnchen · B. Kleinmeier
Technical University of Munich, Department of Informatics, 85748 Garching, Germany

I. Zuriguel et al. (eds.), *Traffic and Granular Flow 2019*,
Springer Proceedings in Physics 252,
https://doi.org/10.1007/978-3-030-55973-1_41

Each simulator uses a limited set of locomotion models [2–5] to describe how agents move through a virtual environment. Each model generates unique simulation results in the form of trajectories which translate into different fundamental diagrams. As a consequence, some models are more suitable for a specific type of scenarios, like evacuations, than others. Before drawing conclusions from simulations, we must decide if they replicate the real-world scenario satisfactorily. This contribution tackles the following research question: how do we find the most accurate microscopic locomotion model to simulate a specific real-world scenario?

41.2 Methods and Materials

We pick a well-defined real-world experiment, reenact it in Vadere and compare the empirical data with the simulation output using metrics we introduce here.

The T-Junction Experiment To compare simulations with the real world, we require a setup which is well studied, accurately documented and offers publicly available data of high quality. The T-junction experiment [6, 7], which was jointly conducted by researchers of the University of Wuppertal, University of Cologne and the Forschungszentrum Jülich fulfills all those needs. For the experiment, researchers set up a T-junction with a corridor width of 2.40 m. See Fig. 41.1. Researchers regulated the density in each run by varying the entrance width b_{in} from 0.50 to 2.40 m. Up to 350 probands mostly students, participated. The trajectories, which can be downloaded from [8], contain the pedestrian positions in 0.0625 s time steps.

The Simulation Framework Vadere We reenact the experiment in the pedestrian simulator Vadere, which has been specifically designed as a software framework to compare different locomotion models [1]. Vadere is packaged with several widely used microscopic locomotion models, including the Behavioral Heuristics Model (BHM) and the Optimal Steps Model (OSM) used in our comparison.

Simulation Models The OSM was introduced in [5] and further developed by [9]. We use its most recent version, which models the personal space of an agent by a

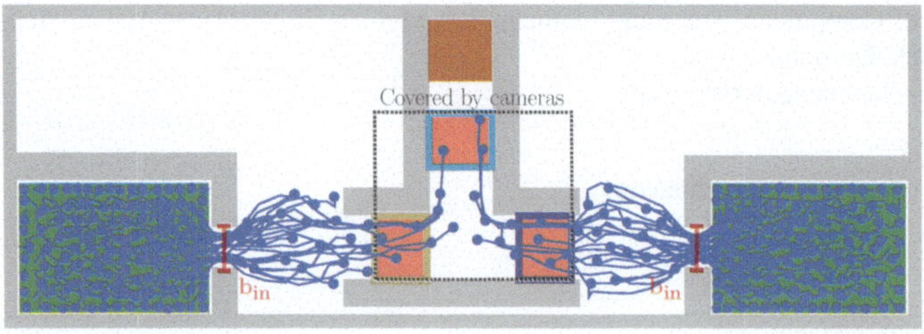

Fig. 41.1 Illustration of the scenario presented in [6, 7]. Agents / pedestrians walk from the green region (left and right) to their brown destination (top) through the red measurement areas. The width of the entrance b_{in} is varied in each run

potential function [9]. The model places an agent's next position at the minimum of the superposition of different potentials within the agent's step circle. An attracting potential, containing the geodetic distance to the target, leads agents to their destinations, whereas repulsive potentials from obstacles and agents prevent collisions. In the BHM, on the other hand, agents use three heuristic strategies to move [10]: to reach their destination, they step in the direction of a target, that is, the negative gradient of the geodetic distance to the target, if that space is free. If not, they try to "evade tangentially". If that fails, the agents wait. In contrast to the OSM, the BHM's agents have a fixed stride length. We used the parameters of former calibrations with two exceptions: We increased the "obstacle potential height" in the OSM to keep agents away from walls, and we reduced the standard deviation of the free-flow speed since the real participants were a homogeneous group.

Metrics By running the simulations in Vadere, we obtained three data sets: experiment, OSM, and BHM. We visualize the trajectories through Vadere. We decided to look with the naked eye instead of automatic image recognition that is still reported to lack robustness [11]. For quantitative analysis, we compared fundamental diagrams, walking times and speeds. Let $T_i = (\mathbf{x}_1, t_1), \ldots, (\mathbf{x}_N, t_N)$ be the observed trajectory, that is, a series of positions and times, of agent, or pedestrian i. We interpolate its position at time t with $t_1 \leq t \leq t_N$ by $\mathrm{pos}_i(t) = \mathbf{x}_k + (\mathbf{x}_{k+1} - \mathbf{x}_k)/(t_{k+1} - t_k) \cdot (t - t_k), t_k < t \leq t_{k+1}$ and its speed $v_i(t)$ by $v_i(t) = ||\mathbf{x}_{k+1} - \mathbf{x}_k|| /(t_{k+1} - t_k), t_k < t \leq t_{k+1}$. The mean speed at time t inside the measurement area A is given by $\overline{v}_A(t) = \sum_{k=1}^{N} v_k(t)/N$. Furthermore, the density at a given time t for the measurement area A is $\rho_A(t) = N/|A|$. N is the number of agents for which $\mathrm{pos}_i(t)$ lies inside area A. The walking time is defined as the time that an agent takes to walk through the region covered by cameras, which is the area depicted in the experiment setup. We measure positions, speeds and densities every 2 s from the start to the end of each simulation run and experiment and generate fundamental diagrams for three measurement areas A_{left}, A_{right}, A_{top} in Fig. 41.1.

41.3 Results and Discussion

In the following, we compare the experiment results with the simulation data based on trajectories, fundamental diagrams, walking times and mean speed. First, we observe that, for all data sets, the variance of walking times increases with the door width b_{in} (Fig. 41.2).

For low densities, agents and pedestrians walk at speed approximately equal to their free-flow speed. An increasing density implies a slower movement. With increasing door width b_{in}, the density spectrum widens and, with it, the velocity spectrum. Both models reproduce this effect.

In the experiment and also in the OSM simulations, pedestrians and virtual agents slow down when the left and right stream merge and accelerate afterward. This effect can be observed in video streams of the data sets generated through Vadere,

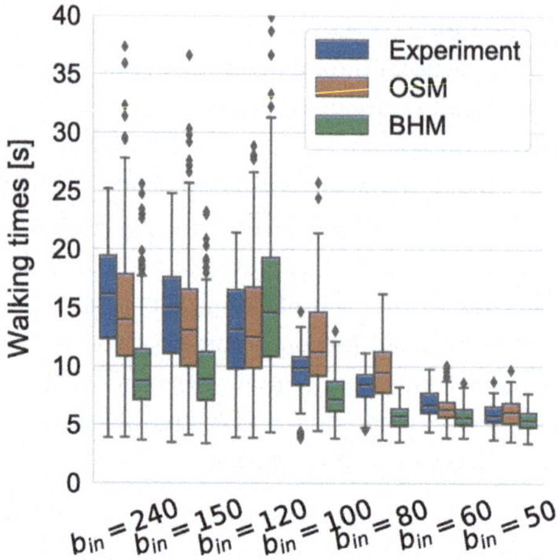

Fig. 41.2 Box plot of the time required by agents and pedestrians to go through the region covered by cameras depicted in the experiment setup. The first box relates to the experiment, the second to the OSM and the third to the BHM. The plot whiskers extend to 1.5 of the interquartile range

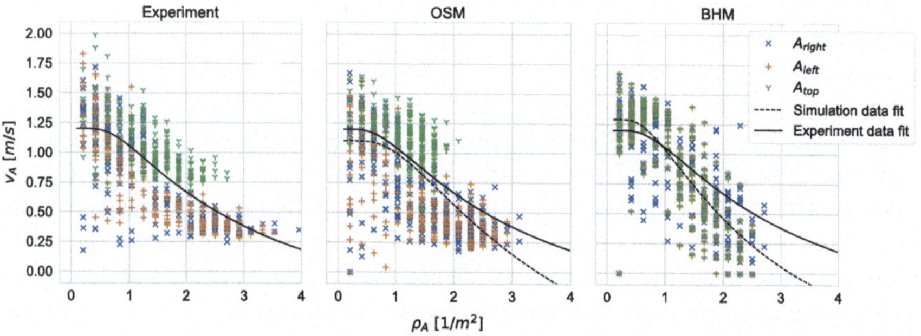

Fig. 41.3 Fundamental diagrams for model and experiment trajectories for the three measurement areas A_{left}, A_{right} and A_{top}. All experiment and simulation runs are merged together. The black line is the best fit of Kladek formula of the experiment data and the dashed line the best fit for the simulation data

but also in the fundamental diagrams in Fig. 41.3, where we see larger velocities for the measurement area A_{top} placed behind the merge point. This behavior was not reproduced by the BHM.

In the experiment, students act cooperatively with almost no pushing or scrambling. Furthermore, walking times and speeds show a smaller variance compared to the simulations (Figs. 41.2 and 41.4). For the OSM, this difference can be explained by the purely mathematical definition of an optimal step. The result is a more 'competitive' behavior, where some agents are overtaken by several others, and with sharper, more zig-zag looking trajectories, shown in Fig. 41.5. In the BHM, the large variance of the walking speed depicted in Fig. 41.4 is a result of the fixed step size, which makes agents either walk fast or wait. Furthermore, BHM agents are more

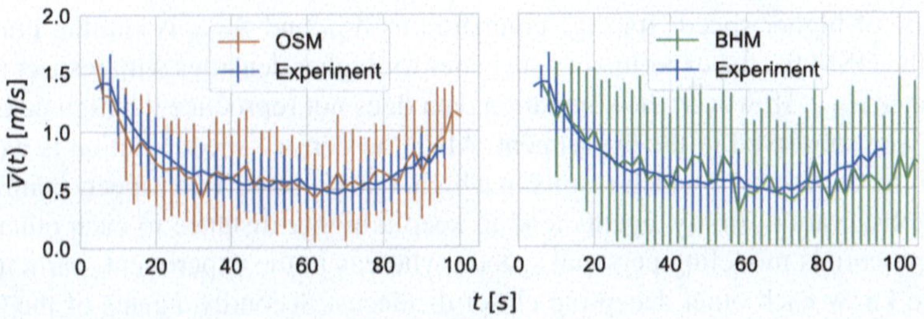

Fig. 41.4 The mean and standard deviation of the speed of all agents and pedestrians at different times for $b_{\text{in}} = 120\,\text{cm}$

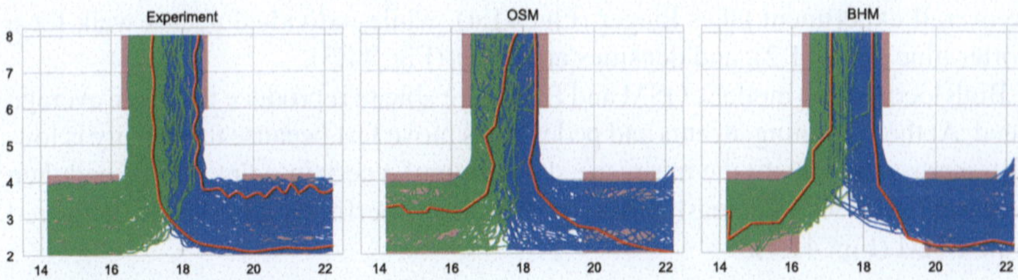

Fig. 41.5 Trajectories for $b_{\text{in}} = 150\,\text{cm}$. The three measurement areas and two example trajectories are highlighted in red. The empirical trajectories are smother than the simulated trajectories. The oscillation from tracking heads are visible

restricted in their step direction (step forward or tangential), so that fewer agents are overtaken, resulting in a smaller variance of walking times (Fig. 41.2).

The fundamental diagrams in Fig. 41.3 show the relationship between speed and density for all runs, that is, for all door widths b_{in} and all measurement areas. We compute a fit based on the Kladek formula which describes the relation between average momentary speed $\overline{v}_m$ and density ρ_A of motorized urban road traffic [12]: $\overline{v}_m = v_{\text{free}} \cdot (1 - \exp(-\gamma \cdot (1/\rho_A - 1/\rho_{\max})))$ with v_{free} as free-flow speed, $\rho_{\max}$ as maximum density where speed becomes zero, and γ a free parameter to be calibrated. Weidmann's parametrization of the Kladek formula for pedestrian dynamics is given in the table. It also shows all parameters for the best fit of the Kladek formula for each data set.

	v_{free}	$\rho_{\max}$	γ
Experiment	1.20	5.42	2.58
OSM	1.10	3.45	3.44
BHM	1.29	3.19	2.36
Weidmann curve	1.34	5.40	1.93

Wider doors mainly lead to higher densities. For all data sets, increasing density causes a decrease in speed. By eye-sight comparison, the OSM reproduces the fundamental diagram of the experiment more accurately than the BHM because the tendency of higher speeds for A_{top} compared to A_{left} and A_{right} is similar. Furthermore, the OSM fits the experiment data better for higher densities with respect to the parameter ρ_{max}. However, the simulation data does not reproduce the high densities which were observed in the experiment. Also, the density-speed relation is weaker than in the experiment, which results in a higher value of γ. Two aspects contribute to this observation: firstly, agents tend to keep a higher distance to each other due to the potentials modeling personal spaces, whereas in the experiment, participants seem to know each other accepting closer distances. Secondly, agents of the OSM step 'selfishly' towards the target resulting in less congestion and higher speeds. The BHM generates lower densities because of the movement restrictions. Therefore, the agents collectively take up less of the available space (Fig. 41.5). For the BHM, the overall experiment takes longer (Fig. 41.4), while individual agents walk for a shorter time (Fig. 41.2), and densities are lower (Fig. 41.3).

Both locomotion models, OSM and BHM, are able to reproduce the mean average speed. At the beginning, agents and pedestrians move fast because the density is low. With increasing density the mean speed drops until a certain point in time and then increases again with decreasing density. The OSM performs a bit better compared to the BHM (Fig. 41.4).

41.4 Conclusion

In this contribution, we presented the pedestrian simulation framework Vadere and how this framework can be used to compare real-world trajectory data with simulated data. We were able to spot and explain important differences in the movement of real-world pedestrians and simulated agents of the OSM and BHM. Using this procedure, we extended our understanding of the locomotion models as well as the real world. We started with a visual comparison to reveal movement patterns that the human eye is trained to recognize. In a second step, we used quantitative metrics, like the fundamental diagram, to underpin our observations. Overall, the OSM reproduces the experiment results better than the BHM. Its performance is good enough to estimate results for situations not covered by the experimental study. The key observation is that participants of the experiment walked more smoothly and homogeneously than our models assume. This behavior may be enabled by cooperation and a sense of fairness.

Both calibrated models were intentionally used without a re-calibration based on the given data set. In future work, the parameters could be varied for both models to get results that are closer to the experiment. For instance, the personal space could be lowered in the OSM to yield higher densities. For the BHM, it would be interesting to adjust step lengths or develop finer heuristics.

Acknowledgements We thank the research office (FORWIN) of the Munich University of Applied Sciences and the Faculty Graduate Center CeDoSIA of TUM Graduate School at Technical University of Munich for their support.B. K. and B. Z. are supported by the BMBF (German Federal Ministry of Education and Research) through the project OPMoPS (grant no. 13N14562) and S^2UCRE (grant no. 13N14463).

References

1. B. Kleinmeier, B. Zönnchen, M. Gödel, G. Köster, Collect. Dyn. **4**, (2019). https://doi.org/10.17815/CD.2019.21
2. S. Wolfram, Nature **311**, 419 (1984). https://doi.org/10.1038/311419a0
3. D. Helbing, P. Molnár, Phys. Rev. E **51**(5), 4282 (1995). https://doi.org/10.1103/PhysRevE.51.4282
4. G. Antonini, A discrete choice modeling framework for pedestrian walking behavior with application to human tracking in video sequences. Ph.D. thesis, École polytechnique fédérale de Lausanne (2005)
5. M.J. Seitz, G. Köster, Phys. Rev. E **86**(4), 046108 (2012). https://doi.org/10.1103/PhysRevE.86.046108
6. J. Zhang, W. Klingsch, A. Schadschneider, A. Seyfried, J. Stat. Mech. Theory Exp. **2011**(06), P06004 (2011). https://doi.org/10.1088/1742-5468/2011/06/P06004
7. M. Boltes, J. Zhang, A. Seyfried, B. Steffen, in *2011 IEEE International Conference on Computer Vision Workshops (ICCV Workshops)* (2011), pp. 158–165. https://doi.org/10.1109/ICCVW.2011.6130238
8. URL http://ped.fz-juelich.de/database/#t-junction
9. I. von Sivers, G. Köster, Transp. Res. Part B Methodol. **74**, 104 (2015). https://doi.org/10.1016/j.trb.2015.01.009
10. M.J. Seitz, N.W.F. Bode, G. Köster, J. R. Soc. Interface **13**(121), 20160439 (2016). https://doi.org/10.1098/rsif.2016.0439
11. R. Geirhos, D.H.J. Janssen, H.H. Schütt, J. Rauber, M. Bethge, F.A. Wichmann, Computing Research Repository (2018). URL http://arxiv.org/abs/1706.06969
12. H. Kladek, Über die Geschwindigkeitscharakterisik auf Stadtstraßenabschnitten. Ph.D. thesis (1966)

Part II
Granular and Active Matter

Chapter 42
First-Order Contributions to the Partial Temperatures in Dilute Binary Granular Suspensions

Rubén Gómez González and Vicente Garzó

Abstract The Boltzmann kinetic equation is considered to evaluate the first-order contributions $T_i^{(1)}$ to the partial temperatures in binary granular suspensions at low density. The influence of the surrounding gas on the solid particles is modeled via a drag force proportional to the particle velocity plus a stochastic Langevin-like term. The Boltzmann equation is solved by means of the Chapman–Enskog expansion around the local version of the reference homogeneous base state. To first-order in spatial gradients, the coefficients $T_i^{(1)}$ are computed by considering the leading terms in a Sonine polynomial expansion. In addition, the influence of $T_i^{(1)}$ on the first-order contribution $\zeta^{(1)}$ to the cooling rate is also assessed. Our results show that the magnitude of both $T_i^{(1)}$ and $\zeta^{(1)}$ can be relevant for some values of the parameter space of the system.

42.1 Introduction

In the last years, the understanding of granular matter under rapid flow conditions has raised the interest of many researchers due not only to its practical applications but also due to the fact that the understanding of its properties is really an exciting challenge. At a more fundamental level, the description of polydisperse granular mixtures (namely, the gaseous state of a mixture of smooth hard spheres with inelastic collisions) has been focused on the modification of the kinetic theory of molecular gases to properly adapt it to the dissipative character of collisions among particles.

R. Gómez González (✉)
Departamento de Física, Universidad de Extremadura, 06006 Badajoz, Spain
e-mail: ruben@unex.es

V. Garzó
Departamento de Física and Instituto de Computación Científica Avanzada (ICCAEx), Universidad de Extremadura, 06006 Badajoz, Spain
e-mail: vicenteg@unex.es
URL: http://www.unex.es/eweb/fisteor/vicente/

I. Zuriguel et al. (eds.), *Traffic and Granular Flow 2019*,
Springer Proceedings in Physics 252,
https://doi.org/10.1007/978-3-030-55973-1_42

Although the influence of inelasticity is reflected in all the transport coefficients, the main new feature (as compared with ordinary or molecular fluids) in the hydrodynamic equations is the presence of the so-called cooling rate ζ in the energy balance equation. The cooling rate (which vanishes for elastic collisions) measures the rate of energy dissipation due to inelastic collisions [1, 2].

One of the most intriguing and surprising effects of inelasticity in granular mixtures is the failure of energy equipartition in the homogeneous cooling state [2]. This means that the partial temperatures T_i of each component are different from the global granular temperature. The energy nonequipartition (which is only due to the inelastic character of collisions) has important effects in problems such thermal diffusion segregation. On the other hand, a new contribution to the breakdown of energy equipartition (additional to the one caused by the inelasticity in collisions) has been reported very recently [3]. Although considered in previous works of ordinary gases [4], this new contribution (which is associated with a nonzero first-order term $T_i^{(1)}$ in the expansion of the partial temperatures in powers of the gradients) had not been accounted in previous works of granular mixtures. Since $T_i^{(1)}$ is proportional to the divergence of the flow velocity $\mathbf{U}$ (i.e., $T_i^{(1)} = \varpi_i \nabla \cdot \mathbf{U}$), the coefficient ϖ_i is also involved in the evaluation of the first-order contribution $\zeta^{(1)}$ to the cooling rate ζ.

Although the coefficients ϖ_i have been computed in the case of *dry* granular mixtures (namely, a granular mixture where the influence of the interstitial fluid on the dynamics of grains is neglected), we are not aware of a similar calculation for binary granular suspensions at low density. The objective of this paper is to evaluate the coefficients ϖ_i in granular suspensions where the effect of the surrounding gas is modeled by means of an effective external force [5] composed by two terms: (i) a viscous drag term that mimics the friction of grains with the surrounding fluid and (ii) a stochastic term representing random and uncorrelated collisions between grains and fluid molecules. Once the first-order $T_i^{(1)}$ contributions to the partial temperatures are evaluated, as a complementary goal, we will also assess the impact of $T_i^{(1)}$ on $\zeta^{(1)}$.

42.2 Boltzmann Kinetic Equation for Binary Granular Suspensions

We consider a granular binary mixture of spheres of masses m_i and diameters σ_i $(i = 1, 2)$. Spheres are assumed to be completely smooth so that, the inelasticity of collisions is characterized by the constant (positive) coefficients of restitution $\alpha_{ij} \leq 1$. The solid particles are immersed in a viscous gas of viscosity η_g. In the low-density regime, the set of Boltzmann equations for the one-particle velocity distribution functions $f_i(\mathbf{r}, \mathbf{v}, t)$ of the component i reads [2]

$$\frac{\partial f_i}{\partial t} + \mathbf{v} \cdot \nabla f_i - \gamma_i \Delta \mathbf{U} \cdot \frac{\partial f_i}{\partial \mathbf{v}} - \gamma_i \frac{\partial}{\partial \mathbf{v}} \cdot \mathbf{V} f_i - \frac{\gamma_i T_{\text{ex}}}{m_i} \frac{\partial^2 f_i}{\partial v^2} = \sum_{j=1}^{2} J_{ij}[f_i, f_j]. \tag{42.1}$$

Here, $\Delta \mathbf{U} = \mathbf{U} - \mathbf{U}_g$, $\mathbf{U}$ and $\mathbf{U}_g$ being the mean flow velocities of the solid particles and the interstitial gas, respectively, $\mathbf{V} = \mathbf{v} - \mathbf{U}$ is the peculiar velocity, and $J_{ij}[f_i, f_j]$ is the Boltzmann collision operator [2].

At low Reynolds numbers, it has been assumed in Eq. (42.1) that the effect of the surrounding molecular gas on the solid particles is modeled by a drag force proportional to $\mathbf{v} - \mathbf{U}_g$ plus a stochastic Langevin force representing Gaussian white noise. While the drag force mimics the friction of grains with the interstitial gas, the stochastic force models the kinetic energy gain of solid particles due to the interaction with the background gas. In addition, γ_i is the drag or friction coefficient associated with the component i and T_{ex} is the temperature of the external gas. Note that the viscosity of the solvent $\eta_g \propto \sqrt{T_{\text{ex}}}$. More details of this kind of Langevin-like models can be found in Refs. [6, 7]. Here, for the sake of simplicity, we consider the coefficients γ_i to be scalars proportional to the viscosity η_g [5]. In this case, according to the results obtained in lattice-Boltzmann simulations [8], γ_i is a function of the partial volume fractions $\phi_i = \pi^{d/2} / \left(2^{d-1} d \Gamma\left(\frac{d}{2}\right)\right) n_i \sigma_i^d$, n_i being the number density of the component i. Based on the restriction that in the dilute limit every particle is only subjected to its respective Stokes drag [8], $\gamma_i = \gamma_0 R_i$ where $\gamma_0 = \left(18\eta_g/\rho\sigma_{12}^2\right)$ and the dimensionless function $R_i = (\rho\sigma_{12}^2/\rho_i\sigma_i^2)\phi_i$. Here, $\sigma_{12} = (\sigma_1 + \sigma_2)/2$, $\rho = \sum_i \rho_i$ is the total mass density, and $\rho_i = m_i n_i$ is the mass density of the component i. Upon deriving the form of γ_i we have considered hard spheres ($d = 3$) and binary mixtures ($i = 1, 2$).

42.3 Homogeneous Steady States

Before analyzing inhomogeneous states, it is desirable to study first the homogeneous case. In this situation, the partial densities n_i are constant, the granular temperature $T = x_1 T_1 + x_2 T_2$ is spatially uniform, and with an appropriate selection of the reference frame, the mean flow velocities vanish ($\mathbf{U} = \mathbf{U}_g = \mathbf{0}$). Here, $x_i = n_i/n$ and $n = n_1 + n_2$ is the total number density. After a transient regime, the system is expected to reach a *steady* state and so, Eq. (42.1) becomes

$$-\gamma_i \frac{\partial}{\partial \mathbf{v}} \cdot \mathbf{v} f_i - \frac{\gamma_i T_{\text{ex}}}{m_i} \frac{\partial^2 f_i}{\partial v^2} = \sum_{j=1}^{2} J_{ij}[f_i, f_j]. \tag{42.2}$$

For elastic collisions ($\alpha_{ij} = 1$), the cooling rate vanishes and in the case that $\gamma_1 = \gamma_2 = \gamma$, Eq. (42.2) admits the Maxwellian solution with a common temperature $T_1 = T_2 = T$. For inelastic collisions ($\alpha_{ij} \neq 1$), $\zeta \neq 0$ and to date the solution of Eq. (42.2) is not known. Thus, one has to consider approximate forms for f_i. Here,

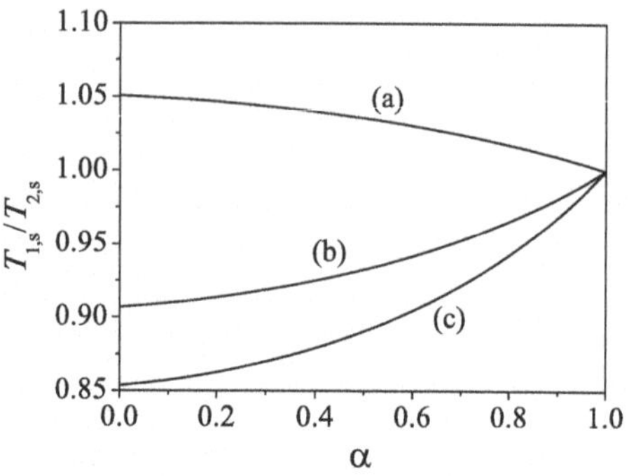

Fig. 42.1 Plot of the temperature ratio $T_{1,s}/T_{2,s}$ as a function of the common coefficient of restitution α for an equimolar mixture ($x_1 = 0.5$) of hard spheres ($d = 3$) with $\sigma_1/\sigma_2 = 1$, and $T_{ex}^* = 0.1$. Three different values of the mass ratio are considered: $m_1/m_2 = 0.5$ (**a**), $m_1/m_2 = 4$ (**b**), and $m_1/m_2 = 10$ (**c**)

for the sake of simplicity and to compute the first velocity moments of f_i, we will replace f_i by the Maxwellian distribution at the temperature T_i:

$$f_i(\mathbf{v}) \to f_{i,\mathrm{M}}(\mathbf{v}) = n_i \left(\frac{m_i}{2\pi T_i}\right)^{d/2} \exp\left(-\frac{m_i v^2}{2T_i}\right), \tag{42.3}$$

where the partial temperatures T_i are defined as

$$T_i = \frac{m_i}{dn_i} \int d\mathbf{v}\; V^2 f_i(\mathbf{v}). \tag{42.4}$$

The (reduced) partial temperatures $\tau_{i,\mathrm{s}} = T_{i,\mathrm{s}}/T_\mathrm{s}$ can be obtained by multiplying both sides of Eq. (42.2) by $m_i v^2$ and integrating over velocity. The result is $2\gamma_{i,\mathrm{s}}^*(\tau_{i,\mathrm{s}} - \theta_\mathrm{s}^{-1}) + \zeta_{i,\mathrm{s}}^* \tau_{i,\mathrm{s}} = 0$, where the subscript s means that all the quantities are evaluated in the steady state. Here, $\gamma_{i,\mathrm{s}}^* = \ell\gamma_{i,\mathrm{s}}/v_{0\mathrm{s}}$, $\theta_\mathrm{s} = T_\mathrm{s}/T_\mathrm{ex}$, $\zeta_{i,\mathrm{s}}^* = \ell\zeta_{i,\mathrm{s}}/v_{0\mathrm{s}}$, $v_{0\mathrm{s}} = \sqrt{2T_\mathrm{s}/\overline{m}}$ is the thermal speed, $\overline{m} = \sum_i m_i/2$, and $\ell = 1/n\sigma_{12}^{d-1}$ is the mean free path of elastic hard spheres. In the Maxwellian approximation (42.3), the (reduced) partial cooling rates $\zeta_{i,\mathrm{s}}^*$ for the partial temperatures $T_{i,\mathrm{s}}$ can be easily evaluated. Its explicit form can be found for instance in Eq. (5.51) of Ref. [2]. The temperature ratio $T_{1,\mathrm{s}}/T_{2,\mathrm{s}}$ is plotted in Fig. 42.1 versus the (common) coefficient of restitution $\alpha \equiv \alpha_{ij}$. As expected, failure of equipartition in the homogeneous steady state is presented when the collisions are inelastic ($\alpha \neq 1$). In addition, it is quite apparent that the extent of the energy violation is greater when the mass disparity is large.

42.4 First-Order Contributions to the Partial Temperatures

We now slightly perturb the homogeneous steady state by small spatial gradients. These gradients give rise to nonzero contributions to the mass, momentum, and heat fluxes. Here, we want to compute the first-order contributions to the partial temperatures. In order to achieve them, we have to solve the Boltzmann equation (42.1) by means of the Chapman–Enskog method [9] adapted to dissipative dynamics. As usual, the Chapman–Enskog method assumes the existence of a *normal* or hydrodynamic solution where all the space and time dependence of the velocity distribution functions $f_i(\mathbf{v}, \mathbf{r}, t)$ occurs via a functional dependence on the hydrodynamic fields. For small spatial gradients, this functional dependence can be made local in space through an expansion of the distribution functions in powers of the spatial gradients: $f_i \to f_i^{(0)} + f_i^{(1)} + \cdots$. Here, only terms up to first order in gradients will be retained (Navier–Stokes hydrodynamic order).

The first-order contributions $T_i^{(1)}$ to the partial temperatures are defined as

$$T_i^{(1)} = \frac{m_i}{dn_i} \int d\mathbf{v}\; V^2 f_i^{(1)}(\mathbf{V}). \tag{42.5}$$

Given that the procedure to determine $T_i^{(1)}$ is relatively large, we display here only the final results. More technical details can be found in Ref. [10]. As said in the Introduction, $T_i^{(1)}$ can be written as $T_i^{(1)} = \varpi_i \nabla \cdot \mathbf{U}$. In the leading Sonine approximation, the coefficients ϖ_i obey the set of algebraic equations

$$\sum_{j=1}^{2} \left[\omega_{ij} + 2\gamma_j x_j \left(\tau_i + \theta \Delta_{\theta,i}\right) - 2\gamma_i \delta_{ij} + \left(T_i^{(0)} + T\theta\Delta_{\theta,i}\right)\xi_j\right]\varpi_j =$$

$$-\frac{2}{d} T\theta\Delta_{\theta,i} - T\sum_{j=1}^{2} n_j \frac{\partial \lambda_1}{\partial n_j} \Delta_{\lambda_1,i}, \tag{42.6}$$

where the coefficients ξ_i are defined by Eq. (25) of Ref. [10] while the collision frequencies ω_{ii} and ω_{ij} are given by Eqs. (26) and (27), respectively, of Ref. [10]. Moreover, the expressions of the derivatives $\Delta_{\theta,i}$, $\partial\lambda_1/\partial n_j$, and $\Delta_{\lambda_1,i}$ are displayed in Ref. [10]. Upon deriving Eq. (42.6), use has been made of the identity $\zeta^{(1)} = \sum_i \xi_i \varpi_i \nabla \cdot \mathbf{U} \equiv \zeta_U \nabla \cdot \mathbf{U}$. The solution to the set of equations (42.6) provides the explicit forms of ϖ_1 and ϖ_2. It is seen that $\varpi_2 = -(x_1/x_2)\varpi_1$, as the solubility conditions of the Chapman–Enskog method requires.

Figure 42.2 shows the α-dependence of $\varpi_1^* \equiv (n\sigma_{12}^2 v_0/T)\varpi_1$ and ζ_U for different systems. We observe that the influence of the inelasticity on both ϖ_1^* and ζ_U is important, specially for strong inelasticity. Thus, both quantities should be taken into account in the kinetic description of binary granular suspensions.

In summary, we have determined the first-order contributions $T_i^{(1)}$ to the partial temperatures in binary granular suspensions. The fact that $T_i^{(1)} \neq 0$ yields a *new* contribution (additional to the one caused by inelasticity in collisions) to the breakdown of energy equipartition. Since this contribution is proportional to the divergence of the flow velocity, it is involved then in the evaluation of the first-order contribution ζ_U to the cooling rate. Our results show that the magnitude of both coefficients $T_i^{(1)}$ and ζ_U can be significant in some regions of the parameter space of the system. This conclusion contrasts with the results obtained for dry granular mixtures [11, 12] where it was shown that both coefficients vanish in the low density limit.

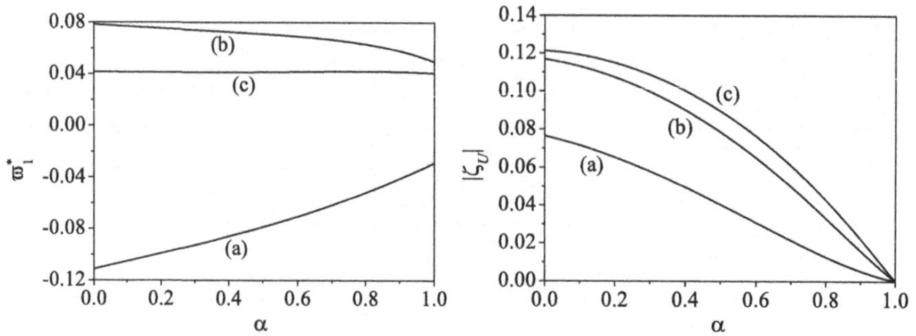

Fig. 42.2 Plot of the (reduced) coefficients ϖ_1^* and ζ_U as a function of the common coefficient of restitution α for an equimolar mixture ($x_1 = 0.5$) of hard spheres ($d = 3$) with $\sigma_1/\sigma_2 = 1$, and $T_{\text{ex}}^* = 0.1$. Three different values of the mass ratio are considered: $m_1/m_2 = 0.5$ (**a**), $m_1/m_2 = 4$ (**b**), and $m_1/m_2 = 10$ (**c**)

Acknowledgements The present work has been supported by the Spanish Government through Grant No. FIS2016- 76359-P and by the Junta de Extremadura (Spain) Grants No. IB16013 (V.G.) and No. GR18079, partially financed by "Fondo Europeo de Desarrollo Regional" funds. The research of R.G.G. has been supported by the predoctoral fellowship BES-2017-079725 from the Spanish Government.

References

1. N.V. Brilliantov, T. Pöschel, *Kinetic Theory of Granular Gases* (Oxford University Press, Oxford, 2010)
2. V. Garzó, *Granular Gaseous Flows* (Springer Nature Switzerland, Basel, 2019)
3. R. Gómez González, V. Garzó, Influence of the first-order contributions to the partial temperatures on transport properties in polydisperse dense granular mixtures. Phys. Rev. E **100**, 032904 (2019). https://doi.org/10.1103/physreve.100.032904
4. J. Karkheck, G. Stell, Bulk viscosity of dense simple fluid mixtures. J. Chem. Phys. **71**, 3636–3639 (1979). https://doi.org/10.1063/1.438806
5. D.L. Koch, R.J. Hill, Inertial effects in suspension and porous-media flows. Ann. Rev. Fluid Mech. **33**, 619–647 (2001). https://doi.org/10.1146/annurev.fluid.33.1.619
6. H. Hayakawa, S. Takada, V. Garzó, Kinetic theory of shear thickening for a moderately dense gas-solid suspension: from discontinuous thickening to continuous thickening. Phys. Rev. E **96**, 042903 (2017). https://doi.org/10.1103/physreve.96.042903
7. T. Kawasaki, A. Ikeda, L. Berthier, Thinning or thickening? multiple rheological regimes in dense suspensions of soft particles. EPL **107**, 28009 (2014). https://doi.org/10.1209/0295-5075/107/28009
8. X. Yin, S. Sundaresan, Fluid-particle drag in low-Reynolds-number polydisperse gas-solid suspensions. AIChE J. **55**, 1352–1368 (2009). https://doi.org/10.1002/aic.11800
9. S. Chapman, T.G. Cowling, *The Mathematical Theory of Non-Uniform Gases* (Cambridge University Press, Cambridge, 1970)

10. R. Gómez González, V. Garzó, First-order contributions to the partial temperatures in binary granular suspensions at low density. Preprint at http://arxiv.org/abs/1910.07490 (2019)
11. V. Garzó, J.W. Dufty, Hydrodynamics for a granular mixture at low density. Phys. Fluids **14**, 1476–1490 (2002). https://doi.org/10.1063/1.1458007
12. V. Garzó, J.M. Montanero, J.W. Dufty, Mass and heat fluxes for a binary granular mixture at low density. Phys. Fluids **18**, 083305 (2006). https://doi.org/10.1063/1.2336755

Chapter 43
Acoustic Resonances in a Confined Set of Disks

Juan F. González-Saavedra, Álvaro Rodríguez-Rivas, Miguel A. López-Castaño, and Francisco Vega Reyes

Abstract We study in this work a system with a binary mixture of disks (both species only differ in their mass) enclosed in a rectangular region. The disks are arranged, at high particle density, in a hexagonal lattice. Specifically, we are interested in the conditions for mechanical signal transmission. Two different potentials are considered. Our results are obtained by means of soft disks computer simulations. We analytically determine the eigenfrequencies spectra of the system from the corresponding Hessian matrix. Alternatively, the eigenfrequencies can also be obtained from the time evolution of molecular dynamics simulations. Previous works have shown that mechanical signal transmission in granular matter can be used to develop acoustic switches. Our results reveal that the working range of the acoustic switch changes dramatically for different interaction models. In particular, strong anisotropies in signal transmission may appear for certain interparticle forces.

43.1 Introduction

Signal transmission devices are ubiquitous in mechanical engineering applications [6, 10]. They are used as vibration insulators [7], acoustic cloaks [20], sound propagation [5], signal rectification [4], etc. In this context, highly dense granular systems can be used to perform these tasks [19]. In particular, granular crystals have proven to be efficient for clean signal transmission [3]. In fact, some recent works show that they can work as switches [19], rectifiers [3] and a range of logic elements [12].

Hence, granular crystals can be used to modify and transmit mechanical energy signals [1]. These applications are analogous to the electrical switches that can control electromagnetic waves propagation [2]. Control of acoustic and thermal information using this type of switches can lead to applications in fields such as thermoelectric systems, communication or ultrasonic imaging [8].

J. F. González-Saavedra (✉) · Á. Rodríguez-Rivas · M. A. López-Castaño · F. Vega Reyes
Departamento de Física and Instituto de Computación Científica Avanzada (ICCAEx), Universidad de Extremadura, 06006 Badajoz, Spain
e-mail: juanfrangs@unex.es

I. Zuriguel et al. (eds.), *Traffic and Granular Flow 2019*, Springer Proceedings in Physics 252,
https://doi.org/10.1007/978-3-030-55973-1_43

As it is known, dense granular systems of identically shaped particles can develop crystalline structures [13]. Usually, the critical density for the appearance of ordering depends strongly on the values of the other relevant physical magnitudes: granular temperature [15], degree of inelasticity in collisions [18], the specifics of the physical processes involved in particle collisions [14], the type of interaction with the boundaries [9], the kind of energy input necessary to thermalize the granular particles [11], etc. Experimental and computational work has put in evidence these effects.

As a first approach, we study how the switch behavior changes for different conservative potentials in the system of macroscopic disks. Our analysis is performed by means of molecular dynamics simulations. In our simulations, we consider two different interaction potentials: a potential that is quadratic in the distance between particles [19], and a Lennard-Jones-type potential [16]. We leave the analysis of the influence of inelasticity in the collisions for a future work.

The paper is organized as follows. The second section is devoted to the system description, including the initial configuration used in this work, together with the computational and theoretical methods used to obtain the results. Finally, Sect. 43.3 presents the results and a discussion, together with a description of possible future extensions of our results.

43.2 Description of the System and Computational Methods

Let us imagine a rectangular region defined by two parallel rigid walls and two parallel periodic boundaries, as sketc.hed in Fig. 43.1. There is no gravitational field acting on the system. Rigid walls-disks interactions are modelled with the same potential as for disk-disk interactions. More specifically, rigid walls are modelled by assuming the existence of a particle of the same size and mass behind the wall, when the wall is within the interaction range of a real particle. The x position of this wall particle coincides at all times with the x position of the corresponding real particle. On the contrary, the y position of the wall particle center is always located at a particle radius away from the wall surface.

Two different types of interaction are used. The first one is parabolic-like, with the form

$$U(r_{ij}) = \begin{cases} \frac{\epsilon}{2}\left(1 - \frac{r_{ij}}{\sigma}\right)^2 & \text{if} \quad r_{ij} \leq \sigma \\ 0 & \text{if} \quad r_{ij} > \sigma \end{cases} \tag{43.1}$$

whereas the second potential is a Lennard-Jones type potential [16]

$$U(r_{ij}) = \begin{cases} 4\epsilon\left[\left(\frac{\sigma}{r_{ij}}\right)^{12} - \left(\frac{\sigma}{r_{ij}}\right)^{6}\right] + \epsilon & \text{if } r_{ij} \leq r_c \\ 0 & \text{if } r_{ij} > r_c\,, \end{cases} \tag{43.2}$$

where ϵ sets the energy scale, σ is the diameter of the disks, r_{ij} is the distance between centers of disks i and j and r_c is the energy cut distance, set in our simulations to $2^{1/6}\sigma$.

In order to obtain the eigenfrequencies of the system, we computed the eigenvalues ω_k^2 of the dynamical matrix $\mathcal{D}$, whose elements are [17, 19]

$$\mathcal{D}_{ij} = \mathcal{M}_{ik}^{-1}\mathcal{H}_{kj} \tag{43.3}$$

with

$$\mathcal{H}_{ij} = \frac{\partial^2 U}{\partial x_i \partial x_j}, \qquad \mathcal{M}_{ij} = m_{L,H}\delta_{ij} \tag{43.4}$$

where $\mathcal{M}_{ij}$, $\mathcal{H}_{ij}$ are the corresponding Hessian matrix [17] and the particle mass diagonal matrices, respectively. The eigenvalues were obtained by: (a) considering an initial equilibrium state of the system; (b) from the average positions of the particles for a simulation run.

In the simulations, a 2D system of particles (disks with equal diameter σ) inside the rectangular box is considered. The dimension of the box is such that an integer number of horizontal layers can be perfectly accommodated into a hexagonal lattice, as it can be seen in Fig. 43.1. The total number of particles is 100, each of them with a diameter σ, which we take as length unit. We consider particles with two different masses, the mass ratio being $m_H/m_L = 10$. The particle distribution is such that we have 25 light disks (with mass m_L) and 75 heavy disks (with mass m_H). The initial configuration is the mechanically stable packing of the disks.

For this work, we only analyze the steady state behavior, and for this reason the initial state does not have an impact on our results. Thus, for simplicity, initial particle velocities are drawn from a Maxwell-Boltzmann distribution whose standard deviation is determined by the system initial temperature. In our case, this temperature is a control parameter in the simulations. The evolution of the system is obtained from molecular dynamics simulations using leap-frog method [16]. The pressure can also be modified by approaching the rigid walls to each other.

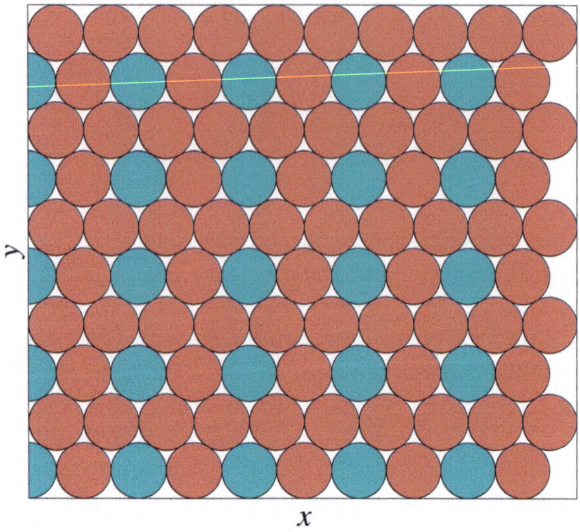

Fig. 43.1 Initial setup of the light (blue) and heavy (red) disks, with 10×10 particle layers. x stands for the horizontal axis and y for the vertical axis

43.3 Results and Discussion

In Fig. 43.2 we can see the corresponding eigenvalues for both parabolic (43.1) and Lennard-Jones (43.2) potentials. The equilibrium positions are considered as $p = 0$ configurations. This state can be modified by approaching the top and bottom walls. In this case, the disks will overlap, and, as a consequence, repulsion forces between them appear. By means of this procedure, we can slightly increase the pressure to $p = 0.1$. As we can see, the eigenvalues display significant discontinuities. This produces frequency gaps which can be tuned by modifying the pressure of the system. It is

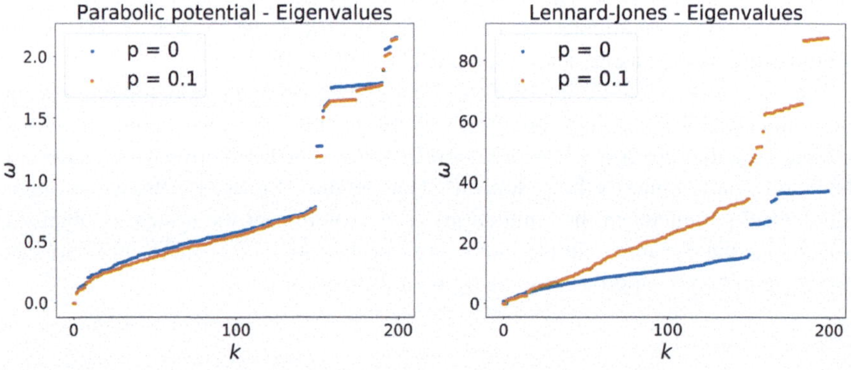

Fig. 43.2 Eigenfrequencies of the system when the parabolic (left panel) and Lennard-Jones (right panel) potentials are considered. The curves represent a system with the particles in their equilibrium positions (blue) and with a pressure $p = 10^{-1}$ (red)

this behavior which allows for the use as a switch. Put in other words, a mechanical signal with a given frequency ω will be transmitted only if it does not lie within a forbidden frequency gap (for a given value of the system pressure p).

A way to look at the behavior of an eventual switch is to calculate the power spectra density

$$D(\omega) = \int_0^\infty dt \frac{\langle \mathbf{v}(t_0 + t) \cdot \mathbf{v}(t_0) \rangle}{\langle \mathbf{v}(t_0) \cdot \mathbf{v}(t_0) \rangle} e^{i\omega t} \tag{43.5}$$

As we can see in Fig. 43.3 the normalized velocity autocorrelation function has finite decay length and is isotropic for the parabolic potential whereas for the Lennard-Jones potential does not decay and there are significant differences between the x and y directions. The peculiar behavior in the case of the Lennard-Jones may be due to its steep repulsive term, as compared to the parabolic potential.

In Fig. 43.4, we can see the corresponding Fourier transform of the panels in Fig. 43.3. It is interesting to note that the isotropy properties of the autocorrelation function show up here as well. More specifically, for the Lennard-Jones potential

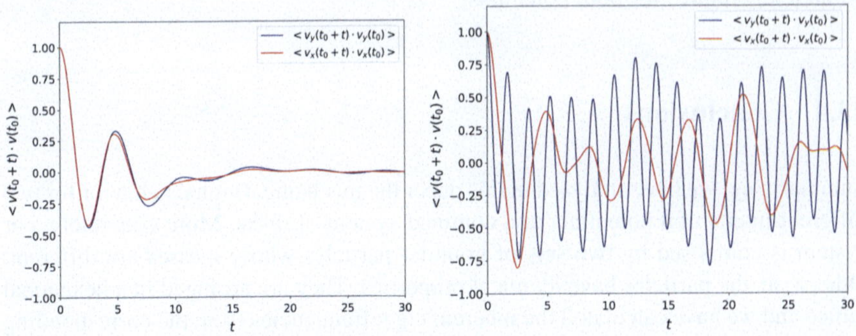

Fig. 43.3 Velocity autocorrelation function in the x and y directions for the parabolic (left panel) and the Lennard-Jones potentials (right panel) at temperature $T = 10^{-4}$

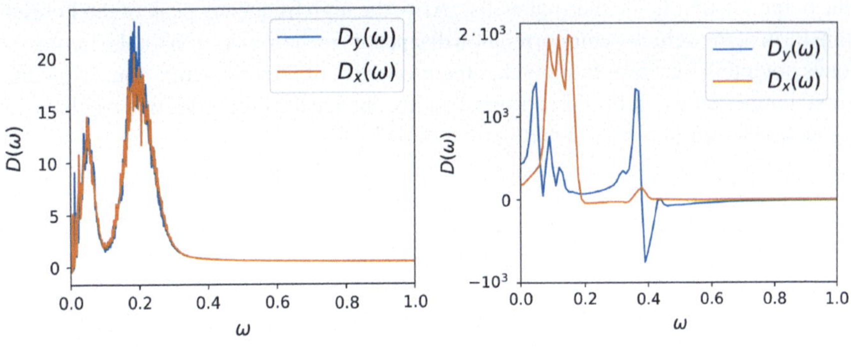

Fig. 43.4 Power spectra distribution (Fourier transform of the velocity autocorrelations) in the x and y directions for the parabolic and the Lennard-Jones potentials

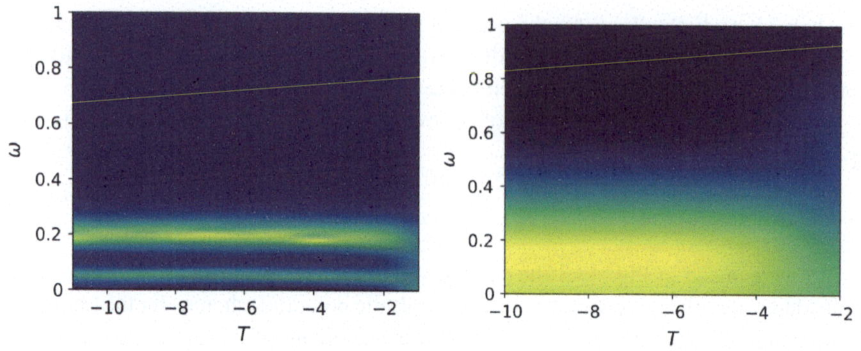

Fig. 43.5 Power spectra distribution (Fourier transform of the velocity autocorrelations) in the x direction for the parabolic and the Lennard-Jones potentials

(right panel), the high frequency peak disappears in the x direction for $T < 10^{-4}$. This is illustrated in more detail in Fig. 43.5, where $D(\omega)$ is displayed for a range of initial temperatures for both potentials.

43.4 Conclusions

Summarizing, we have analysed in this paper the mechanical signal behavior for two different interaction potentials in a confined system of disks. More concretely, our system is composed by two sets of granular particles whose masses are different. Otherwise, the particles have identical properties. They are arranged in a hexagonal lattice and we have calculated the inherent eigenfrequencies from the corresponding Hessian matrix. In addition, we inspected the behavior of the velocity autocorrelation functions, finding important anisotropic behavior in the case of the Lennard-Jones potential. For this potential the transmission is clearly more intense in the y direction (the perpendicular to the thermal walls). Also, the high frequency peak of the Fourier transform of the velocity autocorrelation disappears in the x direction for the Lennard-Jones potential. Our results show that the properties of the interaction forces and the initial temperature conditions can significantly change the properties of a mechanical signal transmission in a confined system of disks.

References

1. S. Alagoz, B. Baykant Alagoz, Sonic crystal acoustic switch device. J. Acoust. Soc. Am. **133**, EL485–EL490 (2013)
2. J. Bardeen, W.H. Brattain, Phys. Rev. **74**, 230 (1948)
3. N. Boechler, G. Theocharis, C. Daraio, Nat. Mater. **10**, 665 (2011)
4. N. Boechler, G. Theocharis, P.G. Kevrekidis, C. Daraio, J. Appl. Phys. **109**, 074,906 (2011)
5. S.A. Cummer, Science **343**, 495 (2014)
6. T. Devaux, A. Cebrecos, O. Richoux, V. Pagneux, V. Tournat, Acoustic radiation pressure for nonreciprocal transmission and switch effects. Nature Commun. **10**, 3292 (2019)
7. G. Gantzounis, M. Serra-Garcia, K. Homma, J.M. Mendoza, C. Daraio, Granular metamaterials for vibration mitigation. J. Appl. Phys. **114**, 093,514 (2013)
8. F. Li, P. Anzel, J. Yang, P.G. Kevrekidis, C. Daraio, Granular acoustic switches and logic elements. Comput. Sci. Eng. **5**, 5311 (2014)
9. A. Lobkovsky, F. Vega Reyes, J. Urbach, Eur. Phys. J. Spec. Top. **179**, 113–122 (2009)
10. Y.J. Lu, Y. Ge, S.Q. Yuan, H.X. Sun, X.J. Liu, Acoustic logic gates by a curved waveguide with ultrathin metasurfaces. J. Phys. D Appl. Phys. **53**, 015,301 (2020)
11. M.A. López-Castaño, J.F. González-Saavedra, A. Rodríguez-Rivas, F. Vega Reyes, Statistical properties of a granular gas fluidized by turbulent air wakes (2019). E-print arXiv:1911.01384
12. I. Mahboob, E. Flurin, K. Nishiguchi, A. Fujiwara, H. Yamaguchi, Nat. Commun. **2**, 198 (2011)
13. P. Melby, F. Vega Reyes, A. Prevost, R. Robertson, P. Kumar, D.A. Egolf, J.S. Urbach, The dynamics of thin vibrated granular layers. J. Phys. Condens. Matter **17**, S2689–S2704 (2005)
14. J.A. Olafsen, J.S. Urbach, Clustering, order and collapse in a driven granular monolayer. Phys. Rev. Lett **81**, 4369–4372 (1998)
15. A. Prevost, P. Melby, D. Egolf, J.S. Urbach, Phys. Rev. E **17**, 050,301(R) (2005)
16. D.C. Rapaport, *The Art of Molecular Dynamics Simulation* (Cambridge University Press, Cambridge, UK, 2004)
17. A. Tanguy, J.P. Wittmer, F. Leonforte, J.L. Barrat, Continuum limit of amorphous elastic bodies: A finite-size study of low-frequency harmonic vibrations. Phys. Rev. B **66**, 174,205 (2002)
18. F. Vega Reyes, J.S. Urbach, Clustering, order and collapse in a driven granular monolayer. Phys. Rev. E **78**, 051,301 (2008)
19. Q. Wu, C. Cui, T. Bertrand, M.D. Shattuck, C.S. O'Hern, Active acoustic switches using two-dimensional granular crystals. Phys. Rev. E **99**, 062,901 (2019)
20. L. Zigoneanu, B.I. Popa, S.A. Cummer, Nat. Mater. **13**, 352 (2014)

Chapter 44
Morphological Response of Clogging Arches to Gentle Vibration

B. V. Guerrero, Iker Zuriguel, and Angel Garcimartín

Abstract We report experiments on the vibration-induced unclogging process in a silo. Arches blocking the orifice are perturbed by a gentle vibration intensity (below the fluidization limit). We explore how the endurance and the morphology of arches are affected by the applied perturbation. By performing a survival analysis of the breaking times we characterize the stability of arches. We use two geometrical variables to describe the arch morphology at two instants: when vibration starts and right before the collapse. Regardless of the morphological descriptor, it is confirmed that the arches that are initially more irregular are more unstable, but even so, it should not be taken for granted that the initial configuration unequivocally determines the endurance of any particular clog. Furthermore, the final configurations reveal that there exists a maximum degree of irregularity that the arches can withstand (before the collapse occurs), which is independent of the shaking strength.

44.1 Introduction

It is known that the gravity-driven outpouring of grains from a silo occurs at a constant rate and increases with the outlet size [1]. When aperture is just a few times wider than particles, the flow may be halted by arches blocking the outlet [2]. Vibration is one effective mechanism to destabilize those arches [3, 4], which will otherwise stand forever once all the kinetic energy in the granular medium is dissipated. Recent works have been devoted to study how the vibration affects the unclogging, focusing only on arch geometrical features. For two-dimensional silos filled with spheres, the angle ϕ (subtended between the centers of three successive beads) has been shown to be a useful variable to describe, in average, the endurance of arches [5–7]. In particular, the maximum of those ϕ angles, ϕ_{max}, is able to account for the mechanical stability of the arch. The higher ϕ_{max} is, the easier it is to break the arch, especially when $\phi_{max} > 180$. Moreover, the standard deviation of the angles $\{\phi\}$ in a single arch, σ,

B. V. Guerrero (✉) · I. Zuriguel · A. Garcimartín
Dep. Física y Mat. Apl., Fac. Ciencias, Universidad de Navarra, 31080 Pamplona, Spain
e-mail: bguerrero.3@alumni.unav.es

I. Zuriguel et al. (eds.), *Traffic and Granular Flow 2019*,
Springer Proceedings in Physics 252,
https://doi.org/10.1007/978-3-030-55973-1_44

quantifies its irregularity [8], which increases along time if the arch is vibrated. Both descriptors reveal that arches evolve to less regular configurations through an aging dynamics, rationalized in terms of some sub-diffusive transport models [9, 10].

In order to extend the morphological analysis performed in [8], here we investigate how the applied vibration affects the morphology and the breakup of the arches.

44.2 Experimental Setup and Methods

The specifications of the experimental setup are reported in [8]. It consists of a flat bottomed two-dimensional silo that can be vibrated vertically (Fig. 44.1). It is made with two transparent polycarbonate sheets (390 mm high × 80 mm wide), separated by a steel frame (1 cm wide and 1.2 mm thick). The container is halved with two metal strips leaving a gap of length D, such that the upper chamber acts as the silo with outlet size D. This design resembles an hourglass; so once the silo is emptied, it can be refilled just by rotating the container upside down. This device is mounted on a electromagnetic shaker that delivers the vibration. The silo is partially filled with stainless steel spheres of diameter $d = 1 \pm 0.01$ mm.

The automated experimental procedure starts with the silo upright and the grains outflowing by gravity. A camera captures the region near the outlet at 25 frames per second. By performing a real time image analysis, the clog is detected when it is formed. Then, the shaker is switched on and the camera starts to record the time-stamped images until the arch breaks or the maximum experimental duration is reached (1200 s). Thereupon, the silo is turned and prepared for the next realization. Finally, for each image, the centers of the spheres are determined (see Fig. 44.1 and [11]) and the arch is identified [12].

The experimental conditions have been chosen seeking to achieve a sizable amount of long lasting arches which, in turn, are vulnerable to the shaking. The chosen outlet width $D = 4.4d$ fulfills this compromise. Concerning the perturbation, we apply three different steady sinusoidal shaking defined by the parameters given in

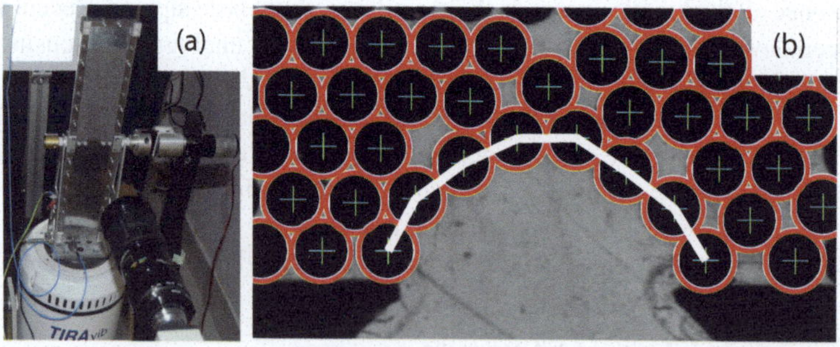

Fig. 44.1 **a** Experimental setup. The silo is placed on top of the shaker and a camera located in front of a backlighted region near the silo outlet. **b** The result of the image processing, where the center of each bead is detected (*green cross marks*) and the clogging arch is identified (*white line*)

Table 44.1 The three vibration intensities applied, their corresponding amplitude A, and the number of arches available for each case. The data for $\{\Gamma = 0.6,\ f = 105\ \text{Hz}\}$ are the same that were used in [8]

Γ	f (Hz)	$A(\mu\text{m})$	Number of arches
0.6	105	13.5	4458
0.6	1000	0.2	783
0.8	105	18.0	2554

Table 44.1. The vibration intensity is characterized by the maximum dimensionless acceleration $\Gamma = A(2\pi f)^2/g$, where g is the gravity acceleration, A the amplitude and f the frequency of vibration. In each cases $\Gamma < 1$, so it is below the fluidization limit ($\Gamma = 1$) but is enough to provoke rearrangements of the arch beads. Two conditions were chosen with same Γ and other two with same f, in order to explore the influence of these parameters separately.

44.3 Results

First, we characterize the stability of arches in terms of the survival function of breaking times P($t > t_b$), which is the probability that an arch lasts longer than t. Experimentally, the breaking time, t_b, is defined as the time elapsed between switching on the vibration and the last image where the arch is identifiable. The survival function is computed as explained in [8]. Figure 44.2 shows the survival functions for the three experimental conditions explored in this work. Notice that in all cases the vibration is capable to shatter long lasting arches although is not enough to break all of them. Indeed, the amount of unbroken arches (i. e. arches that reach the maximum measurement time) is quite high. The heavy tails show an apparent power law decay (note the double logarithmic scale) and then they flatten for long times. Recently, an energy trap model [13] rationalized this behavior reproducing quite well the shape of the survival functions.

Experiments with the same Γ show that the higher the frequency, the less arches are broken. Interestingly arches break even when the amplitude is smaller than the

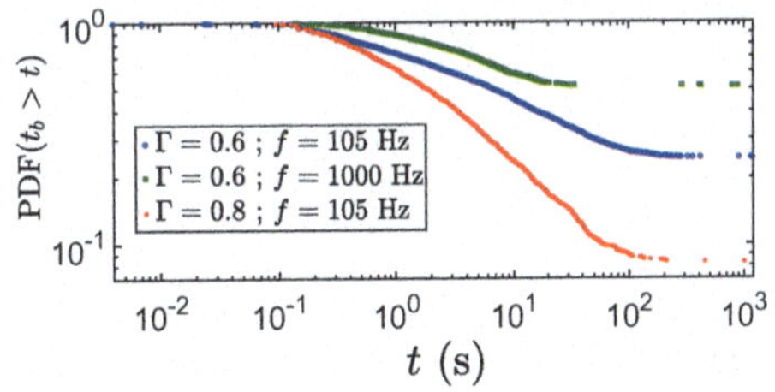

Fig. 44.2 Survival function of the breaking time t_b for three different shaking strengths. Notice that for each perturbation, the point where plateau starts and the proportion of unbroken arches vary significantly

typical roughness of the grains ($\sim$1 µm). Also, the cases with the same f show the expected behavior [13, 14]: higher vibration intensities facilitate the arch breakage. Within this mild vibration scenario, the shaking downward acceleration never overcomes gravity, so the granular medium never detaches from the vibrating base. Therefore, the arch collapse is only possible due to the local rearrangements near the silo outlet, that are capable to overcome friction and geometrical frustration.

We now analyze statistically the effect of the different shaking intensities on the morphology of the arches. The morphological configuration of an arch is characterized in terms of two variables: (i) the maximum angle, ϕ_{max}, and (ii) the standard deviation σ of all $\{\phi\}$ angles in each arch. With the aim of studying the deformation of arches along their lifespan, we focus in two instants: the initial one (when vibration starts) and the final one (before the collapse or at the maximum measurement time). Arches have been classified into unbroken arches and several subsets of broken arches, sorted according to their breaking time.

For the three experimental conditions, the histograms P of the initial maximum angle, $P(\phi_{ini}^{max})$ are shown on the left side of Fig. 44.3. Obviously, the initial state does not depend on forcing because when arches are formed there is no vibration. In addition, these histograms show that, on average, arches with lower ϕ_{ini}^{max} last longer.

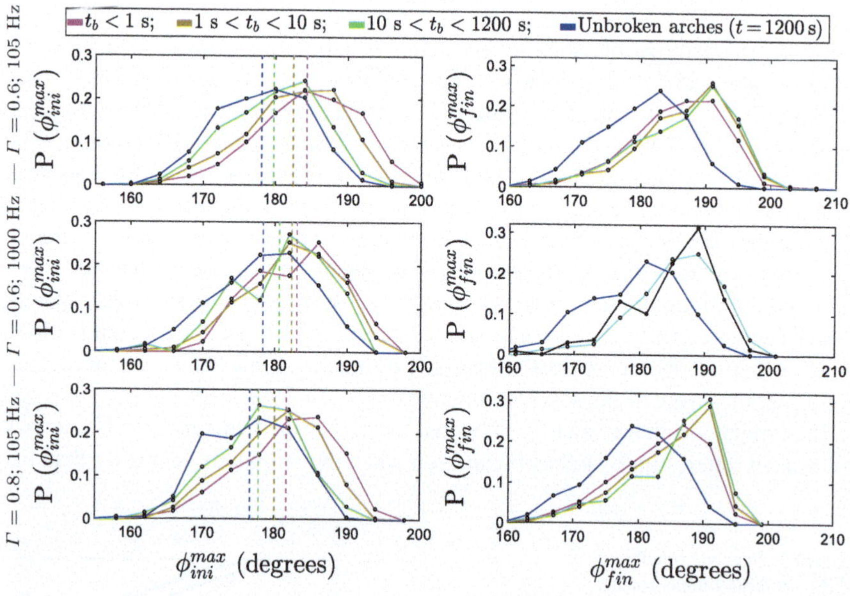

Fig. 44.3 *Left*: Histograms $P(\phi_{ini}^{max})$ of ϕ_{max} just after starting the vibration. Vertical dashed lines represent the average value of the distribution. *Right*: Histograms $P(\phi_{fin}^{max})$ of ϕ_{max} just before the collapse. Colors indicate each subset of arches (sorted by their breaking time as shown in the legend). For $P(\phi_{fin}^{max})$ with $\{\Gamma = 0.6$; f=1000 Hz$\}$, only two subsets of broken arches were taken because of the smaller sample size: cyan, 10 s $< t_b <$ 30 s; and black, 30 s $< t_b <$ 1200 s. The vibration parameters (Γ and f) are given at the left of each histogram

On the right side of the same figure the respective histograms P of the final maximum angle, $P(\phi_{fin}^{max})$, are shown. Note that, unlike the unbroken arches, all histograms for broken arches share the same distribution, regardless their breaking time and the vibration strength. Indeed, the common abrupt falling after $\phi_{fin}^{max} > 190°$ can be associated to an stability threshold, related to the fully mobilized static friction [5].

We also perform this statistical analysis of the arch morphology in terms of σ. Figure 44.4 shows the measured histogram P of the initial and the final σ, $P(\sigma_{ini})$ and $P(\sigma_{fin})$, respectively. The conclusions reached with σ are in agreement with the findings drawn from ϕ_{max}. On average, σ_{ini} is lower for arches that last longer. Besides, the histogram $P(\sigma_{fin})$ corresponding to the unbroken arches clearly differs from the histograms of the broken arches, whereas all the subsets of broken arches exhibit the same distribution (more spread and symmetric than the are obtained for ϕ).

It is worth noting that $P(\phi_{ini}^{max})$ and $P(\sigma_{ini})$ for the different subsets are spread and highly overlapping, so neither ϕ_{ini}^{max} nor σ_{ini} define the stability of any given arch. Nevertheless, the mean values of ϕ and σ (vertical lines in Figs. 44.3 and 44.4) suggest that, on average, arches with lower ϕ_{ini}^{max} or σ_{ini} last longer. Furthermore, our results attest that, on average: (i) the final configuration is quite independent of Γ and f, and (ii) arches evolve to more irregular configurations (until a maximum regularity threshold is reached), without favoring any of them.

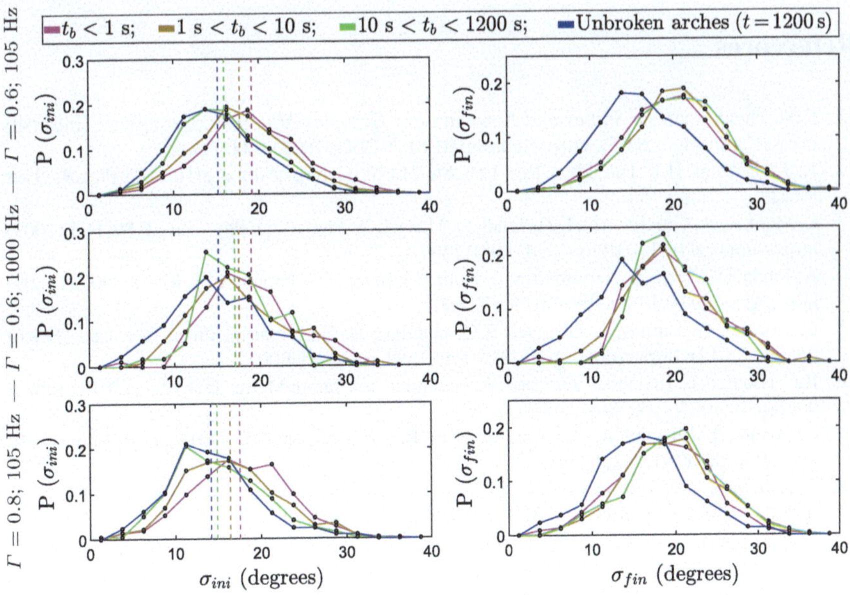

Fig. 44.4 *Left*: Histograms $P(\sigma_{ini})$ of σ just after starting the vibration. Vertical dashed lines represent the average value of the distribution. *Right*: Histograms $P(\sigma_{fin})$ of σ just before the collapse. Colors indicate each subset of arches (sorted by their breaking time, as shown in the legend). The vibration parameters (Γ and f) are given at the left of each histogram

44.4 Conclusions

We checked the validity of our previous results [8] regarding the shape of arches when other shaking intensities are used. We explored cases with: (i) the same Γ and different f and (ii) different Γ and the same f. In each instance, the shaking strength was lower than the fluidization limit, $\Gamma < 1$. The stability of arches against the applied vibration was analyzed by the survival function. Remarkably, arches can be destroyed even when the amplitude is lower than the roughness of the grains, which reveals the complexity of the friction and its role in the silo unclogging. Our findings also confirm that, on average, the final configuration is independent of the chosen vibration parameters (i.e. Γ and f). Finally, these experiments with their respective survival functions provide new evidence which could serve to validate the trap model [13] and to gain a more thorough understanding of the unclogging phenomenon.

Acknowledgements This work was funded by Ministerio de Economía y Competitividad (Spanish Government) through Projects No. FIS2014-57325 and FIS2017-84631. BG thanks Asociación de Amigos de la Universidad de Navarra for a scholarship. We also thank L. F. Urrea for technical help.

References

1. R.M. Nedderman, in *Statics and Kinematics of Granular Materials* (Cambridge University Press, Cambridge, 2005). https://doi.org/10.1017/CBO9780511600043
2. K. To, P.Y. Lai, H.K. Pak, Phys. Rev. Lett. **86**, 71 (2001). https://doi.org/10.1103/PhysRevLett.86.71
3. C. Mankoc, A. Garcimartín, I. Zuriguel, D. Maza, L.A. Pugnaloni, Phys. Rev. E **80**(1), 1 (2009). https://doi.org/10.1103/PhysRevE.80.011309
4. A. Janda, D. Maza, A. Garcimartín, E. Kolb, J. Lanuza, E. Clément, EPL **87**(2), 24002 (2009). https://doi.org/10.1209/0295-5075/87/24002
5. C. Lozano, G. Lumay, I. Zuriguel, R.C. Hidalgo, A. Garcimartín, Phys. Rev. Lett. **109**(6), 068001 (2012). https://doi.org/10.1103/PhysRevLett.109.068001
6. R.C. Hidalgo, C. Lozano, I. Zuriguel, A. Garcimartín, Granul. Matter **15**(6), 841 (2013). https://doi.org/10.1007/s10035-013-0451-7
7. C. Lozano, I. Zuriguel, A. Garcimartín, Phys. Rev. E **91**(6), 062203 (2015). https://doi.org/10.1103/PhysRevE.91.062203
8. B.V. Guerrero, L.A. Pugnaloni, C. Lozano, I. Zuriguel, A. Garcimartín, Phys. Rev. E **97**(4), 042904 (2018). https://doi.org/10.1103/PhysRevE.97.042904
9. C. Merrigan, S.K. Birwa, S. Tewari, B. Chakraborty, Phys. Rev. E **97**(4), 040901 (2018). https://doi.org/10.1103/PhysRevE.97.040901
10. B.V. Guerrero, B. Chakraborty, I. Zuriguel, A. Garcimartín, Phys. Rev. E **100**(3), 032901 (2019). https://doi.org/10.1103/PhysRevE.100.032901
11. M.D. Shattuck. Particle Tracking. http://gibbs.engr.ccny.cuny.edu/technical/Tracking/ChiTrack.php

12. A. Garcimartín, I. Zuriguel, L.A. Pugnaloni, A. Janda, Phys. Rev. E **82**(3), 031306 (2010). https://doi.org/10.1103/PhysRevE.82.031306
13. A. Nicolas, Á. Garcimartín, I. Zuriguel, Phys. Rev. Lett. **120**(19), 198002 (2018). https://doi.org/10.1103/PhysRevLett.120.198002
14. I. Zuriguel, D.R. Parisi, R.C. Hidalgo, C. Lozano, A. Janda, P.A. Gago, J.P. Peralta, L.M. Ferrer, L.A. Pugnaloni, E. Clément, D. Maza, I. Pagonabarraga, A. Garcimartín, Sci. Rep. **4**, 7324 (2014). https://doi.org/10.1038/srep07324

Chapter 45
Gravity-Driven Flow and Clogging in the Presence of an Intruder

Anna Belle Harada and Kerstin Nordstrom

Abstract We present experimental results of the clogging and flow of monodisperse spheres in a silo geometry, with a fixed intruder placed near the exit. Prior work has shown that the placement of such an obstacle can suppress clogging. We extend these observations for a wider array of intruder size and location. Using particle tracking and high-speed video, we analyze the local particle motions and packings to better understand the causes of the clog suppression. We find the mechanisms of clog suppression depend on the intruder size and placement: while the granular temperature appears to be important in all cases, large intruders create spatial disorder that percolates through the system.

45.1 Introduction

The general phenomenology of gravity-driven silo flow is simple. There appears to be a transitional exit aperture size, about five grain diameters. When the exit aperture is larger than this transitional size, particles will flow continuously according to the Beverloo law, with a rate that is independent of filling height, provided the fill height is adequate and the silo width is large enough compared to the aperture and grain diameters [1, 2]. When the aperture is smaller than this transitional size, particles will form a clog with some likelihood [3, 4]. In the infinitely tall silo limit, this means any silo with a small aperture will clog eventually. There is no way to predict when this clog will occur, but larger aperture sizes will have larger average discharge events before a clog [5, 6].

Intriguingly, adding an obstacle near the exit dramatically reduces the clogging probability for silo flow [7–9]. An obstacle can also enhance or suppress the flow rate, though the effect is modest (at most 10%). Prior work has investigated this in several ways in 2D systems, combining clogging measurements with microscopic

A. B. Harada · K. Nordstrom (✉)
Department of Physics, Mount Holyoke College, South Hadley, MA, USA
e-mail: knordstr@mtholyoke.edu

I. Zuriguel et al. (eds.), *Traffic and Granular Flow 2019*,
Springer Proceedings in Physics 252,
https://doi.org/10.1007/978-3-030-55973-1_45

measurements of particle motions. In [7], a circular obstacle of about 10 particle diameters was investigated, and the distance to the exit aperture was varied. In an extension of this work [8], effects of the outlet size were explored. In [9] the shapes of the obstacles were varied. Here we complement prior studies by looking into effects of intruder size and position on clogging.

45.2 Methods

Particles: The particles are clear acrylic spheres (Engineering Labs). They are highly monodisperse, with a size of $D = 3.160 \pm 0.002$ mm, sampled by micrometer measurement. The size distribution skews to the smaller size, with no particles sampled exceeding 3.175 mm. For each experiment, we start with a freshly filled silo, which contains approximately 12,000 particles.

Silo: The front and back of the silo is made with 3/8-inch static dissipative acrylic sheets (McMaster-Carr). Teflon sheeting (McMaster-Carr) was laser cut (Epilog) to make inserts to provide the appropriate sidewall and aperture geometry. An insert is sandwiched between the acrylic sheets, and holes have been tapped into the acrylic sheeting to screw the container together using set screws. The silo is then mounted onto a support frame built from t-slot aluminum, which rests on an optical table. The resulting dimensions of the silo are: 200 mm wide, 475 mm tall, and 3.2 mm thick. In other work, we have verified the system follows the Beverloo law [11] for flow without obstacles [10].

Intruders: Circular intruders are laser cut from Teflon sheeting. We investigate the effect of 5 sizes: $3D$, $5.5D$, $10D$, $17D$ and $30D$. Intruders are fixed to the walls of the silo with double-sided tape, and are centered with respect to the exit. The intruder height H is varied, and it is measured from the exit aperture to the bottom of the obstacle, keeping with the convention in [9].

Measurements: We use a Phantom v1611 camera (Vision Research) to film flow at 1000 fps at a resolution of 1280×800 pixels. The system is illuminated by symmetric LED spotlights to the sides of the camera. This results in bright reflections at the particle centers, used for tracking. Tracking is done via homebuilt code in MATLAB, which is an adaptation of the Crocker-Grier code [13], which gives subpixel accuracy of particle positions.

45.3 Results

In other work [11] we determined our transition from free flow to clogging to be near $R = 5D$, though somewhat higher than $5D$. To study clogging statistics with intruders, we focus on $R = 5D$. This has a clogging probability of 73% with no intruder. We study five intruder heights: $3D$, $6D$, $8D$, $10D$, and $15D$.

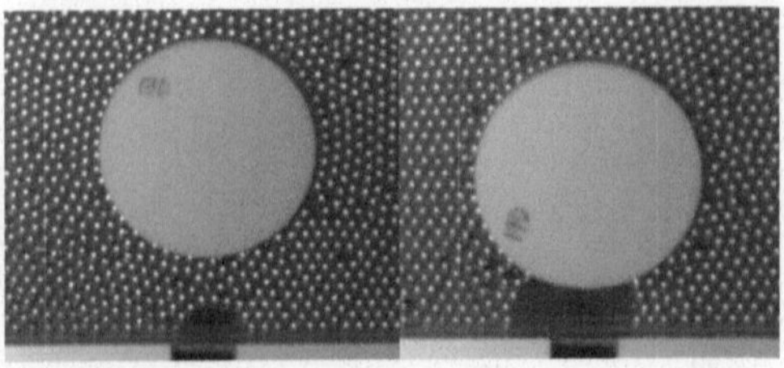

Fig. 45.1 Two modes of permanent clogging in the silo. (left) 17D obstacle, distance above $H = 8D$, shows "normal" clogging via arching over aperture. (right) 17D obstacle, distance above $H = 3D$, shows clogging at the sides

We see two kinds of permanent clogs form. The first kind of clog is the formation of a symmetric arch above the aperture, as shown in Fig. 45.1a. (We do not see asymmetric arches for these particles, though we have in experiments with soft, frictional disks [12]). The second kind of clog we call "side-clogging," the particles form arches between the bottom of the silo and the obstacle, and generally create arches that appear to the sides of the aperture.

The overall results of our clogging trials are presented in Fig. 45.2a, with 10 trials per data point. We find results consistent with prior studies [7, 8]. An obstacle of approximately 10 particle diameters placed closed to the aperture is extremely effective at reducing clogging - in fact we did not see any permanent clogging in this scenario. We emphasize that 10 measurements does not constitute enough statistics for a "true" measurement of probabilities, as there should is a broad distribution of avalanche sizes near the clogging point. Nonetheless the trend is clear. We see an overall trend that in general, larger obstacles are more effective at reducing clogging, sometimes completely reducing the probability to zero. Obstacles placed closer to the exit will also prevent clogging more readily. Both of these results could be anticipated, as bigger and closer obstacles should affect the clogging more. The exception to the "closer and larger" rules occurs for obstacle size $17D$, at a distance of $3D$ above the aperture. In this case, we see a very high likelihood of clogging, where the mechanism is side-clogging.

There are multiple length scales in this system. However, assuming the aperture is fixed, and the particles are small compared to the obstacle, two length scales remain: the obstacle diameter and the height. Systems with the same height:diameter ratio will experience self-similar boundary conditions at the obstacle. As an ansatz, we rescale the obstacle height by its diameter (Fig. 45.2b). The gray line goes through the points where no clogging occurs. We also see that an obstacle placed close to the aperture creates other clogging opportunities at the sides - new bottlenecks are formed. Perplexingly, a very large obstacle placed close to the aperture will create a long bottleneck, which should clog. However, this system (obstacle size $30D$, height $3D$) rarely clogs.

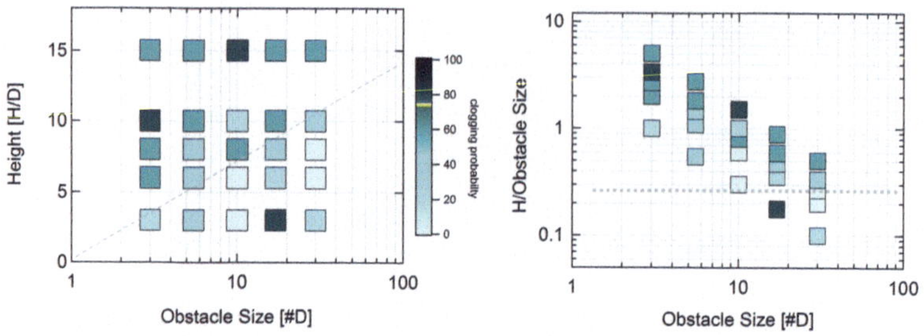

Fig. 45.2 (left) The clogging phase diagram for our system for aperture size $R = 5D$, varying obstacle size and height. The clogging probability for no obstacle is 73%, indicated by the yellow line in the colorbar. We see that clogging is mostly reduced. The grey dashed line is a guide to the eye to divide between regimes where clogging suppression is better or worse. (right) The height data rescaled by obstacle size

By looking at the microscopics we can untangle this mystery. We focus on $10D$ and $30D$ intruders placed near the aperture (bottom row in Fig. 45.2a). Intruder size $17D$ is exceedingly likely to clog. As will be shown, it appears that $10D$ and $30D$ are qualitatively different.

Using particle tracks, we characterize the horizontal granular temperature as a function of time, defined as $T_{gx} = \langle \delta v_x^2 \rangle = \langle (v_x - \langle v_x \rangle)^2 \rangle$ [9]. We consider this quantity for the bottom of the silo as well as for a specific region near the exit, which has a width of $10D$ (centered at the aperture) and a height of $4D$. This region corresponds to particles that are either exiting or would form clogs.

We summarize the results of the granular temperature measurements in the table, reporting the mean granular temperature during developed flow. We find that the exit temperatures are higher than average temperatures for all cases. We find that while the 10D case provides a modest increase in bulk temperature, about 25% compared to no obstacle, the 30D case has a 2.5-fold increase compared to no obstacle, indicating the pronounced effect the large obstacle has on the overall system. The exit temperature is substantially larger than the bulk case (about a 5-fold increase) for 10D, but 30D creates an even larger increase, though similar in magnitude (about 8-fold). Remarkably, the ratio of exit temperatures to bulk temperatures are similar for both obstacles (approximately 40-fold), indicating that obstacles have a similar effect on exit temperature.

Obstacle	$T_{gx}[10^4\ \text{mm}^2/\text{s}^2]$	T_{exit}	T_{gx}/T_{exit}	$T_{exit}/T_{exit,none}$
None	1.31 ± 0.481	14.6 ± 6.06	11.1	1
10D	1.66 ± 0.606	74.1 ± 23.5	44.6	5.08
30D	3.24 ± 1.45	113 ± 42.2	34.8	7.74

We also characterize the packing of the system. However, as this is inherently an inhomogeneous and finite-sized system, the packing fraction itself is not a reliable measurement. Moreover, it would be enlightening to get a map of the spatial structure of the packing, which one does not get by calculating a simple packing fraction. To

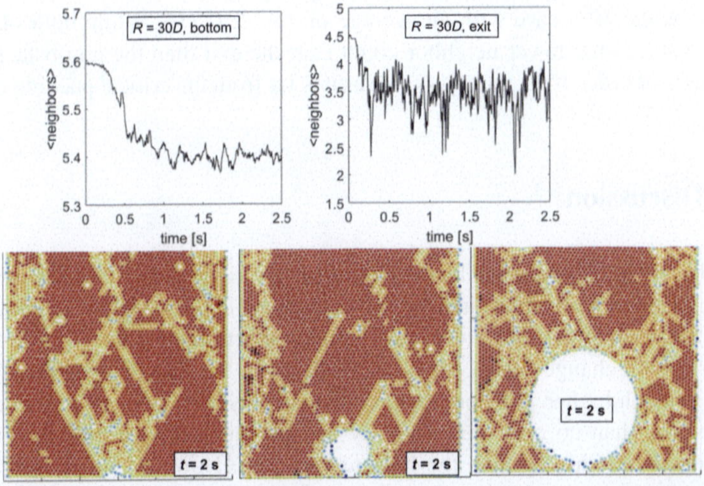

Fig. 45.3 (top) The average neighbor number versus time for the bottom of the silo (left) and the exit (right) for a 30D obstacle. (bottom) The packing for developed flow ($t = 2$ s) with (left to right) no obstacle, 10D, and 30D obstacle, showing the spread of disorder in the system. Red particles have 6 neighbors, other colors have less than six neighbors

solve both problems, we take advantage of the monodispersity of this system and look at the local packing for each particle. For each particle, we search within $1.1D$ and count its neighbors in this region. If a particle has 6 close neighbors, it is most likely closely packed. For any other count the particle is in a configuration with free space; clearly fewer neighbors corresponds to more free space. We measure this for all particles in all frames, and report representative measurements and spatial images as for $30D$ in Fig. 45.3.

For no obstacle, we see that the bottom of the silo has a regular packing initially with some defects. Upon flow, more defects emerge, but they are localized to the exit (Fig. 45.3 bottom, left). The typical number of neighbors is 5 or 6, given an average of 5.60 ± 0.039. In the exit region, the neighbor count is substantially lower, average of 3.93 ± 0.275. For the $30D$ case, we see that the bottom of the silo has some initial defects introduced by the intruder. As the flow develops, the defects spread and eventually span the system. Figure 45.3 (bottom, right) shows the map of neighbor counts for developed flow $t = 2$ s. The typical number of neighbors in the bulk is still 5 or 6, given an average of 5.43 ± 0.0681, the standard deviation is about 2x larger than the no obstacle case. In the exit region, the average neighbor count is 12% lower than the no obstacle case, with an average of 3.49 ± 0.381. For the $10D$ case, we see that the bottom of the silo has some initial defects introduced by the intruder. However, unlike the $30D$ case, these defects remain localized (Fig. 45.3 bottom, center). The typical number of neighbors in the bulk is still 5 or 6, given an average of 5.67 ± 0.0360, which is most similar to the no obstacle case for both mean and standard deviation. In the exit region, the average neighbor count is approximately the same as the $30D$ case with an average of 3.43 ± 0.316. While both obstacles produce a somewhat lower neighbor count near the exit than the no obstacle case, this effect is of order 10%, unremarkable as it is far from the critical packing density.

45.4 Discussion

We see that the horizontal granular temperature is greater near the exit aperture than in the bulk. This is not surprising in itself, as particles are colliding in that region, giving randomness to their motion. The bulk granular temperature for the $10D$ obstacle is not appreciably changed from the no obstacle case, but that the temperature near the exit is much higher. This points to collisions playing a role: $10D$ is much less likely to clog than no obstacle. Similarly the $30D$ temperature near the exit is a similar magnitude higher than the bulk temperature. However, the values for the $30D$ temperatures are a factor of 2 higher than the $10D$ results.

The packing measurements make sense of this. The averages tell a rather uninteresting story: the average neighbor count is smaller for the obstacle cases, but not by much. The average count is smaller near the exit, but again, not changed much from obstacle to none. However, we see a difference in the details for the three cases. Specifically by looking at the map of neighbor counts, we see that disorder is localized in the no obstacle and $10D$ case, while it spreads throughout the system for the $30D$ case. This would explain the higher granular temperature in the bulk for $30D$. The entire system is more fluid for $30D$.

This also explains why we do not see much clogging for $30D$. To recap, we did not see clogging for the $10D$ case, but this is quite consistent with other results [7]. The narrow region that might count as a side bottleneck is infinitesimally small for $10D$, and it is statistically unlikely for a standard clog to occur. However, for $17D$ we see clogging very often, because the lower curvature of the obstacle creates a longer bottleneck. $30D$ has an even longer bottleneck, but we do not see clogging very often. However, $30D$ will disturb the packing order more than $17D$, creating a larger granular temperature. But the particles in $17D$ don't have quite enough temperature to suppress clogging events.

Thus what we see is not exclusively structure or dynamics, it is both: a structural property (the obstacle) makes a structural disturbance in the granular material which affects the particle dynamics. Further, this disturbance can propagate quite far into the material. We have proposed a better metric (the neighbor count) to measure structure than the packing fraction, as disordered and underpacked regions are quite spatially heterogeneous: there is no one packing fraction in these systems, especially near the obstacle and exit.

Acknowledgements This work is partially funded by ACS award PRF 56888-UN19 and by the Research Corporation for Science Advancement.

References

1. R. Nedderman, U. Tuzun, S. Savage, G. Houlsby, J. Chem. Eng. Sci. **37**, 1597 (1982)
2. I. Zuriguel, Pap. Phys. **6**, 060014 (2014)
3. K. To, P.-Y. Lai, H.K. Pak, Phys. Rev. Lett. **86**, 71 (2001)
4. K. To, Phys. Rev. E **71**, 060301 (2005)
5. A. Janda, I. Zuriguel, A. Garcimartin, L.A. Pugnaloni, D. Maza, EPL **84**, 44002 (2008)
6. C.C. Thomas, D.J. Durian, Phys. Rev. Lett. **114**, 178001 (2015)
7. I. Zuriguel, A. Janda, A. Garcimartin, C. Lozano, R. Arevalo, D. Maza, Phys. Rev. Lett. **107** (2011)
8. C. Lozano, A. Janda, A. Garcimartin, D. Maza, I. Zuriguel, PRE **86**, 031306 (2012)
9. K. Endo, K.A. Reddy, H. Katsuragi, Phys. Rev. Fluids **2**, 094302 (2017)
10. C. Mankoc, A. Janda, R. Arevalo, J.M. Pastor, I. Zuriguel, A. Garcimartin, D. Maza, Granul. Matter **9**, 10252 (2007)
11. G. Cai, A.B. Harada, K.N. Nordstrom, submitted to Granular Matter
12. E. Thackray, K. Nordstrom, EPJ Web Conferences, vol. 140, pp. 03087 (2017)
13. J.C. Crocker, D.G. Grier, J. Colloid Interface Sci. **179**, 298 (1996)

Chapter 46
Analysis of the Collective Behavior of Boids

Yoshinari Inomata and Toshiya Takami

Abstract Collective behavior of social organisms emerges by simple interactions between individuals, and the behavior is often found to be chaotic. Automatic classification of such chaotic motion is, in general, difficult because of a large number of degrees of freedom and the stochastic motion of individuals. First, principal component analysis is applied to the instantaneous variables of motion in order to obtain a low dimensional representation of dynamic states and to try to achieve automatic classification. Second, the complex motion of boids is analyzed by dynamic mode decomposition, by which a time-series of dynamical variables is approximated as a linear time-stepping operator. It is shown that the reconstruction error by the linearized operator relates to the transition of complex behavior. Finally, a simple method based on the error analysis is proposed to detect itinerant motion (of the collective behavior).

46.1 Introduction

Swarm intelligence (SI) studies the collective behavior of systems composed of many agents interacting locally with each other and with their environment [1]. In SI, some algorithms are proposed to model social organisms such as birds or fish, and agents in these algorithms generally have some interaction rules. Boids [2], one of the major models, have three interactions, showing emergent behavior of several kinds of collective motion [3], and are also known to exhibit chaotic itinerant behavior. It is important to detect transitions as well as classify the motion because of the chaoticity and itinerancy. While the automatic classification of the collective motion has already been studied [4, 5], these studies do not consider transitions from one pattern to another. Our purpose is to analyze the time-series data of chaotic behavior of boids and to extract characteristics from them.

Y. Inomata (✉) · T. Takami
Oita University, 700 Dannoharu, Oita city, Oita 870-1192, Japan
e-mail: yoshinarhi_i@yahoo.co.jp

I. Zuriguel et al. (eds.), *Traffic and Granular Flow 2019*,
Springer Proceedings in Physics 252,
https://doi.org/10.1007/978-3-030-55973-1_46

In this paper, we examine two methods to analyze the time series. First is the Principal Component Analysis (PCA) in Sect. 46.3, which composes a set of reduced variables to represent the distributed data. It is expected that these reduced representation of data can be used to distinguish the behavior patterns. Second is the Dynamic Mode Decomposition (DMD) in Sect. 46.4, computing a set of modes that includes dynamic information. This is also used to reconstruct the original time-series data. We analyze the magnitude of the difference between the original data and the reconstructed data to find relation to itinerant behavior.

46.2 Boids

Boids is a simulation model of a swarm, proposed by Reynolds in 1987 [2]. Each agent has three local interactions, separation, alignment, and cohesion, with neighboring agents:

- **Separation**: steer not to collide with the nearest agent
- **Alignment**: steer toward the average heading of surrounding agents
- **Cohesion**: adjust the direction to the average position of other agents

Each interaction is adjusted by defining its neighborhood, in which each agent interacts with others, and the magnitudes of forces. In addition, a constant speed is introduced as a property of self-propelled particles. In a study by Couzin et al. [3], with a model similar to Boids with a constant speed, four collective behaviors (swarm, torus, dynamic parallel group and highly parallel group) were observed by changing behavioral zones of the neighborhood.

In a two-dimensional system with N agents, coordinates $\mathbf{r}_t = \{\mathbf{x}_t, \mathbf{y}_t\}$ and velocities $\mathbf{p}_t = \{\mathbf{u}_t, \mathbf{v}_t\}$ are both $2N$ real vectors. The time-stepping calculation is executed by an expression, $(\mathbf{r}_t, \mathbf{p}_t) = F(\mathbf{r}_{t-1}, \mathbf{p}_{t-1})$, with a nonlinear function $F()$. In Fig. 46.1, typical spatial patterns are shown: (a) torus; (b) string torus; (c) random; (d) parallel where agents are in a sufficiently large box with closed boundaries.

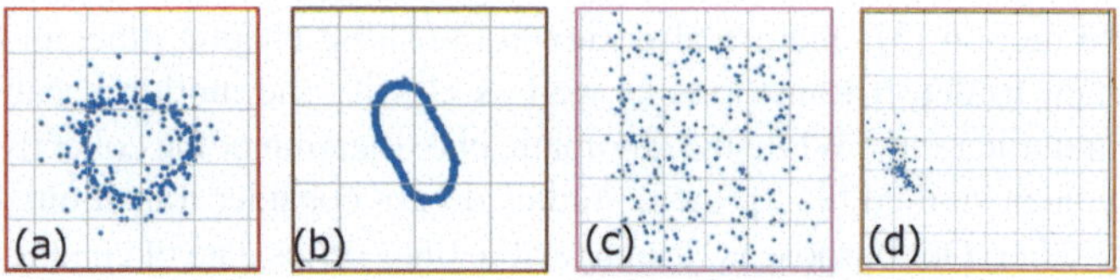

Fig. 46.1 Typical motion patterns of boids: **a** torus; **b** string torus; **c** random; **d** parallel

46.3 PCA

We analyze two types of data by PCA. PCA composes variables, called principal components, to represent the distributed data, and we can introduce higher order principal components. The first one is a set of $4N$ dimensional real vectors obtained by the raw values of dynamical variables, $\{\mathbf{r}_t,\ \mathbf{p}_t\}$. Since the number of dimensions of data is directly determined from the number of boids, N, this raw-data PCA is used for the analysis of a fixed number of boids. We collect 7 different time-series represented in Fig. 46.1 for the number of boids $N = 300$ and the length in time $T = 2000$. Figure 46.2a shows those 2000×7 points projected on the 1st and 2nd principal components, and Fig. 46.2b represents the projection of the same data onto the first three principal components. In these figures, a "string torus" pattern is separated from others, while "torus" and "random" patterns are not distinguished from each other.

The second set of data is represented in a 10 dimensional space created by variances and covariances of elements in the four vectors, $\mathbf{x}_t$, $\mathbf{y}_t$, $\mathbf{u}_t$, $\mathbf{v}_t$, and independent of N. In this case, we use 6 different series with different number of agents containing "torus" and "parallel motions." Fig. 46.3a shows 2000×6 points on two-dimensional space, in which parallel motions, shown in colors of purple and yellow, are distinguished from others. On the other hand, transitions from the initial state to each state are observed in the figure by three dimensional projection Fig. 46.3b.

In summary, distinguishing two types of motion is possible, as shown in Fig. 46.3, but three and more types can only be distinguished via a characteristic motion like Fig. 46.2. Therefore, PCA enables a rough classification of the data, but far from the automatic classification.

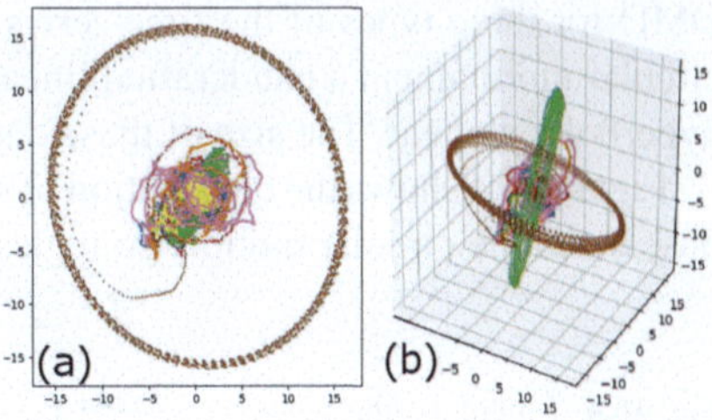

Fig. 46.2 PCA results for the raw data: **a** two and **b** three dimensional representation. The axes are the first 2 or 3 principal components

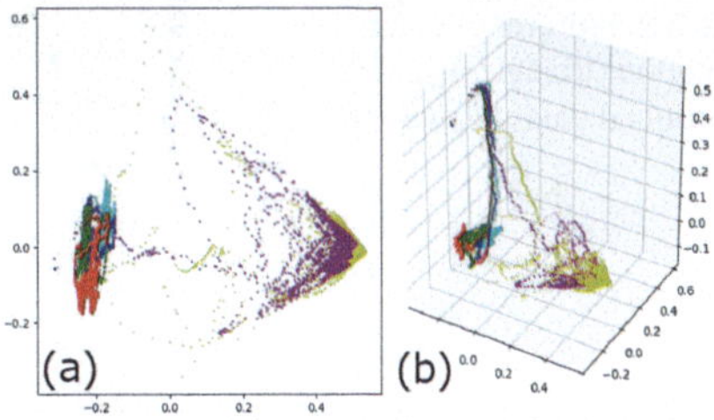

Fig. 46.3 PCA results for the statistical data: **a** two and **b** three dimensional representation

46.4 DMD

Time-series data for N boids are prepared as $N \times T$ complex values, where the coordinate vectors $(\mathbf{x}_t, \ \mathbf{y}_t)$ are represented in a complex vector $\mathbf{z}_t = \mathbf{x}_t + i\mathbf{y}_t$, and T means the length of the time series. At first, a linear approximation $\hat{F}$ of the nonlinear time-stepping function $F()$ is defined by

$$\hat{F} = \underset{\tilde{F} \in \mathbb{C}^{N \times N}}{\arg\min} \frac{1}{T} \sum_{t=0}^{T-1} \left\| \mathbf{z}_{t+1} - \tilde{F}\mathbf{z}_t \right\|^2 . \tag{46.1}$$

Using $N \times T$ complex matrices $Y \equiv \{\mathbf{z}_0, \ \cdots, \ \mathbf{z}_{T-1}\}$ and $Y' \equiv \{\mathbf{z}_1, \ \cdots, \ \mathbf{z}_T\}$, $\hat{F}$ is approximately represented in an expression, $\hat{F} \approx Y'Y^+$, where Y^+ is a pseudo inverse of the rectangular matrix Y with properties $YY^+Y = Y$, $Y^+YY^+ = Y^+$, and YY^+ and Y^+Y are Hermitian. Through singular value decomposition (SVD) with a compact representation, the data matrix Y is represented in the form, $Y = U\Sigma V^*$, where $*$ denotes the conjugate transpose, $U \in \mathbb{C}^{N \times M}$, $M = \min(N, T)$, $\Sigma \in \mathbb{C}^{M \times M}$, and $V \in \mathbb{C}^{T \times M}$. Then, $\hat{F}$ is also represented in a form, $\hat{F} = Y'V\Sigma^{-1}U^*$. The singular values corresponding to dynamic modes are given by the diagonal elements of Σ. There is another representation of SVD, non-compact or full representation, where U and V are chosen an unitary matrices and Σ is a rectangular matrix.

46.4.1 *Singular Values to Represent Boids Dynamics*

First of all, our analysis is focused on singular values obtained in the process of DMD for three types of the time-series data, "torus" shown in Fig. 46.1a, uniform circular motion, and a handcrafted linear series composed of hyperbolic cosine and hyperbolic tangent. The size of these data is fixed at $N = 100$ and $T = 79$.

Figure 46.4 shows the distribution of singular values. For the artificial data (rhombuses) and the circular motion (triangles), only a few numbers of non-zero singular

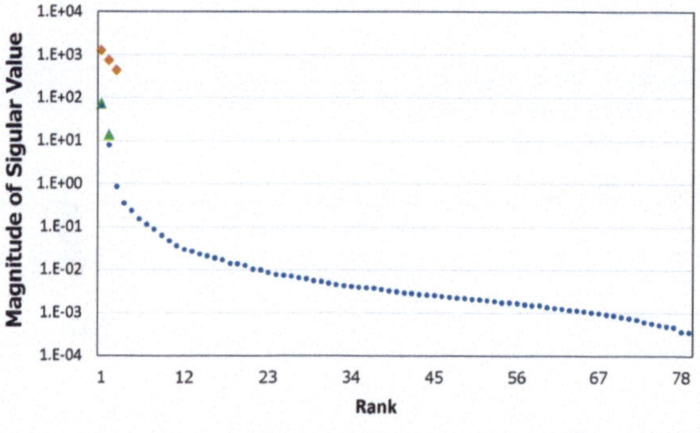

Fig. 46.4 Singular values for "torus" motion of boids (dots), uniform circular motion (triangles), and an artificial dataset (rhombuses)

values are obtained, which means that these artificial data can be reproduced by a small number of modes. On the other hand, for the torus motion of boids (dots), those values are distributed over all ranks. This means that the dynamics of boids contains a large number of modes and is difficult to be reproduced by the linear approximation of the time-stepping function. In this analysis, however, we could not make clear the meaning of the small but non-zero singular values.

46.4.2 Reconstruction Error Analysis

We investigate accuracy of the reconstructed data through DMD. By use of the approximated time-stepping function $\hat{F}$, we can reproduce the time series $\{\hat{\mathbf{z}}_t\}$ from an initial data $\mathbf{z}_0$ as $\hat{\mathbf{z}}_t = \hat{F}^t \mathbf{z}_0$. The total reconstruction error is defined by an average euclidean distance between the original coordinates $\mathbf{z}_t$ and the reproduced ones $\hat{\mathbf{z}}_t$ at t-th step,

$$R_t = \sqrt{\frac{1}{N} \left\| \hat{\mathbf{z}}_t - \mathbf{z}_t \right\|^2}. \tag{46.2}$$

In Fig. 46.5, we show the total reconstruction errors for two time series, where the number of boids is $N = 100$ and the number of time steps is $T = 1000$. The red curve is R_t for torus motion, and the blue curve is for itinerant motion from random state to the parallel state through string-like motion. It is clearly shown that the magnitude of R_t for the itinerant motion is larger than that for the torus motion. Thus, it is supposed that the reconstruction error becomes larger at the time when the transition occurs.

In order to show the relation between reconstruction errors and the transition of motion, we introduce another definition of the reconstruction error by executing DMD for data from slid intervals. We call this error the reconstruction error by time-shifted DMD. The time-shifted DMD is executed for the data $Y = \{\mathbf{z}_{t-\tau}, \cdots, \mathbf{z}_{t-1}\}$ in a relatively narrow interval, i.e., $\tau \ll T$, from $t - \tau$ to $t - 1$. The reconstruction error R'_t by the shifted DMD is the difference between $\hat{\mathbf{z}}_t$ and $\mathbf{z}_t$ calculated by the right-hand-side of the same formula (46.2).

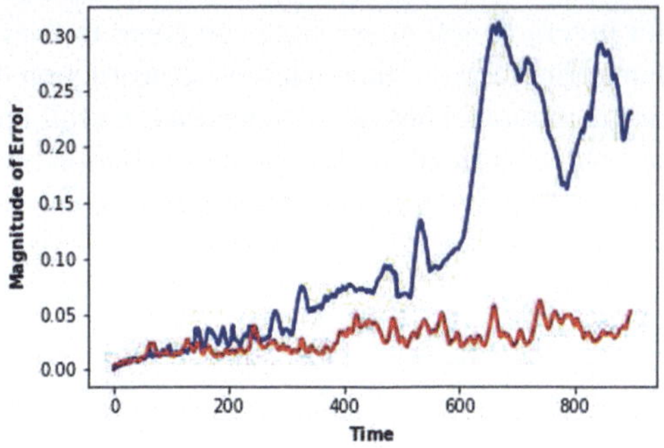

Fig. 46.5 Total reconstruction errors for torus (red) and itinerant motion (blue)

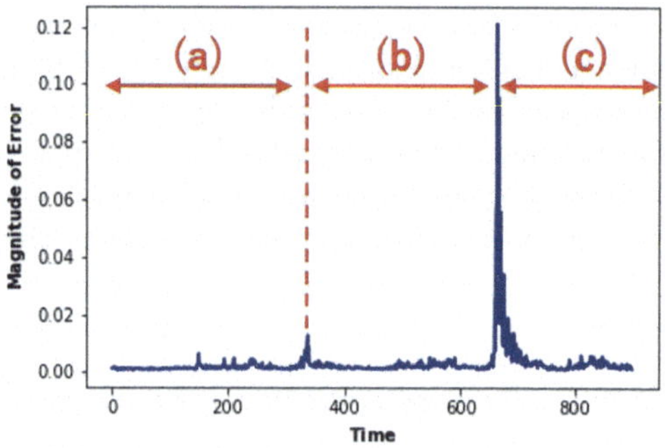

Fig. 46.6 Reconstruction error by the time-shifted DMD ($\tau = 100$)

In Fig. 46.6, we show the reconstruction error by the shifted DMD with $\tau = 100$ for the data exhibiting an itinerant motion. We can observe three types of patterns in this itinerant motion: "random" at $0 < t \lesssim 350$; "string torus" at $350 \lesssim t \lesssim 650$; "parallel" at $650 \lesssim t$. These regions are denoted (a), (b), and (c) in Fig. 46.6, respectively. The reconstruction error by the shifted DMD shows several peaks corresponding to the transition points between these motions. Thus, we can use the reconstruction error to detect transitions of the collective motion of boids.

Through our analysis, it is found that peaks in the reconstruction error by shifted DMD is related to transitions of motion. Thus, we propose the reconstruction error analysis as a simple method to detect itinerant collective motion of complex self-propelled particles.

46.5 Conclusion

In this study, we proposed a simple analysis method to detect changes of complex collective motion of self-propelled particles. While applicability of PCA to the classification was insufficient overall, some specific motion could still be distinguished from others. By the analysis of singular values extracted from the time series of boids, it was shown that the motion is difficult to be reproduced by the linear approximation of the time-stepping function. It is also shown that the reconstruction error for itinerant motion are large compared to the case of simple non-itinerant motion. Furthermore, we found that the reconstruction error relates to the itinerant behavior of the motion. Thus, it is shown that the full reconstruction of the complex motion of boids is difficult by linear approximation of the time-evolving operator, and that this is because the time-series often contains sudden changes of motion.

References

1. D. Martens, B. Baesens, T. Fawcett, Editorial survey: swarm intelligence for data mining. Mach. Learn. **82**, 1–42 (2011)
2. C.W. Reynolds, Flocks, herds, and schools: a distributed behavioral model. Comput. Graph. **21**(4), 25–34 (1987)
3. I.D. Couzin, J. Krause, R. James, G.D. Ruxton, N.R. Franks, Collective memory and spatial sorting in animal groups. J. Theor. Biol. **218**, 1–11 (2002)
4. K. Fujii, T. Kawasaki, Y. Inaba, Y. Kawahara, Prediction and classification in equation-free collective motion dynamics. PLoS Comput. Biol. **14**(11), e1006545 (2018)
5. M. Berger, L.M. Seversky, D.S. Brown, Classifying swarm behavior via compressive subspace learning, in *2016 IEEE International Conference on Robotics and Automation (ICRA)* (2016), pp. 5328–5335

Chapter 47
Modelling the Flow Rate Dip for a Silo with Two Openings

Samuel Irvine, Luke Fullard, Thomasin Lynch, Daniel Holland, Daniel Clarke, and Pierre-Yves Lagrée

Abstract In recent experiments it was shown that a planar silo with two discharge orifices which were separated by a distance, L, displayed counter-intuitive flow rate phenomenon. In contrast to previous studies, the flow rate did not monotonically decrease as the separation between orifices increased. Instead, a rapid decrease in flow rate was observed as the two orifices were separated until, at a critical orifice separation, a minimum flow rate was reached. Upon further separating the two orifices the flow rate steadily increased to the infinite separation limit of two openings. In this work we numerically investigate this so-called 'flow-rate dip' phenomenon. The kinematic and the $\mu(I)$ models are used to examine the two opening silo, with the kinematic model failing to capture any flow rate dynamics and the $\mu(I)$ model capturing the dynamics if appropriately large (yet still physically reasonable) friction parameters are used.

47.1 Introduction

Granular discharge from a silo under the influence of gravity is a complex system to model. In particular, a recent experimental study [1] has shown that for a planar silo with two openings of width W (see Fig. 47.1), the separation distance, L, of these openings causes some unusual flow rate dynamics. For a large separation, the flow rate appears to approach the expected flow rate for two distinct silos with a single opening. When there is zero separation (i.e. the silo has one larger opening), the flow rate is much higher, as predicted by the Beverloo relation. In between these extremes, the

S. Irvine · L. Fullard · T. Lynch (✉)
School of Fundamental Sciences, Massey University, Palmerston North, New Zealand
e-mail: T.A.Lynch@Massey.ac.nz

D. Holland · D. Clarke
Department of Chemical and Process Engineering, University of Canterbury, Christchurch, New Zealand

P.-Y. Lagrée
Sorbonne Université, CNRS, Institut Jean le Rond d'Alembert, Paris, France

I. Zuriguel et al. (eds.), *Traffic and Granular Flow 2019*,
Springer Proceedings in Physics 252,
https://doi.org/10.1007/978-3-030-55973-1_47

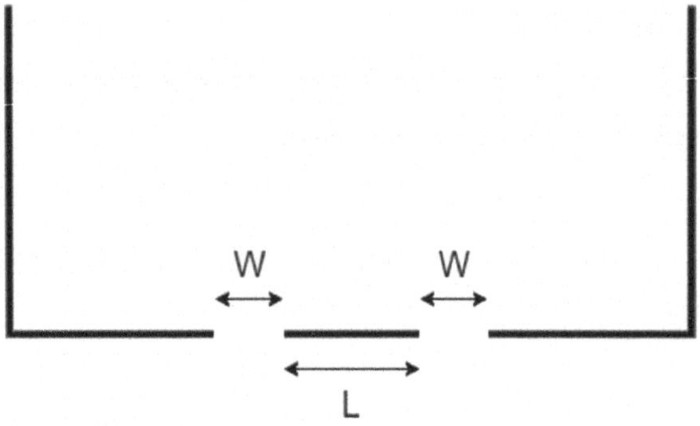

Fig. 47.1 Diagram of the two opening silo

flow rate does not monotonically decrease, it instead dips sharply to a local minimum for a small separation, then slowly increases back to the large separation value. This behaviour had not previously been observed in similar experiments [2, 3] and the new phenomenon was attributed to the larger values of inter-particle friction of the particles used in the study [1].

While it has been shown that discrete element models (DEM) with appropriately large values of inter-particle friction are able to reproduce the observed flow-rate dip [1], as yet, no continuum model of dense granular flow has successfully predicted such a dynamic. As such, the two opening silo is an excellent test of continuum mathematical models of granular flow and may potentially offer insight into the dynamics that are difficult to observe using DEM or experimental methods.

In this work we study two continuum models—the kinematic model [4] and the $\mu(I)$ model [5, 6], and use these to examine the flow rate from a planar dual orifice silo.

47.2 Kinematic Model

The kinematic model [4] can be formulated with the assumption that a gradient in vertical velocity causes horizontal motion, i.e. $u = -B\frac{\partial v}{\partial x}$, where u and v are the horizontal and vertical velocity components respectively and B is some constant parameter. This assumption, combined with that of incompressibility, gives a relatively simple model of granular flow, taking the form of the heat equation [7],

$$\frac{\partial v}{\partial y} = B\frac{\partial^2 v}{\partial x^2}. \tag{47.1}$$

This partial differential equation can be solved exactly across the infinite half-plane with a Dirac delta $v = \delta(x)$ boundary condition at $y = 0$, corresponding to an infinitesimal orifice. The solution of this boundary value problem is

$$v = -\frac{Q}{\sqrt{4\pi By}}\exp\left(-\frac{x^2}{4By}\right), \tag{47.2}$$

where Q is the flow rate. For a silo with two openings separated by some distance L, the boundary condition at $y = 0$ is $v = \delta(x - L/2) + \delta(x + L/2)$, resulting in a solution of the form

$$v = \frac{-Q}{\sqrt{4\pi By}}\left[\exp\left(\frac{(x - L/2)^2}{-4By}\right) + \exp\left(\frac{(x + L/2)^2}{-4By}\right)\right]. \tag{47.3}$$

Note that this model requires scaling by the flow rate, Q. For a single orifice, this can be done using the Beverloo relation [8], although for a smaller orifice a flow rate relation which takes dilatancy into account may be more appropriate [9]. For the two opening case the flow rate for any separation, L, is simply twice the flow rate of a single orifice (i.e. $2Q$) and does not vary with separation. In order to account for orifice interaction a Beverloo-like relation of the form $Q(L)$ would need to be developed. As such, in its original form, this common model of granular flow from a silo is unable to capture the monotonic decrease seen in previous experiments [2, 3] or the flow-rate dip phenomenon. A different model is required to predict flow rate in a silo with more than one opening.

47.3 $\mu(I)$ Model

47.3.1 Derivation

The $\mu(I)$ model defines an effective viscosity η_{eff} which can be used in the incompressible Navier-Stokes equations to model the flow of granular material. In a standard incompressible fluid, shear stress τ can be related to shear rate $\dot{\gamma}$ by the relation $\tau = \eta\dot{\gamma}$, where the viscosity of the fluid η is assumed to be constant. For granular material, the viscosity is determined by the bulk friction μ, which changes with the dimensionless 'inertial number', I, which is given by

$$I = \frac{\dot{\gamma}d}{\sqrt{P/\rho}}, \tag{47.4}$$

where $\dot{\gamma}$ is the shear rate, d is the particle diameter, P is the confining pressure, and ρ is the grain density.

The friction, μ, is the ratio between the shear stress and the pressure, i.e. $\mu = \frac{\tau}{P}$, where τ is the shear stress. It has been found that μ increases with I for granular material in the dense regime [6], with a common form of the relation between μ and I given as

$$\mu(I) = \mu_s + \frac{\Delta\mu I}{I_0 + I}, \tag{47.5}$$

where μ_s, I_0, $\Delta\mu$ are all constant parameters. From this relation, the effective viscosity can be defined as

$$\eta_{\mathrm{eff}} = \frac{\mu(I)P}{|\dot{\gamma}|}, \tag{47.6}$$

with a maximum value, η_{max}, imposed to prevent numerical issues in stationary zones [10]. The effective viscosity together with the incompressible Navier-Stokes equations

$$\frac{\partial \boldsymbol{u}}{\partial t} = -(\boldsymbol{u} \cdot \nabla)\boldsymbol{u} + \nabla \cdot (\eta \nabla \boldsymbol{u}) - \frac{\nabla P}{\rho} + \rho \boldsymbol{g}, \tag{47.7}$$

defines the velocity given initial and boundary conditions. At the side walls and base, no-slip boundary conditions are used, with a free surface at the top and at the orifices. Note that this model is a planar analog, and does not take into account friction at the front and back walls.

This model has previously been used to model a single opening planar silo [11, 12] and the collapse of a granular column [10]. Recently, a selection of conical silos were modelled using the $\mu(I)$ model and compared to experimental data gathered from Magnetic Resonance Imaging experiments [13]. It was shown that the $\mu(I)$ model was able to qualitatively predict the shape of velocity contours in the silo, but was not able to accurately quantify the flow rate. It is possible that accounting for incompressibility may be necessary to accurately capture the flow rate behaviour.

47.3.2 Results

The $\mu(I)$ model was implemented in Basilisk [14] (a computational fluid dynamics software) for a two opening silo for various orifice separations, L. The case where the separation is zero ($L = 0$) is equivalent to a silo with a single orifice twice the orifice width. For each simulation the predicted flow rate is normalised by the flow rate for the zero separation case. To gauge the effect of model constants on the solution, the simulation is also repeated for various different parameters for the $\mu(I)$ rheology, shown in Table 47.1. The results are shown in Fig. 47.2. Basilisk is designed to solve fluid dynamics problems in non-dimensional variables, hence we scale our system such that $W_s/\hat{l} = 10$, where W_s is the silo width and $\hat{l}$ is some reference length. The non-dimensional particle diameter used was $\hat{d} = d/\hat{l} = 0.033$, the scaled orifice width was $\hat{W} = W/\hat{l} = 0.625$ and the initial height of the bed of grains was $\hat{H} = H/\hat{l} = 4.5$. A scaled maximum viscosity of $\hat{\eta}_{max} = 100$ was used to regularise the viscosity at low shear rates.

For low friction values, the model predicts a monotonic smooth decrease from zero separation distance to large separations, as observed in some past studies

Table 47.1 Different parameters for $\mu(I)$ implementation in Basilisk

Low friction	$0.32 + 0.28I/(0.4 + I)$
Medium friction	$0.47 + 0.38I/(0.5 + I)$
High friction	$0.62 + 0.48I/(0.6 + I)$
Extra high friction	$0.77 + 0.58I/(0.7 + I)$

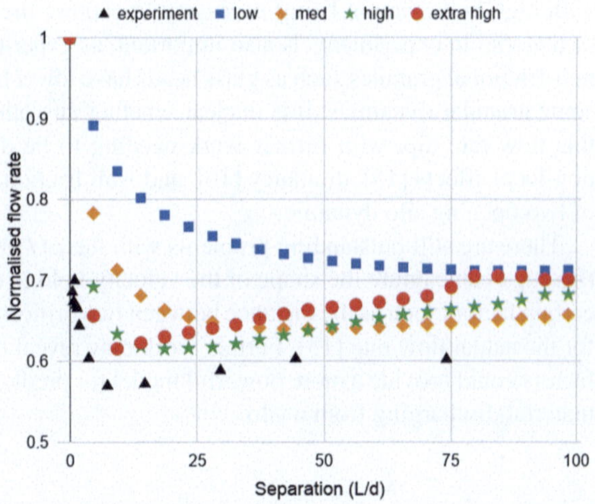

Fig. 47.2 Normalised flow rate for various orifice separations and $\mu(I)$ parameters

[2, 3]. However, for larger friction values the flow rate decreases more sharply as the distance between orifices increases. For large enough friction values the flow rate dips below the large separation limit at a critical value of L, before rising back up to a steady rate. The behaviour for high friction coefficients is qualitatively similar to that observed experimentally.

47.4 Conclusion

In this work we numerically studied whether continuum models of granular flow are able to capture the 'flow rate dip' dynamic in a two opening silo, as observed experimentally. It was shown that kinetic models, such as the kinematic model [4], are unable to produce flow rate interference between the two openings. However, for large enough values of the friction parameters, the continuum model with the $\mu(I)$ granular rheology is able to capture this interference and a type of flow rate dip. For lower values of friction, a monotonic decrease in flow rate was observed as the spacing between openings was increased. This work reports the first time a flow rate dip has been observed in a continuum model of granular flow. Since the $\mu(I)$ model is purely a local-model (as compared with non-local models of granular friction [15]), these results suggests that the flow rate dip is a local effect. The flow rate dip can not be associated with a finite size effect, since, in our continuum model, the particle size enters only into the inertial number definition. Furthermore, since the model is of incompressible type, the dip in flow rate appears to be mainly caused by pressure or shear-rate dynamic interaction. Fully explaining the observed phenomenon is ongoing work.

We note that the values chosen for the $\mu(I)$ parameters greatly affect the dynamics, with high bulk friction being necessary to capture the dip in flow rate. The choice of material in experiments is also important, as experiments with relatively smooth non-frictional granules such as glass beads have lower bulk friction [6] and may miss some granular dynamics. It is unclear whether any other phenomena contributes to this flow rate dip, with further work needing to be done to see if accounting for non-local effects [15], dilatancy [16], and wall friction [17] improves the prediction of two opening silo dynamics.

There are still outstanding problems with the $\mu(I)$ model. In particular, the $\mu(I)$ rheology can capture the shape of the velocity field in a silo, and effect of changing certain factors such as the distance between two orifices, but it gives poor predictions for the actual flow rate [13]. Further work may reveal that accounting for additional factors could provide a more powerful model for predicting the dynamics of granular material discharging from a silo.

References

1. L.A. Fullard, E.C.P. Breard, C.E. Davies, A.J.R. Godfrey, M. Fukuoka, A. Wade, J. Dufek, G. Lube, The dynamics of granular flow from a silo with two symmetric openings. Proc. R. Soc. A **475**(2221), 20180462 (2019)
2. C. Xu, F.L. Wang, L.P. Wang, X.S. Qi, Q.F. Shi, L.S. Li, N. Zheng, Inter-orifice distance dependence of flow rate in a quasi-two-dimensional hopper with dual outlets. Powder Technol **328**, 7–12 (2018)
3. X. Zhang, S. Zhang, G. Yang, P. Lin, Y. Tian, J.F. Wan, L. Yang, Investigation of flow rate in a quasi-2d hopper with two symmetric outlets. Phys. Lett. A **380**(13), 1301–1305 (2016)
4. R.M. Nedderman, U. Tüzün, A kinematic model for the flow of granular materials. Powder Technol. **22**(2), 243–253 (1979)
5. P. Jop, Y. Forterre, O. Pouliquen, A constitutive law for dense granular flows. Nature **441**(7094), 727 (2006)
6. G.D.R. MiDi, On dense granular flows. Eur. Phys. J. E **14**(4), 341–365 (2004)
7. F. Melo, F. Vivanco, C. Fuentes, V. Apablaza, Kinematic model for quasi static granular displacements in block caving: dilatancy effects on drawbody shapes. Int. J. Rock Mech. Min. Sci. **45**(2), 248–259 (2008)
8. W.A. Beverloo, H.A. Leniger, J. Van de Velde, The flow of granular solids through orifices. Chem. Eng. Sci. **15**(3–4), 260–269 (1961)
9. Alvaro Janda, Iker Zuriguel, Diego Maza, Flow rate of particles through apertures obtained from self-similar density and velocity profiles. Phys. Rev. Lett. **108**(24), 248001 (2012)
10. P.-Y. Lagrée, L. Staron, S. Popinet, The granular column collapse as a continuum: validity of a two-dimensional navier-stokes model with a μ (i)-rheology. J. Fluid Mech. **686**, 378–408 (2011)

11. L. Staron, P.-Y. Lagrée, S. Popinet, The granular silo as a continuum plastic flow: the hour-glass vs the clepsydra. Phys. Fluids **24**(10), 103301 (2012)
12. L. Staron, P.-Y. Lagrée, S. Popinet, Continuum simulation of the discharge of the granular silo. Eur. Phys. J. E **37**(1), 5 (2014)
13. L.A. Fullard, D.J. Holland, P. Galvosas, C. Davies, P.-Y. Lagrée, S. Popinet, Quantifying silo flow using mri velocimetry for testing granular flow models. Phys. Rev. Fluids **4**(7), 074302 (2019)
14. J.A. van Hooft, S. Popinet, B.J.H. van de Wiel, Adaptive cartesian meshes for atmospheric single-column models: a study using basilisk 18-02-16. Geosci. Model Dev. **11**(12), 4727–4738 (2018)
15. K. Kamrin, Non-locality in granular flow: phenomenology and modeling approaches. Front. Phys. **7**, 116 (2019)
16. L.A. Fullard, C.E. Davies, G. Lube, A.C. Neather, E.C.P. Breard, B.J. Shepherd, The transient dynamics of dilation waves in granular phase transitions during silo discharge. Granul. Matter **19**(1), 6 (2017)
17. Y. Zhou, P.-Y. Lagrée, S. Popinet, P. Ruyer, P. Aussillous, Experiments on, and discrete and continuum simulations of, the discharge of granular media from silos with a lateral orifice. J. Fluid Mech. **829**, 459–485 (2017)

Chapter 48
Jammed Disks of Two Sizes in a Narrow Channel

Dan Liu and Gerhard Müller

Abstract A granular-matter model is exactly solved, where disks of two sizes and weights in alternating sequence are confined to a narrow channel. The axis of the channel is horizontal and its plane vertical. Disk sizes and channel width are such that under jamming no disks remain loose and all disks touch one wall. Jammed microstates are characterized via statistically interacting particles constructed out of two-disk tiles. Jammed macrostates depend on measures of expansion work, gravitational potential energy, and intensity of random agitations before jamming. The dependence of configurational entropy on excess volume exhibits a critical point.

48.1 Introduction

The research reported here builds on two previous studies [1, 2] employing the same methodology of exact analysis in the framework of configurational statistics. They, in turn, were inspired by work on jammed disks in a narrow channel based on different methods of analysis [3–5] that yielded intriguing results. The focus on disks of two sizes and weights in alternating sequence, for which new exact results are being presented here, is intended to be a first step toward a scenario where the channel contains such disks in a random sequence—a realistic goal in the framework of the same methodology.

D. Liu (✉)
Department of Physics, University of Hartford, West Hartford, CT 06117, USA
e-mail: dliu@hartford.edu

G. Müller
Department of Physics, University of Rhode Island, Kingston, RI 02881, USA
e-mail: gmuller@uri.edu

I. Zuriguel et al. (eds.), *Traffic and Granular Flow 2019*,
Springer Proceedings in Physics 252,
https://doi.org/10.1007/978-3-030-55973-1_48

48.2 Geometry

Disks of two sizes with diameters $\sigma_L \geq \sigma_S$ in alternating sequence are being jammed in a channel of width H. The following two conditions guarantee that (i) the disk sequence remains invariant before jamming, (ii) all jammed disks have wall contact, and (iii) jamming leaves no disks loose:

$$\frac{1}{4} < \frac{\sigma_S}{\sigma_L} \leq 1, \quad 1 < \frac{H}{\sigma_L} < \frac{1}{2}\left[1 + \frac{\sigma_S}{\sigma_L} + \sqrt{\left(1 + \frac{\sigma_S}{\sigma_L}\right)^2 - 1}\,\right]. \tag{48.1}$$

All jammed microstates can be assembled from 8 tiles composed of two disks with one disk overlapping (Table 48.1). Adding a tile to an already existing string must satisfy the successor rule to maintain mechanical stability under jamming forces.

Under mild assumptions the microstate of minimum volume with N (large/small) disk pairs is composed of an alternating sequence, vwvw$\cdots$ v, of just two tiles. We declare it to be the reference state for statistically interacting particles in this application. All other (jammed) microstates can be generated by the activation of quasi-particles from this reference state. We have identified $M = 5$ species of particles that serve this purpose (Table 48.2). Adopting the taxonomy of Ref. [6] we distinguish between the categories of hosts and tags.

Particles from species $m = 1, \ldots, 4$ can be placed directly into the reference state (pseudo-vacuum), meaning that it is possible to add a tile v or w to the left or to the right as follows: w13v, w146w, v253v, v2546w. Particles from species $m = 5$ are parasitic in the sense that they can only be placed inside a particle from species $m =$

Table 48.1 Distinct tiles that constitute jammed microstates of disk sequences subject to the conditions (48.1). Mechanical stability rule: v must be followed by w or 2 etc. Motifs pertain to $\sigma_L = 2$, $\sigma_S = 1.4$, $H = 2.5$. Volume of tiles: $V_c = \frac{1}{2}(\sigma_L + \sigma_S) + \sqrt{H(\sigma_L + \sigma_S - H)}$, $V_f = \frac{1}{2}(\sigma_L + \sigma_S) + \sqrt{\sigma_L \sigma_S}$ (assuming unit cross section of channel)

Motif	ID	rule	vol.	Motif	ID	rule	vol.	Motif	ID	rule	vol.	Motif	ID	rule	vol.
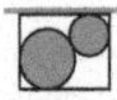	v	w, 2	V_c	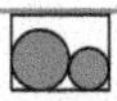	1	3, 4	V_f	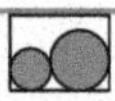	3	v	V_f		5	3, 4	V_c
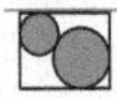	w	v, 1	V_c	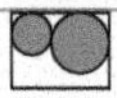	2	5	V_f	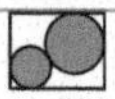	4	5, 6	V_c	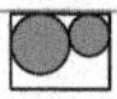	6	w, 2	V_f

Table 48.2 Five species of quasi-particles. The hosts $m = 1, \ldots, 4$ modify the reference state whereas the tag $m = 5$ modifies any one of the hosts. The ID lists the overlapping tiles involved. The activation energy ϵ_m is relevant before jamming and the excess volume ΔV_m after jamming

ID	m	Cat.	ΔV_m	ϵ_m	ID	m	Cat.	ΔV_m	ϵ_m
13	1	Host	$2V_t$	$2pV_t - \gamma_S$	2546	4	Host	$2V_t$	$2pV_t - \gamma_S + 2\gamma_L$
146	2	host	$2V_t$	$2pV_t - \gamma_S + \gamma_L$	45	5	tag	0	$\gamma_L - \gamma_S$
253	3	host	$2V_t$	$2pV_t - \gamma_S + \gamma_L$					

$1,\ldots,4$, at exactly one position and with one disk overlapping as follow: 1|45|3, 1|45|46, 25|45|3, 25|45|46. The number of tag particles that can be added inside the same host is only limited by the size of the number N of disk pairs in the system. For example, two tags inside the first host reads 1|4545|3. The minimum number of tiles v or w between two host can be two as in 13vw146, one as in 13v13, or zero as in 146253.

48.3 Energetics

The activation of every host particle extends the total volume after jamming by the amount $2V_\mathrm{t}$, where $V_\mathrm{t} \doteq V_\mathrm{f} - V_\mathrm{c}$ (see Tables 48.1 and 48.2). Placing a tag does not change the volume. The activation energy ϵ_m assigned to a particle from species m pertains to the state of random agitation before jamming. It consists of work against the ambient pressure exerted by pistons and gravitational potential energy, all relative to the reference state.

We are free to choose the mass density of small and large disks. Therefore, the gravitational potential energies γ_L and γ_S are independent parameters as is the expansion work $2pV_\mathrm{t}$. A fourth parameter is the intensity T_k of random agitations. The jamming protocol is explained in [1]. All results coming out of configurational statistics will only depend on three (dimensionless) ratios of four energy parameters:

$$\beta \doteq \frac{2pV_\mathrm{t}}{T_\mathrm{k}}, \quad \Gamma_\mathrm{L} \doteq \frac{\gamma_\mathrm{L}}{2pV_\mathrm{t}}, \quad \Gamma_\mathrm{S} \doteq \frac{\gamma_\mathrm{S}}{2pV_\mathrm{t}}. \tag{48.2}$$

48.4 Combinatorics

The quasi-particles identified in Table 48.2 are statistically interacting in the sense that activating one particle affects the number d_m of open slots for the activation of further particles from each species. This interaction can be accounted for by a multiplicity expression for jammed microstates involving a generalized Pauli principle [7] in the form, [6, 8, 9],

$$W(\{N_m\}) = \prod_{m=1}^{M} \binom{d_m + N_m - 1}{N_m}, \quad d_m = A_m - \sum_{m'=1}^{M} g_{mm'}(N_{m'} - \delta_{mm'}), \tag{48.3}$$

with capacity constants A_m and statistical interaction coefficients $g_{mm'}$ as listed in Table 48.3, and where N_m is the number of activated particles from species m.

The initial capacity for hosts grows linearly with the number of disks. It is zero for tags, which can only be activated inside hosts. Activating a host ($m' = 1,\ldots,4$) removes one or two slots for activating a further host ($m = 1,\ldots,4$) but adds one

Table 48.3 Capacity constants and statistical interaction coefficients for each particle species

m	A_m
1	$N-2$
2	$N-3$
3	$N-2$
4	$N-3$
5	0

$g_{mm'}$	1	2	3	4	5
1	2	2	1	2	1
2	1	2	1	1	1
3	2	2	2	2	1
4	1	2	1	2	1
5	−1	−1	−1	−1	0

slot for activating a tag ($m = 5$). Activating a tag ($m' = 5$) removes one slot for activating hosts ($m = 1, \ldots, 4$) but leaves the number of slots for activating a further tag ($m = 5$) invariant.

48.5 Statistical Mechanics

We can express the excess volume and the entropy as functions of the average particle content $\{\langle N_m \rangle\}$ of a jammed macrostate as follows [9]:

$$V - V_{\text{ref}} = \sum_{m=1}^{M} \langle N_m \rangle \Delta V_m, \qquad Y_m \doteq A_m - \sum_{m'=1}^{M} g_{mm'} \langle N_{m'} \rangle. \tag{48.4a}$$

$$S = k_B \sum_{m=1}^{M} \Big[\big(\langle N_m \rangle + Y_m\big) \ln \big(\langle N_m \rangle + Y_m\big) - \langle N_m \rangle \ln \langle N_m \rangle - Y_m \ln Y_m \Big]. \tag{48.4b}$$

The average particle numbers are the solutions of the linear equations [8, 9],

$$w_m \langle N_m \rangle + \sum_{m'=1}^{M} g_{mm'} \langle N_{m'} \rangle = A_m. \tag{48.5}$$

The w_m are non-negative solutions of the algebraic equations [6, 8–13],

$$e^{\epsilon_m / T_k} = (1 + w_m) \prod_{m'=1}^{M} \left(1 + w_{m'}^{-1}\right)^{-g_{m'm}}. \tag{48.6}$$

The analytic solution of Eq. (48.6), too unwieldy for display, gives us explicit expressions for the scaled excess volume, $\bar{V} \doteq (V - V_{\text{ref}})/N V_{\text{t}}$, the scaled entropy, $\bar{S} \doteq S/N k_{\text{B}}$, and the particle densities, $\bar{N}_m \doteq \langle N_m \rangle / N$, as functions of β, Γ_{L}, Γ_{S}.

48.6 Results

There is space to present one case, $\Gamma_L = \Gamma_S$, at the border between two regimes, where large and small disks have equal gravitational potential energy when they touch the same wall. Only two of the parameters (48.2) are independent. Increasing β means reducing the intensity of random agitations before jamming and increasing Γ_L means increasing the effects of gravity (e.g. by tilting the plane of the channel from horizontal toward vertical).

In the high-intensity limit, $\beta \to 0$, we have the most disordered macrostate:

$$\bar{N}_1 = \cdots = \bar{N}_4 = \frac{1}{12}, \quad \bar{N}_5 = \frac{1}{6}, \quad \bar{V} = \frac{2}{3}, \quad \bar{S} = \ln 3 \qquad (\beta = 0). \tag{48.7}$$

It is instructive to plot the population densities versus volume (Fig. 48.1a–c). For $\bar{N}_1, \ldots, \bar{N}_4$, $\beta = 0$ is realized at the points where multiple curves merge. For $\bar{N}_5$ that point is at the top of the dotted line. Increasing β toward infinity means moving along any path toward $\bar{V} = 0$ if $\Gamma_L < 1$ or toward $\bar{V} = 1$ if $\Gamma_L > 1$. The path stops at $\bar{V} = \frac{1}{2}$ if $\Gamma_L = 1$, which signals criticality. The following conservation law only holds for the case considered here:

$$\bar{N}_{\rm tot} \doteq \sum_{m=1}^{5} \bar{N}_m = \frac{1}{2} \qquad (\Gamma_L = \Gamma_S). \tag{48.8}$$

Its validity is illustrated by the two diagonal lines in Fig. 48.1a, where the dashed line represents the population density of all hosts combined. The tags $m = 5$ do not contribute to excess volume. Their numbers depend on $\bar{V}$ nevertheless, via the conservation law (48.8). Any increase in volume caused by the activation of one or the other host necessarily crowds out one tag.

The results show that hosts $m = 2, 3, 4$ only contribute significantly at high intensity. The three distinct macrostates associated with $\beta = \infty$ are located at the corners or the center of Fig. 48.1a. All have $\bar{N}_2 = \bar{N}_3 = \bar{N}_4 = 0$.

- The state with $\bar{N}_1 = 0$ and $\bar{N}_5 = \frac{1}{2}$ is realized for $\Gamma_L < 1$ and has $\bar{V} = 0$, $\bar{S} = 0$. It is a doublet consisting of the reference state vwvw$\cdots$vwv and the state 14545$\cdots$4546. The former contains no particles, $\bar{N}_5 = 0$, and the latter one host 2 and a macroscopic number of tags inside, amounting to $\bar{N}_5 = 1$.
- The state with $\bar{N}_1 = \frac{1}{2}$ and $\bar{N}_5 = 0$ is realized for $\Gamma_L > 1$ and has $\bar{V} = 1$, $\bar{S} = 0$. It is a singlet packed with hosts 1: 13vw13vw$\cdots$13v. Hosts 1 proliferate because they have negative activation energies.
- The state with $\bar{N}_1 = \bar{N}_5 = \frac{1}{4}$ is realized for $\Gamma_L = 1$ and has $\bar{V} = \frac{1}{2}$, $\bar{S} = \ln 2$. It is highly degenerate. The hosts 1 are randomly distributed between vacuum tiles with a random number of tags inside. Hosts 1 and tags 5 have zero activation energies whereas the hosts, 2, 3, 4 have positive activation energies.

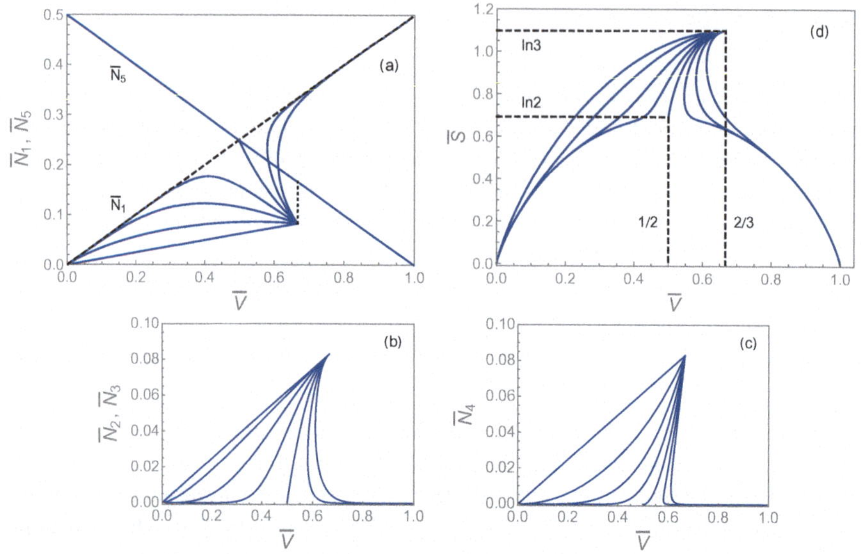

Fig. 48.1 Population densities $\bar{N}_m$, $m = 1, \ldots, 5$ versus excess volume $\bar{V}$ for fixed values 0, 0.5, 0.75, 1.0, 1.25, 1.5 of parameter $\Gamma_L = \Gamma_S$ from bottom up in panel (**a**) and from left to right in panels (**b**) and (**c**). The dotted line connect the points pertaining to $\beta = 0$ of $\bar{N}_1$ and $\bar{N}_5$. The dashed diagonal represents the sum $\bar{N}_1 + \bar{N}_2 + \bar{N}_3 + \bar{N}_4$. **d** Entropy $\bar{S}$ versus excess volume $\bar{V}$ for fixed values 0, 0.5, 0.75, 0.9, 1.0, 1.1, 1.25, 1.5 of parameter $\Gamma_L = \Gamma_S$ (from left to right)

With all this information in place we are ready to interpret a graphical representation of entropy versus excess energy (Fig. 48.1d). The parameter β runs from zero to infinity along each path from the top down. All paths start at coordinates $\bar{V} = \frac{2}{3}$, $\bar{S} = \ln 3$. Paths for $\Gamma_L < 1$ end at $\bar{V} = 0$, $\bar{S} = 0$, and paths for $\Gamma_L > 1$ at $\bar{V} = 1$, $\bar{S} = 0$. At critical gravity, $\Gamma_L = 1$, both volume and entropy decrease monotonically but end at the critical values $\bar{V} = \frac{1}{2}$, $\bar{S} = \ln 2$.

Compact analytic expressions for the curves in Fig. 48.1 pertaining to zero gravity and critical gravity are available. For $\Gamma_L = 0$ we have

$$\bar{N}_1 = \cdots = \bar{N}_4 = \frac{1}{8}\bar{V}, \quad \bar{N}_5 = \frac{1}{2}(1 - \bar{V}), \quad \bar{S} = (\bar{V} - 1)\ln\left(\frac{2}{\bar{V}} - 2\right) + \ln\left(\frac{2}{\bar{V}}\right), \tag{48.9}$$

where the range of volume is $0 \le \bar{V} \le \frac{2}{3}$. For $\Gamma_L = 1$ we have

$$\bar{N}_1 = \frac{(\bar{V} - 1)^2}{2\bar{V}}, \quad \bar{N}_2 = \bar{N}_3 = \frac{3}{2} - \bar{V} - \frac{1}{2\bar{V}}, \quad \bar{N}_4 = 2\bar{V} + \frac{1}{2\bar{V}} - 2, \tag{48.10a}$$

$$\bar{N}_5 = \frac{1}{2}(1 - \bar{V}), \quad \bar{S} = 2(\bar{V} - 1)\ln\left(\frac{1 - \bar{V}}{2\bar{V} - 1}\right) - \ln(2\bar{V} - 1), \tag{48.10b}$$

across the more restricted range $\frac{1}{2} \le \bar{V} \le \frac{2}{3}$ of volume.

48.7 Outlook

Qualitatively different jamming patterns pertain to the regimes $\Gamma_L < \Gamma_S$ of light large disks and $\Gamma_L > \Gamma_S$ of heavy large disks. Yet different jamming patterns are expected when the analysis is generalized to periodic and aperiodic sequences including random sequences of large and small disks (with modified jamming protocols).

References

1. N. Gundlach, M. Karbach, D. Liu, G. Müller, J. Stat. Mech. **P04018** (2013)
2. C. Moore, D. Liu, B. Ballnus, M. Karbach, G. Müller, J. Stat. Mech. **P04008** (2014)
3. R.K. Bowles, I. Saika-Voivod, Phys. Rev. E **73**, 011503 (2006)
4. S.S. Ashwin, R.K. Bowles, Phys. Rev. Lett. **102**, 235701 (2009)
5. R.K. Bowles, S.S. Ashwin, Phys. Rev. E **83**, 031302 (2011)
6. D. Liu, P. Lu, G. Müller, M. Karbach, Phys. Rev. E **84**, 021136 (2011)
7. F.D.M. Haldane, Phys. Rev. Lett. **67**, 937 (1991)
8. Y.-S. Wu, Phys. Rev. Lett. **73**, 922 (1994)
9. S. B. Isakov, Phys. Rev. Lett. **73**, 2150 (1994), Mod. Phys. Lett. B **8**, 319 (1994)
10. D.-V. Anghel, J. Phys. A 40, F1013 (2007), Europhys. Lett. **87**, 60009 (2009)
11. P. Lu, J. Vanasse, C. Piecuch, M. Karbach, G. Müller, J. Phys. A **41**, 265003 (2008)
12. P. Lu, D. Liu, G. Müller, M. Karbach, Condens. Matter Phys. **15**, 13001 (2012)
13. D. Liu, J. Vanasse, G. Müller, M. Karbach, Phys. Rev. E **85**, 011144 (2012)

Chapter 49
Statistical Properties of a Granular Gas Fluidized by Turbulent Air Wakes

Miguel A. López-Castaño, Juan F. González-Saavedra, Álvaro Rodríguez-Rivas, and Francisco Vega Reyes

Abstract We perform experiments with a granular system that consists of a collection of identical hollow spheres (ping-pong balls). Particles rest on a horizontal metallic grid and are confined within a circular region. Fluidization is achieved by means of a turbulent air current coming from below. Air flow is adjusted so that the balls do not elevate over the grid, as an approach to 2D dynamics. With a high-speed camera, we take images of the system. From these images we can infer horizontal particle positions and velocities by means of particle-tracking algorithms. With the obtained data we analyze: (a) the systematic measurement error in the determination of positions and velocities from our digital images; (b) the degree of homogeneity achieved in our experiments (which depends on possible deviations of the grid from the horizontal and on the homogeneity of turbulent air wakes). Interestingly, we have observed evidences of crystallization at high enough densities.

49.1 Introduction

Experimental works on the symmetry properties of nearly close-packed particles were carried out in the early 20th for direct visualization of the crystal structure in laboratory scale model systems (by that time this was technically not possible for real solid crystals). Pioneering work by L. Bragg and J. F. Nye in experiments with soap bubbles monolayers [1], which clearly reported a macroscopic hexagonal crystal structure, was followed by other interesting analogous works on different phases in systems with macroscopic particles [2–4].

The present work was conceived as an approach to this kind of experimental works. Our laboratory set-up is directly inspired in the work by Durian and co-workers, who analyzed the properties of Brownian motion in a macroscopic particle [5]; a ping pong ball fluidized by turbulent air wakes. These wakes are produced

M. A. López-Castaño (✉) · J. F. González-Saavedra · Á. Rodríguez-Rivas · F. Vega Reyes
Departamento de Física and Instituto de Computación Científica Avanzada (ICCAEx), Universidad de Extremadura, 06071 Badajoz, Spain
e-mail: malopez00@unex.es

I. Zuriguel et al. (eds.), *Traffic and Granular Flow 2019*, Springer Proceedings in Physics 252,
https://doi.org/10.1007/978-3-030-55973-1_49

at the Von Kármán streets due to air flow past a spherical particle [6]. We built a very similar set-up, this time using $\sim 10^2$ particles in most of our experiments. The final aim of our series of measurements is to search for eventual phase transitions, in analogy with the observations in thin vibrated layers [7].

49.2 Description of the System

We perform experiments with a variable number N of identical spherical particles. Specifically, our particles are ping-pong balls with diameter $\sigma = 4$ cm (ARTENGO© brand balls, made of ABS plastic, this material having a mass density $\rho \simeq 0.08$ g cm^{-3}). Particles rest on a metallic mesh (circular holes of 3 mm diameter arranged in a triangular lattice) and are enclosed by a circular wall made of polylactic acid (PLA). The diameter of this circular boundary is $D = 72.5$ cm and its height is $h \simeq 4.5$ cm $> \sigma$. Thus, the total area of the system available to the spheres is $A = 0.413$ m$^2 = 328.65 \times \pi(\sigma/2)^2$, which means that up to $N_{\text{max}} = 0.9069 \times 328.65 \simeq 298$ balls can fit in our system, if we ignore boundary effects (which would reduce this number). The grid is mounted and carefully levelled so that it remains as horizontal as possible (i.e.; perpendicular to gravity). In this way, we expect to achieve dynamics in which gravity does not create any anisotropy in the system behaviour.

Steady dynamics of the particulate system is achieved by means of a vertical air current coming from below, as depicted in Fig. 49.1. We use for this a fan, model SODECA© HCT-71-6T-0.75/PL, that is able to produce air current intensities u_{air} passing through the metallic grid plane with rates in the range of [2–5.5] m/s. An intermediate foam (2 cm thick) homogenizes the air current from the fan. The air flow distribution over the grid was measured using a turbine digital anemometer, plugged to a computer to collect data. We observed that the air current flow suffers local deviations over the grid of less than 10%, with respect to the average u_{air}.

The air flow coming from the fan produces turbulent wakes past the spheres [6]. Air current intensity is adjusted so that particles never levitate, and remain in contact with the metallic grid at all times. With this, we achieve a particle dynamics that is

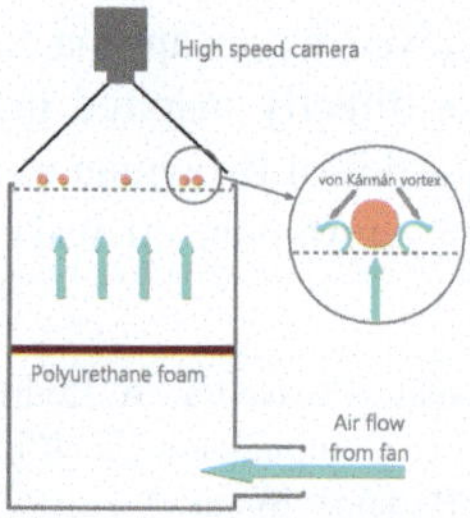

Fig. 49.1 Sketch of the experimental set-up

effectively two-dimensional, since the relevant particle motion is contained within the grid plane.

Summarizing, we built a set-up that has the following properties: (1) It is a many-particle system; (2) the energy input is homogeneous; (3) the dynamics is contained in a horizontal plane (the grid) and as a consequence gravity does not introduce a predominant direction; (4) symmetry-break is observed in the form of a hexagonal arrangement of the particles.

A series of experiments has been carried out, by modifying the values of air flow intensity (2 m/s $\leq u_{air} \leq$ 5.5 m/s) and packing fraction $\varphi \equiv N(\sigma/D)^2$, ($0.03 \leq \varphi \leq 0.79$).

49.3 Particle Tracking

A Phantom VEO 410L high speed camera has been used [8]. This model is capable of recording at 5200 frames per second at its maximum resolution (1280 $\times$ 800 pixels). At this resolution, we have recorded for each experiment a 99.92 s clip at 250 frames/s (well below the maximum operational speed of our camera). For low particle densities ($\varphi < 0.20$) three clips have been taken for each set of parameters (which improves the statistics of the data). Before each take the system has been thermalized for a few minutes, in order to assure steady state conditions (transient regime is very short for granular gas systems in most cases [9]).

We have developed a custom detection algorithm, where we combine the use of the open source library OpenCV [10] for particle detection together with TrackPy [11] (a python version of the Crocker and Grier algorithm [12]) for linking particle positions. Only a central region of interest (ROI) has been used for our subsequent analysis. This is done minimize eventual boundary effects [13, 14]. Our ROI is rectangular and has dimensions $L_x = 9.6\,\sigma$, $L_y = 7.7\,\sigma$.

49.3.1 *Error Estimation*

Static localization error is defined as the standard deviation from the position of a motionless particle [15–17]. In order to quantify this static error, we recorded five videos with a single static sphere (fan switched off), positioned at different points of the system in each one. Measuring the deviation from the mean position, we have found this static error to be $\epsilon_s = 0.059$ pixels or, in units of the particle diameter, $\epsilon_s = 7.6 \times 10^{-4}\,\sigma$. Additionally, there are other factors that can have an effect on the quality of measurements, such as motion blur–also called dynamic error–[18]; this is due to the fact that cameras take a certain amount of time Δt to acquire each frame. In our particular case we used $\Delta t = 1.5 \times 10^{-3}$ s. Therefore, the positions that we obtain are not instantaneous, but averaged over that short period of time. This error has implications, for example, in determining the value of the diffusion coefficient D, and introducing uncertainty in the mean squared displacement (MSD) measurement [15]. Since we do not discuss here dynamical properties with characteristic times shorter than our Δt, this dynamic error can be neglected in the present work.

49.3.2 Isotropy and Homogeneity Conditions

In order to analyze the degree of isotropy in the collective dynamics, we have looked at the position distribution for x and y coordinates. Results are plotted as histograms in Fig. 49.2. We observed that except for very low (Fig. 49.2a) or very high particle densities (Fig. 49.2b), particles are evenly allocated throughout the surface, as seen in Fig. 49.2c, with a standard deviation never greater than 5% [19]. In the case of high packing fractions, evidences of crystallization show up in the form of sharp and evenly distributed peaks (Fig. 49.2b). On the other hand, at very low densities (Fig. 49.2a), granular temperature is so low that the particles rarely move far away from their initial positions, thus skewing the distribution towards those regions.

With respect to the velocity distributions, and contrary to what happens with positions, we did not detect different behaviours depending on the packing fraction. Therefore, we represent in Fig. 49.3a and b x and y velocity distribution functions respectively, combining the data from all of the clips altogether (we analyze in this way the *global isotropy of the set-up* rather than the isotropy of a particular experiment). We have observed that distribution functions in v_x and v_y are not Gaussian, which agrees with previous observations in granular dynamics experiments [20]. In any case, our velocity distributions are rather well centered at zero (mean values $\mu_x = 4.32 \times 10^{-3}\sigma/\text{s}$ and $\mu_y = 8.56 \times 10^{-3}\sigma/\text{s}$) and both have a similar standard deviation ($\sigma_x = 1.06\,\sigma/\text{s}$ and $\sigma_y = 1.07\,\sigma/\text{s}$), which is an indication of a high degree of horizontality. In addition, we represent in Fig. 49.3c the particle speed modulus global distribution function $v = (v_x^2 + v_y^2)^{1/2}$, with mean value $1.22\,\sigma/\text{s}$ and standard deviation $0.88\,\sigma/\text{s}$. It is interesting to notice that a Gamma probability distribution

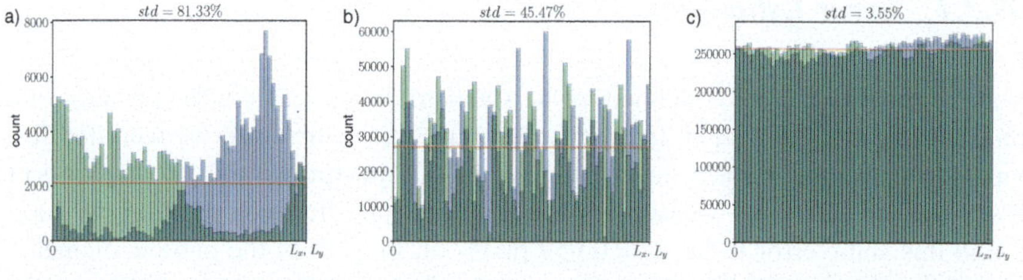

Fig. 49.2 Histogram of particle locations in the x (blue) and y (green) directions, during full-length clips for: **a** a system with a low packing fraction ($\varphi = 0.03$, $u_{air} = 4.63$ m/s), here particles remain near their initial positions; **b** a case in which packing fraction approaches a crystallization regime ($\varphi = 0.69$, $u_{air} = 2.80$ m/s), the reason behind the histogram maxima is that particles organize in a lattice structure. And **c**, combined data for all the experiments recorded at intermediate densities, for which an isotropic liquid state is found ($0.20 \leq \varphi \leq 0.54$)

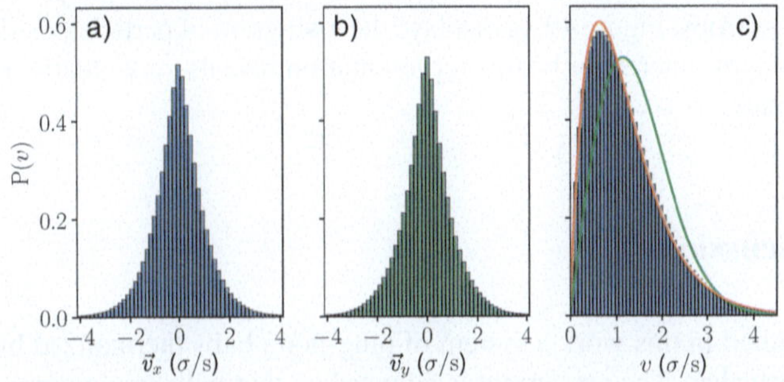

Fig. 49.3 Probability distribution function for the velocity of particles in the whole series of experiments. Displacements in the **a** x direction, **b** y direction, and **c** modulus of the velocity; for reference, gamma (orange) and Maxwellian (green) distribution fits are represented

($f(x) = x^{a-1}e^{-x}/\Gamma(a)$ with shape parameter $a = 1.863$, in this case) provides a much better fit than a 2-dimensional Maxwell-Boltzmann ($f(x) \propto xe^{-x^2}$).

In Fig. 49.2 velocities are deduced from to frame-to-frame displacements. Let us analyze now the distribution of displacements at longer times. For this, we have determined the distribution function for 200-frames displacements, $\mathbf{d}_{200} = \mathbf{r}(t + 200) - \mathbf{r}(t)$, over the entire length of each experiment (of about 10^2 s); i.e., we have of the order of 2.5×10^4 events per particle. Would anisotropy occur, it should show up in a polar coordinate representation (as in Fig. 49.4b) in the form of a well defined maximum in the preferential displacement direction. We have selected a 200 frames window (this is roughly the time at which, at intermediate densities, particles 'forget' in our set-up their initial velocity, as calculated from the velocity autocorrelation function [21]). Figure 49.4b reveals a high degree of dis-

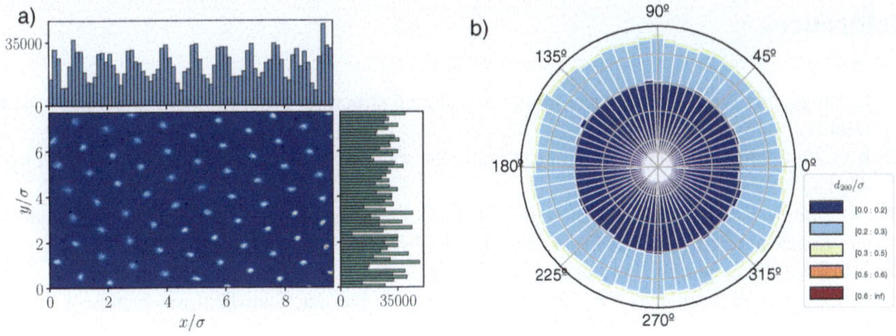

Fig. 49.4 Experiment with packing fraction $\varphi = 0.67$ and air flow velocity $u = 2.80$ m/s. **a** Particle positions histograms (upper and left panels) and plot of particle positions superimposed for all frames (central panel). A crystalline structure is clearly visible. **b** Stacked histogram in polar coordinates of 200 frame displacements ($\mathbf{d}_{200}$) for that same system; different colours represent different displacement lengths (in units of the particle diameter). The histogram strongly suggests that particle dynamics is highly isotropic

placements isotropy. Figure 49.4a displays the histogram of particle positions for the same experiment. Interestingly, this representation reveals very clearly a hexagonal crystal structure.

49.4 Discussion

We have studied in this work a system of ping-pong balls thermalized by means of turbulent air wakes. The air current is adjusted so that balls are in contact with the table at all times (i.e., their movement is primarily two-dimensional). In the first part of this work, we have analyzed the accuracy of our experimental methods for particle tracking. As in previous works by other authors [15–17], we have made distinction between inherent static and dynamic errors, finding that in both cases they are very small. Homogeneity of particle density indicates that we achieved a good degree of horizontality in of our set-up (this work is intended to study only inhomogeneity-free dynamics). For intermediate densities, the degree of isotropy is very high (Figs. 49.2c and 49.3). It is very interesting to remark that the distribution function of v (averaged over all of the series) fits a gamma distribution [22].

At high densities, it is interesting to notice the remnants of what seems to be a hexagonal crystal structure (Figs. 49.2b and 49.4a). Phase transitions that can eventually appear in this system will be studied in more detail in an impending work.

Acknowledgements The authors thank J. S. Urbach, E. Abad and S. B. Yuste for fruitful discussions. We acknowledge funding from the Government of Spain through project No. FIS2016-76359-P and from the regional Extremadura Government through projects No. GR18079 & IB16087, both partially funded by the ERDF.

References

1. L. Bragg, J.F. Nye, A dynamical model of a crystal structure. Proc. R. Soc. Lond. A **190**, 474 (1947)
2. B.G. Eaton, R.G. Finstad, P.D. Lane, Kinetic theory simulator for laboratory use. Am. J. Phys **47**, 132 (1979)
3. J.L. Finney, Random packings and the structure of simple liquids. I. The geometry of random close packing. Proc. R. Soc. Lond. A **319**, 479 (1970)
4. J. Friedel, *Dislocations* (Pergamon, New York, 1964)
5. R.P. Ojha, P.A. Lemieux, P.K. Dixon, A.J. Liu, D.J. Durian, Statistical mechanics of a gas-fluidized particle. Nature **427**, 521 (2004)
6. M. Van Dyke, *An Album of Fluid Motion* (The Parabolic Press, Stanford, CA, USA, 1982)
7. J.A. Olafsen, J.S. Urbach, Clustering, order and collapse in a driven granular monolayer. Phys. Rev. Lett **4369–4372**, 81 (1998)
8. Ametek: Phantom high speed. https://www.phantomhighspeed.com/products/cameras/veo
9. F. Vega Reyes, A. Santos, G.M. Kremer, Role of roughness on the hydrodynamic homogeneous base state of inelastic spheres. Phys. Rev. E **89**, 020,202(R) (2014)
10. OpenCV. https://opencv.org/

11. D. Allan et al., soft-matter/trackpy: Trackpy v0.4.2 (2019). https://doi.org/10.5281/zenodo.3492186.
12. J.C. Crocker, D.G. Grier, Methods of digital video microscopy for colloidal studies. J. Colloid Interface Sci. **179**, 298–310 (1996)
13. A.R. Abate, D.J. Durian, Approach to jamming in an air-fluidized granular bed. Phys. Rev. E **74**(3), 1–12 (2006)
14. Y. Lanoiselée, G. Briand, O. Dauchot, D.S. Grebenkov, Statistical analysis of random trajectories of vibrated disks: towards a macroscopic realization of Brownian motion. Phys. Rev. E **98**(6), 062,112 (2018)
15. A.J. Berglund, Statistics of camera-based single-particle tracking. Phys. Rev. E **82**(1), 1–8 (2010)
16. R.J. Ober, S. Ram, E.S. Ward, Localization Accuracy in Single-Molecule Microscopy. Biophys. J. (2004)
17. R.E. Thompson, D.R. Larson, W.W. Webb, Precise nanometer localization analysis for individual fluorescent probes. Biophys. J. (2002)
18. T. Savin, P.S. Doyle, Static and dynamic errors in particle tracking microrheology. Biophy. J. **88**(1), 623–638 (2005)
19. C. Scholz, T. Pöschel, Velocity distribution of a homogeneously driven two-dimensional granular gas. Phys. Rev. Lett. (2017)
20. J.S. Olafsen, J.S. Urbach, Velocity distributions and density fluctuations in a granular gas. Phys. Rev. E **60**, R2468 (1999)
21. M.A. López-Castaño, J.F. González-Saavedra, A. Rodríguez-Rivas, E. Abad, S.B. Yuste, F. Vega Reyes, Diffusive properties of a system of macroscopic particles thermalized by turbulent air wakes (2019). In preparation
22. R.V. Hogg, A.T. Craig, *Introduction to Mathematical Statistics*, 6th edn. (Macmillan Publishing Co. Inc., New York, 1989)

Chapter 50
Scaling Analysis and CFD Simulations of the Silos Discharge Process

David Méndez-Esteban, R.C. Hidalgo, and Diego Maza

Abstract In this work, we numerically examine granular flows in a silo using Computational Fluid Dynamics (CFD). In particular, we use constitutive models implemented in a commercial ANSYS software package and employ them to simulate the silo discharge process. Here, the analysis is focussed on the velocity and density profiles at the silo exit, while the numerical protocol is validated, comparing with experimental data of mass flow rate. Furthermore, the outcomes obtained for different sizes of the orifice are rationalized employing a theoretical scheme, which is based on the self-similar properties of the density and the velocity profiles.

50.1 Introduction

Particle flows play a relevant role, in many industrial and natural processes. Snow avalanches and powder handling in plenty of chemical engineering processes are only examples of this fact. Thus, the description of granular media in terms of its local particle-particle interactions have deserved important experimental and theoretical efforts in the past.

When examining macroscopic granular systems, particle shape and polydispersity notably limit the derivation of complete theoretical approaches. Moreover, their numerical modeling typically includes many tuning parameters, which require extensive experimental calibration. Therefore, the data obtained from experiments allow a deeper understanding of these systems, as well as the numerical tools calibration. Granular media models are usually introduced through a detailed description of each of its constituent particles (Discrete Element Methods). However, in practical situations, the use of coarse-grained continuous approaches becomes mandatory due to the vast number of particles involved. Very often, these methods also provide reasonable macroscopic predictions.

D. Méndez-Esteban · R.C. Hidalgo · D. Maza (✉)
Dpto. de Física y Matemática Aplicada, Facultad de Ciencias, Universidad de Navarra, Pamplona, Spain
e-mail: dmaza@unav.es

I. Zuriguel et al. (eds.), *Traffic and Granular Flow 2019*,
Springer Proceedings in Physics 252,
https://doi.org/10.1007/978-3-030-55973-1_50

Since the '50s, continuous models of granular media have been developed [1–6]. Typically, these theoretical frameworks use the mass fraction (or solid fraction) as a control parameter. Thus, depending on its local value the samples are considered as gases (collisional regime), liquids (kinetic regime) or solids (frictional regime, based on soil mechanics). Consequently, to deduce a generalized constitutive model that fits well in all circumstances is challenging.

Silo discharge is a paradigmatic example of particle flows, although the theoretical attempts to predict its mass-flow-rate as a function of the aperture size still remain unsuccessful [7].

In this work, the discharge process in a flat bottom silo (or bunker) is investigated. We perform simulations using computational fluid dynamics (CFD) to study the silo flow, examining the kinematic variables that control the mass delivery at the outlet. Previous experimental findings [8] allow the model calibration for a flat bottomed silo. These parameters might be extended to other geometries, such as hoppers of different angles. Based on a theoretical framework introduced in [9], we analyze the velocity and density profiles at the silo exit. The analysis strongly suggests that the outlet size is the only relevant length scale is this process. From that starting point, we introduce a novel methodology for calibration and testing of the CFD approach, which might be useful in other applied scenarios.

50.2 Numerical Protocol and Calibration Procedure

The numerical model rests on the kinetic theory of granular flow (KTGF). It is a continuous approach that combines the momentum balance equation with phenomenological granular constitutive laws [10]. In this continuous approach, the particles behave as a gas with inelastic collisions, which fills the cylinder. The KTGF assumes that particles collide inelastically in binary collisions and that their speed distribution is a Gaussian [10]. From that point onwards, the constitutive equations are deduced for the kinetic and collisional regimes.

Typically, the constitutive laws contain shear and bulk viscosities arising from the mechanisms of particle momentum exchange. On the one hand, the bulk viscosity accounts for the resistance of the system to compression and expansion [5]. Moreover, the total shear viscosity accounts for different viscosity sources, depending on the specific type of behavior (collisional and kinetic) [10]. To obtain more realistic outcomes, the model also includes an additional source of shear viscosity [11], which mimics the macroscopic Mohr-Coulomb friction (frictional part). It is worth to mention that it is very relevant when modeling systems densely packed. When computing the shear frictional viscosity [11], the solid pressure P_s is an important ingredient that includes kinetic-collisional pressure P_k, and frictional pressure P_f. In principle, the kinetic-collisional pressure P_k is an outcome of the original KTGF, where the restitution coefficient of the particles plays a significant role. However, the original KTGF only accounts for the collisional dissipation, due to normal collision between particles. Here, we use the formulation of Chialvo and Sundaresan [12], which also

Table 50.1 Inputs to the CFD code Ansys Fluent

Inputs Granular model	Value	Reference
Packing limit	0.6	
Restitution coefficient	0.97	
Frictional packing limit	0.25	
Angle of internal friction	27°	
Effective restitution coefficient	0.85	[12]

accounts for the tangential collisional dissipation, introducing an effective restitution coefficient e_{eff}, who depends on the particle friction and the normal restitution coefficient e_n. On the other hand, P_f is function of the local volume fraction $\phi(r)$. In particular, in our calculation we use the Jhonson and Jackson formulation for $P_f(\phi)$ [13], which also involves other parameters as the packing limit ϕ_{max}, i.e. the maximum volume fraction that the model can handle, and the frictional packing limit ϕ_{min}, which is volume fraction threshold, above which a frictional viscosity is included. Finally, to mention that the model includes two-phases (gas and solid), where both phases can occupy each point of the space in any proportion. Furthermore, it can also be extended to more components. For the sake of simplicity, in our calculation, the drag force between the particle and the interstitial air is not considered, and the initial condition is a densely packed system with $\phi_o = 0.6$, in all cases.

The simulated bunker is a flat bottomed cylinder, $0.160\,m$ diameter resembling the experimental condition of Ref. [8]. Hence, we set the model parameters, which mimic steel spheres of diameter $d = 1 \cdot 10^{-3}\,m$. The outlets explored (centered at bottom) range between $6 \cdot 10^{-3}$ to $20 \cdot 10^{-3}\,m$ in diameter. We explore different mesh configurations until velocity and vertical acceleration profiles become almost independent of the mesh size. Model relevant parameters are summarized in Table 50.1.

Calibrating the numerical protocol, we adjust the model parameters to reproduce the experimental mass flow rate values, corresponding to the flat silo geometry. Figure 50.1 compares the mass flow rate measured in [8] with the ones obtained via CFD. Although some disagreements can be observed for small outlets, we obtain a reasonable agreement between numerical and experimental values of the mass flow rate. Our intention is to develop a cross-checking procedure, involving the experimental results, the Beverloo's correlation, as well as, a novel theoretical estimation based on scaling arguments, motivated by the shape of the velocity and density profiles, at the orifice. As a result, all the relevant parameters of the solver Ansys-fluent can be set, for reproducing the flow rate values corresponding to each orifice size R. Remark that in all the simulation presented here, only the orifice size R is changed.

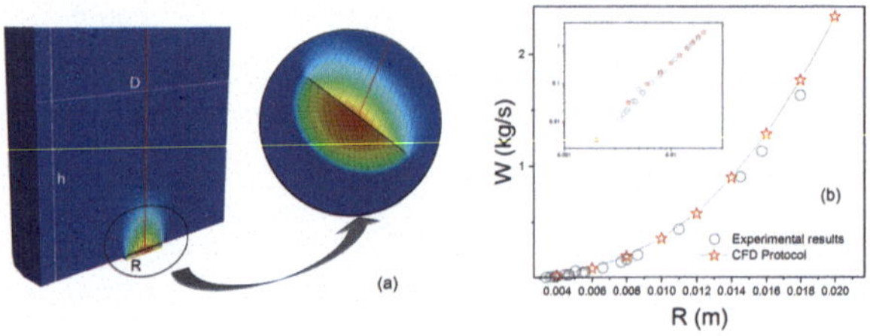

Fig. 50.1 **a** Model geometry and mesh configuration. Colors are proportional to the modulus of vertical velocity **b** Mass flow rate obtained for different outlet sizes (star symbols) compared with the experimental data (open circles). Connecting lines are a guide for the eye. Inset: the same data displayed on a logarithmic scale

50.3 Numerical Results

Here, we focus our attention on the velocity and volume fraction profiles in the orifice. Figure 50.2 illustrates the vertical velocity, $v_z(r)$, obtained just at the exit ($z = 0$), for different orifice sizes R. For comparison, the numerical data is fitted with the function:

$$v_z(r) = v_c(R)[1 - (r/R)^2]^a = \sqrt{2g\gamma R}\,[1 - (r/R)^2]^a,\ \gamma = \int_0^\infty \frac{a(z)}{gR}\,dz \quad (50.1)$$

The relationship $v_c(R) = \sqrt{2g\gamma R}$ can be deduced integrating the normalized advective acceleration profiles (see Fig. 50.2b) along the vertical direction, as was proposed by Rubio-Largo et al [9]. Additionally, the profiles can be scaled by $v_c(R)$ and consequently its self-similar behavior (see inset of Fig. 50.2a) can be specified by a single exponent, $a = 0.35 \pm 0.01$. These outcomes indicate conclusively that the orifice size is the relevant scale of this process.

Following the arguments that were given in [9, 14], the exit velocity profile is fully determined by the kinematic field at the outlet neighborhood. While the origi-

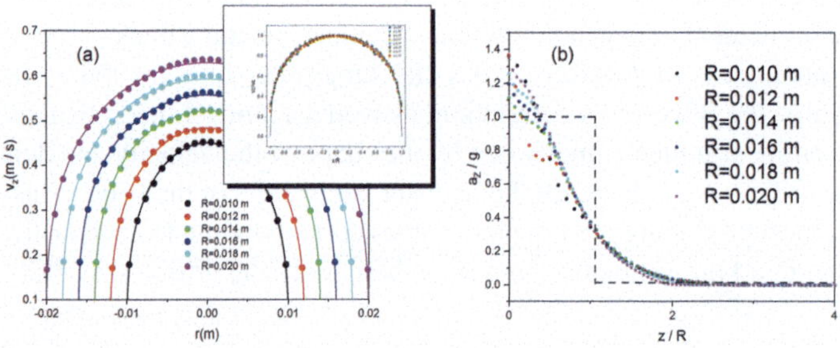

Fig. 50.2 **a** Vertical velocity profiles $V_z(r)$ on the plane of the orifice (in the inset $V_z(r/R)/V_z(R)$) **b** Normalized vertical acceleration $a(z/R)/g$. Dashed line represents the *free fall arch* approach

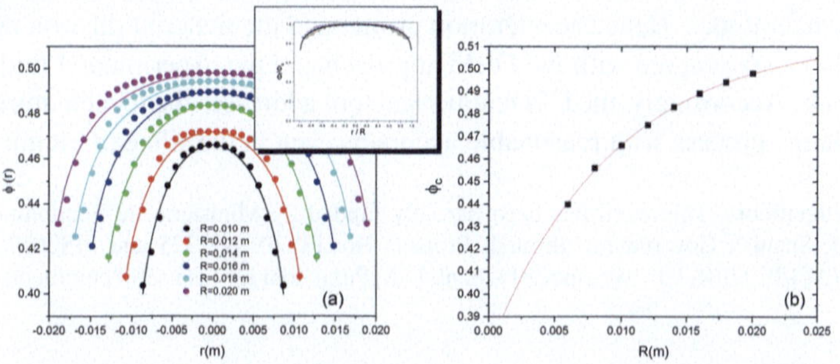

Fig. 50.3 **a** Volume fraction profiles $\phi(R, r)$ (in the inset $\phi(r, R)$ is normalized by its value at the outlet centre, $\phi_c(R)$). **b** $\phi_c(R)$ evolution on R

nal *free fall arch* idea [7] rest on the assumption that vertical particle's acceleration equals gravity for coordinates below R and equal zero anywhere else, the acceleration profiles obtained in our simulations are smooth evolving functions that can be collapsed in a singular curve $a(z)/g = F(z/R)$ as is depicted in Fig. 50.2b. Although some dispersion is evident when $z \simeq 0$, different profiles are very similar. Thus, γ's values range between $\gamma = 0.97$ and $\gamma = 1.01$, providing the arguments to consider $v_c(R) = \sqrt{2gR}$ as the typical scale to collapses the data obtained for different orifices as is evidenced in the inset of Fig. 50.2a.

Figure 50.3 shows the volume fraction profiles $\phi(r)$ obtained numerically, for different orifice sizes. It is noticeable that the volume fraction depends on the exit radius, but with a very flat profile. This fact is also evidenced in the inset of Fig. 50.3a, where the values are scaled with the volume fraction at the outlet center, $\phi_c(R)$. We fit the collapsed profiles using the same expression introduced to analyze the 2D-case [14]:

$$\phi(r) = \phi_c(R)[1 - (x/R)^2]^b = \phi_\infty(1 - \alpha_1 e^{-\frac{R}{\alpha_2}})[1 - (x/R)^2]^b \qquad (50.2)$$

where $\phi_c(R) = \phi_\infty(1 - \alpha_1 e^{-\frac{R}{\alpha_2}})$ accounts the packing fraction dependence on the exit radius and $b = 8.0 \pm 0.1 \cdot 10^{-3}$ fixes the curvature. The values $\phi_\infty = 0.515 \pm 0.003$, $\alpha_1 = 0.261 \pm 0.005$ and $\alpha_2 = 9.5 \pm 0.5 \cdot 10^{-3}$ m correspond to the best fit (see Fig. 50.3b). Two important conclusions can be extracted from the values of the parameters of Eq. 50.2: (a) the asymptotic volume fraction ϕ_∞ is lower than the bulk packing limit, $\phi_B = 0.60$, and (b) the scale imposed by α_2 is around ten times the particle diameter, as it occurs in the 2D-case [14].

Summarizing: we numerically study a silo flow using Computational Fluid Dynamics (CFD) and focus the analysis on velocity and density fields at the silo outlet. The numerical protocol is validated, comparing with previous data of mass flow rates obtained experimentally. We find that CFD protocols provide a valuable tool to describe local coarse-grained fields of the silo discharge process. The resulting velocity and density profiles are self-similar, with R as appropriate length to scale the obtained magnitudes. Hence acceleration profile and the material dilation predicted for small exit sizes agree with the DEM approach and the experimental findings for the 2D-case. Accordingly, the CFD numerical tool addresses the local features of the silo discharge process with reasonable accuracy, even in the silo exit vicinity.

Acknowledgements This work has been partially funded by Ministerio de Economía y Competitividad (Spanish Government) through Projects No. FIS2014-57325 and FIS2017-84631-P, MINECO/AEI/FEDER, UE. We sincerely thank L.A. Pugnaloni for his useful comments and suggestions.

References

1. R.A. Bagnold, Experiments on a gravity-free dispersion of large solid spheres in a newtonian fluid under shear. Proc. R. Soc. A Math. Phys. Eng. Sci. **225**(1160), 49–63 (1954)
2. S. Chapman, T.G. Cowling, in *The Mathematical Theory of Non-Uniform Gases: An Account of the Kinetic Theory of Viscosity, Thermal Conduction and Diffusion in Gases*, 3rd edn. (Cambridge University Press, Cambridge, 1970). (Eng.)
3. J.T. Jenkins, S.B. Savage, A theory for the rapid flow of identical, smooth, nearly elastic, spherical particles. J. Fluid Mech. **130**, 187–202 (1983)
4. P.C. Johnson, R. Jackson, Frictional-collisional constitutive relations for granular materials, with application to plane shearing. J. Fluid Mech. **176**, 67–93 (1987)
5. C.K.K. Lun, S. Br Savage, D.J. Jeffrey, N. Chepurniy. Kinetic theories for granular flow: inelastic particles in couette flow and slightly inelastic particles in a general flowfield. J. Fluid Mech. **140**, 223–256 (1984)
6. S.B. Savage, M. Sayed, Gravity flow of coarse cohesionless granular materials in conical hoppers. Z. Angew. Math. Phys. ZAMP **32**(2), 125–143 (1981)
7. R.M. Nedderman, in *Statics and Kinematics of Granular Materials* (Cambridge University Press, November 1992)
8. C. Mankoc, A. Janda, R. Arévalo, J.M. Pastor, I. Zuriguel, A. Garcimartín, D. Maza, The flow rate of granular materials through an orifice. Granul. Matter **9**(6), 407–414 (2007)
9. S.M. Rubio-Largo, A. Janda, D. Maza, I. Zuriguel, R.C. Hidalgo, Disentangling the free-fall arch paradox in silo discharge. Phys. Rev. Lett. **114**(23), 238002 (2015)
10. D. Gidaspow, in *Multiphase Flow and Fluidization: Continuum and Kinetic Theory Descriptions* (Elsevier Science, 1994)
11. D.G. Schaeffer, Instability in the evolution equations describing incompressible granular flow. J. Differ. Equ. **66**(1), 19–50 (1987)
12. S. Chialvo, S. Sundaresan, A modified kinetic theory for frictional granular flows in dense and dilute regimes. Phys. Fluids **25**(7), 070603 (2013)
13. P.C. Johnson, P. Nott, R. Jackson, Frictional-collisional equations of motion for participate flows and their application to chutes. J. Fluid Mech. **210**, 501–535 (1990)
14. A. Janda, I. Zuriguel, D. Maza, Flow rate of particles through apertures obtained from self-similar density and velocity profiles. Phys. Rev. Lett. **108**(24), 248001 (2012)

Chapter 51
Dense Pedestrian Crowds Versus Granular Packings: An Analogy of Sorts

Alexandre Nicolas

Abstract Analogies between the dynamics of pedestrian crowds and granular media have long been hinted at. They seem all the more promising as the crowd is (very) dense, in which case the mechanical constraints prohibiting overlaps might prevail over the decisional component of pedestrian dynamics. These analogies and their origins are probed in two distinct settings, (i) a flow through a narrow bottleneck and (ii) crossing of a static assembly by an intruder. Several quantitative similarities have been reported for the former setting and are discussed here, while setting (ii) reveals discrepancies in the response pattern, which are ascribed to the pedestrians' ability to perceive, anticipate and self-propel.

Besides its obvious practical relevance, the study of pedestrian dynamics owes much of its interest to its singular positioning at the crossroads between social sciences and mechanics. This is reflected by a dichotomy of approaches.

On the one hand, statistical models inspired from economics, such as discrete-choice models,[1] are often used when it comes to predicting distant destinations or routes chosen by pedestrians [1]. (Nevertheless, it should be mentioned that they have also been applied to stepping dynamics [2]).

On the other hand, mechanical approaches are generally favoured when the dynamics are investigated in finer detail, at a more 'microscopic' scale, in particular when strong interactions or hindrances are expected between pedestrians because of the crowd's density. These approaches are formally based on Newton's equation of motion (or variants thereof), supplemented with a term describing the (externally inferred) desired direction.

How the pedestrians' *decisions* and *choices* of motion in view of their surroudings should enter this mechanical framework remains however unclear, from a fundamen-

[1] These models depend on an abstract utility function, which quantifies the attractiveness of a given destination, direction or trajectory.

A. Nicolas (✉)
Institut Lumière Matière, CNRS and Université Claude Bernard Lyon 1, 69622 Villeurbanne, France
e-mail: alexandre.nicolas@polytechnique.edu

I. Zuriguel et al. (eds.), *Traffic and Granular Flow 2019*,
Springer Proceedings in Physics 252,
https://doi.org/10.1007/978-3-030-55973-1_51

tal viewpoint. *In practice*, the two components (the 'cognitive' one and the purely mechanical one) are often combined in an *ad hoc* way. In the famous social-force model [3], at the heart of various commercial software products, these components are simply put on an equal footing, in that the choices of motion (influenced by the other pedestrians and the built environment) are converted into 'social forces' and summed up with the mechanical forces in Newton's equation of motion. One is left with a mechanical problem, without further justification.

Mechanical descriptions are quite appealing, especially at high density. Indeed, when the crowd is very dense, one may even think that excluded-volume effects (precluding overlaps) restrict the possibilities of motion so much that they stifle the effect of decision-making, thus the singularities of pedestrians as compared to, say, grains of matter. To translate this idea into a reasoning on configuration space, most configurations are admissible for a sparse crowd, because randomly positioned pedestrians seldom overlap; in contrast, at very high density, volume exclusion reduces the available volume of configuration space; the prevalence of areas that are forbidden because of overlaps might even restrict the possible directions of evolution of the assembly in configuration space to such an extent that the former virtually govern its evolution (as if in a maze).

This begs the following question: Is a purely mechanical description satisfactory at very high density? Does the crowd then behave like a granular medium? We shall address this question in two distinct situations, a bottleneck setup in Sect. 51.2 and an intrusion experiment in Sect. 51.3, after setting the theoretical framework in Sect. 51.1.

This manuscript aims to provide a pedagogic discussion. Thus, most quantitative features are skimmed off; references to already published or forthcoming manuscripts will be provided for readers interested in quantitative aspects.

51.1 Theoretical Framework

We start with a brief introduction to the theoretical framework in which the motion of pedestrians is to be handled, within classical mechanics. As a mechanical body, a pedestrian i obeys Newton's equation of motion,

$$m\ddot{\boldsymbol{r}}_i = \boldsymbol{F}_{\text{ground}\to i} + \boldsymbol{F}^{(\text{mech})}_{\text{wall}\to i} + \sum_j \boldsymbol{F}^{(\text{mech})}_{j\to i}.$$

Here, the *mechanical* forces possibly exerted by walls, $\boldsymbol{F}^{(\text{mech})}_{\text{wall}\to i}$, and other pedestrians j, $\boldsymbol{F}^{(\text{mech})}_{j\to i}$, in case of contact and pushes, have been taken into account, even though they seldom arise in practice. The ground force, $\boldsymbol{F}_{\text{ground}\to i}$, is composed of (i) a static (vertical) reaction force that would subsist with no deformation of the pedestrian's body (*passive case*), and (ii) a force $\boldsymbol{F^{(p)}}_i$ that results from changes in the body shape

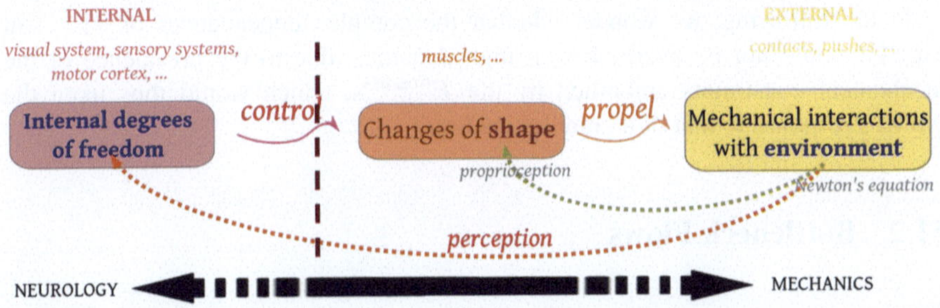

Fig. 51.1 Schematic representation of the interplay between external mechanical forces, the self-propelling force, and non-mechanical cues from the environment

(created by flexing leg muscles) and that propels the pedestrian. In the horizontal plane, one can thus write

$$m\ddot{\boldsymbol{r}}_i = \boldsymbol{F}^{(p)}{}_i + \boldsymbol{F}^{(\text{mech})}_{\text{wall}\to i} + \sum_j \boldsymbol{F}^{(\text{mech})}_{j\to i}.$$

Note that this self-propelling force $\boldsymbol{F}^{(p)}{}_i$ does not require any process akin to decision-making: Active particles such as vibration-driven 'hexbugs' [4, 5] are also propelled by $\boldsymbol{F}^{(p)}{}_i$. Entities that make decisions about their motion on the basis of their perceptions [6, 7] are nonetheless singular in that their (active) propelling force will be a function of *complex dependences on environmental cues, the history of past body shapes*, etc., generically denoted by '(...)'; see Fig. 51.1. Noting that the reaction to external cues cannot be instantaneous, it needs a finite time $\tau_\psi > 0$ to be processed (for humans, voluntary reactions take at least 0.1 s; for instance, the behavioural response to a complex visual stimulus occurs after typically $\tau_\psi = 0.4$ s [8]), one can write

$$\begin{cases} m\ddot{\boldsymbol{r}}_i & = \boldsymbol{F}^{(p)}{}_i + \boldsymbol{F}^{(\text{mech})}_{\text{wall}\to i} + \sum_j \boldsymbol{F}^{(\text{mech})}_{j\to i} \\ \tau_\psi \dot{\boldsymbol{F}}^{(p)}{}_i & = f(\ldots) \end{cases} \quad (51.1)$$

Of course, the foregoing derivation of the singularity of perceptual entities is overly simplified, because $\boldsymbol{F}^{(p)}{}_i$ may in fact also vary with the configuration of the crowd for non-perceptual entities, for instance if a magnetic self-propelled particle has its polarisation axis rotated via magnetic interactions with its neighbours. Nonetheless, it is interesting to note that, even in the case of such an elementary dependence of $\boldsymbol{F}^{(p)}{}_i$ on the environment, the collective behaviour of self-propelled particles may differ from that of their passive driven counterparts, notably in their self-organisation abilities [9]. A more accurate view on the topic is reserved for a forthcoming paper (Nicolas, *in preparation*).

In the following, we wonder whether the complex dependences of $f(\ldots)$ in Eq. 51.1 could not be overlooked at high densities, due to the prevalence of the mechanical constraints embodied by the $\boldsymbol{F}^{(\mathrm{mech})}$'s, which would thus liken the crowd's response to that of a granular medium.

51.2 Bottleneck Flows

The foregoing question is first addressed in the context of competitive flows through a (narrow) bottleneck, insofar as this setting seems to severely constrain the possibilities of motion.

51.2.1 Common Features

In the last few years, several similarities (which had long been hinted at in a tentative way) have been quantitatively demonstrated in experiments with granular materials flowing out of a vibrated hopper, on the one hand, and pedestrian crowds asked to rush through a narrow doorway, on the other hand [10]. These similarities include:

- exponential bursts of egresses in close succession (see [11] for a discussion about the universality of this feature)
- heavy tails in the distribution of time gaps τ between egresses in competitive escapes, which are at least reasonably well described by power laws: $p(\tau) \sim \tau^{-\alpha}$ [12]. Note that this expression entails a divergence of the mean time gap $\langle\tau\rangle$ when $\alpha < 2$, hence a dwindling flow rate as the system size tends to infinity.
- the possibility of a faster-is-slower effect in the evacuation, whereby tuning up the entities' drive to escape (*by tilting the hopper or prescribing a more competitive behaviour to the pedestrians*) may actually delay the evacuation due to longer clogs [12]. It should however be noted that, while the possibility of such an effect has been experimentally demonstrated with crowds, it requires high competitiveness and perhaps also aggressive pushes; for less competitive evacuations, a more trivial faster-is-faster effect is observed [13, 14]

Given that the flow rate J is given by $J = 1/\langle\tau\rangle$, the distribution $p(\tau)$ of time gaps τ is a central feature, which we now study in granular flows and then in their pedestrian counterparts.

51.2.2 Granular Hopper Flows

Long time gaps are caused by the formation of clogs, due to arches blocking the aperture of the vibrated hopper in two dimensions. The ultimate shattering of an existing arch, if it occurs, owes mostly to the vibration-induced destabilisation of the weakest link in the arch, i.e., of the grain that forms the widest angle with its neighbours in the arch [15]. Thus, we were led to study the evolution of the position of this grain under vibrations. Using non-trivial, but reasonable assumptions, we showed that this evolution can be likened to that of a Brownian particle in an energy trap, with an inverse temperature β associated with the acceleration Γ of the vibrations, $\beta \approx \frac{\gamma}{\Gamma^2}$, where γ is a drag coefficient [16]. It follows from this formal mapping that the lifetime of an arch is given by the Kramers' escape time

$$\tau \propto \exp\left(\beta E_b\right), \tag{51.2}$$

where E_b is the trap depth (or energy barrier). The values of E_b were inferred from ramp experiments, in which the intensity of vibrations was gradually increased until an existing arch broke; they were shown to follow a Weibull distribution

$$p(y) = e^{-y} \text{ with } y \propto \sqrt{E_b}$$

Arch stabilities, quantified by E_b, are thus fairly narrowly distributed. But the bottomline is that this fairly narrow disorder is amplified by the exponential dependence on E_b in Kramers' escape time, Eq. 51.2. Were energy barriers exponentially distributed (as they are in the Soft Glassy Rheology model [17]), this would lead to a power-law distribution of escape times τ. Here, the disorder in the arch stabilities is slightly larger, therefore the distribution of arch lifetimes decays even more slowly than a power law as $\tau \to \infty$. The results of the foregoing minimal model attain semi-quantitative agreement with experimental data [16]. Importantly, they lead to the prediction that the mean time gap $\langle\tau\rangle$ may diverge irrespective of the vibration intensity (in the limit of moderate shaking), which casts doubt on the ability of gentle vibrations to restore the hopper flow persistently.

51.2.3 Pedestrian Flows Through a Bottleneck

Turning to pedestrians, temporary clogs are also observed in competitive pedestrian flows through a bottleneck, due to pedestrian 'arches' blocking the doorway. Our initial observation was that cellular automata (CA) provide a computationally very efficient way to simulate them, e.g., to test evacuation scenarios, but that their predictions are seldom validated against experimental measurements, in particular detailed statistics about the time series of egresses [12]. As a matter of fact, it seemed to us that none of the CA that we surveyed in the literature was able to describe the seemingly heavy-tailed distribution of time gaps between egresses reported for highly competitive evacuations [18]. This feature turned out to be difficult to replicate. Indeed, the regular grid typically used in CA precludes geometric disorder in the pedestrian arches, whereas such disorder was central to the replication of heavy tails in granular flows (see the previous section).

It goes without saying that the specific mechanics governing the clogging dynamics differ between pedestrians and grains. However, the broad insight gained from the granular case helped us devise a CA that was finally able to reproduce experimental results on controlled evacuations semi-quantitatively. To this end, after realising (i) the importance of having some intrinsic disorder in the model (and not just random fluctuations) and (ii) the constraints affecting the geometry in a CA, we decided to introduce disorder in the agents' behaviours, more precisely in the impatience displayed by agents close to the exit (the so called 'friction'), which controls the number of conflicting endeavours to escape simultaneously. This enabled us to capture the desired experimental features [18].

All in all, despite the discrepancies in the physical origins of their motion, (competitive) pedestrians and grains flowing through a narrow bottleneck exhibit striking similarities, which hints that basic mechanisms control the dynamics in this setup, in particular the joint presence of a driving 'force' towards the escape and paramount volume exclusion effects (included in $\boldsymbol{F}^{(\mathrm{mech})}_{j\rightarrow i}$ in Eq. 51.1).[2]

51.3 Crossing by an Intruder

To test if the similarities observed in bottleneck flows extend to other setups, we now move on to crossing tests of a static (pedestrian or granular) two-dimensional assembly by an intruder.

51.3.1 Granular Mono-Layer

This kind of test is fairly classical for granular media. In particular, Cixous, Kolb and colleagues studied the passage of a circular intruder in a mono-layer of granular disks below jamming [21] (also see the related works by Séguin et al. [22]). They distinguished different regimes, but in the regime most relevant for the upcoming comparison they observed an average flow field of the form displayed in Fig. 51.2c. It features radial velocities on the fore-side of the intruder, with recirculation eddies on the side. Furthermore, the perturbation is short-ranged and decays exponentially in space (contrary to what happens in a viscous fluid). It should also be stated that a depleted zone appears in the wake of the cylinder, with a size that decreases with increasing packing fraction.

[2]Note that these similarities may perhaps not extend to the effect of placing an obstacle in front of the door [19].

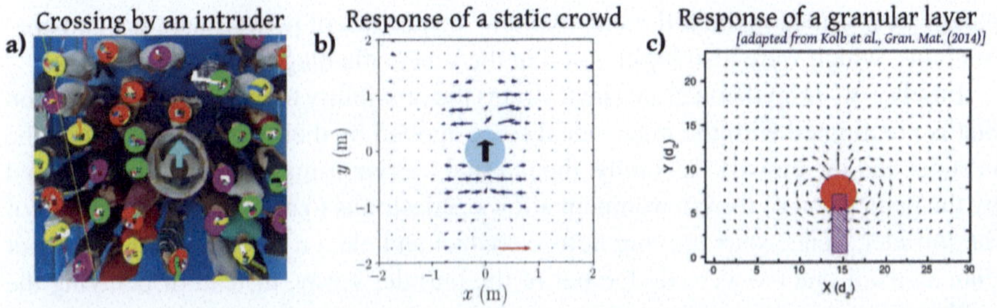

Fig. 51.2 Response to the crossing of an intruder in a static crowd (**a-b**) and a granular layer (**c**). Picture (**a**) is a snapshot taken during a controlled experiment, while panels (**b**) and (**c**) are average flow fields in the co-moving frame of the intruder. The red arrow in (**b**) represents a velocity of 30 cm/s. Adapted from Ref. [20, 21]

51.3.2 Static Pedestrian Crowd

To allow comparison, controlled experiments were performed in Orsay (France) and Bariloche (Argentina), in which a cylinder of around 70 cm in diameter moved linearly through a group of static participants (see Fig. 51.2a). The group's density was varied from $1-2\,\mathrm{ped/m^2}$ to $6-7\,\mathrm{ped/m^2}$ and its orientation was also varied; for further experimental details, refer to [20].

In short, when the participants were asked to behave casually, as if they were on an underground platform, and were facing the incoming intruder, the crowd's response consisted of strictly transverse moves directed outwards downstream from the obstacle and inwards in its wake, as shown in Fig. 51.2b, with only little sensitivity to the crowd's density. This is in stark contrast with the granular response of Fig. 51.2c. On the other hand, similarly to the granular case, the perturbation was short-ranged and left a pedestrian-free zone in the wake of the obstacle, which was refilled after the passage, via the inwards moves mentioned just above.

Only when the participants were asked to refrain from anticipating the intruder's passage *and* had to turn their back to the intruder (which thus arrived from behind), only then did the crowd's response mirror the granular one somewhat more closely, with displacements aligned approximately radially on the fore-side of the intruder (*data not shown here*).

51.3.3 Pedestrian specifics

As in the bottleneck setup, but to a much lesser extent here, some similarities were noticed in the pedestrian and granular responses, which consist of a short-ranged perturbation field (possibly due to the discreteness of the systems) and a depleted zone in the intruder's wake. In contrast, the observed discrepancies in the displacement patern are ascribed to the following two salient specifics of pedestrians, as compared to grains, which were also highlighted in the schematic diagram of Fig. 51.1.

Firstly, self-propulsion grants pedestrians the possibility to move along a direction that is not aligned with the mechanical force applied by their surroundings (i.e., the intruder and the crowd). Secondly, the internal 'decision-making' process informed by the perception of the environment allows pedestrians to anticipate the passage of the intruder, hence start moving before contact and elect a transverse move, rather than an axial one (so as to dodge out of the intruder's way, instead of delaying the collision).

In the formal framework outlined in Sect. 51.1, this is tantamount to saying that an avoidance strategy is encoded in the function $f(\ldots)$ determining the evolution of the self-propelling force in Eq. 51.1; this strategy is not reducible to a simple social or mechanical force inducing radial repulsion. It remains to be checked whether the observed features (beyond the reach of simple social force models) can be captured by some of more elaborate lines of modelling, which provide a more realistic account of anticipating effects, in particular descriptions based on anticipated times to collision [7].

Acknowledgements The authors acknowledges partial funding from PALM (ANR-10-LABX-0039-PALM) through the project PERCEFOULE.

References

1. M. Bierlaire, T. Robin, in *Pedestrian Behavior: Models, Data Collection and Applications* (Emerald Group Publishing Limited, 2009), pp. 1–26
2. G. Antonini, M. Bierlaire, M. Weber, Transp. Res. Part B Methodol. **40**(8), 667 (2006)
3. D. Helbing, P. Molnar, Phys. Rev. E **51**(5), 4282 (1995)
4. T. Barois, J.F. Boudet, N. Lanchon, J.S. Lintuvuori, H. Kellay, Phys. Rev. E **99**(5), 052605 (2019)
5. G.A. Patterson, P.I. Fierens, F.S. Jimka, P. König, A. Garcimartín, I. Zuriguel, L.A. Pugnaloni, D.R. Parisi, Phys. Rev. Lett. **119**(24), 248301 (2017)
6. M. Moussaïd, D. Helbing, G. Theraulaz, Proc. Natl. Acad. Sci. **108**(17), 6884 (2011)
7. I. Karamouzas, B. Skinner, S.J. Guy, Phys. Rev. Lett. **113**(23), 238701 (2014)
8. S. Thorpe, D. Fize, C. Marlot, Nature **381**(6582), 520 (1996)
9. N. Bain, D. Bartolo, Nat. Commun. **8**, 15969 (2017)
10. I. Zuriguel, D.R. Parisi, R.C. Hidalgo, C. Lozano, A. Janda, P.A. Gago, J.P. Peralta, L.M. Ferrer, L.A. Pugnaloni, E. Clément et al., Sci. Rep. **4** (2014)
11. A. Nicolas, I. Touloupas, J. Stat. Mech. Theory Exp. **2018**(1), 013402 (2018)
12. J.M. Pastor, A. Garcimartín, P.A. Gago, J.P. Peralta, C. Martín-Gómez, L.M. Ferrer, D. Maza, D.R. Parisi, L.A. Pugnaloni, I. Zuriguel, Phys. Rev. E **92**(6), 062817 (2015)
13. A. Nicolas, S. Bouzat, M.N. Kuperman, Transp. Res. Part B Methodol. **99**, 30 (2017)
14. M. Haghani, M. Sarvi, Z. Shahhoseini, Transp. Res. Part A Policy Pract. **122**, 51 (2019)
15. C. Lozano, G. Lumay, I. Zuriguel, R. Hidalgo, A. Garcimartín, Phys. Rev. Lett. **109**(6), 068001 (2012)
16. A. Nicolas, Á. Garcimartín, I. Zuriguel, Phys. Rev. Lett. **120**(19), 198002 (2018)
17. P. Sollich, F. Lequeux, P. Hébraud, M. Cates, Phys. Rev. Lett. **78**(10), 2020 (1997)
18. A. Nicolas, S. Bouzat, M.N. Kuperman, Phys. Rev. E **94**, 022313 (2016)

19. Á. Garcimartín, D. Maza, J.M. Pastor, D.R. Parisi, C. Martín-Gómez, I. Zuriguel, New J. Phys. **20**(12), 123025 (2018)
20. A. Nicolas, M. Kuperman, S. Ibañez, S. Bouzat, C. Appert-Rolland, Sci. Rep. **9**(1), 105 (2019)
21. E. Kolb, P. Cixous, J. Charmet, Granul. Matter **16**(2), 223 (2014)
22. A. Seguin, Y. Bertho, F. Martinez, J. Crassous, P. Gondret, Phys. Rev. E **87**(1), 012201 (2013)

Chapter 52
Elongated Self-propelled Particles Roaming a Closed Arena Present Financial Stylized Facts

Germán A. Patterson and Daniel R. Parisi

Abstract We report the existence of financial stylized facts in a system of mechanical vehicles driven by vibration (VDV). The VDVs are restricted to a closed geometry that is composed of two chambers connected by an opening which allows a continuous flow of agents between the two regions. We studied the temporal evolution of the density of particles around the opening and made a statistical comparison with the price evolution of bitcoin (BTC). We found remarkable similarities between these two systems enabling us to study financial systems from a new perspective.

52.1 Introduction

Financial stylized facts are the common statistical properties that can be found in different studies of markets and instruments. Although they are qualitative properties, stylized facts are difficult to find in synthetic stochastic processes generated by artificial models [1]. Recently, it has be shown that the counterflow through a door of simulated pedestrians, with decision-making capacity, presents several of the stylized facts observed in financial markets [2]. This picture of counterflows of particles through a narrow constriction can be interpreted as a flow of two types of orders (buy and sell) that interact in a narrow price interval. In this direction, several physical analogies were proposed to describe different financial properties [3–6].

In this work, we present results of experiments carried out with vehicles driven by vibration [7–9]. The system under study comprises two closed regions linked by a

G. A. Patterson (✉) · D. R. Parisi
Instituto Tecnológico de Buenos Aires (ITBA), CONICET, Buenos Aires, Argentina
e-mail: gpatters@itba.edu.ar

D. R. Parisi
e-mail: dparisi@itba.edu.ar

I. Zuriguel et al. (eds.), *Traffic and Granular Flow 2019*,
Springer Proceedings in Physics 252,
https://doi.org/10.1007/978-3-030-55973-1_52

narrow opening that yields a balanced flow with a bottleneck. We studied the temporal evolution of the density of particles around the opening and made a statistical comparison with the price evolution of the Bitcoin crypto-currency price. Our findings show that the density fluctuations experimentally reproduce the main stylized facts, which allow us to address the study of financial systems from a new perspective.

52.2 Experiment

We built a closed arena comprised of two regions connected by a narrow opening of length $L = 40$ mm using acetate tape as flexible walls and wooden blocks to fix the opening size. This particular design allows us to generate a continuous flow of particles through a bottleneck. Figure 52.1a shows the experimental setup. We used 13 vibration-driven vehicles (VDVs) named Hexbug Nano [10]. The dimensions of these vehicles are 43 mm × 15 mm × 18 mm. As can be seen in Fig. 52.1a, we used a four-color label design for the tracking of the VDVs.

Figure 52.1b shows the estimated probability density function (PDF) of the positions of all agents during the entire experiment. The color scale shows that VDVs tend to be on the edges of the arena. In particular, there is a high probability of finding agents around the opening.

We recorded the experiment using a GoPro camera in a zenith position with a time resolution $1/30$ s. The recording time was ≈ 60 min.

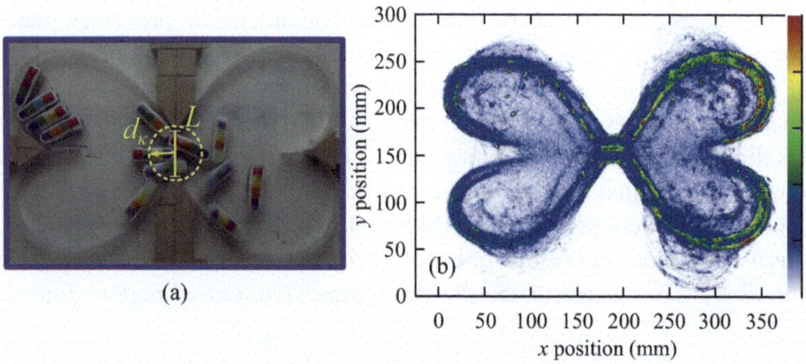

Fig. 52.1 **a** Experimental setup. Two regions are connected by a narrow opening of size $L = 40$ mm. **b** Estimated PDF of the positions of all agents. A linear color scaling represents the probability of finding an agent in a certain position

52.3 Results

We studied the evolution of the density of particles ρ around the opening. For this, we measured the distance of the κ nearest-neighbor d_κ and estimated the density as

$$\rho \propto \frac{\kappa - 1}{d_\kappa^2} \,. \tag{52.1}$$

Note that the missing proportinality constant will have no effect on the results obtained. In this analysis, we used $\kappa = 3$. The analysis of the influence of κ on the emerging stylized facts is beyond the scope of this work but, we found results to be robustly present in the interval $\kappa \in [2, 4]$. Following the analogy put forward by Parisi et al. [2], we calculated 'returns' of the time series Y over a window of size j as

$$R_Y(t_i, j) = Y(t_{i+j}) - Y(t_i) \,, \tag{52.2}$$

where, for the VDV system, Y is the density of particles given by Eq. 52.1. Figure 52.2a shows the time evolution of the density of VDVs for 100 s. Data reveal alternating periods of high and low density. The temporal evolution of R_ρ presents periods of high variability clustering in time as can be seen in Fig. 52.2b. Finally, the probability distribution funcion of R_ρ reveals a peaky distribution with fat tails.

Using the above definition of returns, we computed different statistical properties and compared them to the financial stylized facts. Specifically, we took the time series of the Bitcoin crypto-currency expressed in US dollars (BTC) ranging from 2012/12/31 to 2018/06/30 at a sampling rate of 1 h [11]. In this case, Y is the log-price of BTC.

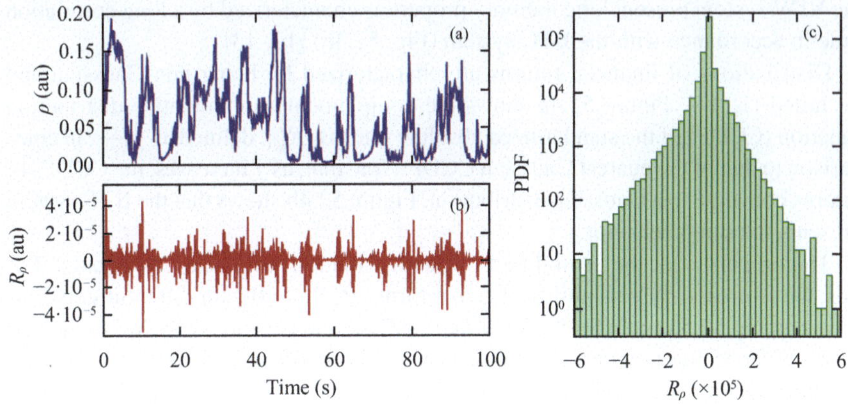

Fig. 52.2 Temporal evolution of **a** the density of particles and **b** the returns R_ρ for an arbitrary time window of 100 s. **c** Probability distribution function of R_ρ for the entire experiment

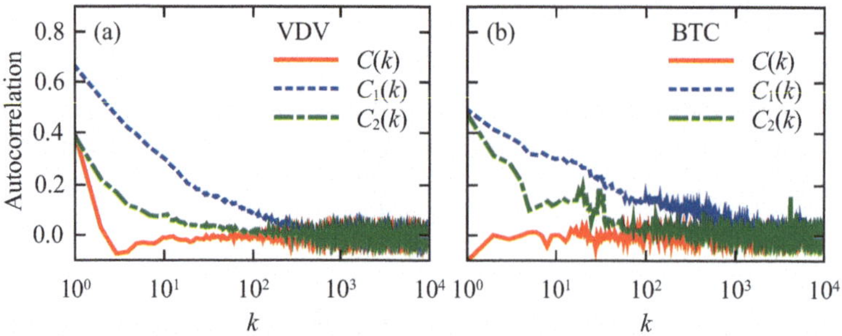

Fig. 52.3 **a** Autocorrelation functions for the variation of density R_ρ, $|R_\rho|$ and $|R_\rho|^2$ of the VDV system. **b** Same results for the BTC system

We first analyzed the correlations of several functions of R_Y. The sample correlation function is defined as

$$C(k) = \text{corr}(R_Y(t_{i+k}), R_Y(t_i)) \,, \tag{52.3}$$

where k is the time lag. Figure 52.3a shows results for the VDV system revealing the absence of linear autocorrelation for times larger than a few lags. The BTC system presents the same behavior as can be seen in Fig. 52.3b.

The volatility clustering of the returns observed in Fig. 52.2b was quantified by the autocorrelation function of an arbitrary power of the absolute returns as

$$C_\alpha(k) = \text{corr}(|R_Y(t_{i+k})|^\alpha, |R_Y(t_i)|^\alpha) \,. \tag{52.4}$$

In contrast to $C(k)$, Fig. 52.3a shows that the autocorrelations of $C_1(k)$ and $C_2(k)$ for the VDV system present long memory properties characterized by a long correlation time in accordance with the BTC system (Fig. 52.3b) [12, 13].

Distributions of financial returns are characterized by being non-Gaussian and fat tailed [1, 14]. Figure 52.4a shows the complementary cumulative distribution function (CCDF) of the standardized absolute returns $|R^*_Y|$, defined as $\frac{|R_Y|}{\langle |R_Y| \rangle}$, in comparison to that of the nearest Gaussian CCDF. Note that, as j increases, the CCDFs of returns converge to a Gaussian distribution. Figure 52.4b shows that the BTC system presents the same behavior.

The properties of the central part of the distribution were also investigated. For this, we estimated the probability of zero returns ($P(R_Y = 0)$) for different steps j as shown in Fig.52.5a and b for the VDV and BTC systems, respectively. We found that $P(R_Y = 0)$ decays as a power law function $j^{-\delta}$ in accordance with results obtained for the S&P500 index [15].

Finally, we studied the long-range dependence of the time series of absolute returns by means of the detrended function analysis (DFA) [16, 17]. Figure 52.6 shows the root-mean-square deviation from the trend $F(n)$ as a function of n. Both, the VDV (a) and BTC (b) systems, present a linear relationship in log-log scale. We fit power

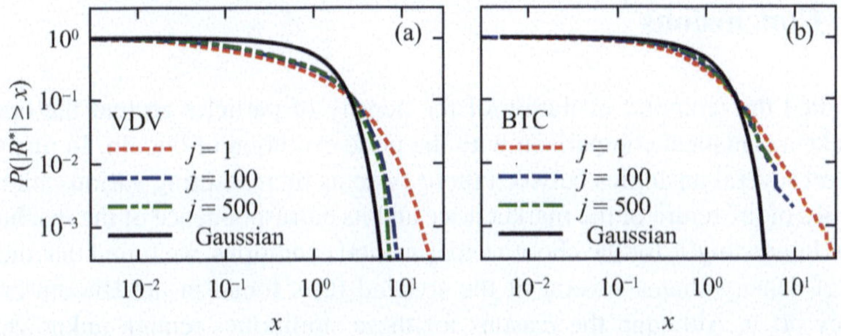

Fig. 52.4 CCDF for normalized absolute returns series R^*_ρ (**a**) and R^*_{BTC} (**b**) for various time windows j. Solid black lines stand for the CCDF of the nearest Gaussian distribution

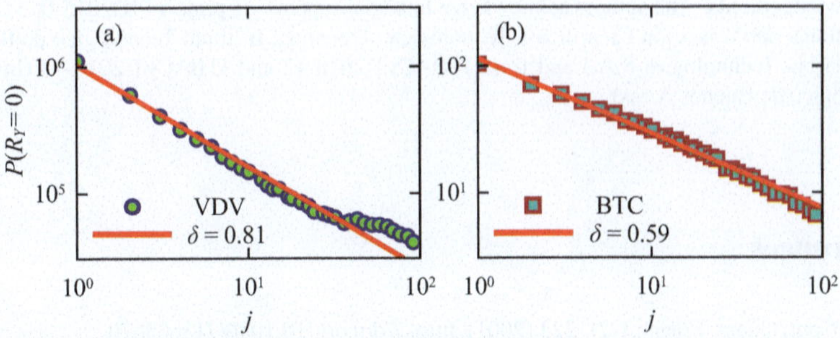

Fig. 52.5 Estimated probability of zero return as a function of the time step j for the VDV (**a**) and BTC. Red solid lines stand for power law fits

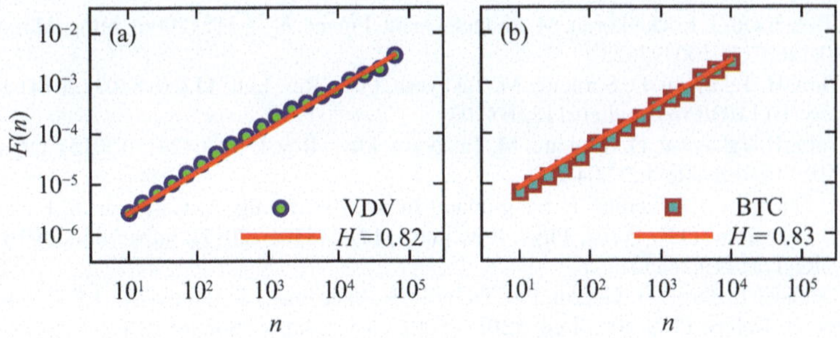

Fig. 52.6 Detrended function analysis for the VDV (**a**) and BTC (**b**) systems. Solid lines stand for power law fits

law functions and estimated Hurst exponents $H = 0.82$ and $H = 0.81$ for the VDV and BTC systems, respectively. Again, both systems are in accordance showing that periods of positive trends are followed by periods with the same trends.

52.4 Conclusions

We studied the temporal evolution of the density of particles around the opening and make a statistical comparison with the price evolution of bitcoin. In this work, we report several analogies between these systems by measuring various statistical properties of the return of the market price and its correspondence of the mechanical system. Interestingly, for the chosen experimental conditions, we found that the time series of density shares several of the stylized facts found in the Bitcoin crypto-currency price. Although the reasons for these similarities remain unknown, this simple mechanical toy-model can be used to get insights into more complex financial systems.

Acknowledgements The authors acknowledge financial support via project PID2015-003 (Agencia Nacional de Promoción Científica y Tecnológica, Argentina; Instituto Tecnológico de Buenos Aires; Urbix Technologies S.A.) and from ITBACyT-2018-42 and ITBACyT-2018-43 (Instituto Tecnológico de Buenos Aires).

References

1. R. Cont, Quant. Financ.**1**(2), 223 (2001). https://doi.org/10.1080/713665670
2. D.R. Parisi, D. Sornette, D. Helbing, Phys. Rev. E **87**, 012804 (2013). https://doi.org/10.1103/PhysRevE.87.012804
3. C. Vamo, N. Suciu, W. Blaj, Phys. A Stat. Mech. Appl. **287**(3), 461 (2000). https://doi.org/10.1016/S0378-4371(00)00385-X
4. J.P. Bouchaud, J. Kockelkoren, M. Potters, Quant. Financ. **6**(2), 115 (2006). https://doi.org/10.1080/14697680500397623
5. Y. Yura, H. Takayasu, D. Sornette, M. Takayasu, Phys. Rev. Lett. **112**, 098703 (2014). https://doi.org/10.1103/PhysRevLett.112.098703
6. Y. Yura, H. Takayasu, D. Sornette, M. Takayasu, Phys. Rev. E **92**, 042811 (2015). https://doi.org/10.1103/PhysRevE.92.042811
7. G.A. Patterson, P.I. Fierens, F. Sangiuliano Jimka, P.G. König, A. Garcimartín, I. Zuriguel, L.A. Pugnaloni, D.R. Parisi, Phys. Rev. Lett. **119**, 248301 (2017). https://doi.org/10.1103/PhysRevLett.119.248301
8. A. Deblais, T. Barois, T. Guerin, P.H. Delville, R. Vaudaine, J.S. Lintuvuori, J.F. Boudet, J.C. Baret, H. Kellay, Phys. Rev. Lett. **120**, 188002 (2018). https://doi.org/10.1103/PhysRevLett.120.188002
9. T. Barois, J.F. Boudet, N. Lanchon, J.S. Lintuvuori, H. Kellay, Phys. Rev. E **99**, 052605 (2019). https://doi.org/10.1103/PhysRevE.99.052605
10. HEXBUG Nano. https://www.hexbug.com/nano. Accessed 18 May 2017
11. Bitcoincharts. https://bitcoincharts.com/charts. Accessed 30 June 2018
12. R. Cont, M. Potters, J.P. Bouchaud, in *Scale Invariance and Beyond*, ed. by B. Dubrulle, F. Graner, D. Sornette (Springer, Berlin Heidelberg, Berlin, Heidelberg, 1997), pp. 75–85
13. Z. Ding, C.W.J. Granger, R.F. Engle, *Chap. A Long Memory Property of Stock Market Returns and a New Model*, (Harvard University Press, Cambridge, MA, USA, 2001), pp. 349–372
14. J.P. Bouchaud, M. Potters, *Theory of Financial Risk and Derivative Pricing: From Statistical Physics to Risk Management* (Cambridge University Press, 2003)

15. R.N. Mantegna, H.E. Stanley, Nature **376**(6535), 46 (1995)
16. C.K. Peng, S.V. Buldyrev, S. Havlin, M. Simons, H.E. Stanley, A.L. Goldberger, Phys. Rev. E **49**, 1685 (1994). https://doi.org/10.1103/PhysRevE.49.1685
17. C. Peng, S. Havlin, H.E. Stanley, A.L. Goldberger, Chaos Interdiscip. J. Nonlinear Sci. **5**(1), 82 (1995). https://doi.org/10.1063/1.166141

Chapter 53
Set Voronoi Tessellation for Particulate Systems in Two Dimensions

Simeon Völkel and Kai Huang

Abstract Given a countable set of points in a continuous space, Voronoi tessellation is an intuitive way of partitioning the space according to the distance to the individual points. As a powerful approach to obtain structural information, it has a long history and widespread applications in diverse disciplines, from astronomy to urban planning. For particulate systems in real life, such as a pile of sand or a crowd of pedestrians, the realization of Voronoi tessellation needs to be modified to accommodate the fact that the particles cannot be simply treated as points. Here, we elucidate the use of Set Voronoi tessellation (i. e., considering for a non-spherical particle a *set of points* on its surface) to extract meaningful local information in a quasi-two-dimensional system of granular rods. In addition, we illustrate how it can be applied to arbitrarily shaped particles such as an assembly of honey bees or pedestrians for obtaining structural information. Details on the implementation of this algorithm with the strategy of balancing computational cost and accuracy are discussed. Furthermore, we provide our python code as open source in order to facilitate Set Voronoi calculations in two dimensions for arbitrarily shaped objects.

53.1 Introduction

Particulate systems are ubiquitous in nature, industry and our daily lives, ranging from active ones like pedestrians or animals, as in Fig. 53.1, to passive ones such as a thermal granules.

S. Völkel (✉) · K. Huang
Experimental Physics V, University of Bayreuth, Universitätsstraße 30, 95447 Bayreuth, Germany
e-mail: simeon.voelkel@uni-bayreuth.de

K. Huang
e-mail: kai.huang186@dukekunshan.edu.cn

K. Huang
Division of Natural and Applied Sciences, Duke Kunshan University, No. 8 Duke Avenue, Kunshan 215316, Jiangsu, China

I. Zuriguel et al. (eds.), *Traffic and Granular Flow 2019*, Springer Proceedings in Physics 252,
https://doi.org/10.1007/978-3-030-55973-1_53

Fig. 53.1 Carniolan honey bees on a partially sealed honeycomb. The Set Voronoi tessellation (black) is based on multiple points (white) per bee

They all share the characteristics that the macroscopic behavior depends on the arrangement of individual particles relative to each other. Prominent collective effects regarding the dynamics of granular materials, such as the solid-liquid-like transition [1–8] and pattern formation [9–11], are typically triggered by the mobility of individual particles. Moreover, there also exists evidence showing that the strain field associated with local rearrangement of particles can be used to obtain the local stress field, which in turn can provide indispensable hints on the establishment of local force networks during jamming transition [12–14]. Investigations on granular systems thus often rely on accurate measurements of properties such as local volume fraction, neighborhood, etc. For defining these quantities, it is expedient to attribute to every particle a portion of the available space, that "belongs to" or "is occupied" by a single particle.

53.2 Limitations of the Classical Voronoi Tessellation

The honeycomb in Fig. 53.1 can be seen as an example of a naturally formed tessellation, optimized to have a fair distribution of space for individual larvae of bees to grow inside. Motivated by such self-organized processes in nature, the concept of spatial tessellation was established long time ago [15–18] in order to analyze structures of various systems in diverse disciplines, from the structure of the universe in astronomy [15] through positioning public schools or post offices in urban planning [19] to characterizing topological aspects of molecular structures [20–22].

A generic, parameterless approach to attribute space based on a single point per object (typically its center) is the Voronoi tessellation (VT), also known as Dirichlet tessellation, Thiessen polygons or Wigner-Seitz cells [16, 17, 23, 24].

For mono-disperse spherical particles (as well as point-like particles), classical VT based on the particle centers is applicable and delivers intuitive results (see Fig. 53.2a). This, however, is not the case for polydisperse packings (see Fig. 53.2b) or non-spherical particles (see Fig. 53.2c). Here, part of the particle on the left lies outside of the Voronoi cell, contributing to an obvious source of error. To overcome this, various weighted Voronoi diagrams have been proposed (see, e. g., [18] for an overview) and implemented in physics and material science [25, 26].

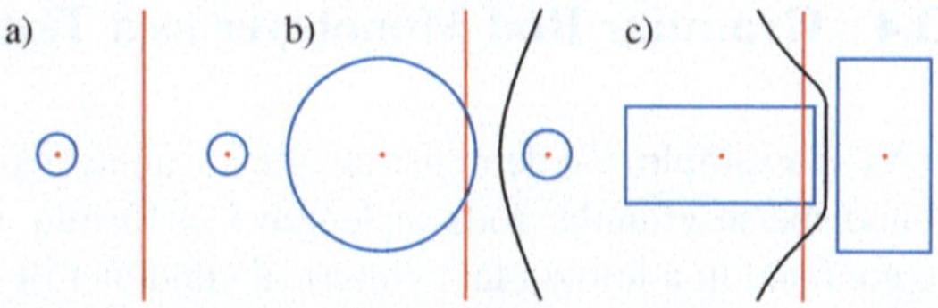

Fig. 53.2 Comparison of classical VT based on the center (red) of particles and Set VT (black) considering the border of the two particles (blue)

Problems due to applying classical VT to a glass composed of *differently sized* (but spherical) atoms have already been discussed [26]. Here, we highlight the problems arising from applying classical VT to equally sized, but *elongated* particles.

53.3 Set Voronoi Algorithm and Implementations

A generic way to attribute space to both differently sized particles and non-spherical particles is to tessellate space according to the distance to the closest *surface* (as opposed to center). This can be seen as the limiting case of a classical VT, when considering for each particle the *set of points* marking its surface. Therefore it is also referred to as *Set Voronoi* diagram [27]. The resulting, intuitive cell-borders of the examples in Fig. 53.2 are depicted in black. Note that the Set Voronoi border for nonoverlapping monodisperse spheres (or circular discs in two dimensions) like in Fig. 53.2a coincides with the classical Voronoi cell border. Voronoi diagrams with lines and arcs as generators have been studied systematically since the late 1970s [18]. Nevertheless, their importance in the realm of granular physics was rarely realized until the past decade [27–30].

Recently, Weis and Schönhöfer [31] provided a program based on the Voro++ library [32] to calculate the Set VT, targeting reconstructed CT-scans of three-dimensional (3D) packings of non-spherical granular particles [30]. However, an urgent demand for a solution in two dimensions remained, as it is computationally not economic to apply the 3D algorithm directly to two-dimensional (2D) systems [33]: Adding dimensions to the problem inevitably increases the computational cost for obtaining the tessellation, letting aside the efforts now required to treat the borders in the added dimension(s). Considering as well the fact that many particulate problems of practical interest (e. g., dynamics of pedestrians, flocks or monolayers of granulates) can be treated (quasi-) two-dimensionally and very often data is acquired using 2D imaging techniques, we implemented the Set VT strategy based on discrete points put forward by Schaller et al. [27] in 2D. Our implementation uses the python scripting language together with scipy/numpy libraries [34], relying on VT routines provided by the Qhull library [35]. It is available as free open source software [36].

53.4 Granular Rod Monolayer as a Test Case

As an example, we demonstrate the advantages of Set VT using a monolayer of monodisperse granular rods of length $l = 15$ mm and diameter $d = 3$ mm. They are confined in a horizontal cylindrical container of diameter $D = 19$ cm, which is subjected to sinusoidal vibrations against gravity with oscillation frequency f and dimensionless acceleration Γ as two control parameters to keep the rods mobilized. More details on the experimental set-up and image analysis procedure can be found in [37]. Depending on the packing density, the rods may organize themselves into an uniaxial nematic state with two-fold rotational symmetry or tetratic state with 4-fold symmetry. For analyzing the disorder-order transition quantitatively, an accurate determination of the local packing density is desired. For amorphous media, such as a random packing of granular particles, (Set) VT provides a direct route to the local area or volume fractions of individual grains [27, 38, 39].

Figure 53.3 compares the outcome of classical and Set VT based on a snapshot reconstructed from the positions and orientations obtained experimentally in [37]. This particular snapshot is chosen, as it consists of both dense and dilute regions. The classical VT does not include walls and leads to cells cutting the walls of a container (red solid lines crossing the gray area in Fig. 53.3a) or even extending to infinity (red dashed lines in Fig. 53.3a). These problems can be easily and consistently avoided using Set VT, as the container can be included as an additional 'particle', naturally limiting the cells of all contained particles (see Fig. 53.3b). Additionally Set VT delivers a much more reasonable tiling in the sense that no particle cuts its cell's border, as the container lid prohibits the 'hard' rods from overlapping.

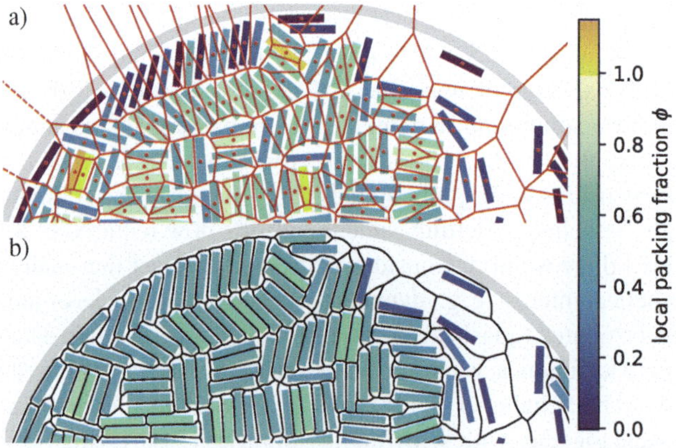

Fig. 53.3 VT (**a**) and Set VT (**b**) of partly ordered granular rods in quasi-two-dimensions. The parameters are: particle number $N = 400$, driving frequency $f = 50$ Hz and dimensionless acceleration $\Gamma = 6.26$. The color code indicates the local packing fraction ϕ according to the different tessellations: classical VT considering the particle centers (red, upper panel) versus Set VT (black, lower panel) including the container rim as additional particle

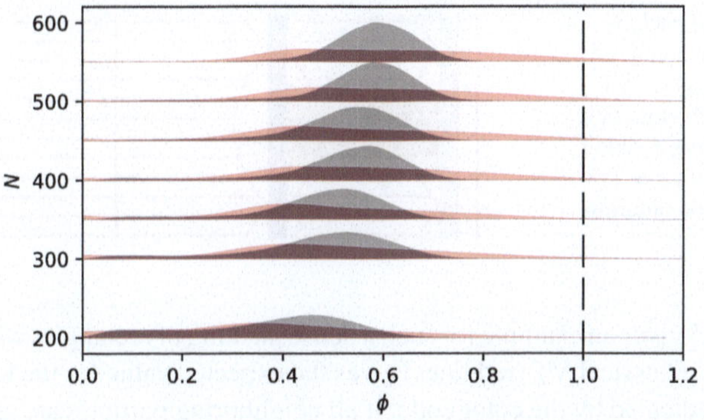

Fig. 53.4 Probability density P_ϕ of the local packing fraction ϕ for different numbers of particles N in the container, obtained from kernel density estimation. Light red and dark gray curves correspond to classical and Set VT, respectively. The peak acceleration is kept in the range $\Gamma \in [1.94; 20.59]$ and $f = 50$ Hz is fixed. Other experimental parameters are the same as in Fig. 53.3. The reflection method [41, 42] has been used to correct for ϕ being non-negative by definition, and the bandwidth of the Gaussian kernel is chosen according to Scotts rule [43] as $\sigma \cdot n^{-1/5}$ proportional to the standard deviation σ of the distribution and dependent on the number of rods n (here 1.7×10^5 to 5.3×10^5 per stacked plot). Captured frames with detection problems (particles (partially) outside of the container, overlapping or of zero size) were skipped to avoid bias

Quantitatively, the local packing density $\phi = A_p/A_c$ derived from the projected area of the particles $A_p = l \cdot d$ and the area of the corresponding cell A_c obtained from the space tiling also demonstrates the difference clearly: Cells with 'impossible' $\phi > 1$ (see color code) disappear and large fluctuations of ϕ for rods aligned with each other in a similar local configuration diminish as Set VT is implemented.

Figure 53.4 compares the local packing density distribution obtained from VT (red) and Set VT (black) for different global packing densities. For Set VT, the probability density P_ϕ shows a clear trend of an increasing local packing density with the global one. This feature is much less obvious for the classical VT. There, the most striking observation is the tail of the distribution towards large ϕ becoming more prominent. The dashed line at $\phi = 1$ marks the upper bound for hard particles and a reasonable tiling. All red curves clearly exceed this limit. This manifests again the importance of observing the applicability of each tessellation technique.

53.5 Why Set Voronoi Is Essential for Elongated Particles

To analyze quantitatively the maximal error introduced by applying the classical VT to elongated particles, we consider a perfect, dense packing of identical "hard" rectangles. With no space left between the non-overlapping rectangles, a constant local packing fraction equal to the global packing fraction of unity is expected.

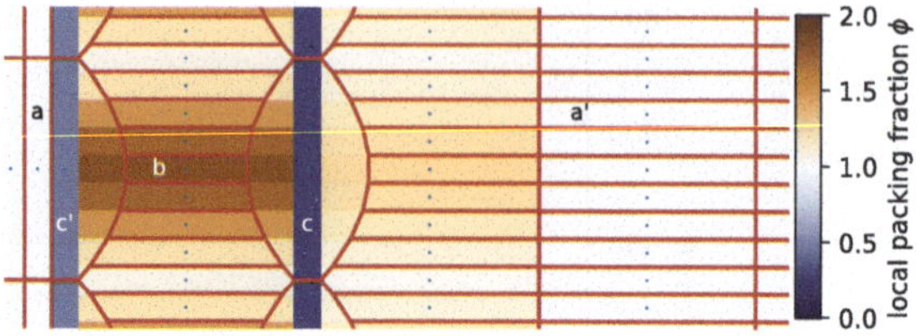

Fig. 53.5 Local packing fraction (see color code) according to classical VT (red lines, based on particle centers) in a dense packing of rectangles with $\varepsilon = 1/8$ exhibiting tetratic ordering

Figure 53.5 gives an example for such a dense packing of rectangles with width-to-length ratio ε. Classical VT (red lines) gives the expected value for the local packing fraction, as indicated by the color code, if all neighboring particles are aligned in the same direction (cf. particles a and a' in Fig. 53.5) but fluctuates significantly where rectangles of different orientation come together. Here, local packing fractions up to $\phi_{\text{VT,max}} = 2/(1+\varepsilon)$ are obtained, e.g., for particle b in Fig. 53.5. This overestimation is thus a first order effect in terms of ε. It can reach a remarkable error of 100% for vanishing width-to-length ratio, even when letting aside any experimental inaccuracies or elasticity of the particles. At the same time, the local packing fraction for neighboring particles c and c' in Fig. 53.5 is underestimated, up to an equally dramatic extent as $\varepsilon \to 0$. Even for a *dense* packing of hard particles (global packing fraction of unity) and without boundary effects, classical VT can indicate *vanishing* local packing fraction, deviating disastrously from the expected value!

On the contrary, Set VT gives consistent local packing fractions. It delivers a value of $\phi_{\text{Set VT}} = 1$ for every rectangle in Fig. 53.5 within the numerical uncertainty arising from the approximation of the surface through a finite number of points.

53.6 Performance Versus quality Trade-Off via Erosion

For closely spaced particles special care is required to identify their surface properly before applying the Set VT. For rounded particles with a finite minimal curvature radius r_c, Schaller et al. [27] show that it is highly beneficial to consider the maximally eroded surfaces, having the constant minimal distance r_c to the original ones: Using the eroded surface, dramatically fewer points suffice for obtaining the same Set VT and accuracy. In addition, erosion can resolve slight particle overlap. This is particularly helpful for densely packed particles close to jamming [44] and is expected to make investigations of deformable particles feasible [27].

For the granular rods discussed above, the rectangles seen by the camera have sharp corners. In other words, $r_c = 0$, preventing a 'lossless' erosion. Accurately capturing the sharp corners commands a vanishing erosion depth, which in turn dictates a very close spacing of the discretization points on the eroded surface. As describing a surface with a larger number of points inevitably increases the computational costs, a compromise between accuracy and speed has to be found. Here, we erode the rods by one pixel to remove slight overlap due to finite experimental resolution, detection accuracy and finite elasticity of the particles. Setting the maximum distance between discretization points to the erosion depth delivers satisfactory results and we recommend this as a rule of thumb.

For many systems, however, the situation is more pleasant, especially if the most important property is the elongation, while the dimensions along the other direction(s) are equal and the exact shape of each particle only plays a tangential role, like in the case of the bees shown in Fig. 53.1. In such a case, it is typically acceptable to approximate the particle as spherocylinder. This approximation paves the way to a very efficient representation of the particle when using Set VT: The maximally eroded 'surface' of each particle is then just its medial axis, a one-dimensional line segment. Furthermore, according to the rule of thumb, one discretization point per radius of the spherocylinder r suffices for satisfactory results even at high packing densities. For the bees on the honeycomb, this translates to a single digit number of discretization points, each depicted as a white dot in Fig. 53.1. Nevertheless, this is sufficient for resolving the most prominent and important effects due to the elongated shape of the individual bee, as the black Set Voronoi cell borders illustrate.

As similarly sized rod-shaped particles represent a diverse class of systems, from liquid crystal molecules at a microscopic scale to pedestrians viewed from the top (i.e., when taking their shoulders into account), the above analysis demonstrates that it is essential to employ Set VT for elongated particles.

53.7 Conclusion

Using dynamics of a granular rod monolayer as an example, we demonstrate that Set VT provides a more meaningful tiling of space in comparison to the classical VT that relies on the center of particles. From polydisperse systems to irregularly shaped or even deformable particles, the Set VT algorithm is expected to be substantially more consistent in characterizing the geometric and topological features of particulate systems, many of which can be approximated as elongated particles.

Note that in addition to obtaining the local packing density, VT can also be used to extract other order parameters, such as determining neighbors of individual particles. The Delaunay triangulation, which connects particle centers to their neighbors, follows the natural definition that neighbors share a part of a Voronoi cell border and is a typical approach after classical VT. The extension of this definition to Set VT is straightforward and can be used in further characterizations, for instance, using the bond orientational order parameter [5]. How to extract more meaningful information from Set VT in addition to the local packing density and the improvements against the classical VT will be a focus of future investigations.

Acknowledgements We gratefully acknowledge helpful discussions with Simon Weis, Matthias Schröter, Ingo Rehberg, Stefan Hartung and Wolfgang Schöpf. We thank Alexander and Valentin Dichtl for the possibility of taking the picture of the bees in Fig. 53.1 during an inspection of the hive. This work is supported by the German Research Foundation (DFG) under Grant No. HU1939/4-1.

References

1. K. Huang, C. Krülle, I. Rehberg, J. Appl. Math. Mech. **90**(12), 911 (2010). https://doi.org/10.1002/zamm.201000110
2. K. Huang et al., Phys. Rev. E **79**(1), 010301 (2009). https://doi.org/10.1103/PhysRevE.79.010301
3. K. Huang, K. Röller, S. Herminghaus, Eur. Phys. J. Special Top. **179**(1), 25 (2009). https://doi.org/10.1140/epjst/e2010-01191-5
4. B. Andreotti, Y. Forterre, O. Pouliquen, *Granular Media: Between Fluid and Solid* (Cambridge University Press, Cambridge, 2013)
5. C. May et al., Phys. Rev. E **88**(6), 062201 (2013). https://doi.org/10.1103/PhysRevE.88.062201
6. S.C. Zhao, M. Schröter, Soft Matter **10**(23), 4208 (2014). https://doi.org/10.1039/C3SM53176G
7. P. Ramming, K. Huang, EPJ Web Conf. **140**, 08003 (2017). https://doi.org/10.1051/epjconf/201714008003
8. M. Baur, K. Huang, Phys. Rev. E **95**(3), 030901 (2017). https://doi.org/10.1103/PhysRevE.95.030901
9. G.H. Ristow, *Pattern Formation in Granular Materials* (Springer, 2000)
10. A. Fortini, K. Huang, Phys. Rev. E **91**(3), 032206 (2015). https://doi.org/10.1103/PhysRevE.91.032206
11. A. Zippelius, K. Huang, Sci. Rep. **7**, 1 (2017). https://doi.org/10.1038/s41598-017-03844-0
12. A.J. Liu, S.R. Nagel, Nature **396**(6706), 21 (1998). https://doi.org/10.1038/23819
13. D. Bi et al., Nature **480**(7377), 355 (2011). https://doi.org/10.1038/nature10667
14. Y. Zhao et al., Phys. Rev. Lett. **123**, 158001 (2019). https://doi.org/10.1103/PhysRevLett.123.158001
15. R. Descartes, *Principles of Philosophy* (Amsterdam, 1644) (3-7873-1697-3)
16. G. Dirichlet, Journal fur die reine und angewandte Mathematik (40), 209 (1850)
17. G. Voronoi, Journal fur die reine und angewandte Mathematik (133), 97 (1907)
18. A. Okabe et al., *Spatial Tessellations: Concepts and Applications of Voronoi Diagrams*, 2nd edn. (Wiley, 2000). https://doi.org/10.1002/9780470317013
19. M. McAllister, D. Kirkpatrick, J. Snoeyink, Discrete Comput. Geom. **15**(1), 73 (1996)
20. A.L. Mackay, J. Mol. Struct. **336**(2), 293 (1995). https://doi.org/10.1016/0166-1280(95)04172-3
21. F. Aurenhammer, A.C.M. Comput. Surv. **23**(3), 345 (1991). https://doi.org/10.1145/116873.116880
22. J. Bernauer et al., Bioinformatics **24**(5), 652 (2008). https://doi.org/10.1093/bioinformatics/btn022
23. A.H. Thiessen, Mon. Weather Rev. **39**(7), 1082 (1911). https://doi.org/10.1175/1520-0493(1911)39<1082b:PAFLA>2.0.CO;2
24. E. Wigner, F. Seitz, Phys. Rev. **43**, 804 (1933). https://doi.org/10.1103/PhysRev.43.804
25. J. Park, Y. Shibutani, Intermetallics **15**(2), 187 (2007). https://doi.org/10.1016/j.intermet.2006.05.005
26. J. Park, Y. Shibutani, Intermetallics **23**, 91 (2012). https://doi.org/10.1016/j.intermet.2011.12.019
27. F.M. Schaller et al., Philos. Mag. **93**(31–33), 3993 (2013). https://doi.org/10.1080/14786435.2013.834389
28. A. Baule et al., Nat. Commun. **4**(1), 2194 (2013). https://doi.org/10.1038/ncomms3194
29. A. Baule, H.A. Makse, Soft Matter **10**(25), 4423 (2014). https://doi.org/10.1039/C3SM52783B
30. S. Weis et al., EPJ Web Conf. **140**, 06007 (2017). https://doi.org/10.1051/epjconf/201714006007
31. S. Weis, P. Schönhöfer, Pomelo: calculate generic set Voronoi diagrams with C++11 (2017). http://theorie1.physik.fau.de/research/pomelo/index.html
32. C. Rycroft, Voro++: a three-dimensional Voronoi cell library in C++. Tech. Rep. LBNL-1432E, 946741 (2009). https://doi.org/10.2172/946741

33. S. Weis, Private communication
34. P. Virtanen et al., (SciPy 1.0 Contributors). (2019). https://urldefense.proofpoint.com/v2/url?u=https-3A__dx.doi.org_10.1038_s41592-2D019-2D0686-2D2&d=DwIDAw&c=vh6FgFnduejNhPPD0fl_yRaSfZy8CWbWnIf4XJhSqx8&r=Rzdb4lgIEtBrxhH5PwQ-9aXsz3dDphMvRIUvnf30PH8DcdqQZMTYhV81QB_xhaSP&m=XWbU6CtLdAQS84WwBUSRvhZgHH58zBe8va1uian1EGM&s=WaJsHPfQKKp7DKs_HkMBwKOEtoRXS51GBKjyhXDW9xo&e=
35. C.B. Barber et al., ACM Trans. Math. Softw. **22**(4), 469 (1996). https://doi.org/10.1145/235815.235821
36. S. Völkel, Setvoronoi2d: calculate the Set Voronoi tessellation in two dimensions. https://doi.org/10.5281/zenodo.3531546 (online)
37. T. Müller et al., Phys. Rev. E **91**(6), 062207 (2015). https://doi.org/10.1103/PhysRevE.91.062207
38. C. Song, P. Wang, H.A. Makse, Nature **453**(7195), 629 (2008). https://doi.org/10.1038/nature06981
39. S. Torquato, F.H. Stillinger, Rev. Mod. Phys. **82**(3), 2633 (2010). https://doi.org/10.1103/RevModPhys.82.2633
40. O. Tange, *GNU Parallel 2018* (Ole Tange, 2018). https://doi.org/10.5281/zenodo.1146014
41. B. Silverman, *Density Estimation for Statistics and Data Analysis*. Monographs on Statistics & Applied Probability (Chapman & Hall, 1986)
42. L.I. Boneva, D. Kendall, I. Stefanov, J. Royal Stat. Soc. B **33**(1), 1 (1971). https://doi.org/10.1111/j.2517-6161.1971.tb00855.x
43. D.W. Scott, *Multivariate density estimation: theory, practice, and visualization*, 2nd edn. (Wiley, Hoboken NJ, 2015). https://doi.org/10.1002/9781118575574
44. I. Zuriguel et al., Sci. Rep. **4**, 7324 (2014). https://doi.org/10.1038/srep07324

Part III
Cities, Vehicular Traffic and Other Transportation Systems

Chapter 54
A Geostatistical Approach to Traffic Flow Reconstruction from Sparse Floating-Car Data

Eduardo del Arco, Mihaela I. Chidean, Inmaculada Mora-Jiménez, Samer H. Hamdar, and Antonio J. Caamaño

Abstract In this work, we consider that the underlying nature of traffic flow is a random field, and the contributions of individual vehicles, the Floating-Car Data, are single realization of those random fields. We have found that Stop-and-Go waves can be characterized by the anisotropic variogram. This function can be used to reconstruct the traffic random field with better performance than the average approach. We have tested the geostatistical interpolations (Triangular Irregular Network and Kriging) with synthetic and real-world data, scoring its performance by means of a cross-validation technique. In the future, this approach may be useful in reducing uncertainty in the estimation of travel times and even decreasing the occurrence of road capacity breakdowns.

54.1 Introduction and Problem Statement

In road traffic, observable phenomena are the product of many interacting physical, sociological and psychological processes. Most of these can be characterized and reproduced using models of road traffic properly calibrated [1]. Except in trivial cases, the interaction among processes is so complex that some physically determined

E. del Arco (✉) · M. I. Chidean · I. Mora-Jiménez · A. J. Caamaño
Rey Juan Carlos University, Móstoles, Spain
e-mail: eduardo.delarco@urjc.es

M. I. Chidean
e-mail: mihaela.chidean@urjc.es

I. Mora-Jiménez
e-mail: inmaculada.mora@urjc.es

A. J. Caamaño
e-mail: antonio.caamano@urjc.es

S. H. Hamdar
The George Washington University, Washington, D.C., USA
e-mail: hamdar@gwu.edu

I. Zuriguel et al. (eds.), *Traffic and Granular Flow 2019*,
Springer Proceedings in Physics 252,
https://doi.org/10.1007/978-3-030-55973-1_54

phenomena appear to be completely random. In this line, we propose to model the road traffic flow as a random field of velocities, and reconstruct it from a sparse subset of vehicles embedded in such a flow, or Floating-Car Data (FCD). The reconstruction will be done by means of interpolations that historically where developed in the context of geostatistics. The main advantage of this approach, compared to using a phenomenological traffic model, e.g. Intelligent Driver Model [2], is that it provides a velocity map with physical characteristics of real-world traffic directly from the time-space structure of the data. An inspiring precedent of our approach is [3], where the traffic patterns are reconstructed from aggregated data of stationary traffic detectors. Our work differs in two fundamental aspects: (i) No *a priori* assumptions are made in the propagation speed of the disturbances, and (ii) the use of data provided by a sparse subset of the vehicles (FCD).

We consider the microscopic flow of traffic as a random field of velocities V over a spatial domain S and temporal domain T, $\Omega = T \times S$. The value of the velocity v at any space and time pair $\mathbf{x} = (t, s)$ is $v(\mathbf{x})$, a single realization of the random variable $V(\mathbf{x})$. The set of all random values at Ω is a random process $V(\mathbf{x})$. For shake of simplicity, we call it the Space-Time-Velocity Field (*STV*-field). The sampling of the STV-field is performed by a subset of the vehicles involved in such a flow, the Sensor Vehicles (SV). This set provides realizations of the STV-field or samples among space and time, the Floating-Car Data. The relation between this sub-flow and the rest of the traffic (Passive Vehicles or PV) for given time-space region Ω is the Fraction of Sensor Vehicles or FSV. So, $FSV = {}^{|SV|}/_{|SV|+|PV|}$, where $|.|$ denotes the cardinality operator of a set.

54.2 Interpolation of STV-Field: Method and Assumptions

The main assumption of this work is the stationary nature of the STV-field. This assumption is valid as long as the spatial extension of the random field is conveniently bounded, e.g. a road segment, a single junction or a single roundabout. The duration of the stationarity is limited, and given by the persistence of the same traffic regime. Thus, for any coordinate $\mathbf{x} \in \Omega$, the value of the STV-field is given by the sum of the expected value of the random process plus a deviation. Under the assumption of intrinsic stationarity [4], the deviation (or fluctuation) from the expected value is characterized by the *semivariance*. It captures the dissimilarity of velocity among space and time pairs, whose relative position is given by the vector $\mathbf{h} \in \Omega$, usually termed *lag*. That is, $\gamma(\mathbf{h})$ does not depend on particular coordinates, but on a time-space oriented interval, and is defined as

$$\gamma(\mathbf{h}) = \frac{1}{2}\mathsf{VAR}\left[V(\mathbf{x}+\mathbf{h}) - V(\mathbf{x})\right] = \frac{1}{2}\mathsf{E}\left[\left(V(\mathbf{x}+\mathbf{h}) - V(\mathbf{x})\right)^2\right] \tag{54.1}$$

The function $\gamma(\mathbf{h})$ for all possible $\mathbf{x}$ and $\mathbf{h}$, is the *theoretical variogram*. This function captures the time-space structure of the field, and is usually considered in geostatistical methods as the Kriging interpolation approach to reconstruct the STV-field. According to Kriging, the estimation of STV-field value at any point $\mathbf{x}_0$ is defined by the weighted sum of local samples $\{v(\mathbf{x}_i)\}_{i=1}^{M}$ provided by SV, in the vicinity of $\mathbf{x}_0$

$$\hat{V}(\mathbf{x}_0) = \sum_{i=1}^{M} w_i v(\mathbf{x}_i) \tag{54.2}$$

where weights $\{w_i\}_{i=1}^{M}$ are found by solving the Kriging System (details provided in [4]). The theoretical variogram requires the perfect knowledge of γ for all $\mathbf{x}$ and $\mathbf{h}$, but that is not possible. Thus, we take from geostatistics the estimation of the semivariance, for a fixed lag module h and orientation θ, i.e. $\mathbf{h} = h_{\angle\theta}$, here adapted to the STV-field:

$$\hat{\gamma}(h, \theta) = \frac{1}{2|N(h \pm \delta, \theta \pm \tau)|} \sum_{(i,j)\in N(h\pm\delta,\theta\pm\tau)} \left[v(\mathbf{x}_i) - v(\mathbf{x}_j)\right]^2 \tag{54.3}$$

where $N(h, \theta)$ is the set of points encompassed by the given h and θ, whilst δ and τ are required to deal with the sparsity of the gathered data.

The physical interpretation of the angle θ is key: from vertical (space dimension), angles around $\theta = 0°$ involves the consideration of velocity pairs with a purely spatial lag over the length of the road, independently of the time. Angles around 90° involves the consideration of velocity pairs for a fixed time lag, independently of the road sections. If the variation of a random field is the same in all its directions, then it is isotropic and only depends on h. However in traffic flow, beyond the instantaneous velocities of the vehicles, there are other observable phenomena characterized by their speed, i.e.: field lines and propagation of the Stop-and-Go Waves (SGW). In geostatistics, the variation in the characteristics of the random field associated to the direction is termed *anisotropy*. We have found that, in a congested traffic regime, the angle linked to the variogram with maximum continuity θ_{max}, is related to the propagation speed of the SGWs v_{sgw}, given by $v_{sgw} = \cot\theta_{max}$.

Note that solving the Kriging System requires the evaluation of semivariances in arbitrary lags. Therefore, it is necessary to fit a variogram model allowing the regression of one theoretical variogram for each angle. Once they are obtained, an ellipse-model is fitted to find the directions (velocities) of propagation of traffic characteristics. By means of the ellipse parameters, we construct an orthonormal basis by reduced distance technique and rotation (invertible affine transformation) [5] to isotropize the STV-field. The underlying idea of the anisotropy correction is similar to that used in the congestion filter [3], with no *a priori* assumptions of v_{sgw} nor the traffic regime. This kind of transformation enables the Kriging reconstruction to be used with isotropic (geometrically corrected) variogram models, e.g.: Gaussian and spherical models. In addition, the invertible change of basis is also valid with methods of reconstruction other than Kriging.

54.3 Experiments and Results

We have considered two data sets, a synthetic data set from a microscopic simulation provided by the Intelligent Driver Model [2], and real vehicle tracking data from I-80 highway [6]. Both sets are composed by samples of velocity, position and time of all the involved vehicles. The considered reconstruction methods are:

1. Triangular Irregular Network (TIN). This is a parameter-free method using a Delaunay mesh of irregular triangles. It has been already tested successfully with vehicle data [7].
2. Isotropic Kriging. It deliberately ignores the time-space correlation structure, imposing an isotropic variogram. The gaussian (KIG) and spherical (KIS) variogram models are tested.
3. TIN with anisotropy correction. We use a linear transformation in order to isotropize the space-time correlation structures. The gaussian (TIN Gauss) and spherical (TIN Sph) variogram models are tested.
4. Kriging with anisotropy correction. The gaussian (KAG) and spherical (KAS) variogram models are tested.

We choose the refined version of Willmott's dimensionless index of agreement [8] as a figure of merit (d_r). The reconstruction methods are compared against the arithmetic mean of the samples, labeled MEAN. By definition of d_r, if the mean of the predictions is equal to the mean of the observations, then $d_r = 0.5$, being the maximum score for this method. In the real world, it is only possible to access the data provided by the SV set. Therefore, the evaluation method uses only these data, and the performance of the different methods is evaluated following a k-fold cross-validation approach [9]. A fair method of evaluation is to divide the samples provided by the SV set in long fragments of sequential samples or full trajectories. In this way, we mitigate the (deceptively beneficial) effect of a vehicle as a predictor of its own movement.

First, we show qualitative results of the different interpolator using the IDM data, for an FSV of 12.5% (Fig. 54.1). Note that isotropic methods (Fig. 54.1b, c) do not characterize the propagation speed of the SGW. Once the anisotropic correction is performed, the propagation waves are reconstructed correctly, obtaining the best results with the spherical variogram model (Fig. 54.1f). Note also that the TIN method does not interpolate coordinates out of the convex-hull, and there are some triangular artifacts in the borders.

Regarding the experiments with I-80, we have taken a space-time fragment where congestion waves are easily distinguishable (see Fig. 54.2a, black rectangle). To provide an statistical analysis about the performance of the interpolators, we perform 100 random selections of SV sets. In Fig. 54.2 we show boxplots of the Willmott's d_r considering three different FSV. The dispersion of d_r shows the differences between realizations: the lower the FSV, the greater the dispersion. When using $FSV = 10\%$, it seems that Kriging with anisotropy correction shows the best average performance. When increasing the FSV, the methods with geometric correction of anisotropy

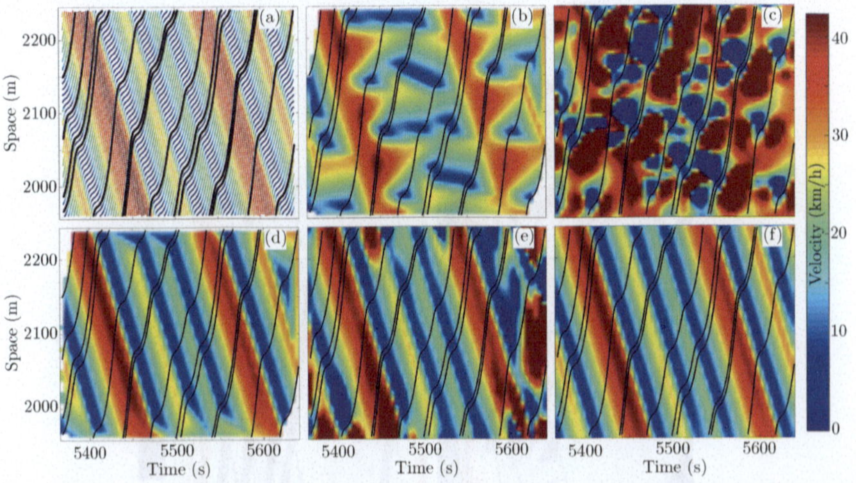

Fig. 54.1 **a** Ground truth and SV trajectories in black lines, with $FSV = 12.5\%$; Results for **b** TIN; **c** KIG; **d** TIN Sph; **e** KAG; and **f** KAS interp. methods

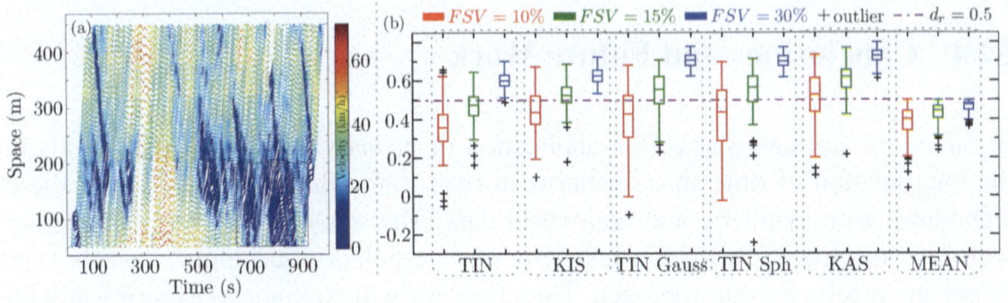

Fig. 54.2 **a** I-80 ground truth. The studied part is highlighted; **b** Willmott's d_r. Results of tested methods for different FSV

(TIN Gauss, TIN Sph and KAS) show better performance than the isotropic ones. Any interpolation method obtaining a score greater than 0.5 is better than using the arithmetic mean of the observations (MEAN method).

The benefits of anisotropy correction are very evident in Fig. 54.3. We show a single realization with $FSV = 12.5\%$ and sensor vehicles distributed very unevenly (note the period between 750 and 800 seconds with no sensor vehicles on the road). Isotropic methods do not correctly detect the SGW's propagation speed. The TIN method (Fig. 54.3a), is giving time-space patterns without physical sense in the context of traffic. In the KIS method (Fig. 54.3b), the absence of sensor vehicles causes that the STV-field is only given by the process average. The reconstruction improves substantially with the geometric correction of anisotropy. Thus, when using TIN Sph (Fig. 54.3c), some waves are recovered, though the absence of vehicles in the aforementioned time interval causes triangular artifacts. In Fig. 54.3d, the Stop-and-Go Waves are reconstructed, even in the period with no sensor vehicles.

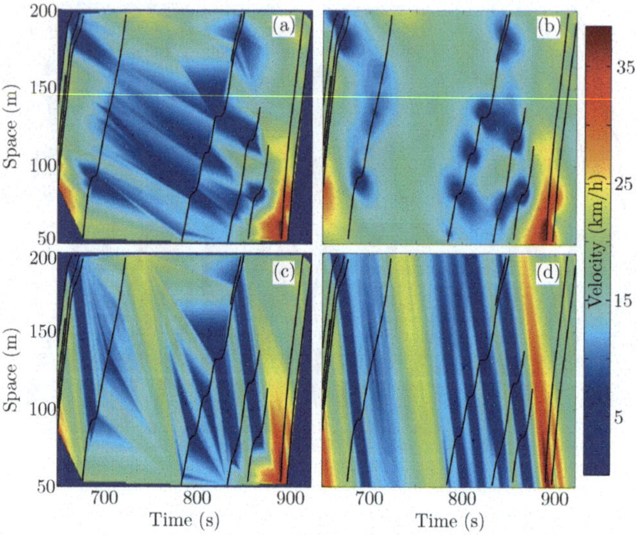

Fig. 54.3 Reconstruction of I-80 STV-field. **a** TIN; **b** KIS; **c** TIN Sph; **d** KAS

54.4 Conclusions and Future Work

In this work we have shown the application of primary geostatistical methods in the interpolation of time-space patterns in road traffic flow. We have tested these techniques with synthetic and real-world data. The random field model, together with the anisotropic variogram and Kriging interpolation technique, reveals very promising results for this approach. Therefore, we will continue to experiment with the recently released HighD dataset [10]. On the other hand, these techniques can be applied to flows of another kind, such as two-dimensional granular flows and pedestrians. In the future, the geostatistical approach may be useful in several areas of road transportation, where some examples arise: Travel time predictions from FCD with less uncertainty, and the anticipation and damping of SGW by means of sensor vehicles equipped with Cooperative Adaptive Cruise Control. Finally, we thought that the geostatistical theoretical framework, very briefly synthesized in this work, is key in the understanding of the time-space underlying structures of road traffic flow. Deepening in these techniques will provide technological resources in the prevention of massive road capacity breakdowns.

References

1. S. Hoogendoorn, R. Hoogendoorn, Calibration of microscopic traffic-flow models using multiple data sources. Philos. Trans. R. Soc. A Math. Phys. Eng. Sci. **368**(1928), 4497–4517 (2010)
2. M. Treiber, A. Hennecke, D. Helbing, Congested traffic states in empirical observations and microscopic simulations. Phys. Rev. E **62**(2), 2000 (1805)

3. M. Treiber, D. Helbing, Reconstructing the spatio-temporal traffic dynamics from stationary detector data. Cooperative Transp. Dyn. **1**, 3.1–3.24 (2002)
4. M.A. Oliver, R. Webster, A tutorial guide to geostatistics: computing and modelling variograms and kriging. Catena **113**, 56–69 (2014)
5. M. Eriksson, P. Siska, Understanding anisotropy computations. Math. Geol. **32**(6), 683–700 (2000)
6. M. Montanino, V. Punzo, Trajectory data reconstruction and simulation-based validation against macroscopic traffic patterns. Transp. Res. Part B Methodol. **80**, 82–106 (2015)
7. E. del Arco et al., Sparse vehicular sensor networks for traffic dynamics reconstruction. IEEE Trans. Intell. Transp. Syst. **16**(5), 2826–2837 (2015)
8. C.J. Willmott, S.M. Robeson, K. Matsuura, A refined index of model performance. Int. J. Climatol. **32**(13), 2088–2094 (2012)
9. R.O. Duda, P.E. Hart, D.G. Stork, *Pattern Classification* (Wiley, 2012)
10. R. Krajewski, J. Bock, L. Kloeker, L. Eckstein, The highD dataset: a drone dataset of naturalistic vehicle trajectories on german highways for validation of highly automated driving systems, in *21st International Conference on Intelligent Transportation Systems (ITSC)* (IEEE, 2018

Chapter 55
Double-Deck Rail Car Egress Experiment: Microscopic Analysis of Pedestrian Time Headways

Marek Bukáček, Veronika Pešková, and Hana Najmanová

Abstract Based on the data from two double-deck rail car egress experiments, presented project aims to estimate the effects of various parameters on the evacuation time. This paper focuses on a microscopic analysis of the time headways showing that several headway-based statistics may illustrate the macroscopic behavior as well as may explain observed phenomena. Applying this approach, we have concluded that the evacuation movement may be decomposed into two phases with different performance and that the efficiency of evacuation is given by their proportion. Moreover, we have observed that the total evacuation time is affected mainly by the frequency of extremely large headways, i.e. the performance significantly depends on few slowest persons.

55.1 Introduction

The level of details applied in analytical studies of pedestrian behavior differs with the complexity of infrastructure. While the planar motion in corridors with bottlenecks [1], counter-flow [2] or even complex city infrastructure [3], has been widely studied, the field of public transport such as passenger train lacks microscopic data.

From a more macroscopic perspective, Marcos and Pollard [4] provided detailed results describing movement characteristics related to required egress time of people drawn from different experimental egress trials and other studies dealt with evacuation processes from trains in hazardous conditions in tunnels [5, 6].

This project focuses on a double-deck electric unit class EPJ 471 (CityElefant) intended for passenger service in the vicinity of city agglomerations. The aim of the project is to quantify the effect of various parameters on the total evacuation time. To

M. Bukáček (✉)
Faculty of Nuclear Sciences and Physical Engineering, CTU in Prague, Trojanova 13, 120 00 Prague, Czech Republic
e-mail: marek.bukacek@fjfi.cvut.cz

V. Pešková · H. Najmanová
Faculty of Civil Engineering, CTU in Prague, Thákurova 7, 166 29 Prague, Czech Republic

I. Zuriguel et al. (eds.), *Traffic and Granular Flow 2019*,
Springer Proceedings in Physics 252,
https://doi.org/10.1007/978-3-030-55973-1_55

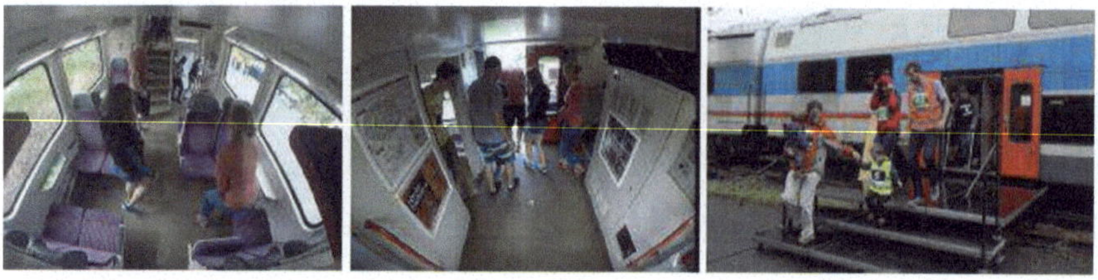

Fig. 55.1 Illustration of both experiments

handle this task, two train egress experiment were carried out, each covering one part of the parametric space—see details in articles [7, 8]. Concerning both experiments, more than 130 volunteers participated in the evacuation trials and more than 10 organizers were engaged.

The train units were equipped by wide angle cameras covering most of the experiment area, see snapshots in Fig. 55.1. The passing times of each pedestrian through the critical bottlenecks were extracted manually.

Such data enables not only to evaluate the total evacuation time, but also the time headways [9]. This article summarizes this "microscopic" point of view, trying to explain the observed total evacuation times by means of individual behavior.

To handle the microscopic behavior, the "headway" defined as the time distance between two consecutive pedestrian was evaluated at the main exit. Its width variation (from 0.6 m to 1.34 m) enabled in some scenarios independent movement of two pedestrian lines. In this case, the mutual orientation of such lines is expressed as the "phase shift" and affects the measurements even there is no direct effect on the evacuation time, see illustration in Fig. 55.2.

When people are moving side by side, the phase shift is equal to zero. This configuration is detected by the saw-like pattern (the side by side walkers have time headways are close to zero, while the next pair keeps the distance). On the other hand, when the lines are shifted by half of the period, the distance to the predecessor is the same as to the follower. In this case, constant trend is observed

To evaluate effects of settings with different time progress, we will compare corresponding phases of the evacuation. Based on measured outflow we will distinguish:

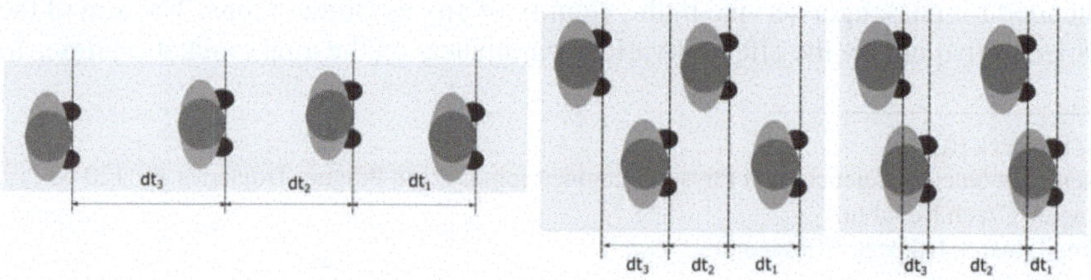

Fig. 55.2 Left: Definition of time headway in narrow corridor. Right: Two possible setups of lines—almost synchronous (phase shift close to zero) and asynchronous (phase shift close to $\pi/2$)

- Main phase—the core of the passengers is leaving. The persons have formed a queue (crowd) thus the outflow fluctuates around the capacity.
- Tail phase—the last passengers are egressing. The pool of remaining people does not saturate the capacity and thus the time headways rise

First, the phase transition (driven by the time development) itself may explain the value of total evacuation time. Second, the statistical evaluation of headway statistics just in given phase will help to focus on relevant data.

55.2 Experiment 1

The first experiment simulating an emergency egress was organized in June 2015, see [7, 8]. It involved 56 volunteers at the age from 21 to 51 years (18 women, 38 men) and focused on binomial parameters (two trials for each combination):

- Exit availability (width 1.34 m): both sides (1)/one side only (2)
- Time to stop effect: no delay (I)/50 s delay (II)

The exit availability affects the size of the crowd in front of the exit as the number of participants is the same in both cases. The time to stop represents a potential delay between start of evacuation and opening the door simulating the time needed to stop.

The Table 55.1 summarizes the macroscopic quantities related to the total evacuation time defined as the leaving time of the last participant (t_{Last}), the total number of participants passing through given exit (N), and the time of first passing participant t_{First} as the scenarios differed by the door-opening time. From these quantities, mean flow and mean headway may be extracted by simple formulas:

$$J_{\text{mean}} = \frac{N}{t_{\text{Last}} - t_{\text{First}}}, \; \mathrm{d}t_{\text{mean}} = \frac{t_{\text{Last}} - t_{\text{First}}}{N - 1} \tag{55.1}$$

From these quantities one can see that the evacuation time slightly differs with the scenario, but the mechanism is hardly visible. Applying the headway approach, we can get more detailed insight, see the progress of a trial visualized in Fig. 55.3. The headway time series may be characterized by mean headway value within main

Table 55.1 Macroscopic results of Experiment 1. Values were averaged over two trials and two main exits, where available

Scenario	t_{First} [s]	t_{Last} [s]	N [ped]	J_{mean} [ped/s]	$\mathrm{d}t_{\text{mean}}$ [s/ped]
A-1-I	7.6	28.5	28	1.34	0.75
A-1-II	52.6	69.7	28	1.64	0.61
A-2-I	11.7	47.0	56	1.59	0.63
A-2-II	52.4	87.8	56	1.59	0.63

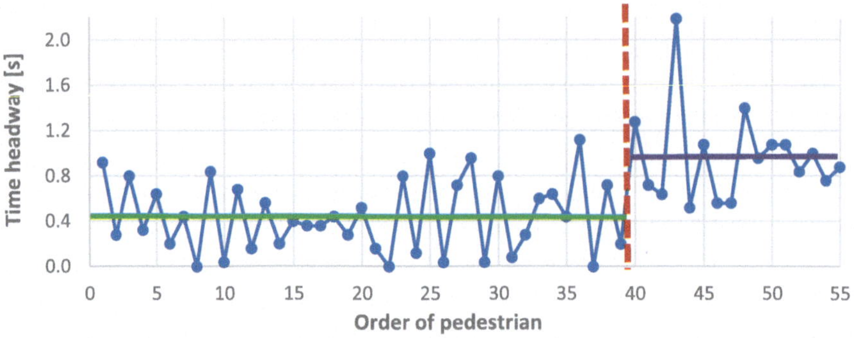

Fig. 55.3 The time development of headways in scenario A-2-II trial 1. The mean headways within both phases and phase separation are visualized by corresponding lines

Table 55.2 Microscopic results of Experiment 1. Values were averaged over two trials and two main exits, where available

Scenario	Main phase mean [s/ped]	Tail mean [s/ped]	Phase prop [%]
A-1-I	0.64	1.35	17
A-1-II	0.50	1.14	19
A-2-I	0.51	1.16	19
A-2-II	0.43	1.07	31

phase, mean headway within tail phase and proportion of phases as summarized in Table 55.2.

Averaging these features among the scenarios brings numbers sufficient to understand the effect of settings. Scenario A-1-I is the slowest one in both main phase and the tail phase. Scenarios A-1-II resp. A-2-I are faster by 20% relatively to A-1-I due to higher motivation caused by the effect of waiting resp. the effect of bigger crowd. Scenario A-2-II indicates even better performance within both phases, but the macroscopic performance is worse due to the higher proportion of tail phase.

The higher proportion of tail phase in A-2-II is in line with the naked eye observations. The combination of closed doors and one exit available caused that lower deck passengers filled the exit room, thus the upper deck passengers had to wait at the stairs. When this room became available again, the upper deck passengers started the evacuation, but due to the other bottlenecks inside the train, only one half of the exit capacity was used. On the other hand the performance of pedestrians was better than during other scenarios thus the total evacuation time was in line with other scenarios.

As seen from Table 55.2, combining the lowest mean time headways with the lowest tail proportion could bring faster evacuation. In the case the upper deck and lower deck passengers would share the exit during the whole evacuation thus the tail proportion would reach 19%, the total evacuation time could be decreased by 5 s (10%).

55.3 Egress Experiment 2

In March 2018, the second experiment extended the parametric space by:

- Passengers heterogeneity: high school students (A)/population sample (B)
- Exit width: 0.60 m / 0.75 m / 0.90 m / 1.10 m / 1.34 m
- Exit type: high platform/stairs/terrain

With respect to larger set of parametric combinations (30), only one trial was realized for each scenario. For details, see [10].

Again, headway analysis brings more light on the mechanism behind the evacuation time dynamics. This time, the changeable exit width dramatically affects the possibility of independent passing of two lines and the exit type together with the heterogeneity affected the mean time headway.

Applied approach goes even beyond the previously presented statistical method; conclusions will be extracted directly from the headway charts visualized in Fig. 55.4. First, observations confirmed the main and the tail phase separation (driven by the same misbalance between lower and upper deck passengers), and hence the performance within the phase is explained.

As expected, the headways within the tail phase are doubled relatively to the main phase when the exit enabled two lines motion, but this difference vanishes in narrow exit cases. Moreover, the behavior inside a phase is quite interesting—it seems that the headways do not hold arbitrary values, but they exist within several levels, i.e. the constant trend along 1 s indicates one line motion, saw like patter alternating 1 s and 0 s level corresponds to two line motion with no phase shift.

The phase transition from one line to two lines motion induced by the exit width is characterized by more frequent 0 s level rather than continuously decreasing headway. Moreover, the negative effects of less efficient exit types affect the headways the same fashion—instead of systematically higher headways, one can observe very high headways more frequently. This indicates the situation when one pedestrian needs

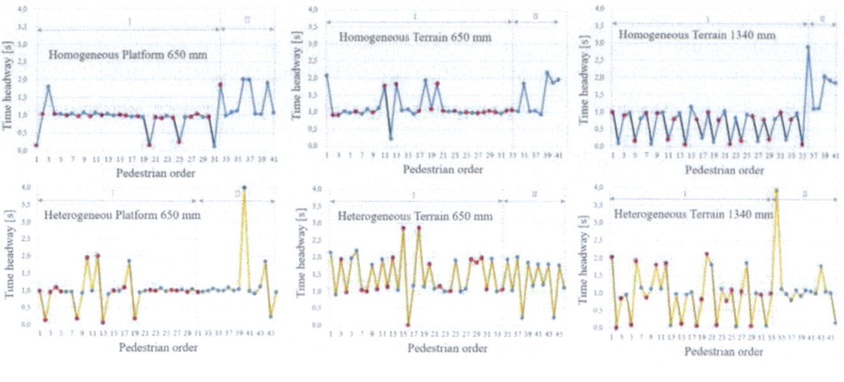

Fig. 55.4 Part of headway study, Experiment 2. The color of dots corresponds to the passenger's initial position (red: lower deck, blue: upper deck)

significantly longer time to leave the train while the behavior of majority passengers is rather the same. This phenomenon was detected mainly for heterogeneous group (population ample) and jump-to-terrain exit type.

55.4 Conclusions

In this article, the travel time analysis of train evacuation using time headways was performed. This approach enables to explain the impact of scenario parameters on evacuation time incorporating the microscopic behavior.

Each evacuation trial was split into the main phase and the tail phase with significantly different performance. Moreover, headways inside the phases depends on the scenarios as well as the motivation of pedestrians. These aspects are not correlated, the scenario with the lowest headways reached the highest proportion of tail phase (upper deck passengers blocked upstairs), which left a space for improvements. Assuming a scenario combining low headways and low proportion of tail phase, the total evacuation time could be 10% lower than observed.

The second experiment extended the observation by the exit type and width effects. The higher flow through wider exit as well as the other phenomena are related to the significant change of individual behavior of several people rather than small change of all group behavior.

In general, the total evacuation time differs with the probability of consecutive egress of two pedestrians or, on the other end of the scale, with the probability that one pedestrian would need more time steps to leave rather than systematically higher mean value. That explains the worst performance of heterogeneous group egressing to the terrain where the high evacuation time was caused by the need of special assistance. Even the majority of the other headways is the same as measured in other scenarios, the total performance was significantly worse. These observations indicate that the effort to decrease evacuation time should focus on the worst performing persons.

Acknowledgements Presented research has been supported by the Grant Agency of the Czech Technical University in Prague, grant SGS18/107/OHK1/2T/11 and SGS18/188/OHK4/3T/14. We would like to thank GAMS team members and other volunteers for significant help with organizing the experiment. All experiment participants approved for the experimental data to be stored, used and published for academic purposes.

References

1. M. Bukáček, P. Hrabák, M. Krbálek, Microscopic travel time analysis of bottleneck experiments, Transportmetrica A: Transp. Sc. **14**(5–6), 375–391 (2018)
2. M. Boltes, J. Zhang, A. Seyfries, B. Steffen, T-junction: experiments, trajectory collection, and analysis, in: ICCV Workshop, pp. 158–165 (2011)

3. J. Zhang, A. Sayfried, Quantification of bottleneck effects for different types of facilities. Transp. Res. Procedia **2**, 39–51 (2014)
4. S. Markos, J.K. Pollard, Passenger train emergency systems: single-level commuter rail car egress experiments, Tech. Rep.DOT/FRA/ORD-15/04, Research and Innovative Technology Administration, John A. Volpe, National Transportation Systems Center, Cambridge, MA (2015)
5. M. Oswald, H. Kirchberger, C. Lebeda, Evacuation of a high floor metro train in a tunnel situation: experimental findings, in *Pedestrian and Evacuation Dynamics 2008*, ed. by W.W.F. Klingsch, C. Rogsch, A. Schadschneider, A. Schreckenberg (Springer, Berlin Heidelberg, Berlin, Heidelberg, 2010), pp. 67–81
6. K. Fridolf, D. Bilsson, H. Frantzich, Evacuation of a metro train in an underground rail transportation system: flow rate capacity of train exits, tunnel walking speeds and exit choice. Fire Technol. **52**(5), 1481–1518 (2016)
7. M. Bukáček, H. Najmanová, P. Hrabák, the effects of synchronization of pedestrian flow through multiple bottlenecks—train egress study, in PED'16. Pp. 105–112 (2017)
8. H. Najmanová, P. Hejtmánek, M. Bukáček, Fire safety of passenger trains: experimental analysis of evacuation from CityElefant double-deck unit. Int. Conf. Fire Saf. **2016**, 294–301 (2016)
9. M. Krbálek, P. Hrabák, M. Bukáček, Pedestrian headway—reflection of territorial social forces. Phys. A **490**(1), 38–49 (2018)
10. H. Najmanová, L. Kuklík, V. Pešková, M. Bukáček, P. Hrabák, D. Vašata, Fire safety of passenger trains: experimental and sensitivity study of evacuation from a double-deck train unit, in progress

Chapter 56
Urban Mobility Observatory

Winnie Daamen, Arjan van Binsbergen, Bart van Arem, and Serge P. Hoogendoorn

Abstract The Urban Mobility Observatory (UMO) will gather, store, and disseminate empirical multi-modal traffic, transport and mobility data, using a well-balanced set of innovative data collection methods. It will make these comprehensive data available for scientific research to develop and test new theories and models to better understand, predict and facilitate multi-modal mobility in large urbanized regions. UMO is located in the Netherlands, being one of the densest urbanized countries in the world, facing severe accessibility problems and environmental pressures from transport.

56.1 Introduction

Urban transport systems are becoming increasingly complex. At the same time, transport technologies and services are developing rapidly, which changes travel behaviour. The Netherlands is one of the densest urbanized countries in the world, facing severe accessibility problems and environmental pressures from transport, while having a strong interdisciplinary and multi-stakeholder collaboration tradition. This combination demands new and integrated observation, analysis and modelling approaches. The Urban Mobility Observatory (UMO) will collect, integrate and store empirical and experimental multi-modal traffic and transport data and make these available for scientific research in order to better understand and facilitate multi-modal mobility in large urbanized regions.

UMO will be realised by a dedicated academic consortium of 6 universities (Delft University of Technology (lead), Vrije Universiteit Amsterdam, Eindhoven University of Technology, University of Twente, University of Groningen and Utrecht University) and 2 research institutes (Centre for Mathematics and Computer Science (CWI) and Amsterdam Institute for Advances Metropolitan Solutions (AMS)) that together with the Dutch Research Council will invest 3.2 million Euro. The

W. Daamen (✉) · A. van Binsbergen · B. van Arem · S. P. Hoogendoorn
Transport & Planning, Delft University of Technology, 2628 CN Delft, The Netherlands
e-mail: w.daamen@tudelft.nl

I. Zuriguel et al. (eds.), *Traffic and Granular Flow 2019*,
Springer Proceedings in Physics 252,
https://doi.org/10.1007/978-3-030-55973-1_56

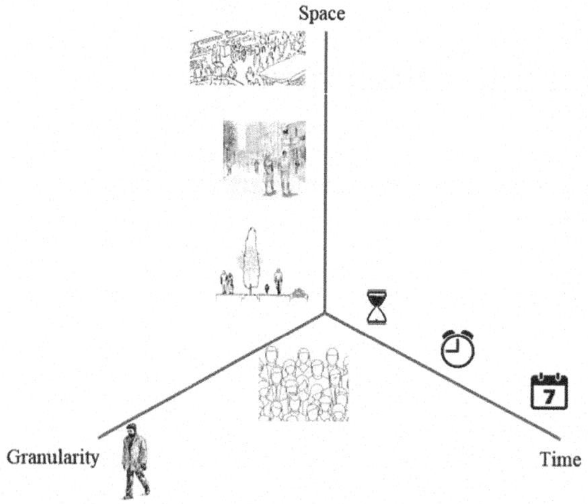

Fig. 56.1 Three dimensions to structure data and data sources: time, space and granularity

development of UMO will be based on experience gained in earlier projects including mass pedestrian movement observations during large events in Amsterdam, experimental observations in the Green Village living lab in Delft and spatial analysis and modelling in the SPINlab in Amsterdam.

In the remainder of this paper, we will first address different types of data, that will be collected by UMO. Then, we show the detailed design of the UMO facility, followed by an example of one of the living labs in de Delft University of Technology Campus (the Leeghwaterstraat). We end with conclusions.

56.2 Data Types

A wide variety of data sources describing mobility, transport and traffic exists. We structure these data sources along three dimensions: time, space and granularity, see Fig. 56.1. The time scale ranges from several microseconds (e.g. delay in propagation of information in autonomous vehicles) to multiple years (e.g. household composition). The spatial scale ranges from a cross-section (e.g. lateral distance between bicycles) to a whole city (e.g. origin-destination patterns). Finally, the granularity scale relates to the information available on the individual travellers. Low granularity implies little information (e.g. for social media data), while high granularity implies a lot of information per person (e.g. for people participating in a driving simulator experiment). In a way, this granularity is linked to the distinction between microscopic (individual) and macroscopic (aggregate or flow) perspectives.

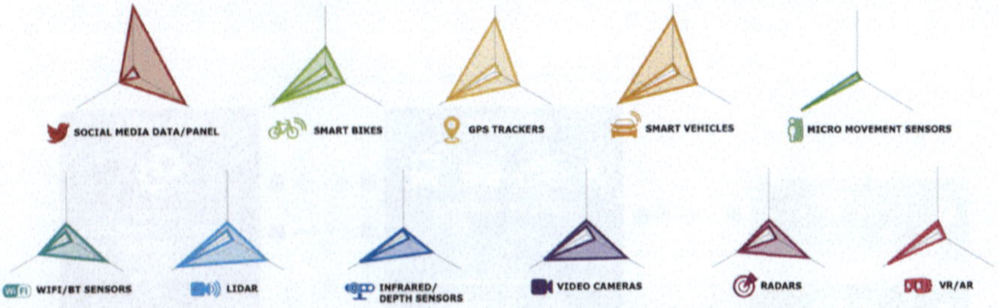

Fig. 56.2 Combinations of different data collection methods, covering the whole range of the three dimensions time, space and granularity

A specific sensor or data collection method collects data within a limited range of each of the three dimensions. To provide data on the whole range of all dimensions, we typically combine different data sources, see Fig. 56.2.

56.3 Components of UMO

The UMO facility consists of five components, namely sensor networks (Sect. 56.3.1), population sampling (Sect. 56.3.2), simulators (Sect. 56.3.3), a coupling module (Sect. 56.3.4) and the data module (Sect. 56.3.5). Figure 56.3 shows the relations between these components. Each component is further detailed in the following sections.

56.3.1 Sensor Networks

Sensor networks support distributed data collection for real life and on-site controlled experiments. These sensor networks consist of instruments to permanently observe real-life situations in a broad range of urban conditions, and functionalities for flexible observations. It also includes a moving sensor network of instrumented vehicles and instrumented bikes and portable devices for additional observations. State-of-the-art sensor technologies include cameras, radars, LIDARs, BT/Wi-Fi trackers, PIR sensors, and cell phone apps. All sensors are used to collect data on, for example, traffic conditions/traffic/driver state, drivers' and vehicle behaviour, bicyclists and pedestrian strain and fitness. Depending on the type of sensors, they collect data on individuals, individual vehicles (human driven, partially or fully automated, or remote controlled) or traffic flows (groups of individuals, vehicles) in a range from very local (area directly surrounding the vehicle, bicyclist, pedestrian) up to spatial and temporal aggregated (area, network) level. In addition, situational data may include weather conditions (temperature, precipitation, visibility, wind, lighting conditions), environmental conditions (air quality, noise) and characteristics of the geographic setting.

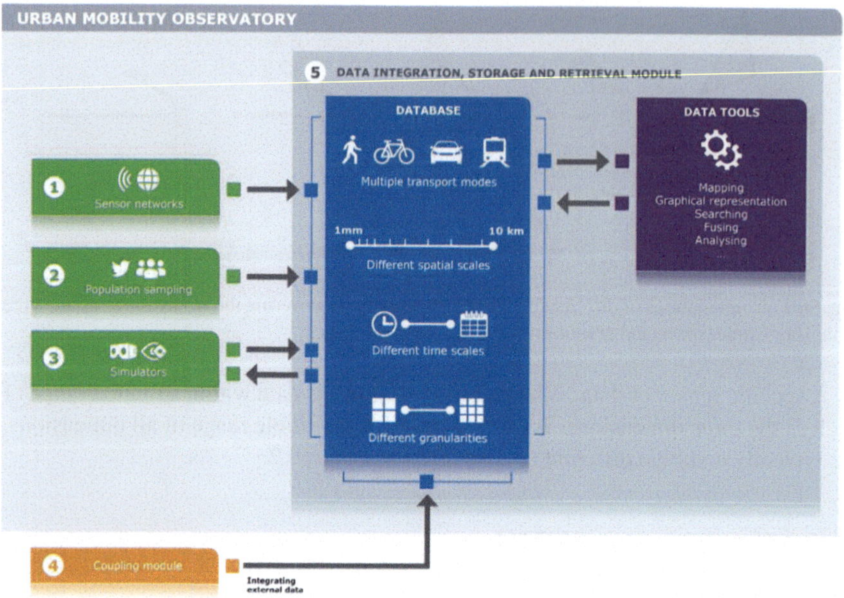

Fig. 56.3 Overview of UMO and its five components

56.3.2 Population Sampling

Traffic observation techniques alone cannot give a complete picture of current travel and mobility behaviour, for example because motivations of behaviour cannot be observed, choice alternatives are not discovered, and the observation systems only cover a part of the complete network/area. Furthermore, behaviour in exceptional and dangerous situations or behaviour in interaction with new technologies and systems, such as automated vehicles and Mobility-as-a-Service, is difficult or even impossible to observe in real life. This module therefore consists of crowd sensing and panel data gathering abilities. The panel data collection platform collects data from a representative sample of the Dutch population. It gathers self-reported behaviour and preferences of individuals, combining tested and validated questionnaires and app-based polling approaches. Crowd sensing is performed via crowd sourcing, social media and public fora data harvesting.

56.3.3 Simulators

The virtual reality (VR), augmented reality (AR) and simulator environments aim to closely observe (individual) human behaviour and experiences in specific experimental traffic conditions, such as potentially dangerous, not-yet existing, or rarely occurring situations With the VR, AR and simulator environment interactions between different traffic participants can be observed, including human-human as well as human-machine interactions (in cases of semi- or fully automated vehicles). Also, reactions of drivers to infrastructure modifications, various methods of information provision, and driver assistance approaches can be tested. This part of the facility is used in an iterative way. First, the collected data are used to generate realistic scenarios, which can be used to construct a virtual world: a VR/AR/simulator experimental setting. Second, the reactions and responses (behaviour or their evaluations of travel contexts) of the participants are observed, and as such provide input to the database.

56.3.4 Coupling Module

The Urban Mobility Observatory will functionally connect to existing data sources and living labs such as MobiliteitsPanel Nederland (MPN; set up by the Netherlands Institute for Transport Policy Analysis), National Datawarehouse (NDW; road traffic information sourced from Dutch road authorities), TomTom data, KPN data, Google data, OV Chipcard data (i.e., public transport data sourced from the Dutch national public transport payment and tariffing system). UMO will also link to existing facilities operated by consortium partners such as the datalabs DiTTlab, SPINlab and Urban Mobility Lab, and data collection facilities including Living Lab Smart Mobility, RADD and Social Glass6. This will, among others, enable a full integration of traffic and mobility behaviour on the one hand, and spatial data on for example land use, urban densities, land and real-estate prices, locational choice on the other.

56.3.5 Data Integration, Storage and Retrieval Module

The data gathered through the sensor networks, population sampling and simulators are all conveyed to different, spatially separated, but—as far as possible—federated databases, together with associated metadata (and essential tools) to keep the data retrievable and usable for a long period. These data have a variety of time and spatial characteristics, as well as different granularity, see Sect. 56.2. The backbone of UMO will complement the data with contextual information, either derived from the observation systems or from external, coupled data sources.

56.4 Leeghwaterstraat: Example of a Living Lab in UMO

To collect data within the UMO project a number of living labs will be initiated. One of these living labs will be on the campus of the Delft University of Technology, starting with the Leeghwaterstraat. It is a 450 long so-called bicycle street ('fietsstraat' in Dutch), where mixed traffic exists between cyclists and cars, while cyclists have priority. As this type of street is relatively new, it provides an excellent opportunity to investigate this co-existence of cars and (large) bicycle flows: when do car drivers overtake cyclists, who gives priority at an intersection, how many cyclists cycle next to each other, where do pedestrians cross the bicycle street? Eight poles in the Leeghwaterstraat have been equipped with cameras. These cameras focus on the road, and do not reveal any information on the entrances and the neighbouring buildings. The images are only stored the moment experiments takes place. This is announced through traffic signs mounted at the Leeghwaterstraat entrances. Two types of cameras have been installed, see Fig. 56.4. The oversight cameras are suitable to observe the manoeuvres of the cyclists, while the Star light cameras can be used to derive trajectories.

Figure 56.5 shows a screenshot of camera A, an oversight camera. Persons and license plates have been blurred for privacy reasons. We see that the street is wide enough to let two cars pass, and it also gives sufficient space for cars to overtake cyclists.

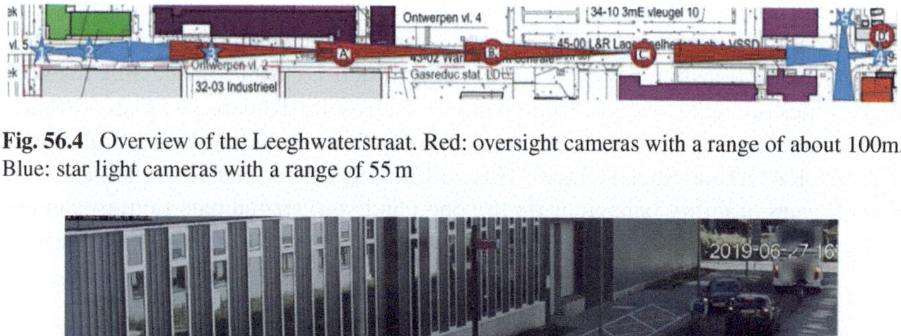

Fig. 56.4 Overview of the Leeghwaterstraat. Red: oversight cameras with a range of about 100m. Blue: star light cameras with a range of 55 m

Fig. 56.5 Screenshot of camera A on the Leeghwaterstraat

56.5 Conclusions

UMO will offer the possibility to collect, store and disseminate data from a wide variety of sensors and data sources representing different time and geographic scales, transport modes, and different settings (real life, experimental, virtual). UMO will provide accurate, quantitative, real-time as well as off-line, large scale tracking opportunities of people and vehicles. The data will be acquired in research projects covering a range of research methodologies, including exploratory research, theory development, hypotheses testing, model identification, model calibration, and model validation. At the same time, UMO will provide enriched (real time) data, as the basis for insights and models to unveil possibilities for the development of new transport services, transport related product-market combinations, traffic management concepts, transport policy options, transport mode and infrastructure development options and even land-use strategies.

Acknowledgements This project is funded by the Dutch Research Council (NWO), grant number 175.2017.017.

Chapter 57
Investigating the Role of Network Morphology in the Underpinning of a Network Fundamental Diagram

Alonso Espinosa Mireles de Villafranca, Richard D. Connors, and R. Eddie Wilson

Abstract We investigate the role that network morphology plays in setting up the routing interactions that govern the emergence of a network fundamental diagram (NFD) in homogeneous road network *patches*. We generate ensembles of synthetic networks and find solutions to the *static traffic assignment* (STAP) for a range of (equivalent) global demands and trip distribution structures. The network ensembles are generated by building β-skeletons of randomised lattices according to the methodology presented by the authors in TGF 2017. We find that the total road length of the network largely determines network performance in the uncongested regime. The clustering coefficient is a secondary factor.

57.1 Introduction

Macroscopic behaviour of traffic has been studied for many decades, for example Mahmassani et al. [10] show that clear relationships between macroscopic traffic variables can be observed in simulations on gridded networks. More recently, empirical evidence for the existence a *macroscopic fundamental diagram* (MFD) [4] in Yokohama spurred the study of conditions for the emergence of an MFD [2, 5] as well as into ways of deriving [7] and approximating it, for example [9]. A given network may be partitioned into 'homogeneous-enough' sub-regions each of which has its own MFD. Network traffic may thus be modelled by a set of reservoirs that exchange flows [6].

For this paper, we take as a starting point that networks partitioned into homogeneous regions can lead to well defined individual MFDs and take the opposite approach to previous works by asking the question: Does starting with homoge-

A. Espinosa Mireles de Villafranca (✉) · R. E. Wilson
University of Bristol, Merchant Venturers Building, Woodland Road, Bristol BS8 1UB, UK
e-mail: alonso.espinosa@bristol.ac.uk

R. D. Connors
University of Leeds, Leeds LS2 9JT, UK
e-mail: R.D.Connors@its.leeds.ac.uk

I. Zuriguel et al. (eds.), *Traffic and Granular Flow 2019*,
Springer Proceedings in Physics 252,
https://doi.org/10.1007/978-3-030-55973-1_57

neous network regions or patches lead naturally to the emergence of a macroscopic fundamental diagram for the network? We will adopt the term *network fundamental diagram* (NFD) to highlight the central role of the network itself.

In Sect. 57.2 we define static traffic assignment and cast it as a convex optimisation programme; this is the form it will be used in for our numerical investigations. In Sect. 57.3 we describe the network model used for generating the morphological ensembles and then proceed to describe the experimental set-up and results in Sect. 57.4. Finally we conclude in Sect. 57.5.

57.2 Static Traffic Assignment

The static (time-independent) traffic assignment problem is concerned with allocating volumes (per unit time) of users to routes through a network. A traffic pattern or flow in a network is a *user equilibrium* assignment if it satisfies Wardropian equilibrium: the cost of all used routes are all the same; any unused routes have equal or greater cost.

For a network whose edges are characterised by cost functions that are monotonous, convex and separable (i.e., $c_i(\mathbf{x}) = c_i(x_i)$), then the STAP can be expressed by the following convex optimisation programme [11],

$$\min_{\mathbf{x}} \quad F_{\mathrm{UE}}(\mathbf{x}) \tag{57.1a}$$

$$\text{subject to} \quad \mathbf{E}\,\mathbf{x} = \mathbf{d}_\omega, \quad \forall\, \omega \in \mathcal{W}, \tag{57.1b}$$

$$\mathbf{x} \geq 0. \tag{57.1c}$$

Here F_{UE} is the Beckmann functional [1], $\mathbf{E}$ is the node-link incidence matrix, and $\mathbf{d}_\omega$ is an OD-specific demand vector that vanishes at all nodes unless they are the origin or destination of OD pair ω. That is, $(\mathbf{d}_\omega)_{\text{origin}} = -(\mathbf{d}_\omega)_{\text{destin}} = -d_\omega$, where d_ω the total demand associated to ω. Expression (57.1b) captures flow conservation at intersections.

57.3 Network Model

To generate the ensembles we use the model introduced in [3]. It consists of three steps to generate a random network: first, randomising a regular lattice to generate the node set; second, assembling the β-skeleton [8] for the set of nodes obtained from the previous step; third, allocating parameters for the cost function of each edge, the a_i and b_i in the affine function $a_i + b_i x_i$. Special care is taken to make definitions endogenous to the network model.

The *morphological* parameters of our model, which are also our experimental variables, α and β, capture the *griddedness* of the network and the *connection density*,

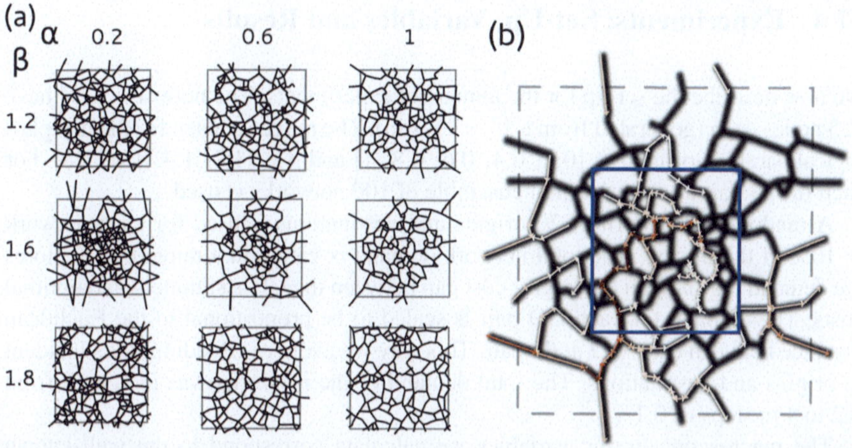

Fig. 57.1 **a** Examples of the different network morphologies. **b** A user equilibrium flow pattern. 20 OD pairs are used, redder (darker) corresponds to higher flows and whiter (lighter) shades are lower flows. The central quadrant (blue square) is the region *A* for which macroscopic traffic variables are calculated

respectively. The griddedness α is defined in the interval $\alpha \in [0, 1]$ and changes the node distribution between a regular lattice ($\alpha = 0$) and uniform random points ($\alpha = 1$). The effect of increasing β is to decrease the overall density of edges in the network, it also has topological consequences on the network. The number of triangles decreases as β increases ($\beta \in [1, 2]$), thus so does the clustering coefficient of the networks. The clustering coefficient is therefore a good measure for describing the effects of β on morphology (more detail on the network model in [3]).

Conceptually we are trying to construct a set of networks which is dense around real-world networks, from which we will take a finite experimental sample of 'substrates' (network patches) on which the STAP will be solved. Figure 57.1 shows example networks from different ensembles, as well as a resulting equilibrium traffic pattern.

To be able to compare the networks (with different number of nodes and morphologies) the parameters of the cost functions, are defined implicitly by using the following 'constant infrastructure supply' heuristic. We consider all intersections to have equal constant capacities that are evenly split among the incoming roads. This results in all roads incoming to a node having the same b_i. The a_i are set equal to the euclidean length of the edges.

Throughout this paper we distinguish between the terms *topology* and *morphology* to highlight that both the adjacency structure (topology) and the geometric structure (morphology) are important.

57.4 Experiments: Set-Up, Variables and Results

We now describe the set-up for the numerical experiments. The networks used have 225 nodes each (generated from a 15 × 15 grid). The ranges chosen for the morphological parameters are $\alpha \in \{0.2, 0.4, 0.6, 0.8, 1\}$ and $\beta \in \{1.2, 1.4, 1.6, 1.8\}$. For each (α, β) pair an experimental ensemble of 100 networks is used.

A random OD pattern of 20 origins and destinations is used for each network (∼10% of the nodes). The total network demand is varied in a range that explores the demand region with the largest cost gap between the UE assignment and optimal costs. The demand for each OD pair is scaled to be proportional to the Euclidean distance between each OD node-pair. This way we avoid overloading links incident to origins and destinations. The total demand of the networks was increased from 0.2 to 4 in steps of 0.1.

The macroscopic traffic variables we calculate correspond to the well-known microscopic ones involved in the relationships referred to as the (classical, i.e. link-based) fundamental diagram.

The network level definitions of variables we use are *trip production* P_A and region density ρ_A. The production is calculated as

$$P_A = \frac{\sum_{i \in A} v_i x_i}{L_A}, \tag{57.2}$$

where L_A is the total road length of all roads fully contained in A and v_i is the speed of vehicles on edge i. Trip production is the total veh · km travelled in region A per unit time.

The region density is given by

$$\rho_A = \frac{\sum_{i \in A} x_i c_i(x_i)}{L_A} = \frac{k_A}{L_A}, \tag{57.3}$$

where k_A is the number of vehicles in the region, or more specifically, the aggregated vehicle occupation of all the links completely contained in A.

In Fig. 57.2a the trip production (P_A) versus density (ρ_A) is shown for 20 ensembles. Each data point is an average of the density and production over a whole ensemble subjected to the same global demand. The curves of fit are of the form $P_A = C\sqrt{\rho_A}$, with fit coefficient C. They are coloured according to the mean observation region length L_A to show the trend of the family of P_A versus ρ_A curves with network length. Each data point in Fig. 57.2a corresponds to the ensemble average of P_A over 100 networks.

There is a near-monotonic relationship between region length and performance, which is a sense-check on the results, as infrastructure supply increases so does trip production. However, the effects of network topology can be seen more clearly in Fig. 57.2b. For mid-α values the spread of the NFD curves due to β means ensembles with similar α can be hard to distinguish by performance alone. Near the turning point

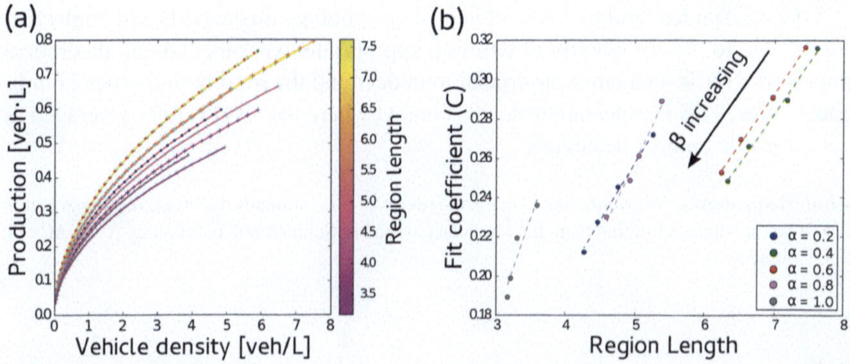

Fig. 57.2 **a** The increase of trip production P_A with respect to vehicle density ρ_A. The marker colours blue, green, red, cyan, grey correspond to $\alpha = 0.2, 0.4, 0.6, 0.8, 1$, respectively. The different shapes correspond to different values of β. The fitted curves are coloured according to the region length L_A and are of the form $P_A = C\sqrt{\rho_A}$. In general L_A is larger for networks that perform better, however the trend is not smooth. **b** The fit coefficient C is plotted against the L_A. The coefficients depend linearly on the region length that is strongly modulated by α. Note the cross-over of the $\alpha = 0.2$ (blue) and $\alpha = 0.8$ (violet) lines reflects the staggering of the production curves in (**a**)

$\alpha \simeq 0.5$, the coefficient lines are parallel and close. As α increases the production curves for ensembles with very different griddedness (see lines for $\alpha = 0.2$ and $\alpha = 0.8$ in Fig. 57.2b yield similar production curves but can be distinguished due to their values of β).

For $\alpha = 1$ the linear trend of the fit coefficient is not as good. In these ensembles, due to the interplay between clustering of nodes and the edge density, the networks show more diversity, thus exhibiting more complex performance behaviour.

57.5 Conclusions

Through computational experiments on network ensembles we confirm that the key factor determining the NFD relationship (the uncongested branch only) between trip production and density is the infrastructure supply, captured mainly in our model through total observation region road-length. That is the physical properties of the network, play a more important role than morphological features. As a first approximation, the morphological properties are important in the way they indirectly affect the total road length. This is seen in how the fit coefficients of the NFD depends on the clustering coefficient (captured by our β).

Even with affine cost functions the different ensembles show distinct behaviour. This is slightly surprising considering the nuanced qualitative change in morphologies presented by the different ensembles.

A detailed understanding of the effects of morphology on the NFD, although challenging due to the complexity of defining appropriate experimental and theoretical frameworks for investigation, is crucial to understand the routing properties of individual cities. This is especially true in a context where the NFD is to be leveraged as tool for traffic control strategies.

Acknowledgements We would like to thank the reviewer for comments that made this paper better. AEMV acknowledges funding from the Mexican Council of Science and Technology (CONACyT), CVU: 538874.

References

1. M. Beckmann, C. McGuire, C. Winsten, *Studies in the Economics of Transportation* (Yale University Press, 1956)
2. C. Daganzo, N. Geroliminis, An analytical approximation for the macroscopic fundamental diagram of urban traffic. Transp. Res. Part B Methodol. **42**(9), 771–781 (2008)
3. A. Espinosa Mireles de Villafranca, R. Connors, R. Wilson, Static traffic assignment on ensembles of synthetic road networks, in *Traffic and Granular Flow '17* (Springer, 2019), pp. 29–36
4. N. Geroliminis, C. Daganzo, Existence of urban-scale macroscopic fundamental diagrams: some experimental findings. Transp. Res. Part B Methodol. **42**(9), 759–770 (2008)
5. N. Geroliminis, J. Sun, Properties of a well-defined macroscopic fundamental diagram for urban traffic. Transp. Res. Part B Methodol. **45**(3), 605–617 (2011)
6. J. Haddad, N. Geroliminis, On the stability of traffic perimeter control in two-region urban cities. Transp. Res. Part B Methodol. **46**(9), 1159–1176 (2012)
7. D. Helbing, Derivation of a fundamental diagram for urban traffic flow. Eur. Phys. J. B **70**(2), 229–241 (2009)
8. J. Jaromczyk, G. Toussaint, Relative neighborhood graphs and their relatives. Proc. IEEE **80**(9), 1502–1517 (1992)
9. L. Leclercq, N. Geroliminis, Estimating MFDs in simple networks with route choice. Transp. Res. Part B Methodol. **57**, 468–484 (2013)
10. H. Mahmassani, J. Williams, R. Herman, Performance of urban traffic networks, in *Proceedings of the 10th International Symposium on Transportation and Traffic Theory* (Elsevier, 1987), pp. 1–20
11. M. Patriksson, *The Traffic Assignment Problem: Models and Methods* (Dover, 2015)

Chapter 58
Experiments and Usability Tests of a VR-Based Driving Simulator to Evaluate Driving Behavior in the Presence of Crossing Pedestrians

Claudio Feliciani, Luca Crociani, Andrea Gorrini, Akihito Nagahama, Katsuhiro Nishinari, and Stefania Bandini

Abstract Road accidents are among the leading cause of death worldwide, with developing countries paying the highest toll. Among the different road users, pedestrians are the most vulnerable. A better understanding of the relation occurring between car drivers and pedestrians when they interact is necessary for the development of new research approaches and their fall in education/prevention. Given the risks involved in the study of car-pedestrian interactions, experimental studies are difficult and only simulators represent a feasible alternative. However, the high cost of driving simulators prevent developing countries from investing on these tools. In this research, we considered the possibility to use a simple low-cost (below 500 $) virtual reality (VR) based driving simulator to study the interaction between cars and pedestrians at unsignalized crosswalks. Young adults and elderlies were asked to drive in a virtual environment where simulated pedestrians would pass randomly on crosswalks. Driving behavior was numerically determined for both age groups using speed, time gap and driving accuracy. The obtained results fit with the behavior observed in real conditions, suggesting the empowerment of the application of low-cost VR-based systems for traffic safety research. Results on usability indicated the existence of

C. Feliciani (✉) · K. Nishinari · S. Bandini
Research Center for Advanced Science and Technology, The University of Tokyo, 4-6-1 Komaba, Meguro-ku, Tokyo 153-8904, Japan
e-mail: feliciani@jamology.rcast.u-tokyo.ac.jp

L. Crociani · A. Gorrini · S. Bandini
Complex Systems and Artificial Intelligence Research Center, University of Milano-Bicocca, Viale Sarca 336/14, 20126 Milan, Italy

A. Nagahama
Ritsumeikan Global Innovation Research Organization, Ritsumeikan University, 1-1-1 Noji-higashi, Kusatsu, Shiga 525-8577, Japan

K. Nishinari
Department of Aeronautics and Astronautics, The University of Tokyo, 7-3-1 Hongo, Bunkyo-ku, Tokyo 113-8656, Japan

I. Zuriguel et al. (eds.), *Traffic and Granular Flow 2019*, Springer Proceedings in Physics 252,
https://doi.org/10.1007/978-3-030-55973-1_58

discomfort during the use of VR tools, suggesting deeper studies on digital-divide issues.

58.1 Introduction and Background

As highlighted by the WHO [16], road accidents represent the eighth leading cause of death in the world population (and the leading one for children and young adults): 1.35 million people are killed on roads every year, with most fatalities occurring in developing countries. In particular, pedestrians and cyclists are some of the most vulnerable road users, with a percentage of fatalities corresponding to 26% of the overall victims.

To contrast pedestrian-car accidents it is necessary to identify which factors influence traffic safety to support public institutions in the design of transportation infrastructures and traffic policies. Risky driving behavior is essentially caused by human factors and environmental factors. On one hand, traffic accidents are linked to the lack of coordination among motor, perceptive and attentional skills and, on the other side, on individual psychological disposition towards hazardous situations [6]. As a consequence, very often these factors could affect non-compliance of drivers to traffic norms.

The study and data collection of human factors (in a selected road environment) is difficult in real conditions. Supervised experiments are limited due to ethical and safety concerns, and data obtained from observation are often biased by several intrinsic variables. Driving on traffic simulators could be a viable alternative, but two main aspects come into effect upon their use: cost and reliability of the results. As Table 58.1 shows, the regions which would need more investment on education and research on traffic safety lack the financial resources. On the other hand, it is also questionable whether expensive driving simulators are necessarily needed to perform such studies. A report by the Swiss Army [11] (which employed truck simulators as part of the training of its soldiers) concluded that one hour in the simulator would

Table 58.1 Road traffic fatalities and expenditure on R&D (research and development) and education by population for different countries (selection is based on the significance for this study and availability of reliable data) [1, 2, 10, 12–16]

Indicator	Japan	Italy	Turkey	China	Thailand
Road traffic related fatalities per 100,000 people (year)	4.1 (2016)	5.6 (2016)	12.3 (2016)	18.8 (2013)	32.7 (2016)
Expenditure on R&D per capita (year)	1349 $ (2017)	557 $ (2017)	271 $ (2017)	357 $ (2017)	48 $ (2017)
Education expenditure per capita (year)	1372 $ (2014)	1172 $ (2016)	470 $ (2015)	491 $ (2018)	253 $ (2013)

cost about 420$, while the same time would cost 150$ in driving school (with real trucks). Thanks to the recent trend in relation to virtual reality (VR) technologies, costs have decreased and it is now possible to consider the development of driving simulators at an accessible price for research using commercial components.

Within this framework, we present a set of preliminary experimental results on driving behavior achieved through a low-cost VR driving simulator. The obtained results allowed assessing potentials and shortcomings, and considering a more systematically use of VR tools to study the impact of considered human factors in driving behavior.

58.2 Experimental Equipment, Participants and Procedure

The developed VR system consists of a driving simulator in an ad hoc designed urban scenario (see Fig. 58.1, the virtual environment) representing a slightly asymmetrical 8-shaped road intersecting multiple unsignalized pedestrian crosswalks (see Fig. 58.2). Standard commercial devices were used (a Oculus Go headset, a Thrustmaster T80 wheel and pedal set and a low-cost router and PC for communication between the devices). The virtual environment has been developed with the Unity software package, freely available for academic research and education.

The selection of the scenario has been based on the combination of multiple factors: scientific relevance, relation with the real world, technical aspects and accumulated scientific experience within this topic. In particular, a previous study on pedestrian-driver interaction at unsignalized crosswalk allowed to gain empirical knowledge on the decision-making process related to the crossing task performed by pedestrians [6]. Previous research also led to the creation of different simula-

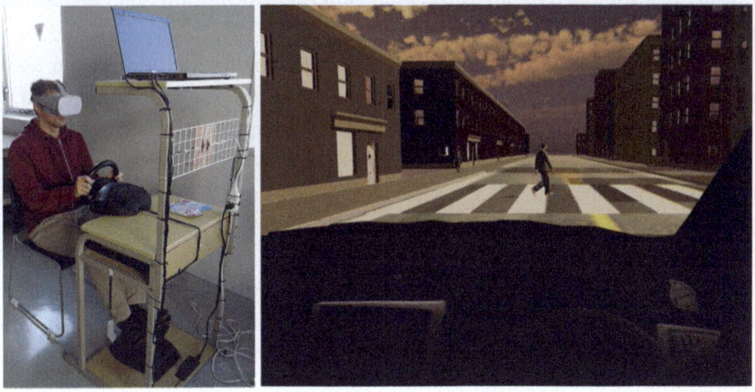

Fig. 58.1 Left: picture of the VR driving simulator including headset, steering wheel, pedals and the PC used to transfer data between them. Total cost of the whole equipment was below 500 $. Right: screenshot of the scenario represented in VR; a pedestrian is seen crossing the road from inside the car

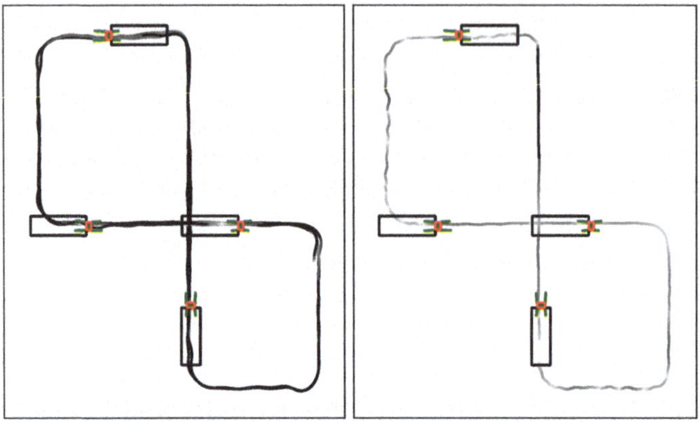

Fig. 58.2 Trajectories representing the driving behavior of two participants during the experiment. Car trajectories are given in gray gradations with the darkness representing the speed (the darker the fastest). Green trajectories are the paths used by crossing pedestrians, with the crosswalk represented in red. The black area before the crosswalk represents the region where drivers and pedestrians interact

tion models, capable to estimate time-gap and fatalities occurring at unsignalized crosswalks based on a variety of parameters (speed limit, vehicular and pedestrian flow, etc.) [4, 5]. Psychological studies in relation to decision-making by pedestrians where also performed using a VR approach [3].

A sample of 16 participants (composed by 12 young adults and 4 elderly drivers) were recruited for the experiments and received a remuneration for the time spent. Before actually having them wearing the VR headset, a questionnaire was handed them to eventually exclude people with a medical history presenting risks in relation with VR and those who came by car or bicycle (one young participant was excluded). In the initial questionnaire we also collected general information (age, gender, driving history, etc.) and asked on previous experiences with VR.

After completing the initial questionnaire, participants had to drive in a straight road where they would practice on accelerating/stopping and lane change. To reduce discomfort, participants had to rest for about 5–10 min before actually starting the real experiments (whose results are presented below) where they would drive through the urban scenario (see Fig. 58.1 on the right) for about 3–5 min. We did not give particular instructions beside telling them to drive safely as they would in the real world.

After the experimental session, to more fruitfully evaluate the performance of the VR tool in terms of user experience, we administered to participants two validated psychological measures, focusing on: usability [8] and feeling of sickness due VR by itself [9].

58.3 Results

In this preliminary analysis, we focused on the driving behavior in the presence of crossing pedestrians, collecting data on aspects such as: driving compliance to pedestrian right-of-way, driving speed and driving accuracy. Driving accuracy was evaluated by computing the standard deviation of the yaw angle error from the road direction in the straight sections of the circuit. Under these conditions, 0 would represent a completely straight perfect driving. The time gap (the ratio between the distance to the crosswalk and the speed of the car) was evaluated for every frame when both car and pedestrians were inside the interaction region given in black in Fig. 58.2 (the minimum was later used). Data collected from the questionnaires were also used to study limitations of VR in relation to user experience.

Some initial qualitative results can be obtained by inspecting Fig. 58.2. It appears clear that each participant has a different driving style, some are confident and fast (left side of Fig. 58.2), sometimes leading to imperfections when turning; others are less confident (right) and reduce their speed to avoid collisions and stay well within the roadway.

When results are converted into numerical values, our analysis show that driving behavior in the VR environment is in line with experimental observations [7]. In particular, young adults are more prone at showing a risky behavior by driving at higher speed and having a lower time gap compared to elderly (see Table 58.2). However, the lower driving accuracy by elderly could also be related to their difficulties in adapting with the VR system as shown in the results from the questionnaire.

Concerning user experience, the results achieved through the usability and sickness measures show, from a qualitative point of view, the satisfactory usability of the VR system (especially considering the very low budget) and the influence of participants' age. However, the low scores obtained also show that there is still a long way to go until VR systems could actually reproduce driving behavior with fidelity.

Table 58.2 Results relative to the driving behavior of participants depending on the age group and most important indicators from the questionnaires (normalized to a 0–1 scale)

Age group	Average speed	Maximum speed	Minimum time gap	Driving accuracy
Adults (19–33)	48.86 km/h	76.12 km/h	0.351 s	10.05°
Elderlies (67–70)	20.78 km/h	59.09 km/h	0.528 s	20.15°
Age group	Nausea	Oculomotor symptoms	Simulation fidelity	Overall system usability
Adults (19–33)	0.412	0.484	0.570	0.554
Elderlies (67–70)	0.500	0.491	0.450	0.436

58.4 Conclusions

In this paper some preliminary studies on the use of VR low-cost commercial technologies to observe driving behavior have been presented. A VR driving simulator has been developed and tested using subjects belonging to different age groups. Results show that driving behavior difference among age groups was comparable between the virtual environment and reality. Although a more complete analysis is necessary to exclude, for example, if this is due to the digital literacy of younger generations (and to understand the digital-divide more in general), results are also promising as they show that simple low-cost solution could be used to study traffic safety, thus paving the path to developing countries with limited financial resources.

Acknowledgements The authors wish to thank the wonderful secretary-team of the Nishinari laboratory (Erika Shiihara, Yuko Tsumori and En Fujimoto) not only for their administrative support but also for the perfect translation of the questionnaires into Japanese language. We also want to thank the students who helped organizing the experiments. This work was supported by JST-Mirai Program Grant Number JPMJMI17D4, Japan and partially by FONDAZIONE CARIPLO (Italy) "LONGEVICITY-Social Inclusion for a Elderly through Walkability", Rif. 2017-0938 and the JSPS grant FY2019/L19513.

References

1. Bangkok Post, State sets R&D milestone of 1.5% of GDP (2018). https://www.bangkokpost.com/tech/1435590/state-sets-rd-milestone-of-1-5-of-gdp
2. China Daily, China's spending on education in 2018 increases (2019). http://www.chinadaily.com.cn/a/201905/02/WS5cca77e6a3104842260b986d.html
3. N.T. Dang, V. Cavallo, J. Thomson, A. Gorrini, A. Piwowarczyk, F. Vienne, S. Bandini, G. Saint Pierre, Crossing the street in dyads: Social influence on pedestrian behaviour in an immersive virtual environment, in *Road Safety and Simulation International Conference* (RSS2017) (2017)
4. C. Feliciani, L. Crociani, A. Gorrini, G. Vizzari, S. Bandini, K. Nishinari, A simulation model for non-signalized pedestrian crosswalks based on evidence from on field observation. Intelligenza Artificiale **11**(2), 117–138 (2017). https://doi.org/10.3233/IA-170110
5. C. Feliciani, A. Gorrini, L. Crociani, G. Vizzari, K. Nishinari, S. Bandini, Calibration and validation of a simulation model for predicting pedestrian fatalities at unsignalized crosswalks by means of statistical traffic data. J. Traffic Transp. Eng. (English Edition) (2019). https://doi.org/10.1016/j.jtte.2019.01.004
6. A. Gorrini, L. Crociani, G. Vizzari, S. Bandini, Observation results on pedestrian-vehicle interactions at non-signalized intersections towards simulation. Transp. Res. Part F Traffic Psychol. Behav. **59**, 269–285 (2018). https://doi.org/10.1016/j.trf.2018.09.016
7. B.A. Jonah, Age differences in risky driving. Health Educ. Res. **5**(2), 139–149 (1990). https://doi.org/10.1093/her/5.2.139
8. R.S. Kalawsky, VRUSE—a computerised diagnostic tool: for usability evaluation of virtual/synthetic environment systems. Appl. Ergon. **30**(1), 11–25 (1999). https://doi.org/10.1016/S0003-6870(98)00047-7
9. R.S. Kennedy, N.E. Lane, K.S. Berbaum, M.G. Lilienthal, Simulator sickness questionnaire: an enhanced method for quantifying simulator sickness. Int. J. Aviat. Psychol. **3**(3), 203–220 (1993). https://doi.org/10.1207/s15327108ijap0303_3

10. Organisation for Economic Co-operation and Development (OECD), OECD database (2018). https://data.oecd.org
11. Swiss Federal Audit Office, Einsatz von Simulatoren bei der Schweizer Armee (2013) (in German)
12. The United Nations Educational, Scientific and Cultural Organization (UNESCO): UIS.Stat (2019). http://data.uis.unesco.org
13. The World Bank, DataBank—World Development Indicators (2019). https://databank.worldbank.org
14. Trading Economics, Thailand GDP per capita (2017). https://tradingeconomics.com/thailand/gdp-per-capita
15. World Health Organization, Global status report on road safety 2015 (2015)
16. World Health Organization, Global status report on road safety 2018 (2018)

Chapter 59
Bicycle Parking Choice Behaviour at Train Stations. A Case Study in Delft, the Netherlands

Alexandra Gavriilidou, Laura Pardini Susacasa, Nagarjun Reddy, and Winnie Daamen

Abstract Due to the increasing use of the bicycle as access and egress mode to public transport hubs, bicycle parking facilities are being constructed and extended. The provision of appropriate infrastructure requires the understanding of the needs and preferences of the cyclists. In order to gain insights into the parking spot choices of the cyclists, we have collected data at a bicycle parking facility in Delft train station (the Netherlands). This dataset is used to estimate discrete choice models that distinguish between an uncongested (i.e. many empty spots) and a congested (i.e. almost full) state of the facility. The findings reveal that cyclists want to park close to the exit of the parking facility which promotes the design in tiers. At the same time, they want to exert minimal effort, which calls for an easier mechanism to park in the top tier. Last but not least, providing real-time information signs is advisable, as it facilitates the parking spot choice as by reducing searching time.

59.1 Introduction

The interest in and use of bicycles is increasing in urban areas, also as access and egress mode for longer distance trips. The latter requires the provision of seamless connections to other modes. Bicycle parking facilities at train stations, for example, are of vital importance to guarantee this seamless connection. In order to provide appropriate bicycle parking infrastructure, it is important to understand and predict the bicycle flows and time spent in the facility. One of the main determinants of the

---

A. Gavriilidou (✉) · L. Pardini Susacasa · N. Reddy · W. Daamen
Delft University of Technology, Stevinweg 1, 2628 CN Delft, The Netherlands
e-mail: a.gavriilidou@tudelft.nl

L. Pardini Susacasa
e-mail: L.PardiniSusacasa@student.tudelft.nl

N. Reddy
e-mail: N.Reddy@student.tudelft.nl

W. Daamen
e-mail: w.daamen@tudelft.nl

I. Zuriguel et al. (eds.), *Traffic and Granular Flow 2019*,
Springer Proceedings in Physics 252,
https://doi.org/10.1007/978-3-030-55973-1_59

underlying model is the parking spot choice behaviour of cyclists. To the best of the authors' knowledge, there is no study investigating this choice behaviour from the perspective of the cyclists. The contribution of this paper is, thus, to fill this gap and develop a corresponding choice model.

As a starting point, literature related to car parking choice has been consulted and transferability to bicycle parking choice is discussed. It has been found that car drivers tend to choose spots that are closer to their destination [3]. This effect is especially present for male drivers [5, 6]. The factor age is only significant when a parking fee is applicable [4]. van der Waerden et al. [6] found a difference between parking behaviour during quiet and busy times. This difference has been observed to particularly affect the usage of real-time information signs that display the availability of spots. Drivers do not use the signs when they drive to their first preferred spot. When that is not available, they make use of the signs to find an available spot [2]. In another study, however, the information signs were found to be used and trusted by the drivers leading to time savings [1]. Even though the precise use of the information signs is inconclusive, it becomes clear that they do play a role in the parking choice.

By transferring these car-related findings to bicycles, we hypothesise that the bicycle parking choice follows the flow diagram of Fig. 59.1. A cyclists entering the parking facility needs to weigh two elements, namely the walking distance to the exit and the parking effort when there is an upper and a lower tier. Moreover, we hypothesise that the occupancy of the facility matters. In an uncongested parking facility, cyclists are free to park in their desired spot, while in a congested state (i.e., when the majority of the parking places is occupied), the availability of spots is also weighed in, by considering the real-time information (RTI) signs. Personal characteristics, such as age and gender, are also believed to play a role in this choice. Gender is expected to influence the importance of walking distance, while age affects the significance of parking effort. As we expect that the state of the parking facility

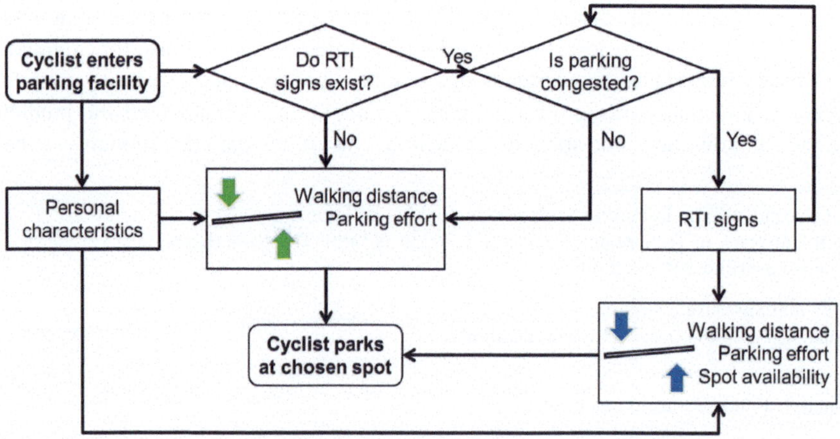

Fig. 59.1 Flow diagram of bicycle parking choice behaviour

affects the parking choice behaviour of cyclists, we develop two models, one for each state, using the random utility maximisation principle and discrete choice theory.

The paper is structured as follows. Section 59.2 describes the data collection of bicycle parking choices, which are used to estimate the behavioural models developed in Sect. 59.3. In Sect. 59.4 the results of the model estimation are presented and discussed, leading up to the conclusions in Sect. 59.5.

59.2 Data Collection

The data collection took place at the central train station of Delft, the Netherlands. There are currently two underground bicycle parking facilities at this station, with a total of 7700 spots. Both have two tier parking spots and RTI signs. The second parking facility of about 2000 spots was chosen for the data collection, because the RTI signs show the spot availability per tier of each row, instead of per row which is the case in the first parking. The layout is sketched in Fig. 59.2, along with pictures taken inside the parking to show the RTI signs and the two tier design within a row.

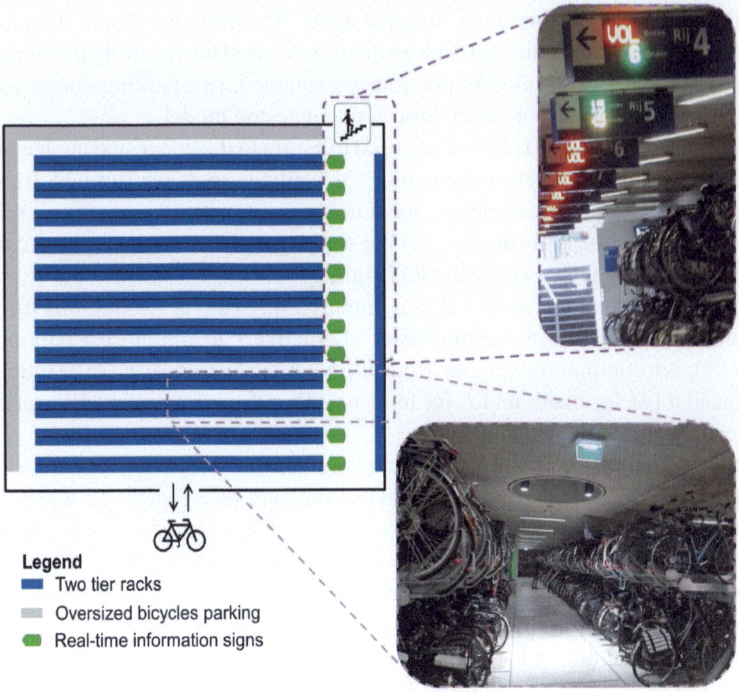

Fig. 59.2 Layout of observed bicycle parking facility in Delft train station, The Netherlands. The photos show the RTI signs and the arrangement of spots within a row in two tiers

Revealed and stated preference data were collected during the morning peak hours of two working days in December 2018. The revealed data consist of observations of the parking choices made by incoming cyclists and of recordings of the number of available spots as indicated by the RTI signs. In total, 1167 parking choices were observed over 13 rows with 2 tiers each. The RTI signs have an updating frequency of 1 min and the recordings were made with an interval of 5 min.

The stated preference data were collected by means of a survey, filled in by users of the parking facility. The questionnaire visualised four parking choice dilemmas, trading off the parking effort (top/bottom tier) and the distance to the parking exit (far/close). The respondents had to choose one alternative per case, which led to 460 stated choices. Moreover, the gender and age group (youngster, adult, elderly) were collected as personal characteristics in both datasets. The majority of the persons, however, belongs to the adult age group, which does not allow the further consideration of age as a factor affecting the bicycle parking choice.

59.3 Model Estimation Approach

The collected datasets cover aggregated spot choices, i.e., segments in the facility with similar characteristics containing multiple spots. These segments are discrete and finite and comprise the choice set. The estimation is based on the utility maximization principle as the decision rule. As previously explained, two parking choice models are developed, namely a congested and an uncongested model.

The observed (revealed) choices are used to estimate the congested choice model, since during the data collection the number of available spots was limited. Based on descriptive statistics, three attributes seem to have a statistically significant relation with the parking choice: (i) the tier, (ii) the walking distance to the parking exit and (iii) the availability indicated by the RTI signs.

The tier of alternative i is coded as a dummy D_{Top_i} (top is 1, bottom is 0), while the walking distance from the middle of alternative i to the closest parking exit $WalkDist_i$ is a continuous variable. Both attributes are hypothesised to be influenced by the gender (M for males and F for females). Therefore, four interaction terms are considered in the utility function.

With respect to the availability signs, the information is captured in two attributes. The first represents the colouring scheme of the sign (see Fig. 59.2), which is red when the tier of a row is full, orange when there are less than 5 spots available and green otherwise. The hypothesis is that cyclists would opt to go towards a choice alternative that has a green sign and, therefore, the dummy D_{Green_i} is created (green is 1, not green is 0). Moreover, if cyclists upon entering would inspect all the signs, they are considered more likely to choose an alternative that has a higher availability compared to the others. The dummy D_{HAV_i} takes the value 1 when the availability of alternative i is greater than the average of all alternatives , and otherwise it is 0.

The utility function for alternative i of the congested model is:

$$V_{\text{congested}_i} = (\beta_{\text{TopF}} \cdot F + \beta_{\text{TopM}} \cdot M) \cdot D_{\text{Top}_i} + \beta_{\text{Green}} \cdot D_{\text{Green}_i} + (\beta_{\text{WalkF}} \cdot F + \beta_{\text{WalkM}} \cdot M) \cdot WalkDist_i + \beta_{\text{HAV}} \cdot D_{\text{HAV}_i} \tag{59.1}$$

The estimation of the uncongested choice model is based on the survey (stated choices). The attributes of this model are the two parking spot properties that were traded off in the survey. The tier attribute is similar to the congested model, including the two interaction terms with the gender. The distance to the parking exit in this case is a dummy D_{Close_i}, indicating whether alternative i is close to or far from the exit (close is 1, far is 0).

The utility function for an alternative i of the uncongested model is:

$$V_{\text{uncongested}_i} = \beta_{\text{Close}} \cdot D_{\text{Close}_i} + (\beta_{\text{TopF}} \cdot F + \beta_{\text{TopM}} \cdot M) \cdot D_{\text{Top}_i} \tag{59.2}$$

59.4 Results

The estimation results of the two models are summarised in Table 59.1. Negative parameter values correspond to a decrease in utility, while positive values indicate an increase. All considered attributes prove to be statistically significant, and hence they affect the parking behaviour choice.

The negative parameters for the top tier indicate that cyclists dislike parking in the top tier. Top tier spots are for both models four to five times less attractive for female cyclists compared to male cyclists.

Regarding the walking distance to the exit, cyclists prefer to minimise it by parking closer to the parking exit. In congestion, male cyclists penalise walking distance twice as much as female cyclists. This is in line with their lower disutility for top tier spots. It means that male cyclists prefer to park as close to the exit as possible, even if it requires greater parking effort, while female cyclists would rather walk further to find a spot in the bottom tier. The break-even point in this trade-off is 56.5 m for women, in comparison to 6.5 m for men.

With respect to the real-time information signs, a green indication increases the utility of that alternative. High availability also has a positive influence, though a bit more conservative. The reason might be that it is rather easy for incoming cyclists to check the colour of all signs, but comparing the actual numbers is more involved and they are less willing to perform the calculations. In any case, it is proven that the RTI signs are appreciated by cyclists when the parking is at a congested state and the effort to find a spot is higher because of the scarcity of spots.

Table 59.1 Model estimation results

Model ($\bar{\rho}^2$)	Coefficient name	Description	Value	Robust standard error	Robust t-test
Uncongested (0.16)	β_{TopF}	Top tier & Female	−1.38	0.18	−7.64
	β_{TopM}	Top tier & Male	−0.29	0.13	−2.19
	β_{Close}	Close to exit	0.91	0.16	5.68
Congested (0.10)	β_{TopF}	Top tier & Female	−2.26	0.14	−16.54
	β_{TopM}	Top tier & Male	−0.52	0.11	−4.81
	β_{WalkF}	Walking distance & Female	−0.04	0.01	−5.57
	β_{WalkM}	Walking distance & Male	−0.08	0.01	−2.19
	β_{Green}	Green number in RTI sign	1.05	0.14	7.58
	β_{HAV}	High availability	0.64	0.09	7.10

59.5 Conclusions

The present study investigated the bicycle parking choice behaviour at parking facilities present at public transport hubs. Data were collected at a bicycle parking facility in Delft, the Netherlands, and used to estimate two discrete choice models, which capture the different behaviour under a uncongested and a congested state of the facility.

The model estimation results reveal that cyclists prefer to park close to the exit of the parking facility to reduce their walking distance. At the same time, they dislike exerting effort when parking. This attitude is stronger for women, who would rather walk more than 50 m to find a spot in the bottom tier, while the corresponding threshold for men is lower than 10 m. The information signs were found to have a positive contribution to the utility function, especially with respect to the colouring scheme.

These insights have implications for the design guidelines of parking facilities. In order to reduce walking distances, the tier design is promoted as it decreases the size of the facility. However, this increases the parking effort, which is especially disliked by women and may call for the installation of an easier mechanism to park in the top tier in the rows furthest away from an exit of the facility. Last but not least, providing real-time information signs with colours for different occupancy levels is advisable, as it facilitates the parking spot choice by reducing the search time.

It is important to check the transferability of these findings before design guidelines can be developed. Future studies should also further investigate the effect of age, as it might have an influence on the parking effort.

Acknowledgements The authors thank the 29 M. Sc. students of the CIE4831-18 course who performed the data collection. Moreover, the authors are grateful to NS Stations for granting permission for the data collection within their facilities. This research was supported by the ALLEGRO project (no. 669792), which is financed by the European Research Council and Amsterdam Institute for Advanced Metropolitan Solutions.

References

1. K.W. Axhausen, J. Polak, M. Boltze, *Effectiveness of Parking Guidance and Information Systems: Recent Evidence from Nottingham and Frankfurt/main*. Transport Studies Unit Oxford University-publications-TSU-all series (1993)
2. K.W. Axhausen, J.W. Polak, M. Boltze, J. Puzicha, Effectiveness of the parking guidance information system in Frankfurt am main. Traffic Eng. Control **35**(5), 304–309 (1994)
3. P. Bonsall, I. Palmer, Modelling drivers' car parking behaviour using data from a travel choice simulator. Transp. Res. Part C Emerg. Technol. **12**(5), 321–347 (2004)
4. D.A. Tsamboulas, Parking fare thresholds: a policy tool. Transp. Policy **8**(2), 115–124 (2001)
5. T.T.A. Vo, P. van Der Waerden, G. Wets, Micro-simulation of car drivers' movements at parking lots. Proc. Eng. **142**, 100–107 (2016)
6. P. van der Waerden, A. Borgers, H. Timmermans, Travelers micro-behavior at parking lots-a model of parking choice behavior, in *Proceedings of the 82nd Annual Meeting of the Transportation Research Board* (2003)

Chapter 60
Exploring the Potential of Neural Networks for Bicycle Travel Time Estimation

Giulia Reggiani, Azita Dabiri, Winnie Daamen, and Serge P. Hoogendoorn

Abstract A tool for travel time estimation of cyclists approaching a traffic light can monitor level of service of intersections in bike crowded cities. This work represents a first step in developing such a tool. Neural Network models are evaluated on how they perform in estimating individual travel time of cyclists approaching a signalized intersection. Based on simulated scenarios, in cities with low bicycle levels (deterministic scenario), Neural Networks are good travel time estimators whereas, in places with high bike volumes (where cyclists depart with a discharge rate) information on queued cyclists is crucial for travel time information.

60.1 Introduction

While some cities are struggling to increase bicycle usage, others are successful in encouraging adoption of cycling but become victims of their own success. Such 'cycling cities' struggle with high levels of bike flows, long queues at traffic lights and discontent cyclists due to the delay in their travel time. Traffic management solutions can mitigate the situation by reducing delay using adaptive traffic controllers or rerouting users to intersections with short delays. In order to deploy such systems, a tool that estimates cyclists travel times, as proxy for bike level of service at intersections, is crucial.

To develop a tool that serves the needs of a bike travel time monitoring system at intersections the following data requirements are set: (1) enabling to derive travel

G. Reggiani (✉) · A. Dabiri · W. Daamen · S. P. Hoogendoorn
Department of Transport and Planning, Technical University of Delft, Delft, The Netherlands
e-mail: g.reggiani@tudelft.nl; giulia.regg@gmail.com

A. Dabiri
e-mail: a.dabiri@tudelft.nl

W. Daamen
e-mail: w.daamen@tudelft.nl

S. P. Hoogendoorn
e-mail: s.p.hoogendoorn@tudelft.nl

I. Zuriguel et al. (eds.), *Traffic and Granular Flow 2019*,
Springer Proceedings in Physics 252,
https://doi.org/10.1007/978-3-030-55973-1_60

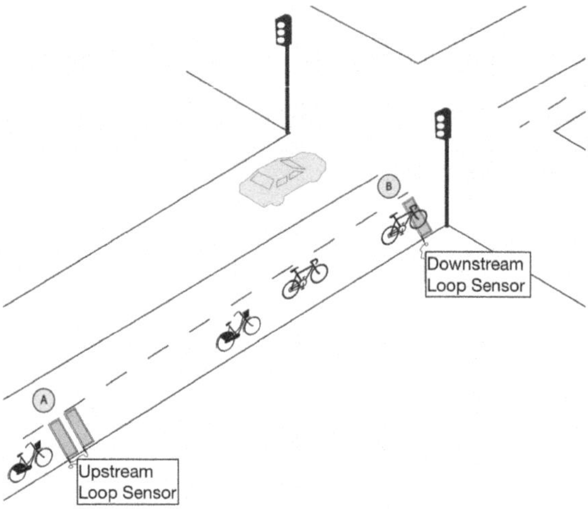

Fig. 60.1 Position of bicycle loop sensors at intersections

time (2) collected over an extensive time frame, (3) representative of user population, (4) readily and real-time available and (5) privacy proof. Some studies investigated the potential of GPS to measure delays (see [3] and references therein). However, GPS only fits the first of the five data requirements: GPS data are collected either via sport apps which enable collection over extensive time frames, but can only represent the "sport" trips, or it is collected via expensive data collection methods, which can equip a representative sample of the population with GPS trackers, but for a limited amount of time (thus not satisfying condition 2). In addition GPS data is not readily available from municipalities and stores sensitive user information. Therefore, this research will use a data set, potentially available from an intersection equipped with loop sensors, a traffic light and a bike queue measurement system because such simulated data set has the potential of meeting all data requirements. Part of this data is readily available to dutch municipalities, due to the extensive deployment of loop sensors on signalised intersections in the Netherlands. Loops are usually installed as shown in Fig. 60.1: 2 upstream of the traffic light (for direction measurement) and one downstream at the stop-line.

Within this work we investigate the properties of a Neural Network (NN) model, when estimating individual cyclists' travel times. Previous studies have explored the possibility of extracting travel times with more "easy to interpret" odels like regressions but were not successful [1]. Our work will go one step further by exploring the potential of a more complex models like NNs.

The wide applicability of NN in the transport domain [4] and scalability would allow these models to easily scale up to incorporate more variables from the same intersection but also from other intersections in a network-oriented approach. We train and test the model on simulated data because it allows for the evaluation of both the model and its input variables.

Section 60.2 presents the methodology describing the simulation settings, the input features and the model. Section 60.3 contains the performances of NNs, in order to investigate if the deployment of these models in reality is effective. Finally, the conclusions are reported in Sect. 60.4.

60.2 Research Methodology

Our research methodology consists of 3 major steps: (1) Simulation of the arrival-departure process of cyclists at the traffic light, (2) Identification of variables to extract from the simulation to use as features for the NN, and (3) Computation of estimation error. Figure 60.2 shows how the research steps interrelate. We decide to test different feature combinations on each scenario (see Fig. 60.3), in order to investigate which feature variables carry more information depending on the simulated setting.

60.2.1 Simulation for Data Generation

We use simulation and not real data from a signalised intersection with sensors because of 4 main reasons. (1) Simulation allows a controlled environment to measure the performances of NNs as complexity is added. (2) We can simulate data not yet available (like queue of cyclists) and assess if collecting such data is valuable for a NN. (3) It allows us to train the model on correct ground truth data that loop sensors alone are not able to deliver, due to occlusion error (see [2] for the definition) in the downstream loop. (4) Based on the NN performance on simulated data, it will be clear if to pursue testings on real data.

Data is simulated based on four scenarios which vary depending on the cycling time, queuing model of cyclists, and high demand of cyclists. Hereafter the four scenarios simulating arrival-departure process of cyclists are described (from simple to more complex):

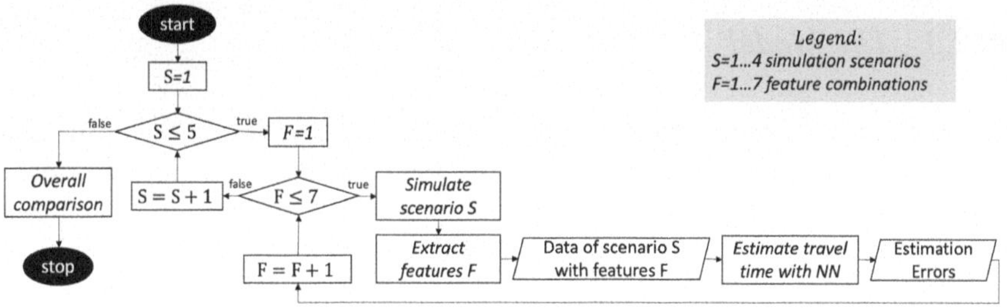

Fig. 60.2 Flow of the research steps

- Deterministic scenario: cycling time is the same for all individuals and cyclists depart from the stop line as soon as the traffic light turns green.
- Discharge Rate scenario: similar to the deterministic scenario with the added complexity that cyclists do not depart from the downstream sensor all at the same time but with a discharge rate.
- Stand over queues scenario: road capacity constraints are considered and arrival rate is modelled to simulate high cyclists demand, so more cyclists are in the queue than can be discharged in one cycle. Cyclists may stay in the queue for more than one red light cycle.
- Stochastic scenario: based on a random arrival-departure process, cycling time between 2 loops is not fixed but modelled according to a normal distribution.

60.2.2 Selection of Feature Variables

Hereafter, a set of five features has been defined based on the data potentially available, the moment a cyclist approaches the upstream senors of bicycle intersections, in the Netherlands.

- Arrival time: this variable contains date-time information of the moment a cyclist reaches the upstream loop sensor.
- Upstream traffic light: carries a 0–1 information to represent green (0) and red light (1) state when the cyclist reaches the upstream sensor.
- Downstream traffic light: carries a 0–1 information to represent green (0) and red (1) light state when the cyclist reaches the upstream sensor (this data might not be available in real settings, but is used as a check).
- Elapsed time from traffic light change: defines, at the arrival time of the cyclist, the seconds passed since the last change in state of the traffic controller.
- Bike queue: defines the number of cyclists waiting for a green light.

Seven combinations of these five features define the data-sets used for the different experimental scenarios (see Fig. 60.3).

60.2.3 The Model

NN models have shown to be extremely versatile and perform well even without a priori assumption on the variable distribution. Their generalization properties make NNs suitable for our purpose. Like all data driven models that learn by minimising the predicted error, NNs need labels (i.e. travel time) of past observations in order to learn how to estimate future ones. Once the NN is trained on past travel time observations (in this case simulated), it will be able to estimate travel time of never seen before observations.

Feature Legend

- Arrival time
- Upstream traffic light
- Downstream traffic light
- Elapsed time from traffic light change
- Bike Queue

Feature combinations \ Scenario	Deterministic	Discharge Rate	Stand Over Queue	Stochastic
1)	0.15	5.64	396.34	19.19
2)	0.00	6.10	300.91	1.58
3)	0.16	5.08	306.26	19.34
4)	0.13	5.00	391.82	18.66
5)	0.14	6.70	315.25	19.18
6)		1.13	2.23	
7)		1.28	2.08	

Fig. 60.3 Mean square error for bicycle travel time estimation: Estimation performance (on test data) of the NN on the 4 scenarios, tested on different feature combinations. Feature combinations with queue information are not considered in the Deterministic and Stochastic case, since these scenarios are simulated without cyclists' discharge rate

60.3 Numerical Results

In this section we report results form the numerical experimentations. For reproducibility, we first describe the structure and parameters of the NN used, as resulting form the numerical experiments. Follows, a description of the NN performance, based on mean square errors in the different scenarios tested on the various feature variable combinations.

60.3.1 Neural Network Structure

Throughout the numerical experiments, a shallow Feed-forward Neural Network, with 6 neurons, emerged as the architecture with the smallest validation error. The NN was implemented in MATLAB software. A structured investigation indicated that the network architecture is adequate, because increasing the number of layers or neurons per layer on average did not improve test performances. Where performances were measured through the mean square error (MSE) as performance function.

60.3.2 Model Performance

Via simulation, a data set of 7200 instances is generated, 70% of which is used for training, 15% for validation and 15% for testing the NN model. Estimation performance of the NN on the different scenarios is reported in Fig. 60.3.

The Deterministic process is the one the NN can estimate better, as expected, since the process is simple. This is deduced by the very small test error, of tenth of a second, on all the scenarios, compared to the other three processes. The data in the feature combination 2 will not be available in real cases; we use it as a check case to see how well the model can predict if we provide it with signal of the traffic light at the time the cyclist would arrive at the downstream (in real life the time the cyclist arrives downstream is not known). The second lowest error in the deterministic scenario is with features: arrival time and elapsed time from traffic light change. This means that, the model estimates better when knowing how many seconds have passed since the change in traffic light state, because it is a FIFO based scenario.

If the process incorporates a queue discharge rate of cyclists, as in the Discharge Rate and Stand Over Queue scenario, feature combinations with bike queue have the smallest estimation error. Having the queue as feature reduces the estimation mean square error up to 2 orders of magnitude. The main reason is that the queue feature incorporates the dynamic information of the arrival-departure process at signalised intersections (i.e. a cyclist has to wait for the queue ahead to discharge, before it can depart again). The Stand Over Queue scenario, incorporated high peak of cyclists arriving at the intersection and Fig. 60.4 shows how accurately the model can estimate travel times in high peak (longer waiting time) when queue information is provided and how it would perform without it. Without queue information the NN can not reproduce the longer travel times that occur when cyclists need to stand in the queue for more than one traffic light cycle. Among the feature combinations that have queue information including the elapsed time form traffic light change improves estimation error in the Stand Over Queue but not in the Discharge rate process. The reason being that in the former process a cyclist is always discharged within the first traffic light cycle, thus elapsed time does not provide as useful information as when the cyclist stays more than one cycle.

Overall, in the Stochastic process, the reached performances are not sufficiently accurate. As information used in the second data-set is not available, on average the error reached by the NN is of 19 s. This indicated that, as the process is more complex, the information considered is not enough for the NN to reproduce the underlying process that generated the data.

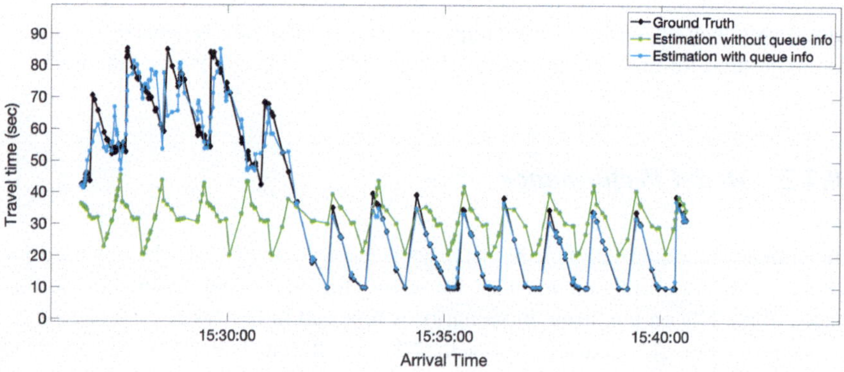

Fig. 60.4 Visualization of model performance in the Stand Over Queue scenario, considering feature combination 3 (without queue info) and 7 (with queue info)

60.4 Conclusions

This work provides a structured investigation, based on a simulation methodology, on how Neural Networks perform for individual travel time estimation. This is the first study on bicycle travel time estimation at intersections, in order to develop real-time bike level of service measures. The investigation of effectiveness of Neural Networks made clear the potentials and limitations of these models. In cities with low bicycle levels (deterministic scenario), NNs are good travel time estimators since with all data sets the reached error is of tenth of a second. Whereas, in places with high bike volumes (where cyclists depart with a discharge rate), only data sets with information on queued cyclists lead to acceptable error of 1–2 s. The main limitation of using NN models to estimate individual bicycle travel time is the availability and richness of the data.

The results enable us to quantify the estimation error in the four scenarios with the different input data. As a consequence, this quantitatively encourages us in future research to develop queue estimation algorithms (of cyclists) that can improve overall travel time estimation. Moreover, future steps should look into the opportunity to cover more complex processes, and more intersections with this methodology.

References

1. D. Duives, W. Daamen, V. Knoop, Berekening voor fietsverliesuren. Technical report (2017)
2. R. Frank Proulx, J. Robert Schneider, F. Luis Miranda-Moreno, Performance evaluation and correction functions for automated pedestrian and bicycle counting technologies. J. Trans. Eng. **142**(3) (2016). Content ID 04016002
3. J. Strauss, L.F. Miranda-Moreno, Speed, travel time and delay for intersections and road segments in the Montreal network using cyclist Smartphone GPS data. Trans. Res. Part D: Trans. Environ. **57**(September), 155–171 (2017)
4. D. Teodorovič, K. Vukadinovič, *Applications of Artificial Neural Networks in Transportation*, Traffic co edn. (Springer, Dordrecht, 1998)

Chapter 61
Extended Longitudinal Motion Planning for Autonomous Vehicles on Highways Including Lane Changing Prediction

Basma Khelfa and Antoine Tordeux

Abstract We propose the concept of multi-lane adaptive cruise control system through an extended algorithm for the longitudinal motion planning of autonomous vehicles on highways including the prediction of lane changing events. The main idea is to couple a hidden Markov model for the detection of lane changes to the adaptive time gap car-following model for a full automation of the driving on multi-lane roads. The aim of the coupled algorithm is to adapt the pursuit according to prediction of the lane change events. Examples of coupling are relaxing the pursuit in case of cut-in to get a smooth vehicle insertion, or reducing temporally the time gap in case of overtaking. We analyse empirically the lane changes in a trajectory database and propose coupling solutions of the car-following model to the identified lane changing events.

61.1 Introduction

More than 90% of road accidents are due to human errors [1]. On highways, the main factor of accident is lane changes [2]. Predicting the trajectories and the lane changing intention of other drivers is one of the crucial issues for the safety of an autonomous vehicle. Although planning trajectories is not a deterministic process, it is possible to specify the most probable trajectory and lane changing events [3]. Furthermore, it has been observed that drivers adapt the pursuit behavior in case of cutting or overtaking events [4]. Basic Adaptive Cruise Control (ACC) systems can provide unappropriated responses, for instance unsafe over-braking in case of cut-in [5].

B. Khelfa (✉) · A. Tordeux
Division for Traffic Safety and Reliability, University of Wuppertal, Wuppertal, Germany
e-mail: khelfa@uni-wuppertal.de

A. Tordeux
e-mail: tordeux@uni-wuppertal.de

I. Zuriguel et al. (eds.), *Traffic and Granular Flow 2019*,
Springer Proceedings in Physics 252,
https://doi.org/10.1007/978-3-030-55973-1_61

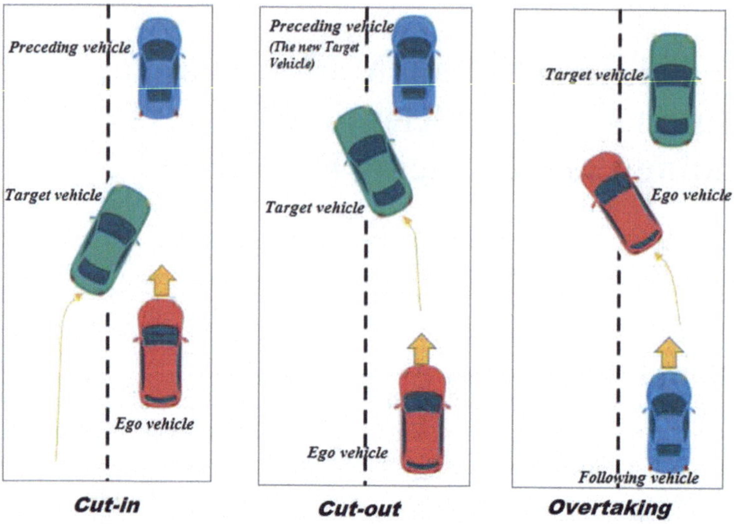

Fig. 61.1 Different cases of lane changes maneuvers (cut-in, cut-out and overtaking) during a pursuit behavior

In this contribution, we propose an extended algorithm for the longitudinal motion planning of autonomous vehicles including the detection of cut-in/cut-out of the target vehicle and other lane changing maneuvers. The extended longitudinal model is based on a Hidden Markov Model (HMM) to predict and detect the lane changing events coupled to the adaptive time gap (ATG) car-following model [6] for the pursuit behavior. The HMM algorithm consists in the estimation of the probability that the target vehicle (the green vehicle in the Fig. 61.1) changes the lane (cut-out) in the middle panel, or the insertion of a new vehicle (cut-in) in the left panel, as well as overtaking maneuvers (the red vehicle in the right panel). Specific pursuit regulations in the ATG are provided in case of cut-in, cut-out and overtaking, in order to adapt and to smooth the behaviors as observed in real situations.

61.2 Lane Change Prediction by Means of Hidden Markov Model

Several models and algorithms are used for the lane changes prediction of autonomous cars. Artificial neural networks (ANN) have attracted the interest of several researchers [7]. Mandalia et al. proposed a vector machine algorithm (SVM) using lateral position data which are measurable by distance sensors to categorize driver intention of changing lanes [8]. Both methods are embraced to predict the practicability and feasibility of lane changing under some conditions w.r.t. spacing distances or speed differences [9]. Even if the methods mentioned above have a high accuracy of lane change detection, they can not recognize slow lane changes unlike hidden Markov chains (HMM) do. Indeed HMM have a higher accuracy of lane change detection [10].

A Markov chain is a series of random variables that can be used to model the dynamic evolution of a random system. A hidden Markov model is based on a Markov model, except that one can not directly observe the sequence of states as the states are "hidden". Each state issues "observations" which are measurable. We are not working on the sequence of states, but on the sequence of observations generated by the states. The graphical model for an HMM is presented in Fig. 61.2. The random variables Z_1, Z_2, ..., Z_n are the hidden variables which represent the hidden states, while X_1, X_2, ..., X_n are the observed variables which represent the observed states. Here n represents the number of states.

The observed variables X_t at time t are the distances Δx and speed differences Δv between the ego vehicle and the target/following vehicles in the current lane c, and in the adjacent lanes (the right r and left l lanes) by translating the ego-vehicle,

$$X_t = \{(\Delta x_t(l), \Delta v_t(l), \Delta x_f(l), \Delta v_f(l)),\ l = c, r, l\}.$$

The hidden variables Z_t at time t are the lane changing intention, namely lane keeping, cut-in left/cut-in right, cut-out left/cut-out right, overtaking,

$$Z_t = \{LK, CIL/CIR, COL/COR, OV\}.$$

The structure of a hidden Markov model satisfies the joint distribution as follows:

$$\mathbb{P}(x_1, x_2, \ldots, x_n, z_1, z_2, \ldots, z_n) = \mathbb{P}(z_1)\mathbb{P}(x_1|z_1)\prod_{k=2}^{n}\mathbb{P}(z_k|z_{k-1})\mathbb{P}(x_k|z_k),$$

where $\mathbb{P}(z_1)$ is the parameter for the initial probability (the Prior), $\mathbb{P}(z_k|z_{k-1})$ is the parameter for the transition probability, and $\mathbb{P}(x_k|z_k)$ is the emission probability parameter which can be calculated using a special case of the Expectation-Maximization (EM) algorithm related as the Welch-Baum algorithm.

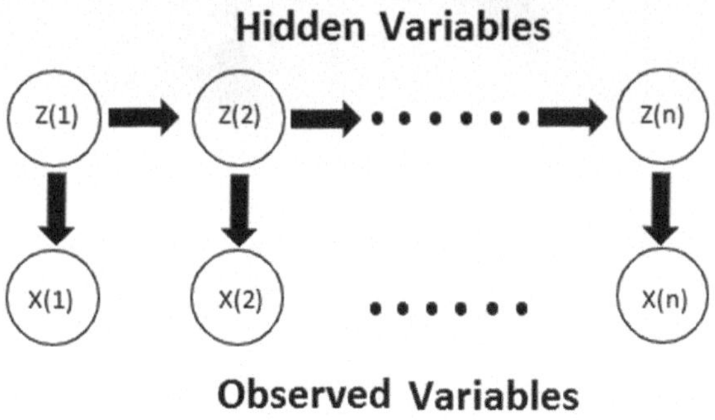

Fig. 61.2 Structure of a Hidden Markov Model (HMM) (Trellis Diagram)

61.3 Empirical Analysis of Lane Changes in Highways

The Next Generation Simulation (NGSIM) data are one of the few published data-sets of microscopic trajectories of road vehicles [11]. The NGSIM project provides several samples of trajectories at our disposal. We analyse the lane change events on a 500 m long segment of the Interstate 80 dataset in Emeryville, San Francisco, California (I-80 dataset), and on a 640 m long segment of the 101 highway (Hollywood Freeway) in Los Angeles, California (101 dataset). Both dataset are trajectories of vehicles in 5 lanes highway with an on-ramp. The data have been respectively collected on April 13, and on June 15, 2005.

The observations (see Fig. 61.3) show in average that the mean speed and mean spacing during the 2 s after the lane change are higher than the performances measured during the 2 s before the lane change (left and middle panels). However, it is not systematically the case. Furthermore, the speed difference with the vehicle in front is generally positive before a lane change (right panel). Such preliminary analysis show that, as expected, no deterministic rules apply for the decision of changing the lane. However, some correlations exist with the speed difference and spacing, suggesting using such variables as inputs of algorithms for lane changing prediction. Some trajectories of vehicles while lane change are presented in Fig. 61.4. The proportions of lane change behaviors for the observed vehicles over the segment (namely lane keeping—LK, lane changing—LC, lane changing left—LCL, and lane changing right—LCR) are given in the Table 61.1.

The input data to the HMM algorithm are the distance Δx_i and the speed differences Δv_i, where $i = 1, \ldots, 6$ between the ego vehicle (the red vehicle in the Fig. 61.5) and the six surrounding vehicles (the blue vehicles in the current lane and neighboring lanes). To calculate the Δx_i and the Δv_i, we need the positions and the velocity of the six vehicle surrounding the ego vehicle. The output data are the changing-lane events, namely

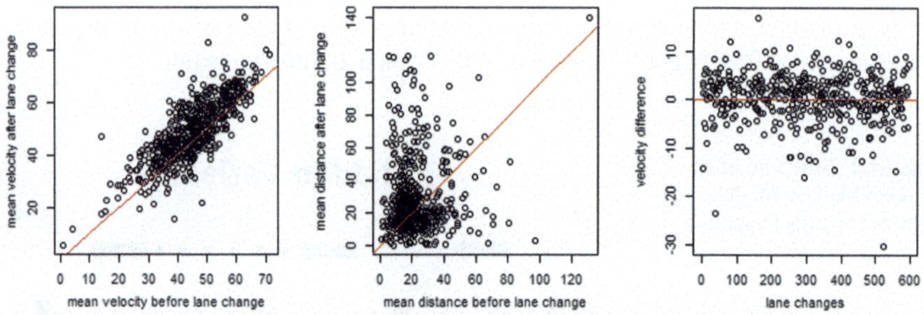

Fig. 61.3 Mean speed in km/h and distance spacing in m 2 s before and after a lane change (left and right panels), and speed difference in km/h with the vehicle in front before a lane change (right panel)

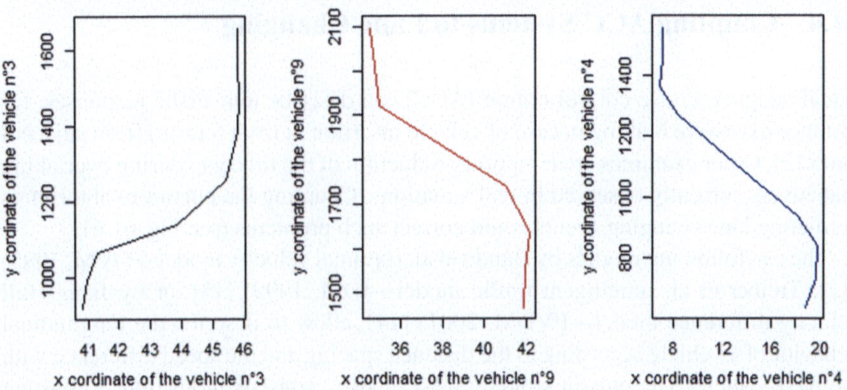

Fig. 61.4 Trajectories of vehicles while lane change (position in meter 3, 5, 7 and 9 s before and after the lane change)

Table 61.1 Probabilities of lane change behavior

States	LK	LC	LCL	LCR
Percent	0.72	0.28	0.16	0.11

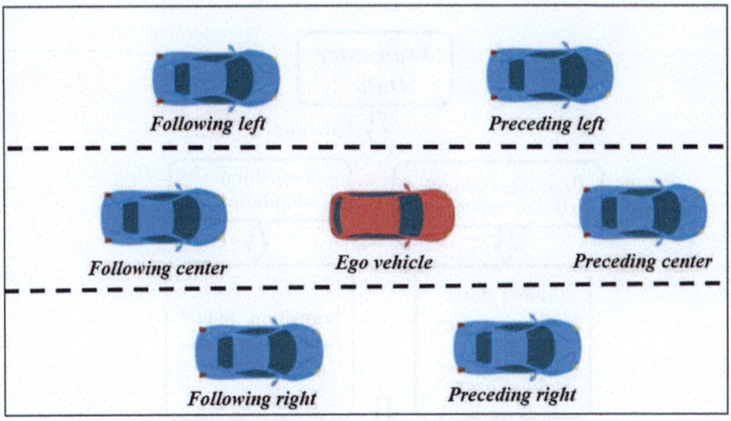

Fig. 61.5 Visualisation of the six vehicles around the ego vehicle that we use to calculate the input data in our algorithm

- LK = lane keeping,
- CIL/CIR = Cut-In Left/ Cut-In Right,
- COL/COR = Cut-Out Left/ Cut-Out Right,
- OV = Overtaking

61.4 Coupling ACC Systems to Lane Changing

Usual adaptive cruise control option (ACC) can describe unrealistic responses, for instance excessive braking in case of vehicle insertion in front (cut-in) from adjacent lanes [5]. Other examples are temporary reduction of the time gap during overtaking maneuvers currently observed in real situations. Coupling the pursuit to algorithms predicting lane changing events could correct such problems (see Fig. 61.6).

The car-following models by Bando et al. (optimal velocity model—OVM, 1995) [12], Treiber et al. (intelligent traffic model—IDM, 1999) [13], or by Jiang (full velocity difference model—FVDM, 2001) [14], allow to describe the longitudinal behavior of a vehicle according to the distance spacing and the speed difference with the next vehicle. The pursuit strategy used in ACC consists in keeping a constant safety time gap T_s with the predecessor. The constant time gap regulation policy is notably the one recommended by the ISO norm 15622:2018 with a safety time gap varying between 0.8 and 2.2 s.

Oppositely to classical car-following models for which the speed, the speed difference, or the distance spacing are relaxed, one assumes relaxation of the time gap towards the equilibrium function with the adaptive time gap ATG model. We can resume it by:

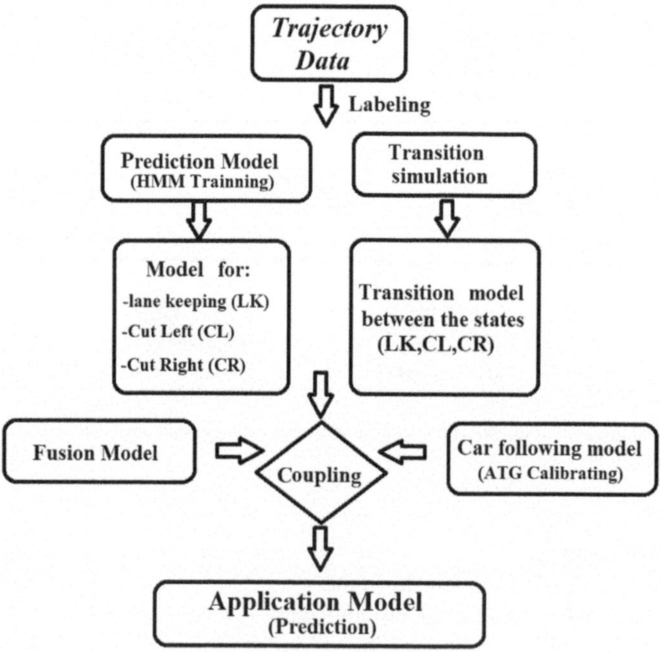

Fig. 61.6 Learning model schema for the lane change prediction method

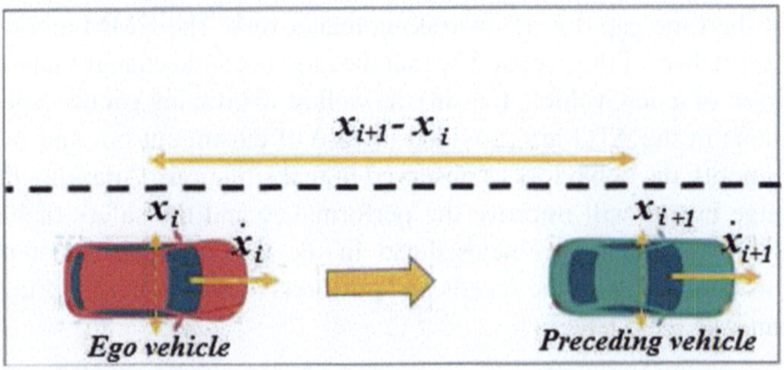

Fig. 61.7 Car-following illustration

$$\dot{T}_i(t) = \frac{1}{T_r}[T_s - T_i(t)],$$

where (the notation is given in Fig. 61.7)

$$T_i(t) = \frac{x_{i+1}(t) - x_i(t) - \ell}{\dot{x}_i(t)}$$

is the time gap of the vehicle i at time t, ℓ being the length of the vehicle including a minimal spacing. Here T_r is a relaxation time parameter and T_s is the constant desired time gap. The corresponding acceleration model is

$$\ddot{x}_i(t) = \frac{\dot{x}_i(t)}{T_r}[1 - \frac{T}{T_i(t)}] + \frac{1}{T_i(t)}[\dot{x}_{i+1}(t) - \dot{x}_i(t)].$$

Robust ACC systems should be based on stable car-following models [15]. The ATG car-following remains stable for any positive value of the parameters T_s and T_r. Consequently, one may change the values of the parameters in case of lane changing events without braking the stability. For instance, the relaxation parameter T_r could be increased in case of insertion of a new vehicle into the lane (cut-in). This could allow to avoid uncomfortable and unsafe over-braking [5]. On the other hand, the safety time gap T_s could be temporary reduced during overtaking maneuvers. This could allow to describe realistic overtaking driving.

61.5 Conclusion

This paper presents an extended longitudinal motion planning for autonomous vehicle including lane changing prediction of the neighboring vehicle. Coupling the pursuit behavior to predicting algorithms for lane changing events could correct a lot of problems such as excessive braking in case of vehicle insertion or temporary reduction of the time gap during overtaking maneuvers. The HMM algorithm consists in the estimation of the probability that the target vehicle changes lane (cut-out), or the insertion of a new vehicle (cut-in), as well as overtaking events. Specific pursuit regulations in the ATG are provided in case of cut-in, cut-out and overtaking, in order to smooth the behaviors as observed in real situations. Adapting the pursuit to lane-change events will improve the performance and the safety of basic ACC systems solely based on next vehicles ahead. In fine, the concept relies on multi-lane ACC systems depending on the speeds and positions of the six neighboring vehicles on current and adjacent lanes.

References

1. S. Singh, Critical reasons for crashes investigated in the national motors. Technical report No. DOT HS 812 115, NHTSA, 11222–022315, vol. 3 (2015)
2. U.S. Department of Transportation, Lane-change/merge crashes: problems size assessment and statistical description, U.S. DOT, Washington, DC, USA, Rep. DOT HS 808 075, (1994)/H. Lum, J.A. Reagan, Interactive highway safety design model: accident predictive module, in *Public Roads*, vol. 58, no. 3 (1995)
3. I. Dagli, G. Breuel, H. Schittenhelm and A. Schanz, Cutting-in vehicle recognition for ACC systems—towards feasible situation analysis methodologies, in *IEEE Intelligent Vehicles Symposium*, Parma, Italy, pp. 925–930 (2004)
4. A. Carvalho, A. Williams, S. Lefvre, F. Borrelli Autonomous cruise control with cut-in target vehicle detection, in *13th International Symposium on Advanced Vehicle Control, IEEE*, pp. 93–98 (2016)
5. Y. Zhang, Q. Lin, J. Wang, S. Verwer, J.M. Dolan, Lane-Change intention estimation for car-following control in autonomous driving. IEEE Trans. Intell. Veh. **3**(3), 276–286 (2018)
6. A. Tordeux, S. Lassarre, M. Roussignol, An adaptive time gap car-following model. Transp. Res. B: Methodol. **44**(8–9), 1115–1131 (2010)
7. Y. Zhang, S. Li, M. Zhang, *Lane-Changing Prediction Modeling on Highway Ramps: Approaches and Analysis, Stanford School of Engineering* (2017). http://cs229.stanford.edu/proj2017/final-reports/5206073.pdf
8. H.M. Mandalia, DD.D. Salvucci, Using support vector machine for lane-change detection, in *Proceedings of the Human Factors and Ergonomics Society Annual Meeting*, vol. 49(22), pp. 1965–1969 (2005)
9. Y. Dou, F. Yan, D. Feng, Lane changing prediction at highway lane drops using support vector machine and artificial neural network classifiers, in*IEEE International Conference on Advanced Intelligent Mechatronics (AIM)*, pp. 901–906 (2016)
10. H. Woo, Y. Ji, H. Kono, Y. Tamura, A. Yamashita, H. Asama, Detection method of lane-change intentions in other drivers using hidden markov models, in *International Conference on Advanced Mechatronics (ICAM): Toward Evolutionary Fusion of IT and Mechatronics*, pp. 253–254 (2015)
11. US Department of Transportation, NGSIM—Next Generation Simulation (2007). https://ops.fhwa.dot.gov/trafficanalysistools/ngsim.htm
12. M. Bando, K. Hasebe, A. Nakayama, A. Shibata, Y. Sugiyama, Dynamical model of traffic congestion and numerical simulation. Phys. Rev. E **51**(2), 1035–1042 (1995)
13. M. Treiber, A. Hennecke, D. Helbing, Microscopic simulation of congested traffic in *Traffic and Granular Flow* vol. 99, pp. 365–376 (2000)

14. R. Jiang, Q. Wu, Z. Zhu, Full velocity difference model for a car-following theory. Phys. Rev E **64**, 017101 (2001)
15. I.A. Ntousakis, I.K. Nikolos, M. Papageorgiou, On microscopic modelling of adaptive cruise control systems, in *4th International Symposium of Transport Simulation-ISTS14*, vol. 6, pp. 111–127 (2015)

Chapter 62
Continuum Traffic Flow Modelling: Network Approximation, Flow Approximation

Megan M. Khoshyaran and Jean-Patrick Lebacque

Abstract Continuum traffic modeling is designed for very large networks and relies on a double approximation. The network is assumed to be dense and is described as a two-dimensional medium, whereas the traffic in the network is described as a bi-dimensional fluid. Traffic on major arterials is described separately using the GSOM (Generic second order model) approach. The dense network is divided into macro-cells, typically one to 10 km in size. The dynamics of traffic in the two-dimensional medium result from the interplay of supply and demand inside and between macro-cells. The paper addresses important modeling issues such as how to estimate the average densities, how to determine the equilibrium traffic supply and demand functions, and how to evaluate traffic supplies and demands.

62.1 Introduction

In order to model traffic over very large areas, typically a metropolitan region, and to solve problems such as reactive or predictive DTA (dynamic traffic assignment), or such as global management problems, a new approach is emerging, based on the idea of continuum approach. The continuum approach relies on a double approximation process: the approximation of the underlying network as a two-dimensional continuous medium, and the approximation of traffic flow in the network by the flow of a two-dimensional fluid. This idea has been applied both to vehicular and pedestrian traffic (for instance the Hughes model [2, 4] of pedestrian flow is well-known).

Taguchi and Iri [17] initiated these ideas by proposing a description of the network as a medium with anisotropic capacity and velocity constraints. Further research

M. M. Khoshyaran
ETC Economics Traffic Clinic, 34 avenue des Champs-Elysées, 75008 Paris, France
e-mail: etclinic@wanadoo.fr

J.-P. Lebacque (✉)
UGE-IFSTTAR-COSYS-GRETTIA, 14-20 Boulevard Newton, Cité Descartes, Champs-sur-Marne, 77447 Marne-la-Vallée, France
e-mail: jean-patrick.lebacque@univ-eiffel.fr

I. Zuriguel et al. (eds.), *Traffic and Granular Flow 2019*,
Springer Proceedings in Physics 252,
https://doi.org/10.1007/978-3-030-55973-1_62

(notably [3, 5, 11]) centered on an isotropic description of the underlying network and a single density for describing a many to one (or one to many) flow. This line of research started with static assignment problems and was extended to DTA. Aghamohammadi and Laval [1] consider applying the concept of MFD (macroscopic fundamental diagram) to dynamic continuum modelling of traffic on large networks.

Saumtally et al [13], Sossoe [14] approximate the underlying network as an anisotropic medium with privileged directions of propagation, and traffic flow is disaggregated per direction or mode, thus allowing for many to many dynamic flows. Sossoe et al. [15] analyzes the limiting system of conservation laws. Khoshyaran and Lebacque [6] extends the model to networks with isotropic and anisotropic areas, using a generalized traffic supply/demand approach (ABTM: anisotropic bidimensional traffic model).

Major arterials should not be integrated into the continuum description of the dense surface road network. Khoshyaran and Lebacque [6] proposes to take into account this fact and to model arterials using a GSOM approach [9, 10]. The GSOM model can be interfaced with the continuum model since both rely on the supply demand approach. This aspect will not be elaborated on in this paper.

For applications, the fundamental problem to be solved is to recover the physical characteristics of the medium representing the network, from the geometric and regulatory properties of the network, as well as from the properties of intersections. These physical characteristics combined with the traffic state of the network determine the local traffic supplies and demands and allow us to evaluate the mean traffic densities. The paper addresses this problem.

62.2 Discretized Continuum Model

62.2.1 General Principles

In this section we describe the continuum model in a phenomenological way, based on a spatial discretization of the network into macro-cells (c), $c \in \mathcal{C}$. The size of a macro-cell ranges from 1–10 km, with a population 10,000–100,000, and a traversal time bounded from below by Δt of the order of 15 mn. The description follows [6].

The **state variables** are the number of vehicles per macro-cell. These variables are updated at each time-step using **flux variables**. Flux variables are of two main types: flows between macro-cells (inter-cell flows) and flows inside macro-cells (intra-flows). Flows are evaluated using the min formula [7], or some generalization thereof as in [8]. Other flows must be taken into account: the travel demand (originating in macro-cells or outside of the network), the exit flows (travelers leaving the network in macro-cells or at the network boundary) and the flows between macro-cells and the arterials.

The **min formula** states that the traffic flow is equal to the minimum between the downstream traffic supply and the upstream traffic demand. Thus the essential ingredients of the model are the traffic demands and supplies which express the physical properties of the network.

62.2.2 Notations

They follow from [6]. Refer also to Fig. 62.1 right.

- Macro-cells (c), $c \in \mathcal{C}$. (ω) constitutes a special macro-cell which represents the outside of the network.
- Time-steps $(t) \stackrel{def}{=} [t\Delta t, (t+1)\Delta t]$ with $t \in \mathcal{T}$.
- Successor macro-cells of a cell $c \in \mathcal{C}$. A macro-cell (g) is a successor of cell (c) if traffic can flow from (c) to (g): $(g) \in \mathrm{Succ}(c)$. The interface between these two macro-cells is denoted (c, g). Physically it is a line separating the macro-cells.
- Predecessor macro-cells of a cell $c \in \mathcal{C}$. A macro-cell (h) is a predecessor of cell (c) if traffic can flow from (h) to (c): $(h) \in \mathrm{Pred}(c)$.
- state and flux variables in macro-cells:

  – $N^t_{c,cg}$ = vehicles in macro-cell (c) at $t\Delta t$, heading for macro-cell $(g) \in \mathrm{Succ}(c)$;
  – $N^t_{c,hc}$ = vehicles in macro-cell (c) at $t\Delta t$, which entered cell (c) through interface (h, c), with $(h) \in \mathrm{Pred}(c)$.

  The total number of vehicles in macro-cell (c) is given by

$$N^t_c \stackrel{def}{=} \sum_{(g)\in\mathrm{Succ}(c)} N^t_{c,cg} + \sum_{(h)\in\mathrm{Pred}(c)} N^t_{c,hc} \tag{62.1}$$

  – inter-cell flows: p^t_{hc} is the flow through interface (h, c) during time-step (t).
  – intra-cell flows $(h) \to (g)$ in (c), where $(h) \in \mathrm{Pred}(c)$ and $(g) \in \mathrm{Succ}(c)$:

    · $q^t_{c,hc}$ flow leaving $N^t_{c,hc}$ during the time-step (t), and $q^t_{c,cc}$ the flow entering the network in macro-cell (c) (for instance parking),
    · $r^t_{c,cg}$ the flow joining $N^t_{c,cg}$ during the time-step (t), and $r^t_{c,cc}$ the flow exiting the network in macro-cell (c) (for instance parking).

Let us now introduce variables pertaining to travel demand and to assignment:

– Θ^t_c: travel demand in cell (c) during time-step (t) ((c) may be equal to (ω));
– Υ^t_c: travel supply in cell (c) during time-step (t);
– $\gamma^t_{c,hg}$ the assignment coefficients for intra-cell flows: fraction of traffic entering (c) from (h) and bound to (g) during time-step (t). Note that $(h) \in \mathrm{Pred}(c)$ but that $(g) \in \mathrm{Succ}(c) \cup \{(c)\}$;
– $\theta^t_{c,cg}$ the assignment coefficients for flows entering the network in macro-cell (c): fraction of traffic entering (c) and bound to (g) during time-step (t). Note that $(g) \in \mathrm{Succ}(c) \cup \{(c)\}$.

These variables constitute exogenous data at the level we consider here. The path choice is outside the scope of the model.

62.2.3 Phenomenological Equations

In order to calculate the dynamics of the network, we will need the following dynamic variables:

- $\delta^t_{c,cg}$: the traffic demand of macro-cell (c) at the (c, g) interface during time-step (t), with $(g) \in \mathrm{Succ}(c)$, during time-step (t). Note that $\delta^t_{\omega,\omega c}$ is the demand from outside the network for macro-cell (c);
- $\sigma^t_{c,hc}$: the traffic supply of macro-cell (c) at the (h, c) interface for traffic entering from (h) during time-step (t), with $(h) \in \mathrm{Pred}(c)$, during time-step (t). Note that if $(\omega) \in \mathrm{Pred}(c)$, then $\sigma^t_{\omega,c\omega}$ denotes the supply outside the network for macro-cell (c);
- $\Sigma^t_{c,cg}$: the internal macro-cell supply for traffic bound to $(g) \in \mathrm{Succ}(c)$, during time-step (t);
- $\Delta^t_{c,hc}$: the internal macro-cell demand of traffic inside macro-cell (c) having come from macro-cell $(h) \in \mathrm{Pred}(c)$, during time-step (t).

We also need functions which describe how conflicts between flows competing for shared resources are arbitrated, the **arbitrage functions**

- Φ^c_{hc} the concave increasing function for flows inside (c) having a common origin $(h) \in \mathrm{Pred}(c)$. These flows share links close to the (h, c) interface;
- Ψ^c_{cg} the concave increasing function for flows inside (c) having a common destination $(g) \in \mathrm{Succ}(c)$. These flows share links close to the (c, g) interface.

The calculation of inter-cell flows is trivial and results from the min principle [7], for all $(c), (h) \in \mathcal{C}$:

$$p^t_{cg} = \min\left[\delta^t_{c,cg}, \sigma^t_{g,cg}\right] \tag{62.2}$$

The intra-flows satisfy supply and demand constraints, for all $(c) \in \mathcal{C}$:

$$\left| \begin{array}{l} 0 \le q^t_{c,hc} \le \Delta^t_{c,hc} \ \forall h \in \mathrm{Pred}(c) \\ 0 \le q^t_{c,cc} \le \Theta^t_c \\ 0 \le r^t_{c,cg} \le \Sigma^t_{c,cg} \ \forall g \in \mathrm{Succ}(c) \\ 0 \le r^t_{c,cg} \le \Upsilon^t_c \end{array} \right. \tag{62.3}$$

They also satisfy conservation equations, for all $(c) \in \mathcal{C}$:

$$r^t_{c,cg} = \sum_{h \in \mathrm{Pred}(c)} \gamma^t_{c,hg}\, q^t_{c,hc} + \theta^t_{c,cg} q^t_{c,cc} \qquad \forall g \in \mathrm{Succ}(c) \cup \{(c)\} \tag{62.4}$$

Finally, following [6, 8], the intra-flows $q^t_{c,hc}$ and $r^t_{c,cg}$ satisfy an optimality principle expressing the arbitrage between resources subject to all constraints (62.3) and (62.4), for all $(c) \in \mathcal{C}$:

$$\underset{\begin{pmatrix}(q_{hc})_{h\in \text{Pred}(c)} \\ (r_{cg})_{g\in \text{Succ}(c)}\end{pmatrix}}{\text{Max}} \left[\sum_{h\in \text{Pred}(c)} \Phi^c_{hc}(q_{hc}) + \sum_{g\in \text{Succ}(c)} \Psi^c_{cg}(r_{cg}) \right] \tag{62.5}$$

The conservation equation in macro-cell (c) for the state variables can be expressed as follows, for all $h \in \text{Pred}(c)$ and all $g \in \text{Succ}(c)$:

$$\left| \begin{array}{l} N^{t+1}_{c,hc} = N^t_{c,hc} + \Delta t \left(p^t_{c,hc} - q^t_{c,hc}\right) \\ N^{t+1}_{c,cg} = N^t_{c,cg} + \Delta t \left(r^t_{c,cg} - p^t_{c,cg}\right) \end{array} \right. \tag{62.6}$$

62.3 Determining the Physical Constants

The object of this section is to analyze how to evaluate supplies and demands in (62.2) and (62.4). The main idea is to derive these quantities from the properties of links.

62.3.1 Equilibrium Functions

Traffic dynamics on links are described by a GSOM model [9, 10], the basic ingredient of which is the **fundamental diagram** (FD) which connects flow to density. We assume traffic attributes to be neutral, i.e. they do not impact the FD (relevant examples of such attributes are route choice attributes). In the absence of calibrated FDs (usually collected traffic data is insufficient to calibrate FDs outside of major arteries), we assume a truncated triangular FD as proposed by Newell:

$$Q^c_e(\rho) = \min \left[V^c_{max}\rho,\, W^c_{max}(\rho_{max} - \rho),\, Q^c_{max}\right] \tag{62.7}$$

normalized for a single lane and specified for a macro-zone (c). In (62.7), Q^c_e denotes the equilibrium flow density function per lane, ρ the density per lane, and the parameters are:

- V^c_{max}: maximum speed, usually resulting from local traffic regulation (between 20 and 45 mph in a urban setting);
- W^c_{max}: maximum congestion wave speed, typically 11 mph;
- ρ_{max}: maximum density per lane (typically 0.18 vh/m i.e. 180 vehicles per km);
- Q^c_{max}: maximum flow (capacity) per lane. If this information is not available, a recommended default value is $\rho_{max} V^c_{max} W^c_{max} / (V^c_{max} + W^c_{max})$.

For a link with ν lanes the maximum flow becomes νQ^c_{max}, the maximum density $\nu \rho_{max}$, the speed parameters are unchanged. The FD parameters can thus be inferred from readily available data.

The equilibrium demand and supply functions [7] of macro-zone (c), normalized for a lane, are given by

$$\begin{bmatrix} \Delta^c_e \stackrel{def}{=} \min \left[V^c_{max}\rho, Q^c_{max}\right] & \text{equilibrium demand} \\ \Sigma^c_e \stackrel{def}{=} \min \left[W^c_{max}(\rho_{max} - \rho), Q^c_{max}\right] & \text{equilibrium supply} \end{bmatrix} \tag{62.8}$$

62.3.2 Effective Evaluation of Mean Densities, Supplies and Demands

In order to use these equilibrium functions we need to estimate traffic densities. Let us consider for instance the density $\rho^t_{c,hc}$ associated to $N^t_{c,hc}$, the number of vehicles in macro-cell (c) at $t\Delta t$, which entered cell (c) through interface (h, c), with $(h) \in \text{Pred}(c)$.

In order to obtain a relevant estimate of this quantity, we consider an auxiliary max-flow problem $(\mathcal{P})$: find the maximum flow from (h) into (c) through the (h, c) interface to all other interfaces $(c, g), \forall(g) \in \text{Succ}(c)$. The flow on each link say (i) is bounded by the capacity $Q_{c,max}$ equal to $\nu_i Q^c_{max}$ with ν_i the number of lanes of (i). The solution to this linear program yields two informations:

- path flows on paths joining (h, c) to all other interfaces $(c, g), \forall(g) \in \text{Succ}(c)$. These paths span a set $\mathcal{A}_{h,c}$ of links in (c). Each of this paths is divided into two parts and we select the links in the path-half closest to (h) in order to form a subset $\mathcal{A}_{c,hc}$ of the set $\mathcal{A}_{h,c}$ of links in (c).
- the minimum cut, along which the maximum flow is achieved. Let $\mathcal{M}_{c,hc}$ be the set of links belonging to this cut.

Let us note that the calculations of $\mathcal{A}_{c,hc}$ and $\mathcal{M}_{c,hc}$ are carried out off-line. These elements allow us to define the quantities necessary to estimate the supplies and demands (Fig. 62.1 right):

- $\ell_{c,hc}$ = total length of lanes in cell (c) which are available for traffic $(h) \to (c)$, if $(h) \in \text{Pred}(c)$. $\ell_{c,hc}$ is the total length of lanes of the links belonging to $\mathcal{A}_{c,hc}$. Note that the length of a link $(i) \in \mathcal{A}_{c,hc}$ is counted ν_i times.
- The density $\rho^t_{c,hc}$ in cell (c) is now defined as (remember it is a density per lane):

$$\rho^t_{c,hc} \stackrel{def}{=} N^t_{c,hc}/\ell_{c,hc} \tag{62.9}$$

- Let μ^c_{hc} be the number of lanes of the links of $\mathcal{A}_{c,hc}$ crossing the cut $\mathcal{M}_{c,hc}$. The demand $\Delta^t_{c,hc}$ is given by:

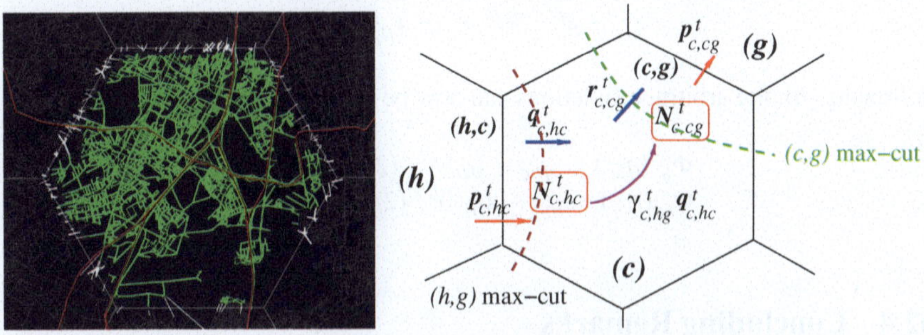

Fig. 62.1 An anisotropic cell.
Left: in white the links involved in inter-cell flows, in green the links involved in intra-cell flows. (R. Belaroussi)
Right: schematic representation of the various quantities involved in calculations

$$\Delta^t_{c,hc} \stackrel{def}{=} \mu^c_{hc} \Delta^c_e \left(\rho^t_{c,hc}\right) \tag{62.10}$$

- Let ν^c_{cg} be the number of lanes available inside cell (c) for traffic flowing $(c) \to (g)$ across (h, c) (Fig. 62.1, left). Then the macro-cell supply for inter-cell flows is given by

$$\sigma^t_{c,hc} \stackrel{def}{=} \nu^c_{hc}\, \Sigma^c_e \left(\rho^t_{c,hc}\right) \tag{62.11}$$

In order to estimate the symmetric elements for $(c) \to (g)$ flows, it is necessary to solve a max-flow problem from all interfaces (h, c), $\forall (h) \in \mathrm{Pred}(c)$ to an interface (c, g) (with $(g) \in \mathrm{Succ}(c)$). The solution of this max-flow problem yields a set $\mathcal{B}_{c,cg}$ of links in (c) which carry the vehicles $N^t_{c,cg}$, and a minimum cut, along which the maximum flow towards (c, g) is achieved. Let $\mathcal{N}_{c,g}$ be the set of links belonging to this cut. These elements allow us to define the quantities necessary to estimate the supplies and demands:

- $\ell_{c,cg}$ = total length of lanes in cell (c) which are available for traffic $(c) \to (g)$. $\ell_{c,cg}$ is the total length of lanes of the links belonging to $\mathcal{B}_{c,cg}$.
- The density $\rho^t_{c,cg}$ in cell (c) is now defined as:

$$\rho^t_{c,cg} \stackrel{def}{=} N^t_{c,cg} / \ell_{c,cg} \tag{62.12}$$

- Let μ^c_{cg} be the number of lanes of the links of $\mathcal{B}_{c,cg}$ crossing the cut $\mathcal{N}_{c,cg}$. The supply $\Sigma^t_{c,cg}$ is given by:

$$\Sigma^t_{c,cg} \stackrel{def}{=} \mu^c_{cg} \Sigma^c_e \left(\rho^t_{c,cg}\right) \tag{62.13}$$

- Let ν^c_{cg} be the number of lanes available inside cell (c) for traffic flowing across (c, g). The macro-cell demand for inter-cell flows is given by

$$\delta^t_{c,cg} \stackrel{def}{=} \nu^c_{cg}\, \Delta^c_e \left(\rho^t_{c,cg}\right) \tag{62.14}$$

Following [6], the arbitrage functions can now be defined:

$$\begin{aligned} \Phi^c_{hc}(q_{hc}) &= q_{hc} - (q_{hc})^2 / \left(2\mu^c_{c,hc} Q^c_{max}\right) \\ \Psi^c_{cg}(r_{cg}) &= q_{cg} - (q_{cg})^2 / \left(2\mu^c_{c,cg} Q^c_{max}\right) \end{aligned} \tag{62.15}$$

62.4 Concluding Remarks

The model presented in this paper completes the ABTM model (anisotropic bidimensional traffic model) previously introduced in [6]. The main contribution is to provide a systematic and simple way to calculate the various equilibrium functions, network characteristics (notably the μ^c_{hc} and μ^c_{cg}), the cuts $\mathcal{M}_{c,hc}$ and $\mathcal{N}_{c,cg}$, and the various supplies and demands.

Further work will address the ABTM-GSOM connection taking into account the main arteries, and DTA in both predictive and reactive context. Much of the work described in this paper is carried out with the intent to model the Ile-de-France area which is a major Metropolitan area 12 M inhabitants, 12,000 km^2, to be divided into some 130 macro-zones.

References

1. R. Aghamohammadi, J.A. Laval, *Dynamic Traffic Assignment using the Macroscopic Fundamental Diagram: A Review of Vehicular and Pedestrian Flow Models* (2018). arXiv:1801.02130
2. J.A. Carrillo, S. Martin, M.T. Wolfram, An improved version of the Hughes model for pedestrian flow. Math. Mod. Methods Appl. Sci. **26**(04), 671–697 (2016)
3. H.W. Ho, S.C. Wong, Two-dimensional continuum modeling approach to transportation problems. J. Trans. Syst. Eng. Inf. Technol. **6**(6), 53–68 (2006)
4. L. Huang, S.C. Wong, M. Zhang, C.W. Shu, W.H. Lam, Revisiting Hughes dynamic continuum model for pedestrian flow and the development of an efficient solution algorithm. Trans. Res. Part B: Methodol. **43**(1), 127–141 (2009)
5. Y. Jiang, S.C. Wong, H.W. Ho, P. Zhang, R. Liu, A. Sumalee, A dynamic traffic assignment model for a continuum transportation system. Trans. Res. Part B: Methodol. **45**(2), 343–363 (2011)
6. M.M. Khoshyaran, J.P. Lebacque, Homogeneous bidimensional traffic flow model. IFAC-PapersOnLine **51**(9), 61–66 (2018)
7. J.P. Lebacque, The Godunov scheme and what it means for first order traffic flow models, in *Transportation and Traffic Theory. Proceedings of the 13th International Symposium on Transportation and Traffic Theory, Lyon, France* (1996)
8. J.P. Lebacque, M.M. Khoshyaran, First-order macroscopic traffic flow models: intersection modeling, network modeling, in *Transportation and Traffic Theory. Flow, Dynamics and Human Interaction. Proceedings of the 16th International Symposium on Transportation and Traffic Theory University of Maryland, College Park* (2005)
9. J.P. Lebacque, M.M. Khoshyaran, A variational formulation for higher order macroscopic traffic flow models of the GSOM family. Trans. Res. Part B **57**, 245–265 (2013)

10. J.P. Lebacque, S. Mammar, H.H. Salem, Generic second order traffic flow modelling, in *Transportation and Traffic Theory 2007. Proceedings of the 17th International Symposium on Transportation and Traffic Theory* (2007)
11. S. Mollier, M.L. Delle Monache, C. Canudas-De-Wit, B. Seibold, Two-dimensional macroscopic model for large scale traffic networks (2018)
12. T. Saumtally, J.P. Lebacque, H. Haj-Salem, Static traffic assignment with side constraints in a dense orthotropic network. Procedia-Soc. Behav. Sci. **20**, 465–474 (2011)
13. T. Saumtally, J-P. Lebacque, H. Haj-Salem, A dynamical two-dimensional traffic model in an anisotropic network. Net. Heterog. Media 663–684 (2013)
14. K. Sossoe, Modeling of multimodal transportation systems of large networks (Doctoral dissertation, Université Paris-Est) (2017)
15. K.S. Sossoe, J.-P. Lebacque, A. Mokrani, H. Haj-Salem, Traffic flow within a two-dimensional continuum anisotropic network. Trans. Res. Procedia **10**, 217–225 (2015)
16. K.S. Sossoe, J.P. Lebacque, Reactive dynamic assignment for a bi-dimensional traffic flow model. In *International Conference on Systems Science* (Springer, Cham, 2016), pp. 179–188
17. A. Taguchi, M. Iri, Continuum approximation to dense networks and its application to the analysis of urban road networks, in *Applications* (Springer, Berlin, Heidelberg, 1982), pp. 178–217

Chapter 63
Voronoi Densities for Bicylists: Adaptation for Finite Object Size and Speed

Victor L. Knoop, Flurin Hänseler, Marie-Jette Wierbos, Alexandra Gavriilidou, Winnie Daamen, and Serge P. Hoogendoorn

Abstract Density is one of the most relevant variables in a traffic flow description. For objects in 2 dimensions, density can be determined by the space that is allocated to each of the objects. This paper introduces a new way of computing the space available for a bicyclist, accounting for speed and accounting for the non-zero size of a bicycle. This changes local densities. The proposed method modifies the Voronoi densities, and assigns space to a bicycle. We assign space to bicycle A if it has a closer proximity to any point of bicycle A than any point of any other bicycle. The proximity is determined by the distance and the angle in relation to velocity of the bicycle. Specific proximity functions need to be formulated and calibrated to match cyclist behavior. This method helps to define a density level for cyclists, which in turn can for instance lead to a better indication of a Level of Service.

63.1 Introduction

To properly describe bicycle traffic in a macroscopic way, the macroscopic variables flow, density and speed need to be defined. We want to define local density, i.e. the density perceived by a cyclist since this is relevant for traffic operations. This quantity explains how bicyclists behave. Densities form the basis for a dynamic macroscopic traffic models, hence a good representation of densities is essential.

Whereas for one-dimensional traffic (e.g., car traffic), the definition of density is relatively straightforward, for 2-dimensional traffic it is more complex. We refer to [4] for an overview of applying different densities to pedestrian movements.

For cyclists, this concept is different. Often, pedestrians are considered as point particles. The shape of a cyclist is—compared to their spacing—so irregular that it substantially differs from a circular (or point) particle. For instance, it would be possible to be very close to the side of a cyclist, but being at a same distance in front

V. L. Knoop (✉) · F. Hänseler · M.-J. Wierbos · A. Gavriilidou · W. Daamen · S. P. Hoogendoorn
Delft University of Technology, Delft, The Netherlands
e-mail: v.l.knoop@tudelft.nl
URL: http://www.transport.citg.tudelft.nl

I. Zuriguel et al. (eds.), *Traffic and Granular Flow 2019*,
Springer Proceedings in Physics 252,
https://doi.org/10.1007/978-3-030-55973-1_63

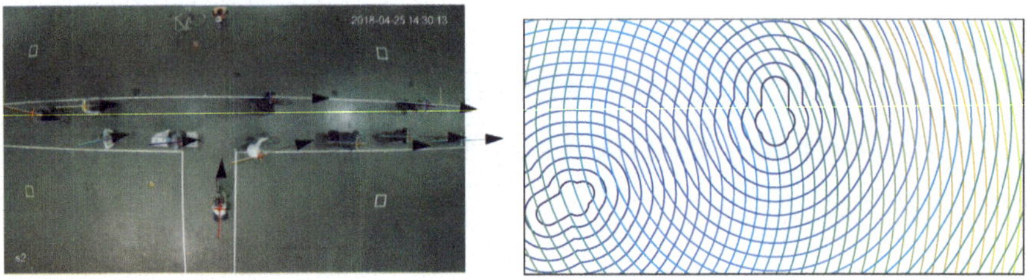

(a) Marked hats (b) Interacting cost fields

Fig. 63.1 The data used for illustration of the method

of the midpoint would be impossible due to its size. Also speed plays a role in the dynamics. A cyclist is expected to adapt his speed more on what is happening in front than what is happening behind, particularly for higher speeds. Following that reasoning, space should not simply be allocated to the cyclist which is closest by, but also speed and direction should be taken into account.

This paper formulates a concept on how the space can be allocated, to in the end get densities. Two concepts are proposed: first, we include the size of particles, which is presented in earlier works as set Voronoi diagrams, e.g. [5], but is now for the first time applied to road traffic. Second, an effect of speed is introduced. As illustration bike sizes and anisotropic cost functions are posed. The aim of the paper is to present this concept. For this, simplified the cost functions and bike sizes are being assumed. Having said so, we do illustrate the ideas presented here with real-world data, collected in an experiment with cyclists [2]. From the data, we use the position of the head, as well as the heading direction and the speed; we use average size of a bike, and simplify its shape to a cross. Figure 63.1a gives an impression of the data used.

The remainder of the paper is organised as follows. First, we give a background on the Voronoi densities. Then, Sect. 63.3 discusses the proposed methodology. The paper finishes with a discussion and conclusions.

63.2 Background: Voronoi Densities

Voronoi diagrams can be used in determining densities. Earlier work [1] extensively discussed the differences between local densities and global densities. Global density is the overall density, for instance determined by dividing the number of people over the area. This might vary on a smaller scale. For this reason, the concept of Voronoi diagrams is applied to pedestrian traffic. This computes the space which is available to everyone; space is allocated to the closest pedestrian. As addition, one can also have a maximum area of influence, and not assign space if it is further from any pedestrian than for instance maximum radius. As said, the concept does not deal with objects of extended size or anisotropy, which is what we introduce in this paper.

63.3 Proposed Method

This section introduces the proposed method, first commenting on the physical space, and then on the velocities.

63.3.1 Accounting for Physical Space

The Voronoi diagrams use the distance to a (mid)point as basis for their space allocation. If we apply this to the center of the bikes, we get the allocation of space as shown in Fig. 63.2a. However, cyclists have a physical size, and we can account for that. The base principle that we apply is that we no longer see a cyclist as one point, but instead take its physical size into account. Let's consider an example in which the space is allocated to the closest object, while considering *any point* of that object. For this, we need a distance function, or as we will call it in the remainder of the paper, a *cost function*. In this section, we consider the Euclidean distance. We now allocate space as follows. For a point in space, we find all costs to all points of all bikes. We identify the bike to which the point with the lowest overall cost belongs. We then allocate the space to that bike.

The example is of two interacting bikes as shown in Fig. 63.1b. The lines show the isocost lines to a bike. They hence already include the concept of getting to the closest point of a bike. For any point in space, it is now considered which bike is closest. In the figure, that translates into checking for which bike the cost of getting to it is lowest. The space is then allocated to that bike.

To clarify the difference with an ordinary Voronoi assignment let's consider the following. The point in space $\mathcal{P}$ has a distance to the midpoint of cyclist a and b,

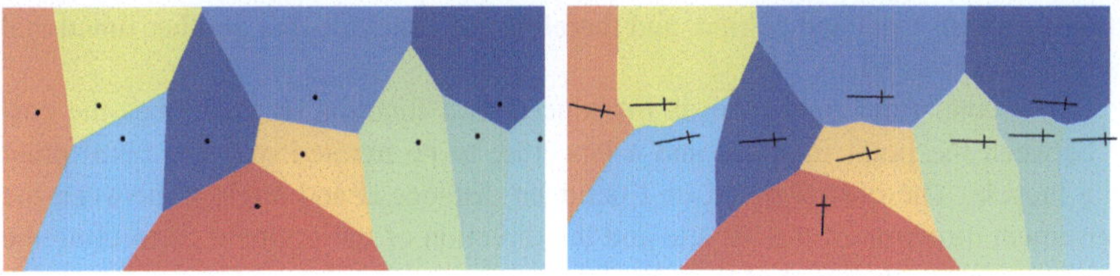

(a) Using one center point (b) Using the full physical size

Fig. 63.2 Extended Voronoi spaces: space allocated to the bicycles based on proximity denoted d^a_{mid} and d^b_{mid}. If $d^a_{\text{mid}} < d^b_{\text{mid}}$, the space would under normal space allocation be assigned to bike a. However, it can occur that the shortest distance to *any point* of bike a (for instance a's handle bar) is further than the shortest distance to *any point* of bike b (for instance b's rear wheel), so $d^a_{\text{shortest}} > d^b_{\text{shortest}}$.

Applying this concept to the data set (see Fig. 63.2b), we see that the boundaries have changed compared to Fig. 63.2a. The boundaries between the space of the various cyclists follow the shape of the bicycles, and are no longer straight. This is due to the non-linearity in the distance. We also see that all areas are "continuous", i.e. there are no islands or concavities. The proof is as follows.

Suppose a point in space a is closest to cyclist c, then all points on the line a-c (being the shortest line from a to c) are closer to cyclist c than to another cyclist c'. Namely, if there were a point b on the line a-c which is closer to c' than to c, then $d_{ac} = d_{ab} + d_{bc} > d_{ab} + d_{bc'} > d_{ac'}$ (due to triangular inequality). This is inconsistent with the assumption, and hence it is proven by contradiction that the areas are "continuous". Following the same proof, it can be concluded that all resulting areas are convex.

63.3.2 Accounting for Speed

While cycling, bicyclists are likely to react more on what is happening in front of them than what is behind and this effect is higher for increasing speed. Therefore, the cost function as defined in the previous section might be different to objects in front of the bicycle than behind. We hence define a cost function which is anisotropic, with lower cost rates (i.e., cost per unit of distance) in front of the bicyclist (hence objects feel as if they were closer) and higher cost rates behind (hence the objects feel as if they were farther away). This cost function is speed and direction dependent; at zero speed the effect is negligible, whereas at high speed, the effect will be stronger.

We will illustrate this principle with two different cost functions. First, an ellipse cost function is considered. Recall that for all points on an ellipsoid the sum of the distance to two (center) points is equal. In this case, we take as center points the mid point of the cyclist and a point ahead of the cyclist, with a distance depending on the cyclist's speed. Figure 63.3a shows isocost lines which are further away in front of the bike than behind. However, they merely are shifted and the costs increase at a same level in front and behind, and therefore we conclude that another functional form is better suited.

The second functional form is introduced as a function. It prescribes the cost C between the point in space and a (specific, to be iterated over the bike) point of a bicycle. The cost is based on Euclidean distance d and angle ϑ between the movement direction of the bicycle and the direction of the segment connecting the point in space to the point of the bicycle:

$$C = \frac{d}{\alpha + \cos(\vartheta)} \tag{63.1}$$

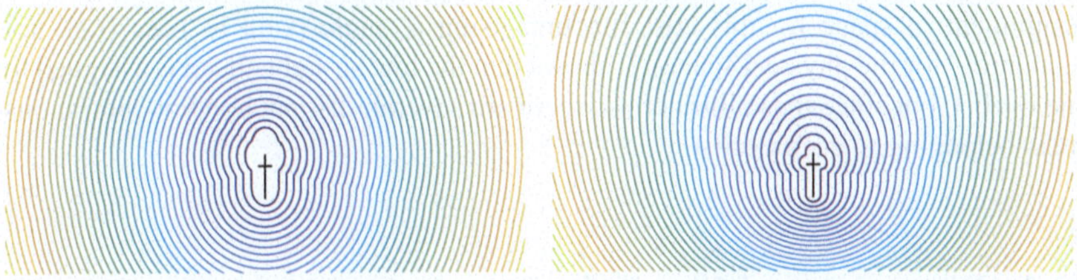

(a) Ellipse cost function (b) Function defined in equation 1

Fig. 63.3 Isocost lines incorporating physical size and speed effects

This function has a lower cost if objects are in the line of travel. Parameter α indicates the strength of the directional component, and should be tuned with speed. Qualitatively, we reason for low speeds, α is expected large, giving equal importance to objects in front and behind. For higher speeds, α can reduce asymptotically to 1, giving more importance to objects in front. This equation is in line with the equation indicating the relative effect of objects for the social force model [3].

Figure 63.3b shows isocost lines for the function (for simplicity, we fixed $\alpha = 3$), still accounting for the the physical size of a bicycle. Note that the lines in the rear are much closer together and hence other objects at the rear are perceived with lower cost. Figure 63.4a shows the resulting areas assigned to each of the bikes. For instance the boundary between the orange bicycle at the rear and the blue bicycle at the bottom curves heavily, allowing much more space to orange bicycle than previously. This is in line with the concepts that the space behind a cyclist is less relevant, hence the blue bicyclist has less use of the space than the orange one.

For the adjusted Voronoi space which accounts for physical space, we proved in Sect. 63.3 that the areas should be "continuous". This cannot be proven without loss of generality for the Voronoi space which accounts for motion due to the angle dependent function, as the next example will show. Consider for instance that all distances straight in front of a bicycle are very low cost (the closer the lower the cost), and always lower than any cost which are off the straight line. For all other angles, an Euclidean distance is taken as cost function. Figure 63.4b shows the

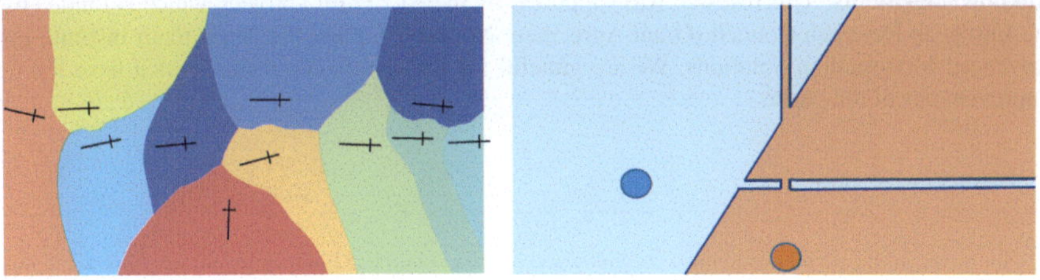

(a) Voronoi areas (b) Example of non continuous areas

Fig. 63.4 Cost fields of two bicycles

assignment of space for this cost function using a single point as base (i.e., not accounting for the physical space). It shows that the line straight ahead is assigned to one cyclist, but since these lines cross, there are several, distinct and not continuous areas belonging to each cyclist.

63.4 Discussion and Conclusion

We addressed the issue of assigning space to individual travellers in order to come to density measures. Earlier work showed differences between local and global densities. The method using Voronoi diagrams turns out to be relevant in determining densities. In order to make this method more accurate for cyclists, in this paper, we have extended this method to include (1) physical size of the objects and (2) speed of cyclists. For the speeds, we explicitly formulated a way to account for the effect that cyclist might adapt their behavior more to objects in front than to objects behind.

The paper proposes a mathematical concept of assigning space to individuals. It is flexible for the specific cost functions, indicating what cyclists consider relevant distances and angles for their objects. The ultimate proof of the functions is in the test whether the allocated space is in line with the speed cyclist choose. Alternatively, (virtual reality) experiments can be done to test where the attention of cyclists is to find what determines and/or limits their speed.

The paper showed some mathematical properties of the areas assigned to each of the cyclist. If anisotropy is included, these areas do not need to be continuous. One might expect that a cyclist adapts his behavior to the space that he expects to be able to use. It is questionable whether space which is shielded by another cyclist is considered accessible—it can be in a dynamic environment. If one wants to exclude this space, one can still use the cost principles introduced here, and combine this with a different assignment methodology. Instead of simply assigning space to the cyclist which has the lowest cost for that area, one can gradually expand the area assigned to each cyclist. Then, one can assign space to a cyclist only if there is a direct line-of-sight between the cyclist and the point in space. Consequently, there can also be areas in space which are not assigned to any cyclist.

Acknowledgements This research was supported by the ALLEGRO project, which is funded by the European Research Council (Grant Agreement No. 669792) and the Amsterdam Institute for Advanced Metropolitan Solutions. We are grateful for the review comments, which have led to improvements of the paper.

References

1. W. Daamen, J. Van den Heuvel, V. Knoop, S. Hoogendoorn, Macroscopic fundamental diagram for train platforms, in *Proceedings of TGF'17* (2019)
2. A. Gavriilidou, M.J. Wierbos, W. Daamen, Y. Yuan, V.L. Knoop, S.P. Hoogendoorn, Large-scale bicycle flow experiment: setup and implementation. Trans. Res. Rec. **2673**(5) (2019)
3. A. Johansson, D. Helbing, P.K. Shukla, Specification of the social force pedestrian model by evolutionary adjustment to video tracking data. Adv. Complex Syst. **10**(supp02), 271–288 (2007)
4. M. Nikoli, Data-driven fundamental models for pedestrian movements. Ph.D. thesis (2017). https://doi.org/10.5075/epfl-thesis-7613
5. F.M. Schaller, S.C. Kapfer, M.E. Evans, M.J. Hoffmann, T. Aste, M. Saadatfar, K. Mecke, G.W. Delaney, G.E. Schröder-Turk, Set voronoi diagrams of 3d assemblies of aspherical particles. Phil. Mag. **93**(31–33), 3993–4017 (2013)

Chapter 64
The HighD Dataset: Is This Dataset Suitable for Calibration of Vehicular Traffic Models?

Valentina Kurtc

Abstract A large-scale naturalistic vehicle trajectory dataset from German highways called highD is used to investigate the car-following behavior of individual drivers. These data include trajectories of 1,10,000 vehicles with the total length of 16.5 h. Solving a nonlinear optimization problem, the Intelligent Driver Model is calibrated by minimizing the deviations between observed and simulated gaps, when following the prescribed leading vehicle. The averaged calibration error is 7.6%, which is a little bit lower compared to previous findings (NGSIM I-80). It can be explaind by the shorter highD trajectories, predominantly free flow traffic and good precision metrics of this dataset. The ratio between inter-driver and intra-driver variability is inversigated by performing global and platoon calibration. Inter-driver variation accounts for a larger part of the calibration errors than intra-driver variation does.

64.1 Introduction

Nowadays, microscopic traffic data have become more available. As a result, the problem of analyzing and comparing microscopic traffic flow models with real microscopic data has become more actual [1, 2]. In this paper, we consider the dataset of naturalistic vehicle trajectories from German highways—the highD dataset [3] for calibration of car-following models. These data include trajectories of 1,10,000 vehicles at a high temporal resolution with the total length of 16.5 h. We compare these data with the well-known NGSIM I-80 trajectory dataset recorded on April 13, 2005 in USA [4]. We demonstrate that the highD data are suitable for calibration of the car-following models. In this paper the Intelligent Driver Model (IDM) [5] is

V. Kurtc (✉)
Peter the Great St. Petersburg Polytechnic University, St. Petersburg 195251, Russia
e-mail: kurts_vv@spbstu.ru

I. Zuriguel et al. (eds.), *Traffic and Granular Flow 2019*,
Springer Proceedings in Physics 252,
https://doi.org/10.1007/978-3-030-55973-1_64

calibrated according to the highD trajectory data. Intra-driver and inter-driver variation are investigated by means of two calibration methods—global calibration and platoon calibration.

The paper organized as follows. Section 64.2 briefly describes the car-following model under investigation. The highD dataset and filtering procedure applied are presented in Sect. 64.3. Section 64.4 describes two calibration methods used within this research. Results are presented in Sect. 64.5. The paper will finish with discussions and conclusion.

64.2 Car-Following Model

In this paper we calibrate the Intelligent Driver model (IDM) [5] with the highD dataset. This model is formulated as coupled ordinary differential equations and characterized by an acceleration function which depends on the actual speed $v(t)$, the approaching rate $\Delta v(t) = v - v_l$ to the leader, and the gap $s(t)$

$$\dot{v}_{\mathrm{IDM}}(v, \Delta v, s) = a \left[1 - \left(\frac{v}{V_0} \right)^4 - \left(\frac{s^*(v, \Delta v)}{s} \right)^2 \right], \qquad (64.1)$$

where a is the maximum acceleration, parameter δ controls the acceleration decrease with the speed increase, V_0 denotes the desired speed. Acceleration function combines the driver's strategy to reach the desired speed with a following behavior to keep the dynamically desired gap $s^*(v, \Delta v) = s_0 + \max(0, vT + v\,\Delta v/(2\sqrt{ab}))$. Parameter s_0 is a minimum bumper-to-bumper gap, T denotes the time gap to the leading vehicle and b is a comfortable deceleration. This model contains five parameters to identify within the calibration procedure $\theta = (a, V_0, s_0, T, b)^T$.

64.3 Dataset of Naruralistic Vehicle Trajectories

The highD dataset is considered for calibration of the IDM. Data contain information about 110000 vehicle trajectories, which were extracted from 60 drone video recordings [3]. The experiments were performed at German highways close to Colonge at six different locations. The recording area is about 420 m length. For each vehicle trajectory we have its ID, longitudinal coordinate, speed, acceleration, ID of its current leader, distance to the leader and etc. The highD trajectory data have high temporal resolution with a discretization time step equal to 0.04 s.

The most similar dataset to the highD dataset is the NGSIM [4], which have been widely used by researchers in the field of traffic flow theory [6, 7]. Recording sites of NGSIM I-80 contain trajectory data collected at the Interstate 80 in Emeryville, USA. Raw NGSIM trajectories contain false trajectory collisions as well as unrealistic vehicle speeds and accelerations present in this dataset [6, 8]. To eliminate these drawbacks, the internal and platoon inconsistencies as well as the noise from original data measurements have been eliminated [9]. Calculation of derived quantities was performed and a smoothing algorithm was proposed in [10]. Nevertheless, [6] claims that it is not the solution, that is, vehicles have to be manually re-extracted from the recordings to get improved longitudinal trajectories.

To perform calibration of car-following models one needs to extract trajectory pairs of vehicles moving one after another in the same lane. The consecutive trajectories are extracted by the procedure described in [7].

64.4 Methods

To find the optimal parameter values of a car-following model with a non-linear acceleration function such as Eq. 64.1, one need to solve a non-linear optimization problem numerically. We initialize the microscopic model with the empirically given speed $v^{sim}(t_0) = v^{data}(t_0)$ and gap $s^{sim}(t_0) = s^{data}(t_0)$, and compute the trajectory of the following car using the ballistic update method (see [11, Sect. 10.2]). Then it is directly compared to the gaps s_i^{data} provided by the highD dataset. To assess quantitatively the deviation between measured and simulated data, we use the absolute error measure in terms of gaps

$$S^{abs} = \frac{\sum_{i=1}^{n}(s_i^{sim} - s_i^{data})^2}{\sum_{i=1}^{n}(s_i^{data})^2} \tag{64.2}$$

where n is a number of data points, s_i^{sim} and s_i^{data} are simulated and empirical gap at time t_i, respectively. We restrict the solution space for optimization to reasonable parameter values without excluding possible solutions by means of box constraints. Two calibration methods are considered: global calibration and platoon calibration [7].

64.5 Results and Discussion

Speed analysis of the highD data reveals that there is mostly free traffic, what results in short trajectory durations (Fig. 64.1, left). Relating to the traffic-state variety, the NGSIM I-80 [4] is superior. Nevertheless the highD contains a few data sets showing indeed impeded traffic or even jams with stop-and-go waves (Fig. 64.1, right). It is worth to mention that data with lower speeds also provide longer trajectories.

For the global approach, 539 trajectory pairs are filtered out, what is comparable to the NGSIM I-80 [7]. For the platoon calibration we consider three vehicles following each other and obtain 388 trajectory triplets.

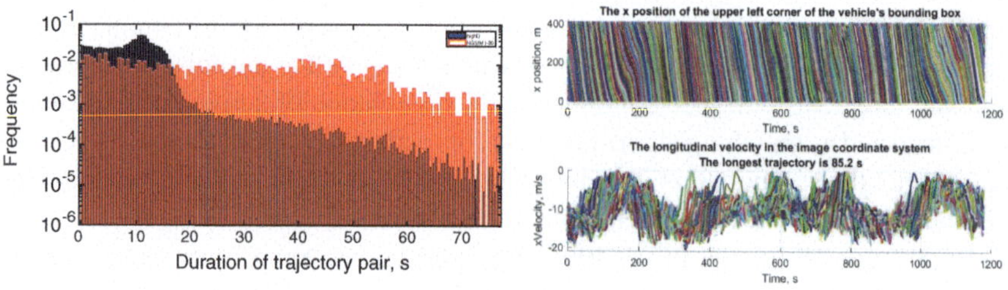

Fig. 64.1 Durations of trajectory pair—highD and NGSIM I-80 datasets (left). Longitudinal positions and speeds of one of the 60 highD video recordings (right)

64.5.1 Global Calibration

To assess the model's fitting power we consider the value of the objective function (Eq. 64.2) at the found optimal solution. Besides that, we consider the absolute gap error $\epsilon_i = s_i^{sim} - s_i^{data}$ for the specific trajectory at time t_i and calculate the variance of this error considering that the mean value is equal to zero. Figure 64.2 visualizes the calibration results of the IDM obtained with highD and NGSIM I-80 datasets [7]. First row represents the distributions of the model parameter values and values of the objective function (Eq. 64.2) subjected to minimization.

To determine the estimation errors we use the variance-covariance matrix of the estimates of the parameter vector θ, which is proportional to the inverse of the Hesse

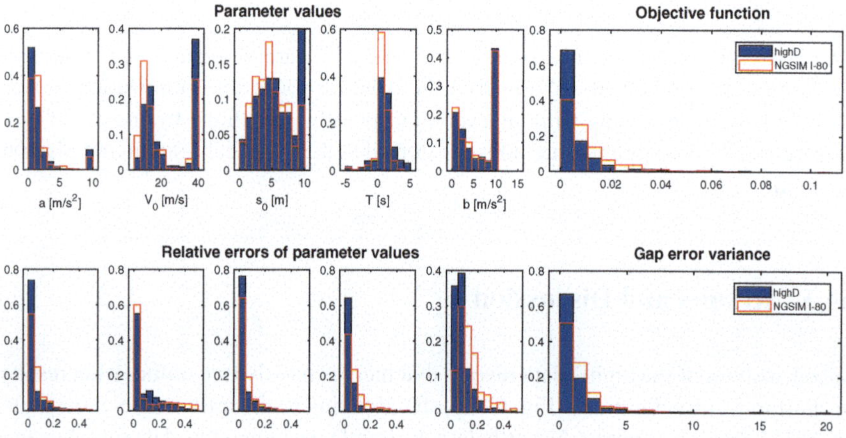

Fig. 64.2 Global calibration of the IDM using highD and NGSIM I-80 datasets. Distributions (relative frequencies) of the parameter values (first row, panels from 1 to 5), estimation errors (second row, panels from 1 to 5), objective function values (first row, panel 6) and gap error variances (second row, panel 6)

matrix of the sum of the squared errors. Distributions of the estimated relative errors and values of the gap variance are depicted at the second row of Fig. 64.2. Only estimates with relative errors below 50% are considered.

64.5.2 Inter-driver and Intra-driver Variations

For platoon approach we use trajectory sets, which contain three vehicles following each other, what results in 388 trajectory triplets. Figure 64.3 presents the distributions of the IDM gap error variance obtained with estimated model parameter values, which estimated errors are less or equal to 50%. The mean value of the gap error variance in case of global calibration is equal to 1.17 m^2, whereas platoon calibration results in 6.87 m^2.

It is obvious that variations in driving behavior come in two forms—inter-driver (different driving styles) and intra-driver (a non-constant driving style of human drivers) variations [7, 12]. In case of the global approach the trajectory of one vehicle is calibrated and, thus only intra-driver variation is considered, that is, $\epsilon^{global} = \epsilon_{intra}$. The platoon method incorporates several driver styles simultaneously and, as a result takes into account both types of variation: $\epsilon^{platoon} = \epsilon_{intra} + \epsilon_{inter}$. We assume that there is no correlation between these two types of errors, and derive the inter-driver variation as follows

$$Var(\epsilon_{inter}) = Var(\epsilon_{platoon}) - Var(\epsilon_{intra}) \tag{64.3}$$

Table 64.1 contains the gap varience values as well as summarizes calibration results.

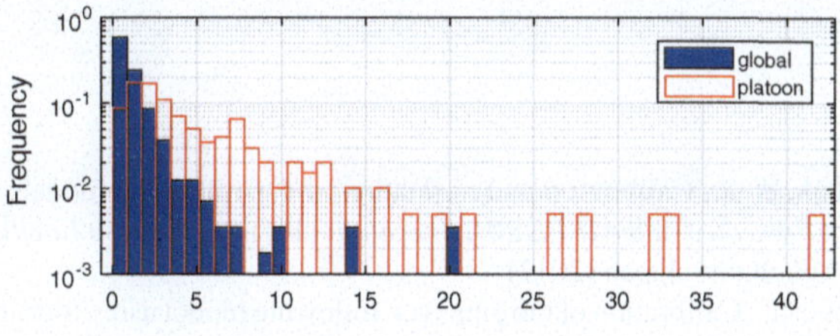

Fig. 64.3 Global and platoon calibration. Distributions of the gap error variance

Table 64.1 Averaged calibration errors and gap variances

	Global	Platoon
$\sqrt{S^{abs}}$, [1]	0.076	0.147
$Var(\epsilon_i)$, $[m^2]$	1.17	6.87
$\epsilon_{inter}/\epsilon_{intra}$, [1]	4.9	

64.6 Discussion and Conclusions

The Intelligent Driver model was considered to reproduce naturalistic vehicle trajectories recorded on the German highways (highD dataset). Calibration error is 7.6% on average, what is lower than the typical error values obtained in previous studies [13–15]. First reason is that the highD trajectories are shorter at the average (Fig. 64.1). Secondly, in contrast to the NGSIM I-80 dataset, the highD data reveal mostly free flow traffic. In particular, highD trajectories often contain no significant accelerations during the time, that is why it is easier to calibrate and leads to lower calibration errors than the NGSIM trajectories do.

We compare the results obtained by calibrating the IDM to the highD and NGSIM I-80 data (Fig. 64.2). Distributions of the parameter values differ from each other for these datasets. The inter-driver variability reflects the variety of the model parameter values, and it is natural that these distributions are different for American and German trajectories. Besides that, relative estimation errors, values of objective function and error gap variance are lower for the highD trajectories (Fig. 64.2).

Calibration errors can be attributed to two types of variability—inter-driver (different driver-vehicle units) and intra-driver (nonconstant driving style) variations. According to previous investigations [7, 12] inter-driver variation prevails the intra-driver variation. Calibrating the IDM to the highD dataset we obtain the same results qualitatively. Two types of calibration methods are applied. The global method incorporates only intra-driver variability, because it considers only one vehicle following its leader. Conversely, the platoon approach includes several drivers following each other with a single data-driven leader. As a result, the inter-driver variation is incorporated as well. Calibrating the IDM to the highD naturalistic trajectories we find out that inter-driver variability is about 5 times more than the intra-driver one is (Table 64.1).

References

1. M. Pourabdollah, et al., Calibration and evaluation of car-following models using real-world driving data, in *Presented at the IEEE 20st International Conference on Intelligent Transportation Systems*, Yokohama, Japan (2017)
2. L. Vasconcelos et al., Calibration of the gipps car-following model using trajectory data. Transp. Res. Proc. **3**, 952–961 (2014)
3. R. Krajewski, J. Bock, L. Kloeker, L. Eckstein, The highD dataset: a drone dataset of naturalistic vehicle trajectories on German highways for validation of highly automated driving systems, in *Presented at the IEEE 21st International Conference on Intelligent Transportation Systems*, Maui, Hawaii, USA (2018)
4. NGSIM, Next Generation Simulation. FHWA, U.S. Department of Transportation. www.ngsim.fhwa.dot.gov. Accessed 5 May 2007
5. M. Treiber, A. Hennecke, D. Helbing, Congested traffic states in empirical observations and microscopic simulations. Phys. Rev. E **62**(2), 1805–1824 (2000)
6. B. Coifman, L. Li, A critical evaluation of the next generation simulation (NGSIM) vehicle trajectory dataset. Transp. Res. Part B Methodol. **105**, 362–377 (2017)

7. V. Kurtc, M. Treiber, Calibrating the local and platoon dynamics of car-following models on the reconstructed NGSIM data, in *Traffic and Granular Flow'15*, pp. 515–522. Springer, Switzerland (2016)
8. A. Duret, C. Buisson, N. Chiabaut, Estimating individual speed-spacing relationship and assessing ability of Newell's car-following model to reproduce trajectories. Transp. Res. Record J. Transp. Res. Board **2088**, 188–197 (2008)
9. M. Montanino, V. Punzo, Trajectory data reconstruction and simulation-based validation against macroscopic traffic patterns. Transp. Res. Part B Methodol. **80**, 82–106 (2015)
10. C. Thiemann, M. Treiber, A. Kesting, Estimating acceleration and lane-changing dynamics from next generation simulation trajectory data. Transp. Res. Record J. Transp. Res. Board **2088**, 90–101 (2008)
11. M. Treiber, A. Kesting, *Traffic Flow Dynamics* (Springer, Heidelberg, 2013)
12. S. Ossen, S.P. Hoogendoorn, B.G. Gorte, Inter-driver differences in car-following: a vehicle trajectory based study. Transp. Res. Record J. Transp. Res. Board **1965**, 121–129 (2007)
13. E. Brockfeld, D. Kuhne, P. Wagner, Calibration and validation of microscopic traffic flow models. Transp. Res. Record J. Transp. Res. Board **1876**, 62–70 (2004)
14. P. Ranjitkar, T. Nakatsuji, M. Asano, Performance evaluation of microscopic flow models with test track data. Transp. Res. Record J. Transp. Res. Board **1876**, 90–100 (2004)
15. V. Punzo, F. Simonelli, Analysis and comparison of microscopic flow models with real traffic microscopic data. Transp. Res. Record J. Transp. Res. Board **1934**, 53–63 (2005)

Chapter 65
Single-File Dynamics of Cyclists: Two Experiments and Two Microscopic Models

Valentina Kurtc and Matrin Treiber

Abstract Studies of bicycle flow are important for design and operation of bicycle facilities. We conjecture that there is no qualitative difference between vehicular and bicycle traffic flow dynamics in the single-file case, so the latter can be described by reparameterized car-following models. To test this hypothesis, we reproduce German and Chinese ring-road bicycle experiments with the Intelligent-Driver Model (IDM) and with the Necessary-Deceleration-Model (NDM), which is specifically designed for bike traffic. Comparing their fit quality (calibration) and predictive power (validation), we find similar quality metrics for both models, so the above hypothesis of a qualitative equivalence cannot be rejected. The NDM represents significant calibration errors for high flow densities, which correspond to flow states with stop-and-go waves. By means of two validation methods, we discover that inter-driver variation is much higher than the intra-driver variation for bicycle traffic. It coincides with the results obtained from vehicular traffic experiments.

65.1 Introduction

Nowadays, pollution and serious congestions lead to a gradual increase of bicycle trips [1]. As a result, studies of bicycle flow are important for design and operation of bicycle facilities. In spite of its growing relevance, past research on bicycle traffic operations in experiments and models [2–5] is remarkably scarce compared to investigations for vehicular traffic [6] and for pedestrian flows [7, 8]. Therefore, it is natural to ask whether there is a significant qualitative difference between vehicular and bicycle traffic flow at all. In other words, if one can use the well-developed car-following models for the simulation of bicycle traffic instead of creating new bicycle

V. Kurtc (✉)
Peter the Great State Polytechnic University, St. Petersburg 195251, Russia
e-mail: kurts_vv@spbstu.ru

M. Treiber
Technische Universität Dresden, 01062 Dresden, Germany
e-mail: martin@mtreiber.de

I. Zuriguel et al. (eds.), *Traffic and Granular Flow 2019*,
Springer Proceedings in Physics 252,
https://doi.org/10.1007/978-3-030-55973-1_65

models. In this paper, we test the Intelligent Driver Models (IDM) [9] as a typical representative of car-following models against the "ring-road" bicycle traffic experiments [2, 4]. We compare the IDM fit quality with that of a specifically designed "bicycle-following model", the Necessary-Deceleration-Model (NDM) [2]. In the following two sections, we shortly describe the models and the experiments. Section 65.4 specifies the calibration procedure before we present and discuss our main results in Sect. 65.5. The paper finishes with a conclusion in Sect. 65.6.

65.2 Models

Two microscopic car-following models are considered—the IDM [9] and the NDM [2]. Both of them are formulated as coupled ordinary differential equations and characterized by an acceleration function which depends on the actual speed $v(t)$, the approaching rate $\Delta v(t) = v - v_l$ to the leader, and the gap $s(t)$.

The IDM is defined by the acceleration function [9]

$$\dot{v}_{\text{IDM}}(v, \Delta v, s) = a \left[1 - \left(\frac{v}{v_0} \right)^4 - \left(\frac{s^*(v, \Delta v)}{s} \right)^2 \right], \tag{65.1}$$

where $s^*(v, \Delta v) = s_0 + \max(0, vT + v\,\Delta v/(2\sqrt{ab}))$ is the dynamically desired gap. The IDM contains 5 parameters to identify via calibration—a, v_0, s_0, T, b.

The acceleration function of the NDM [2] is defined as follows

$$\dot{v}_{\text{NDM}}(v, \Delta v, s) = acc - \min(dec_1 + dec_2, b_{\text{max}}), \tag{65.2}$$

where definition of acc, dec_1 and dec_2 can be found in [2]. The NDM has also 5 parameters to calibrate—τ, v_0, s_0, T, and b_{max}.

65.3 Ring-Road Experiments

Data of two independent experiments with cyclists were used for calibration and validation of two models. The first set of experiments was conducted by the University of Wuppertal in cooperation with Jülich Forschungszentrum on the 6th of May, 2012 [2]. Cyclists aged 7–66 years old were following each other without overtakings along the oval track of 86 86 m length. Two cameras overlooked the whole area, so all trajectories of cyclists were captured. However, because of technical reasons, the extracted trajectories correspond to a straight line of 20 20 m length only. Group experiments were performed for several density levels corresponding to 5, 7, 10, 18, 20 and 33 participants, respectively.

Another set of experiments has been performed with 39, 48 and 63 cyclists [4] accounting for different global densities (similarly to the vehicle flow experiments [10], and the pedestrian flow experiments [11, 12]), different weather conditions (sunny, sunny/cloudy and a little rainy) and both male and female riders. The geometry of the track was an oval with 29 29 m long straight sections joined by two semi-circular curves of radius 14 14 m resulting in a circumference of 146 146 m. The riders were filmed by one video camera mounted on the top of an 18-story building. In contrast to the German experiments, the complete trajectories were extracted. The trajectories cover a longer time interval compared to the German experiments [2]. The Chinese experiments [4] demonstrated even more pronounced stop-and-go waves for the higher densities compared to the German ones.

65.4 Methodology

The German dataset provides the directly recorded three-dimensional positional data x, y and z with the sampling step $\Delta t = 0.04$ s. Using the geometry of the track, we transformed these measurements into arc lengths l. In case of the Chinese data, arc lengths have already been provided with the time step $\Delta t = 1$ s. Since the original data of both experiments contained a significant amount of noise due to measurement and rounding errors, and the noise is aggravated when deriving the speed, a smoothing procedure is necessary. For this purpose, we have applied exponential smoothing of the arc-length data with a symmetric exponential kernel $w(t) \propto exp(\frac{-|t|}{T})$ assuming a smoothing width $T = 0.2$ s and $T = 1$ s for the German and Chinese datasets respectively.

Pairs of consecutive trajectories were used for calibration and validation according to the global approach [6, 13]. The calculated gaps $s^{sim}(t)$ were compared with the experimentally observed gaps $s^{data}(t)$ by means of absolute objective functions

$$S^{abs} = \frac{\sum_{i=1}^{n}(s_i^{sim} - s_i^{data})^2}{\sum_{i=1}^{n}(s_i^{data})^2} \tag{65.3}$$

where n is the number of observation points. We also calibrated according to the relative and mixed error [6, Chap. 16], and found no significant differences.

65.5 Results and Discussion

65.5.1 *Collective Driving Behavior and Calibration*

The first five panels of Fig. 65.1 display the cumulative distribution functions of the calibrated NDM parameters for the Chinese experiments with 39, 48, and 63 bikers corresponding to a global density $\rho = 0.267\,\mathrm{m}^{-1}$, $\rho = 0.329\,\mathrm{m}^{-1}$, and $\rho = 0.432\,\mathrm{m}^{-1}$, respectively. The degree of fitting is comparatively high (more than 70 % of the trajectory pairs of the first two experiments result in an absolute error measure $S_{abs} < 0.1$). Figure 65.2 displays the calibration results to the same data for the IDM. Generally, the IDM fitting quality is slightly better with more than 90 % of all trajectory pairs of the first two experiments having an error measure $S_{abs} < 0.1$. Interestingly, the IDM is better suited to describe the high-density data and the associated stop-and-go waves.

65.5.2 *Inter-Driver Variation and Validation*

Validation by cross comparison implies determining the error measures for a certain test data set by simulating the model with the parameters calibrated to the disjunct "training" data set [6]. For each German experiment, we have separately calculated the calibration-validation matrix M, whose elements M_{ij} give the error measure $\sqrt{S^{abs}}$ for the trajectory pair j as obtained from the model calibrated to the trajectory pair i. The diagonal elements M_{ii} are the calibration errors containing the intra-driver

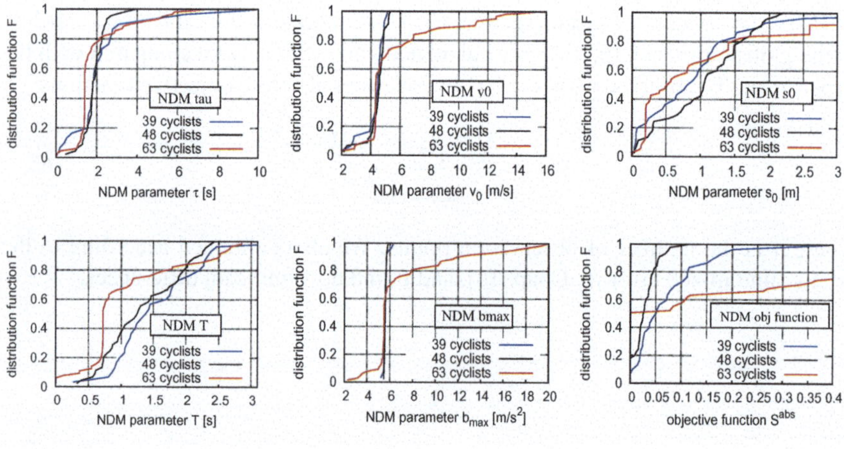

Fig. 65.1 CDFs for the NDM parameters and objective function S^{abs} as obtained from the calibration of all 39, 48, and 63 bicyclist pairs of the Chinese experiments

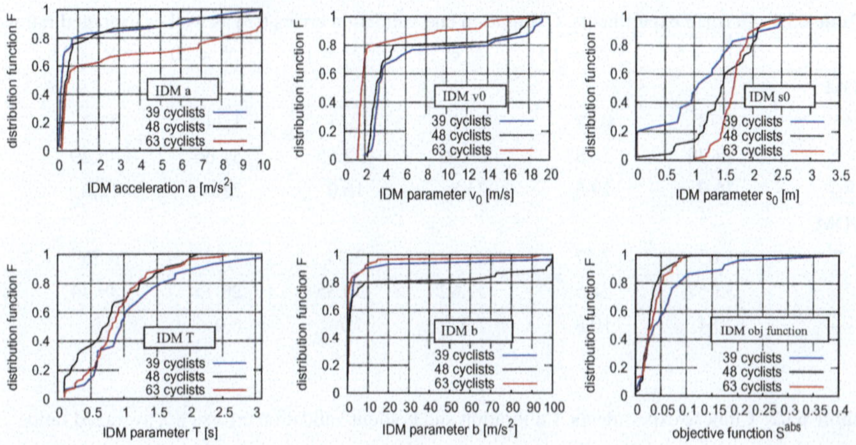

Fig. 65.2 CDFs for the IDM parameters and objective function S^{abs} as obtained from the calibration of all 39, 48, and 63 bicyclist pairs of the Chinese experiments

variation and the model error (or the intra-driver variations alone for a hypothetical "perfect" model). In contrast, the off-diagonal elements M_{ij}, $i \neq j$, give the total error as a superposition of the validation error and the inter-driver variation for the leader-follower pair j with respect to the leader-follower pair i. The average total error ϵ^{tot} and calibration error ϵ^{cal} are given by

$$\epsilon^{tot} = \frac{1}{n(n-1)} \sum_{i=1}^{n} \sum_{j=1, j \neq i}^{n} M_{ij}, \ \epsilon^{cal} = \frac{1}{n} \sum_{i=1}^{n} M_{ii} \qquad (65.4)$$

where n is a number of trajectory pairs. To obtain measures for the overall fitting quality and the predictive power plus inter-driver variations, we have calculated the ratio of the average validation error to the calibration error (Table 65.1). Analyzing the German experiments, we have found unrealistically low calibration errors and large validation-calibration ratios, both for the NDM and the IDM, particularly for the lower densities. Obviously, the trajectories were too short to avoid overfitting, so no conclusions about model properties could be obtained with our procedure from these data.

In order to get rid of the confounding effect of the different driving characteristics of different leaders thereby separating the validation error and the inter-driver variation, we used disjunct parts of the same trajectory for calibration and validation. This is possible for the trajectories from Jiang et al. since they are sufficiently long. The first half of each trajectory was selected for calibration, whereas the second half was used for validation. Table 65.2 represents fitting power (calibration error) and predictive power (validation error) as well as the ratio between these two measures for each experiment (38, 48 and 63 cyclists) and both models (IDM and NDM). In contrast, the low ratios obtained from the longer trajectories of the experiments of

Table 65.1 German experiments. Calibration and validation errors (%) as well as averaged ratios

	$N = 5$	$N = 10$	$N = 15$	$N = 18$	$N = 20$	$N = 33$
IDM						
ϵ^{cal}	0.78	0.98	1.11	1.11	1.34	3.43
ϵ^{tot}	28.70	28.81	28.60	32.55	31.66	43.29
$\frac{\epsilon^{tot}}{\epsilon^{cal}}$	36.9	29.5	25.8	16.0	23.6	12.6
NDM						
ϵ^{cal}	1.65	2.27	3.59	3.94	4.78	6.75
ϵ^{tot}	33.57	24.55	27.82	30.35	28.75	39.74
$\frac{\epsilon^{tot}}{\epsilon^{cal}}$	20.4	10.8	7.7	7.7	6.0	4.5

Table 65.2 Chinese experiments. Calibration and holdout validation errors (%), averaged ratios

	$N = 39$	$N = 48$	$N = 63$
IDM			
ϵ^{cal}	18.27	15.43	32.15
ϵ^{val}	26.71	22.18	34.13
$\frac{\epsilon^{val}}{\epsilon^{cal}}$	1.5	1.4	1.1
NDM			
ϵ^{cal}	23.83	18.75	62.95
ϵ^{val}	28.11	21.77	58.35
$\frac{\epsilon^{val}}{\epsilon^{cal}}$	1.2	1.2	0.9

Jiang et al. (not even significantly greater than one) indicate that overfitting is not a problem there. Therefore, we will restrict our further analysis to these experiments.

65.6 Conclusion

We conclude that the IDM, which has a similar underlying heuristics as the NDM, can not only describe car but also bicycle-following traffic. For the German experiments we obtain, on average, error measures of 1.92 and 4.53 % for IDM and NDM respectively, whereas the Chinese experiments result in error of 25.64 % for the IDM and 23.40 % for the NDM. The NDM demonstrates the highest errors for the highest density, when the stop-and-go waves occure. The different error levels in Tables 65.1 and 65.2 stem from the different lengths of the trajectories. These cover less than 20 20 m in the German experiment and often there are no significant accelerations during this time. In contrast, the Chinese trajectory pairs include several stop-and-go phases. Finally, we should highlight, that the short trajectories of the German experiment are not really useful for calibration and validation.

Validation results show that the predictive power of the NDM is better than that of the IDM. The error from the German experiments is a superposition of the validation error and the inter-driver variation. The ratio of this all-inclusive validation error to the calibration error is in range from 12.6 to 36.9 for IDM and from 4.5 to 20.4 for NDM. Trajectories from Jiang et al. experiment are longer, which enabled hold-out validation and exclusion of the inter-driver variation from validation error. As a result, ratios are significantly lower: 1.1–1.5 for IDM and 0.9–1.2 for NDM. One can conclude that for bicycle traffic the inter-driver variation is higher than the intra-driver variation is. The same results were obtained for vehicular traffic [13].

We conclude that the dynamics of bicycle traffic differs only quantitatively from vehicular traffic and reparameterized car-following models such as the IDM work at least as well as dedicated "bike-following" models. No qualitative difference between vehicular and bicycle traffic flow dynamics for the case of single-lane traffic was found.

References

1. J. Pucher, R. Buehler, Making cycling irresistible: lessons from The Netherlands, Denmark and Germany. Transp. Rev. **28**(4), 495–528 (2008)
2. E. Andresen et al., Basic driving dynamics of cyclists, *Simulation of Urban Mobility: First International Conference* (Springer, Berlin Heidelberg, 2013), pp. 18–32
3. S. Jin et al., An improved multi-value cellular automata model for heterogeneous bicycle traffic flow. Phys. Lett. A **379**(39), 2409–2416 (2015)
4. R. Jiang et al., Traffic dynamics of bicycle flow: experiment and modeling. Transp. Sci. **51**(3), 998–1008 (2017)
5. A. Gavriliidou et al., Modelling cyclist queue formation using a two-layer gramework for operational cycling behaviour. Transp. Res. Part C Emer. Technol. **105**, 468–484 (2019)
6. M. Treiber, A. Kesting, *Traffic Flow Dynamics* (Springer, Heidelberg, 2013)
7. A. Seyfried et al., New insights into pedestrian flow through bottlenecks. Transp. Sci. **43**, 395–406 (2009)
8. A. Tordeux et al., Prediction of pedestrian dynamics in complex architectures with artificial neural networks. J. Intell. Transp. Syst. 1–13 (2019)
9. M. Treiber et al., Congested traffic states in empirical observations and microscopic simulations. Phys. Rev. E **62**(2), 1805–1824 (2000)
10. Y. Sugiyama et al., Traffic jams without bottlenecks-experimental evidence for the physical mechanism of the formation of a jam. New J. Phys. **10**(3), 033001 (2008)
11. D. Helbing et al., Self-organized pedestrian crowd dynamics: experiments, simulations, and design solutions. Transp. Sci. **39**(1), 1–24 (2005)
12. A. Seyfried et al., The fundamental diagram of pedestrian movement revisited. Journal of Statistical Mechanics: Theory and Experiment **2005**(10), P10002–P10002 (2005)
13. V. Kurtc, M. Treiber, Calibrating the local and platoon dynamics of car-following models on the reconstructed NGSIM data, in *Traffic and Granular Flow'15*, pp. 515–522. Springer, Switzerland (2016)

Chapter 66
Response Time and Deceleration Affected by Lateral Shift of Leaders in Vehicular Traffic with Weak Lane Discipline

Akihito Nagahama and Takahiro Wada

Abstract In this study, we focus on the effects of the lateral shift of leading vehicles (leaders) and the steering operation of following vehicles (followers) on the response time and deceleration of the followers. A physical investigation on heterogeneous traffic with weak lane discipline is required because of heavy congestion in certain developing countries. Two-dimensional microscopic traffic models enabled researchers to simulate vehicles' acceleration in traffic with weak lane discipline. However, to the best of our knowledge, there is a lack of research that discusses acceleration considering the variation of response time caused by lateral shifting. We conducted experiments with a driving simulator and observed the followers' behaviors when the leaders suddenly commenced deceleration. The distributions of the deceleration and response time depending on the leaders' position were different under respective conditions of steering operation and relative positions.

66.1 Introduction

Congestion of vehicular traffic is becoming a serious problem in certain developing countries. The traffic observed in these countries often exhibits weak lane discipline and comprises various types of vehicles. We call such traffic "two-dimensional (2D) mixed traffic." Although many models for the lateral maneuvering of vehicles have been proposed [1, 4, 6], the number of models for 2D mixed traffic is relatively small. Against this backdrop, Oketch provided parameters of a microscopic model for nine types of vehicles in 2D mixed traffic [5]. He expressed the arbitrary but discrete lateral position of vehicles utilizing fuzzy logic. Gunay confirmed that the time headway was varied by the lateral shift of a leading vehicle (leader) from a

A. Nagahama (✉)
Ritsumeikan Global Innovation Research Organization, Ritsumeikan University,
1-1-1 Noji-higashi, Kusatsu, Shiga 525-8577, Japan
e-mail: naga0862@fc.ritsumei.ac.jp

T. Wada
College of Information Science and Engineering, Ritsumeikan University,
1-1-1 Noji-higashi, Kusatsu, Shiga 525-8577, Japan

I. Zuriguel et al. (eds.), *Traffic and Granular Flow 2019*,
Springer Proceedings in Physics 252,
https://doi.org/10.1007/978-3-030-55973-1_66

following vehicle (follower) in field 2D mixed traffic [2]. The observation results implied that the lateral shifting of leaders had to be considered when developing a microscopic model. On the other hand, 2D continuous models have been proposed for vehicular traffic. For example, Kanagaraj et al. proposed a 2D vehicular model and estimated model parameters [3]. A car-following model whose sensitivity was varied by the lateral shifting of leaders was applied.

However, to the best of our knowledge, although response time is one of the main factors affecting traffic stability, there is a lack of research that discusses acceleration considering the variation of response time caused by the lateral shifting. Therefore, we first aim to clarify whether the response time is affected by a varying lateral shift. In addition, we also aim to propose a function formulation for the response time and sensitivity to improve the performance of 2D simulations. In this study, the ratio of a follower's acceleration to acceleration without any lateral shift is termed sensitivity. To achieve these objectives, we measure both the response time and acceleration sensitivity with respect to a varying lateral shift and attempt to fit them to certain functions. In Sect. 66.2, we present an experiment conducted using a driving simulator (DS). In Sect. 66.3, we attempt to fit the measured result to certain functions. In Sect. 66.4, we discuss the stability of 2D traffic using the results from Sect. 66.3.

66.2 Experiment to Measure the Response Time and Deceleration

66.2.1 Experiment Configuration

A DS with a fixed base and five monitors was utilized for the experiment. A button was located at the steering wheel, which a driver could press while holding the steering wheel.

The experiment setup is shown in Fig. 66.1. In the experiment, a leader (right, indicated in orange) and a follower (left, indicated in yellow) travel at a constant velocity of 17 m/s on a straight road. The subject is the follower, i.e., the follower is the ego vehicle. At a certain time, the leader suddenly decreases his/her velocity, and the velocity difference instantly reaches −5 m/s. The driver of the vehicle presses the button when he/she notices the leader's deceleration. Here, we define the response time $\tau = t_{\text{button}} - t_{\text{dec}}$, where t_{button} is the time the button is pressed, and t_{dec} is the time the leader starts to decelerate. The subject applies the brake or brake and steering to avoid a collision. It should be noted that the subject cannot start an avoiding maneuver before he/she presses the button.

As indicated in Fig. 66.1, the road has a width of 10.4 m and no lanes. The leader and follower both have a width of 1.9 m. In the experiment, the relative position of the leader was randomly presented from each dx and dy. The longitudinal position dy was selected from $dy_1 = 14$ m, $dy_2 = 34$ m, and $dy_3 = 54$ m from the center of the

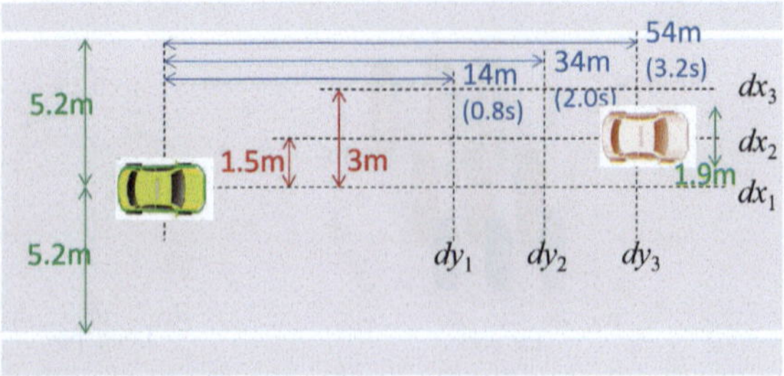

Fig. 66.1 Situation of the experiment and location of the leader

Table 66.1 Information regarding the subjects

Age (year)	Avg. = 22.5, SD = 1.2
Driving experience (year)	Avg. = 2.6, SD = 1.6
Sex (%)	Male = 69.2, Female = 30.8

ego vehicle. In Fig. 66.1, we also indicated the time headways for these longitudinal positions. The lateral position, dx, was selected from $dx_1 = 0$ m, $dx_2 = 1.5$ m, and $dx_3 = 3.0$ m from the center of the ego vehicle toward its left side. We prepared two conditions for the steering operation in order to replicate various traffic situations. The first was the no-steering (NS) condition, in which the subjects are not allowed to steer in avoidance and need to avoid a collision only by braking. The other was the with-steering (WS) condition, in which the subjects could steer and decelerate to avoid a collision. The NS condition replicates the traffic situation in which the ego vehicle cannot move laterally owing to other vehicles. After the subjects familiarized themselves with the DS, we randomly presented the combinations of respective positions and steering conditions to them. In total, the subjects experienced each combination thrice. We measured the positions of both the leader and follower and the timing of the button press for every trial.

Thirteen Japanese students participated in the experiment. In Table 66.1, we present information regarding the subjects. Their driving frequency ranged from once or twice per week to less than once per half year.

66.2.2 Results

Fig. 66.2a indicates the measured response times τ for the leader's positions in the NS case. Each bar indicates the averaged values among all subjects. The leftmost boxes marked in a checkered pattern are the averaged response times among all dx positions for each dy. The outermost yellow boxes are the averaged values among all dy positions for each dx. Figure 66.2b, c highlight the average response times from Fig. 66.2a, with error bars indicating the standard deviations. We first applied a three-way analysis of variance (ANOVA) for factors of the longitudinal and lateral

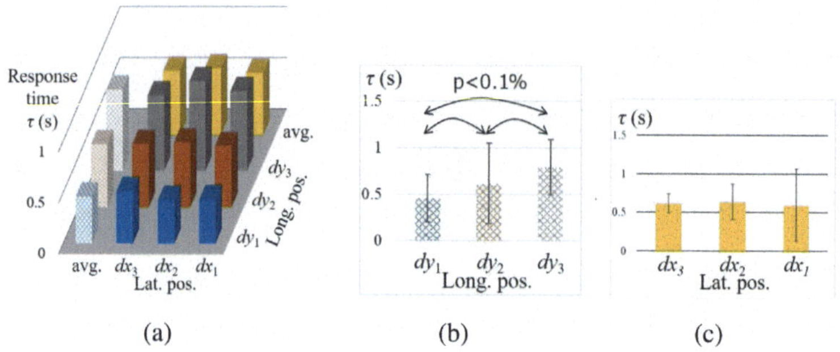

Fig. 66.2 Distribution of the response times in the no-steering case: **a** All the response times for respective positions. **b** Response times averaged among lateral positions. **c** Response times averaged among longitudinal positions. "Lat.", "long.", and "pos." mean lateral, longitudinal, and position, respectively

positions, and the repetition of trials. Because the effect of the repetition was not significant, we could ignore the learning effects of the subjects. Note that we applied the Greenhouse–Geisser correction and all the factors were within-subject factors. We confirmed that the significant effect of the longitudinal position at the significance level was less than 0.1%. No significant effects of the lateral position or interaction of these positions were confirmed. Then, we applied a multi-comparison to the factor of the longitudinal position. Note that here we utilized the Wilcoxon signed-rank test with the Bonferroni correction. We confirmed that a significant difference existed between all pairs as the p-values were less than 0.1%. From Fig. 66.2b, we can conclude that the response time increases with increasing longitudinal distance in the NS condition.

In Fig. 66.3, τ in the WS case is indicated. In contrast to the NS case, a significant difference for the lateral position was also identified. We summarize the results of the statistical tests for the response time and deceleration sensitivity in the NS and WS cases in Table 66.2. Note that we define the deceleration sensitivity w at the position of "$dx = l, dy = m$" as follows.

$$w_{lm} = \frac{a_{lm}}{a_{1m}} \tag{66.1}$$

$$a_{lm} = \frac{\sum \{v_{lm}(t_{\text{button}} + T) - v_{lm}(t_{\text{button}})\}}{T} \tag{66.2}$$

where a, v, and t_{button} are the acceleration, velocity, and the time when the subject pressed the button, respectively. T is a time constant. Here, we determined T as 0.5 s. The deceleration sensitivity is the acceleration ratio at each relative position (a_{lm}) compared to the acceleration when the leader is located in front of the ego vehicle without a lateral shift (a_{1m}).

From Table 66.2, we confirmed that the lateral shift affects the response time in the WS case and the deceleration sensitivity in both the NS and WS cases. In addition,

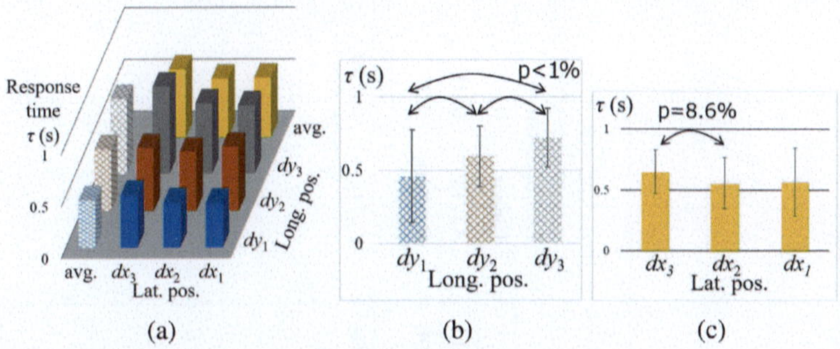

Fig. 66.3 Distribution of the response times in the with-steering case

Table 66.2 Results of statistical tests of the response time and deceleration sensitivity. "NA" indicates that the post-hoc test cannot be applied because of the results of the ANOVA. "n.s." indicates that the difference is not significant. "◇," "‡," "†," and "•" mean that the significance level is less than 0.1%, 1%, 5%, and 10%, respectively

	ANOVA			Multiple comparison	
	Lateral	Longitudinal	Interaction	Lateral	Longitudinal
τ in NS	n.s.	◇	n.s.	NA	$dy_1 < dy_2 < dy_3$◇
τ in WS	†	◇	n.s.	$dx_2 < dx_3$•	$dy_1 < dy_2 < dy_3$◇
w in NS	◇	n.s.	†	NA	NA
w in WS	◇	n.s.	n.s.	$dx_3 < dx_2 < dx_1$†	NA

we confirmed an interaction effect of dx and dy for the deceleration sensitivity in the NS case.

66.3 Function Fitting

We assumed that the response time and the deceleration sensitivity could be determined by the sum of the effects of dx, dy, and the interaction. As candidates of the fitting function, we assumed increasing functions for all the effects except for the interaction, because these effects did not have maximum or minimum values. In these equations, w or τ is in Z, and dx or dy is in q.

Table 66.3 Fitted functions and parameters for the sensitivity and response time

Variable	w in NS				w in WS		τ in NS		τ in WS			
R^2	0.11				0.16		0.16		0.21			
Effect	Lateral		Interaction		Lateral		Longitudinal		Lateral		Longitudinal	
Function	atan		Normal distribution		atan		atan		exp		Linear	
Constant	A_{dx} (-)	−0.16	A_{INT} (-)	0.38	A_{dx} (-)	−32.19	A_{dy} (s)	217.22	A_{dx} (s)	0.11	A_{dy} (s/m)	0.01
	B_{dx} (-)	6.90×10^{11}	ρ (-)	0.00	B_{dx} (-)	0.01	B_{dy} (-)	0.00	B_{dx} (m)	2.96	B_{dy} (m)	2.03
	C_{dx} (m)	1.47	σ_{dx} (m)	0.82	C_{dx} (m)	−4.27	C_{dy} (m)	77.36	C_{dx} (s)	1.08	C_{dy} (s)	−0.79
	D_{dx} (-)	0.68	σ_{dy} (m)	5.30	D_{dx} (-)	1.92	D_{dy} (s)	0.97				
			dx_c (m)	1.56								
			dy_c (m)	35.01								

$$Z_q = A_q(q - B_q) + C_q \tag{66.3}$$

$$Z_q = A_q \exp(q - B_q) + C_q \tag{66.4}$$

$$Z_q = A_q \log(q - B_q) + C_q \tag{66.5}$$

$$Z_q = A_q \mathrm{atan}(B_q(q - C_q)) + D_q \tag{66.6}$$

For the interaction effect on the sensitivity, we utilized the normal distribution function because it had a maximum value, i.e.,

$$w_{\mathrm{INT}} = A_{\mathrm{INT}} \exp\left[-\frac{1}{2(1-\rho^2)}\left\{\frac{(dx - dx_{\mathrm{c}})^2}{\sigma_{\mathrm{dx}}^2} + \frac{(dy - dy_{\mathrm{c}})^2}{\sigma_{\mathrm{dy}}^2} + \frac{2\rho(dx - dx_{\mathrm{c}})(dy - dy_{\mathrm{c}})}{\sigma_{\mathrm{dx}}\sigma_{\mathrm{dy}}}\right\}\right]. \tag{66.7}$$

Table 66.3 indicates the fitted parameters. The functions with the best fitting performances were selected. Note that the normal distribution function for the sensitivity in the NS case had the maximum value at $dx = 1.56$ m, $dy = 35.01$ m.

66.4 Discussion

In Sect. 66.3, we found that both the longitudinal and lateral distances between the leader and ego vehicle affected the deceleration sensitivity and response time. Zhang et al. investigated the stability of car-following models with delay [7], and clarified that both the sensitivity and delay affected the stability of traffic without a lateral shift. The stability condition is $I = w \cdot \tau \cdot v / h_{\mathrm{e}} < \pi/2$, where v is the velocity of the ego vehicle and h_{e} is the equilibrium headway distance. Because I is proportional to the delay or response time [7], the stability of traffic with weak lane discipline would be underestimated if the response time were not considered.

66.5 Conclusion

For better replication of traffic with weak lane discipline, we evaluated the variation of the response time and deceleration sensitivity with the relative position of the leader including a lateral shift. Through an experiment with a DS, we clarified that the relative position of the leader affected both the response time and deceleration sensitivity. In particular, the lateral shift affected both the response time and sensitivity when drivers could steer in traffic. Based on the result in [7], it is implied that the stability of traffic with weak lane discipline is underestimated if these effects are not considered.

For future work, the stability analysis of 2D traffic flow with a delay is essential to evaluate the degree of these effects. In addition, as we utilized a DS, the observed effects would not be quantitatively accurate in actual traffic with real vehicles. An experiment and field observations that detect the response time of each vehicle are required.

Acknowledgements This work was supported by JSPS KAKENHI Grant Numbers 18H05923 and 19K15246, and the Suzuki Foundation.

References

1. K. Ahmed, M. Ben-Akiva, H. Koutsopoulos, R. Mishalani, Models of freeway lane changing and gap acceptance behavior. Transp. Traffic Theory **13**, 501–515 (1996)
2. B. Gunay, Car following theory with lateral discomfort. Transp. Res. Part B Methodol. **41**(7), 722–735 (2007)
3. V. Kanagaraj, M. Treiber, Self-driven particle model for mixed traffic and other disordered flows. Phys. A Stat. Mech. Appl. **509**, 1–11 (2018)
4. A. Kesting, M. Treiber, D. Helbing, General lane-changing model MOBIL for car-following models. Transp. Res. Record **1999**(1), 86–94 (2007)
5. T. Oketch, New modeling approach for mixed-traffic streams with nonmotorized vehicles. Transp. Res. Record J. Transp. Res. Board **1705**, 61–69 (2000)
6. J. Tang, F. Liu, W. Zhang, R. Ke, Y. Zou, Lane-changes prediction based on adaptive fuzzy neural network. Expert Syst. Appl. **91**, 452–463 (2018)
7. X. Zhang, D.F. Jarrett, Stability analysis of the classical car-following model. Transp. Res. Part B Methodol. **31**(6), 441–462 (1997)

Chapter 67
Multi-lane Traffic Flow Model: Speed Versus Density Difference as Lane Change Incentive and Effect of Lateral Flow Transfer on Traffic Flow Variables

Hari Hara Sharan Nagalur Subraveti, Victor L. Knoop, and Bart van Arem

Abstract For better and more efficient motorway traffic control strategies on a lane level, accurate modelling and prediction of traffic conditions is essential. Existing real-time traffic state estimation methods aggregate traffic across lanes. Hence, there is a need for lane-specific traffic flow models. The majority of the existing macroscopic lane-level traffic flow models use speed difference among lanes to explain lane change decisions. One of the major disadvantages of using speed difference as motivation for lane changing is the different speeds at which the lanes on a motorway operate. These models which are usually extensions of the Cell Transmission Model (CTM) also suffer in the manner in which lateral flow is transferred among cells in different lanes. In order to overcome these limitations, a model is proposed using density difference as an incentive to compute lane change rates rather than speed. A two-step transfer of lateral flow among cell segments is also considered where the lateral demand of a cell is dependent upon the receiving capacity of the adjacent and downstream cells in the target lane. The advantages of the presented approach is shown by comparing the proposed model with a conventional model. It was observed that the proposed model performed better in terms estimating densities across lanes. The proposed model is a step in the development of more accurate lane-specific traffic flow models.

67.1 Introduction

With rising congestion on motorways across the world, there is an urgent need to come up with active and efficient traffic management strategies to tackle this problem. For effective traffic management, accurate traffic flow estimation and prediction models are essential. Existing real-time estimation and prediction methods aggregate traffic across lanes without taking into account the difference among lanes. But studies [1, 2] have shown that on multi-lane motorways, lanes operate differently. Depending upon

---

H. H. S. Nagalur Subraveti (✉) · V. L. Knoop · B. van Arem
Delft University of Technology, Stevinweg 1, 2628 CN Delft, The Netherlands
e-mail: h.h.s.nagalursubraveti@tudelft.nl

I. Zuriguel et al. (eds.), *Traffic and Granular Flow 2019*,
Springer Proceedings in Physics 252,
https://doi.org/10.1007/978-3-030-55973-1_67

the conditions, proportion of flows on different lanes vary. This can lead to unbalanced lane usage. Knowledge of lane specific traffic conditions are important to achieve a more balanced distribution of traffic across the lanes and improve efficiency of motorways. Hence traffic flow models which include the dynamics of lane changing are needed for lane specific control.

The majority of the existing macroscopic lane-level traffic flow models use speed difference among lanes to explain lane change decisions. One of the major disadvantages of using speed difference as motivation for lane changing is the different speeds at which the lanes on a motorway operate. These models which are usually extensions of the Cell Transmission Model (CTM) also suffer in the manner in which lateral flow is transferred among cells in different lanes. In order to address these concerns, a model is proposed which uses density difference among lanes as an incentive to compute lane change rates rather than speed. The model also differs from existing models in the manner in which the computed lateral flows are transferred among cells. The proposed model is compared to an existing model by using real world data.

The remainder of the paper is organized as follows: Sect. 67.2 describes state-of-the-art on multilane traffic flow models. Section 67.3 describes the proposed model including the computation and transfer of lateral flows and lane change rates. In Sect. 67.4, the proposed model is tested against real world data and compared to a conventional model and finally in Sect. 67.5, we conclude the paper.

67.2 State-of-the-Art on Multilane Traffic Flow Models

Macroscopic models describe aggregate driving behavior and typically involve a relationship between the density and flow of a network. Earlier works on multi-lane macroscopic models such as [3, 4] were formulated in a continuous space-time domain. Laval and Daganzo [5] expanded on earlier works and presented a hybrid model considering the lane changers as particles with bounded acceleration rates. However, these models aggregated traffic flow variables across lanes. Further models such as [6, 7], and [8] were developed which considered individual lanes as separate entities. Most of these models (except [8]) consider speed difference lanes as an incentive to compute the percentage of flow that might change lane. But on multi-lane motorways, different lanes operate at different speeds. These models use a single fundamental diagram (FD) for the motorway.

A single FD for all lanes implies that lane changes always happen among the lanes as the speeds on lanes differ (especially in free-flow) which is not the case. Using a single FD for all the lanes might not lead to accurate results. Speeds are computed a posteriori if the flow-density relationship is used. If density-speed relationship is used, one of the disadvantages of using speed as a motivation to explain lane changing is the constant relationship between speed and density in the free-flow state (assuming a triangular FD). This implies that no lane changes take place in the free flow state which is not a realistic assumption. Considering different shapes

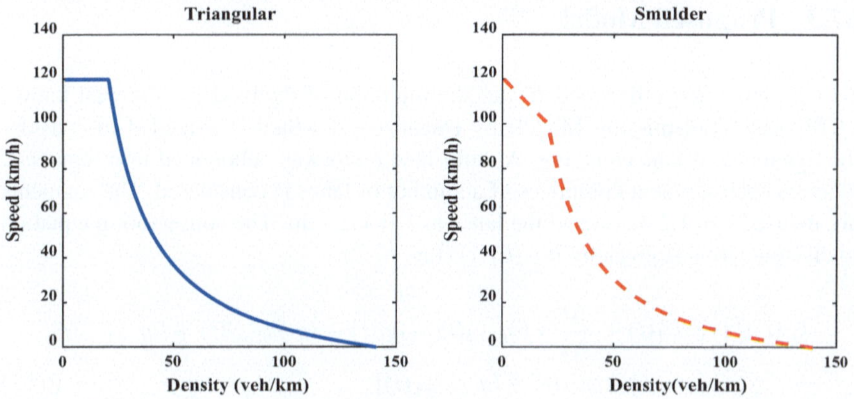

Fig. 67.1 Speed-Density relationship for different FDs

of FD to allow for lane changes in the free-flow state can increase the complexity of the model in terms of the computation of demand and supply capacities due to the introduction of nonlinear terms. As can be seen in Fig. 67.1, if a triangular FD is used, the speed does not vary in the free-flow state. This would imply that there are no lane changes. If a different shape of FD (like Smulder) is used, the number of FD parameters increases because there is a need to estimate both the free-flow and critical speed. This increases the number of parameters as well as introduces a non-linear relationship between flow and density in the free-flow state which affects the computation of demand and supply and consequently transfer of flows. Using density difference among lanes presents us with an advantage that lane changes are possible even in the free-flow state and densities are generally the state variables of traffic flow models. Although there are a number of other factors such as the current location, prevailing traffic conditions, destinations of the vehicles etc. which affect the lane changing flow, in order to keep a simple framework for lateral flows, density difference is used which can capture the basic situations.

A major element of most of the existing models is the transfer of lateral flow among cell segments in the discretization scheme. The transfer of lateral flow among cell segments in different lanes is considered to be diagonal i.e. the lateral flow from a cell depends on the supply capacity of the adjacent downstream cell. While this works in the free-flow state, it may not work in congested conditions where the lateral demand is generally dependent on the receiving capacity of both the adjacent and downstream cell in the target lane (especially when the length of the cell segments are not too long). Considering a diagonal transfer of flow among cell segments can lead to the under-estimation of the distance over which the congestion propagates as well as the erroneous estimation of the strength of congestion. In this study, the motivation behind lane changing is formulated using density as the explanatory variable. The transfer of lateral flows is also based on the assumption that the lateral demand is dependent upon the receiving capacity of the adjacent and downstream cell segments on the target lanes.

67.3 Proposed Model

As with most first order models, the starting point of the model is the well-known CTM (Cell Transmission Model) by Daganzo [9] which is extended to consider the dynamics of lane changing. A multi-lane motorway subdivided into segments, wherein each segment comprises of a number of lanes is considered. The segments are indexed i = 1,2,3....n and the lanes as l = 1,2....m. The conservation equation in discrete terms is given by Eq. (67.1) (Fig. 67.2).

$$\begin{aligned} k_{il}(t+1) =& k_{il}(t) + \frac{\Delta t}{\Delta x} * [q_{i-1,l}(t) - q_{il}(t) + lq_{i,l-1\rightarrow l}(t) + lq_{i,l+1\rightarrow l}(t) \\ & - lq_{i,l\rightarrow l-1}(t) - lq_{i,l\rightarrow l+1}(t)] \end{aligned} \tag{67.1}$$

In (67.1), k and q represent the density and flow at the boundary of the cell-segments respectively. Δt is the size of the time step and Δx is the length of the cell segment. lq denotes the lateral flow between the cell segments. t denotes the simulation horizon where t = 1,2,3.....,T and the total simulation time is given as $t_{\text{sim}} = \text{T}^* \ \Delta t$. The terms $lq_{i,l-1\rightarrow l}$ & $lq_{i,l+1\rightarrow l}(t)$ represent the lateral flow entering cell-segment i from the adjacent lanes $l-1$ and $l+1$ respectively and the terms $lq_{i,l\rightarrow l-1}$ & $lq_{i,l\rightarrow l+1}(t)$ represent the lateral flow exiting cell-segment i to the adjacent lanes $l-1$ and $l+1$ respectively. The lateral flow in (67.1) is considered to be a function of the density difference (directly proportional) rather than the speed difference and is given by Eq. (67.2) for the proposed model.

$$lq_{i,l'\rightarrow l} \propto (k_{il'} - k_{il}) \tag{67.2}$$

where, l' and l represent the original and target lane respectively. The longitudinal flows are computed as the minimum of the sending and receiving capacities of the cell. The transfer of lateral flows is considered as a 2-step process and is shown in Fig. 67.3. The figure on the left shows the conventional transfer while the figure on the right is the proposed transfer method. The proposed method is implemented in the model by modifying the sending and receiving functions of the cell to include the effect of lateral flow and instead utilize effective sending and receiving capacities to compute longitudinal flow among the cell-segments. In the 2-step transfer, the lateral

Fig. 67.2 Representation of the discretized motorway

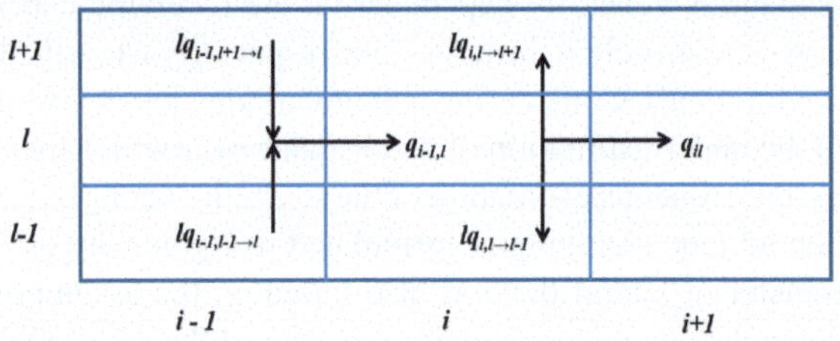

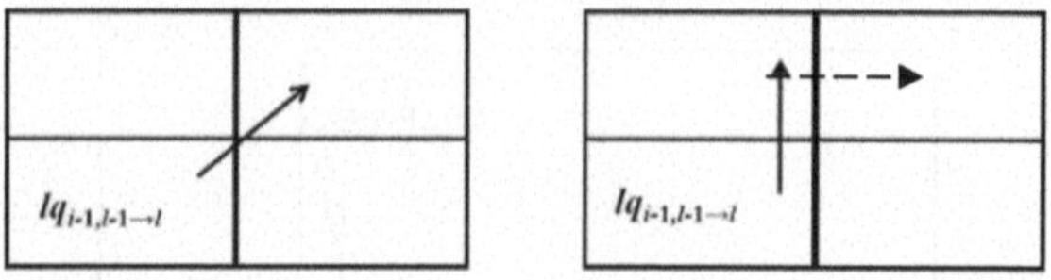

Fig. 67.3 Transfer of lateral flow among cells

demand of a cell is dependent on the receiving capacity of the adjacent as well the downstream cell of the target lane.

67.4 Results

The proposed model is tested against real world data and compared to a conventional model which uses a diagonal transfer of lateral flow. As the speeds at which the lanes are operating are different, speed difference has not been used in the conventional model as this would be lead to obvious errors. Thus only the method of lateral flow transfer is varied between the two models. Data from a lane drop bottleneck on the A12 motorway in Netherlands is selected for this purpose. 1-minute aggregated speed and flow information for the month of May, 2008 from stationary lane based detectors were used to compute densities. The chosen section is a 4-3 lane drop shown in Fig. 67.4.

The performance indicator used to compare the models is the RMSE of lane densities and is given by Eq. (67.3).

$$\text{RMSE} = \left(\sqrt{\sum_{i}^{n}\sum_{j}^{L}\left(k_{est,l}^{i} - k_{obs,l}^{i}\right)^{2}}\right) \tag{67.3}$$

The results for a particular day is shown in Fig. 67.5. The results are shown for detectors upstream of the lane drop.

It can be observed that the 2-step process leads to lower error values on lanes 1 and 2 as compared to a diagonal transfer of flow. There is not much difference on lanes 3 and 4. This can be attributed to the low lane changing activity on these lanes

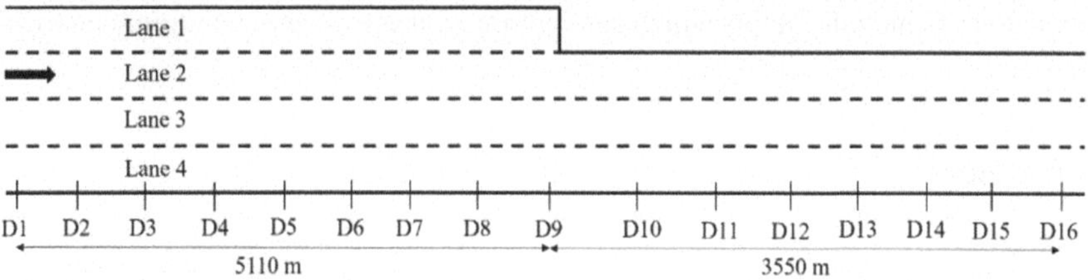

Fig. 67.4 A12 lane drop section

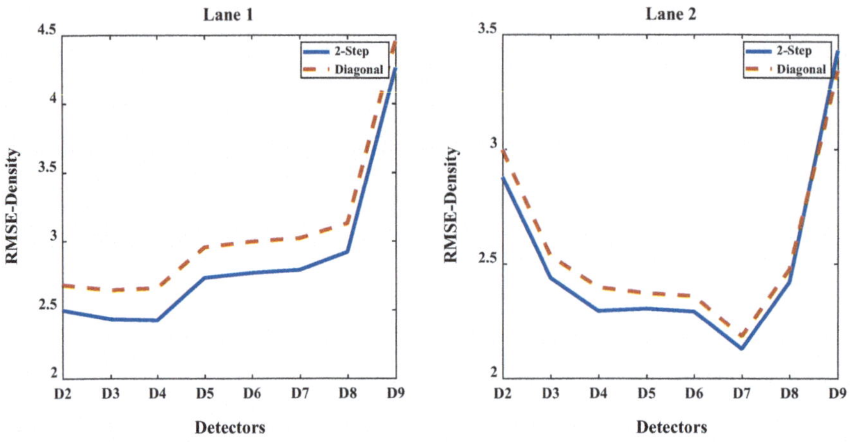

Fig. 67.5 Comparison of the proposed model with conventional model

as compared to lanes 1 and 2 where the lane changing activity is high due to the merging of vehicles. Hence it can inferred that a 2-step lateral flow transfer among cells leads to better lane-level estimations of traffic flow variables when compared a diagonal transfer of flow.

67.5 Conclusions

A multi-lane traffic flow model is proposed where the incentive to change lanes is expressed as a function of the density difference between lanes and it is argued why this suits better than using a speed difference among lanes. A 2-step process of lateral flow transfer is proposed where the lateral flows are dependent upon the receiving capacity of not only the downstream cell of the target lane but also on the adjacent cell of the target lane. The proposed model is then compared to a conventional model using real world data from a lane drop bottleneck. It was observed that the proposed changes lead to better estimation of lane densities as compared to the conventional model. Future works include further testing against real-world/synthetic data for more robust comparison as well as more robust formulation for the computation of lateral flows to include factors which can capture a variety of lane changing scenarios.

References

1. V.L. Knoop, A. Duret, C. Buisson, B. Van Arem, Lane distribution of traffic near merging zones influence of variable speed limits, in Proceedings of the 13th International IEEE Conference on Intelligent Transportation Systems (2010)

2. N. Wu, Equilibrium of lane flow distribution on motorways. Transp. Res. Rec. **1965**(1), 48–59 (2005)
3. P.K. Munjal, L.A. Pipes, Propagation of on-ramp density waves on uniform unidirectional multilane freeways. Transp. Sci. **5**(4), 390–402 (1971)
4. P.G. Michalopoulos, D.E. Beskos, Y. Yamauchi, Multilane traffic flow dynamics: some macroscopic considerations. Transp. Res. Part B: Methodol. **18**(4–5), 377–395 (1984)
5. J.A. Laval, C.F. Daganzo, Lane-changing in traffic streams. Transp. Res. Part B: Methodol. **40**(3), 251–264 (2006)
6. Y. Shiomi, T. Taniguchi, N. Uno, H. Shimamoto, T. Nakamura, Multilane first-order traffic flow model with endogenous representation of lane-flow equilibrium. Transp. Res. Part C: Emerg. Technol. **59**, 198–215 (2015)
7. M. Seraj, Y. Bie, T.Z. Qiu, A macroscopic lane-changing model for freeway considering different incentives and traffic states, in Proceedings of the 96th Annual Meeting of the Transportation Research Board (2017)
8. C. Roncoli, M. Papageorgiou, I. Papamichail, Traffic flow optimisation in presence of vehicle automation and communication systems–Part I: a first-order multi-lane model for motorway traffic. Transp. Res. Part C: Emerg. Technol. **57**, 241–259 (2015)
9. C.F. Daganzo, The cell transmission model: A dynamic representation of highway traffic consistent with the hydrodynamic theory. Transp. Res. Part B: Methodol. **28**(4), 269–287 (1994)

Chapter 68
Diversity Analysis of the Brazilian Air Transportation Network

Izabela M. Oliveira, Laura C. Carpi, and A. P. F. Atman

Abstract In recent years, air transportation networks have been studied with tools from complex networks theory. In this work, the Brazilian domestic air transportation network is analyzed in the period 2010–2018, considering a multiplex approach. Through a recently proposed quantifier, the Multiplex Efficiency Index $\mathcal{E}$, the diversity of Brazilian air transportation routes is studied. Changes in the efficiency of the system are analyzed and discussed, especially considering airport privatization processes and events such as the exclusion of companies, with a case study of a company withdraw with real data. Alternatives for reorganizing routes between the remaining companies are considered to find the most efficient combination. Correlations between the $\mathcal{E}$-index, aircraft seat occupancy rate and airline ticket prices are investigated, highlighting topological features influenced by real factors.

I. M. Oliveira (✉)
Departamento de Matemática, Centro Federal de Educação Tecnológica de Minas Gerais, CEFET-MG. Av. Amazonas, 7675 Belo Horizonte, MG, Brazil
e-mail: izabelamarques@cefetmg.br

I. M. Oliveira · L. C. Carpi · A. P. F. Atman
Programa de Pós-Graduação em Modelagem Matemática e Computacional, PPGMMC, CEFET-MG, Belo Horizonte, Brazil

L. C. Carpi · A. P. F. Atman
Instituto Nacional de Ciência e Tecnologia de Sistemas Complexos, INCT-SC, CEFET-MG, Belo Horizonte, Brazil

A. P. F. Atman
Departamento de Física, CEFET-MG, Belo Horizonte, Brazil

I. Zuriguel et al. (eds.), *Traffic and Granular Flow 2019*,
Springer Proceedings in Physics 252,
https://doi.org/10.1007/978-3-030-55973-1_68

68.1 Introduction

The Brazilian air transportation network has undergone major changes in the last fifteen years. Important airlines have closed their operations, while several low-cost companies have started operating in the country, changing the global scenario [1]. Airports privatizations and financial investments in infrastructure, have also influenced the air transportation market [2]. As a consequence, several new polices have been implemented by the National Civil Aviation Agency (ANAC) [3], to improve the performance of the system. The air transportation assumes a capital importance in Brazil for two main reasons: the continental dimensions of the country and the lack of an efficient rail transportation system [4].

The FIFA World Cup in 2014, brought thousands of people to Brazil, requiring transportation strategies to reach the cities where the games took place [5]. In a previous work [6], we evaluated the evolution of the Brazilian air transportation network, analysing the coverage of the national territory and the overlap of airline routes between 2010 and 2018. For this, the Brazilian Air Transportation Multiplex Network (BATMN) was built and the Multiplex Efficiency Index $\mathcal{E}$ was used to quantify the network efficiency during the studied period. The $\mathcal{E}$-index allows to measure the level of heterogeneity of the system connections in a multiplex approach, on a scale of 0–1, in such a way that a more "efficient" network is one that has more diverse connections between its layers.

This work investigates other factors that may have influenced $\mathcal{E}$ for BATMN, as is the case of ticket prices and the aircraft occupancy ratio. Exploring the values of $\mathcal{E}$ at node levels, the ranking of the most efficient airports of the period is showed, as well as the evolution of the airports efficiency that were privatized up to 2018.

In addition, in 2019, a BATMN airline (Avianca Brasil, ICAO code: ONE) halted operations, concentrating the network in just 4 major airlines, reducing the options offered to the population. Several routes were extinguished, and the slots (landing and takeoff rights) operated by ONE at *Congonhas* airport, which is one of Brazil's main airports, were distributed between two competing airlines. Simulations using real data related to this new slot distribution and route rearrangements were performed to investigate changes in the Multiplex Efficiency Index $\mathcal{E}$ for BATMN.

68.2 Methodology

BATMN was built with a multiplex network approach, so that the nodes represent the 154 Brazilian airports that participated in domestic passenger transportation between 2010 and 2018. The links connecting nodes are the different airline routes with a share of the Brazilian domestic air market exceeding 0.5% of the total number of annual passengers, according to real ANAC data [7]. Thus, each airline has its own network, which is represented as a layer of the multiplex structure. BATMN is composed by 8 layers during the period 2010–2012, 6 layers in 2013, and 5 layers from 2014 to

2018. The airlines considered are: AZU, GLO, ONE, PTB, TAM, TIB, PTN and WEB, so 3 companies were merged with others over the period. For more details about the participating airlines, see Oliveira et al. [6].

To calculate the Multiplex Efficiency Index $\mathcal{E}$ we consider a multiplex network with M layers, each one containing the same set of N nodes. It is possible to obtain the local $\mathcal{E}$-index, at the node level, and global, which is at the network layer level. For this, two probability distributions are extracted from the network, as proposed in a recent work [8]:

1. *Node Distance Distribution*: for each node i in layer p, $\mathcal{N}_i^p(d)$ corresponds to the fraction of nodes reachable from node i, by the shortest path d;
2. *Transition Matrix*: $\mathcal{T}_i^p(j)$ is the probability that node j will be reached in one step by a random walker located at node i, at the layer p.

Then, through the Jensen-Shannon divergence, the node distance (ND) in layers p and q is measured:

$$\mathcal{D}_i(p,q) = \frac{\sqrt{\mathcal{J}(\mathcal{N}_i^p, \mathcal{N}_i^q)} + \sqrt{\mathcal{J}(T_i^p, T_i^q)}}{2\sqrt{\log(2)}} \tag{68.1}$$

for each node i from 1 to N. The difference in connectivity paths between two layers p and q (layer distance—LD) is obtained by averaging over all values of $\mathcal{D}_i$ obtained by Eq. 68.1: $\mathcal{D}(p,q) = \langle \mathcal{D}_i(p,q) \rangle_i$. Then the distance between an element g and a set S is used, defined as the shortest distance between g, not belonging to S, and some element of S:

$$\mathcal{D}(g, S) = \min_{s_i \in S} \mathcal{D}(g, s_i). \tag{68.2}$$

When the elements considered are the network nodes, the ND measure (Eq. 68.1) is used to obtain the distance referred to by Eq. 68.2. In another situation, for the distance between the layers of the network, the LD measure is used. Next, the system diversity value $U(S)$ is calculated by the recursive equation $U(S) = \max_{s_i \in S}\{U(S \setminus s_i) + \mathcal{D}(s_i, S \setminus s_i)\}$. The code for the diversity calculation of a system is freely available in [9]. Finally, we obtain the value of the Multiplex Efficiency Index $\mathcal{E}$, which relates to the value of the system diversity, considering the number of layers present in the network:

$$\mathcal{E} = \frac{U(S)}{M-1}, \tag{68.3}$$

so that $0 \le \mathcal{E} \le 1$. The greater the network edge overlap, the closer to 0 will be the value of $\mathcal{E}$. On the other hand, in networks where links are more diversified between layers, the value of $\mathcal{E}$ will approach 1 [6, 10].

In the case of air transportation, at the layer level, $\mathcal{E}$-values close to 0 mean that airlines present on the network operate with very similar routes, while $\mathcal{E}$-values close to 1 indicate that airlines work with very different routes. At the node level, airports with $\mathcal{E}$-values close to 0 have routes very similar connecting them to other airports considering each airline. Airports with higher $\mathcal{E}$-values are linked in a diversified manner by the routes of the airlines that serve them.

68.3 Results and Discussions

We calculated the value of the Multiplex Efficiency Index for BATMN from 2010 to 2018, which is shown in Fig. 68.1. We found that from 2013 to 2018, $\mathcal{E}$ reaches a plateau and remains around it, with a peak in 2014, which coincides with the year of the FIFA World Cup in Brazil, in agreement with Oliveira et al [6]. With the termination of airline ONE's operations in 2019, its routes are excluded and its 41 slots at *Congonhas* airport, that were used with 9 routes, were redistributed to AZU and PTB, according to the current rules from ANAC [11]. Slots are arrival and departure times available or authorized for flights at busy airports, which are distributed by a regulatory agency [3]. Each airline uses the slots according to its business, creating routes between the airports in which it operates and, thus, the same route can occupy several slots, at different times [7]. The 15 slots received by AZU were all allocated on a single route: the *Rio/São Paulo* air bridge, which connects the airports *Congonhas* and *Santos Dumont*, and is one of the most profitable routes in the country [12]. PTB received 26 slots (14 for PTB and 12 for MAP, which now belongs to PTB [11]) starting operations at *Congonhas* airport with 7 routes, in which 5 of them correspond to links not present in BATMN in 2018. Five experiments were performed using data from 2018, as shown in Fig. 68.1, with 4 layers left in 2019, taking into account the exclusion of ONE from BATMN, which had 91 links in the network, of which only 4 were exclusive to the ONE layer.

- **Experiment 1**: Just the ONE layer is excluded, and the links are not added in any other layer;

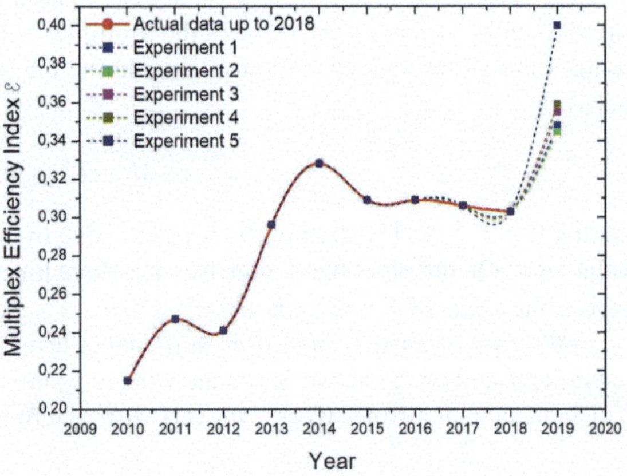

Fig. 68.1 Evolution of the multiplex efficiency index for BATMN, using real data between 2010 and 2018 (*full line*). Results of the 5 experiments conducted from airline ONE's departure from the network in 2019 (*dotted lines*)

- **Experiment 2**: All links belonging to ONE in 2018, are distributed to the 4 remaining BATMN airlines (3 airlines get 23 links each, an done one gets 22);
- **Experiment 3**: ONE is deleted and a link between *Congonhas/Santos Dumont* is added to the AZU and the 7 links that were created by PTB in 2019, are added to the PTB layer, from *Congonhas* airport. This is the situation most similar to what actually happened with the network in 2019, with the ONE exclusion.
- **Experiment 4**: The 82 links belonging to ONE in 2018, except those at the *Congonhas* airport, are equally distributed in the network of experiment 3 among GLO and TAM.
- **Experiment 5**: Excludes ONE and creates 41 random links from *Congonhas* airport (equivalent number of redistributed slots), 15 to AZU and 26 to PTB, to airports that previously did not receive routes linked to *Congonhas*.

It is possible to observe that the $\mathcal{E}$-values found are similar, with small variations in the second or third decimal place, except in the last experiment. In all cases, the value of $\mathcal{E}$ increases with the exclusion of ONE, as it is the airline that has the most overlapping routes with other airlines in the last 5 years [6]. In other words, the 5 airlines present in BATMN in 2018 had more route overlapping than the 4 airlines present in 2019, according to the simulations.

The $\mathcal{E}$-values found in experiments 1 and 2 are equivalent, showing that excluding ONE or redistributing its routes to other companies does not modify the diversification of network links, since 96% of its links already exist in the other layers. In Experiment 3, the creation of new links to PTB, which did not previously operate at *Congonhas* airport, slightly increases the value of $\mathcal{E}$, compared to Experiment 2, due to the creation of 5 new routes in its layer. The addition of the *Congonhas/Santos Dumont* route of AZU does not significantly increase the value of $\mathcal{E}$, since this link already existed in other layers. The $\mathcal{E}$-value of experiment 4 is equivalent to that of experiment 3, due to the high overlap of the 82 links that belonged to ONE. The highest growth, 32% above $\mathcal{E}$ in 2018, can be seen in experiment 5, when 41 different routes are created for PTB and AZU, which are the most diversified layers of the network [6]. Comparing the scenarios of experiments 5 and 1, there is an increase of 15%, whereas from experiment 3 to 1 the increase is only 2%. It can be concluded that releasing slots from one airline to another has no significant effect on the network efficiency as it is not synonymous to route diversification.

To investigate other causes of fluctuations in the Multiplex Efficiency Index, we studied possible correlations between the $\mathcal{E}$ values, the average airfare price (in US dollars) and the load factor, which represents the utilization rate of the seats offered on the aircraft [7]. The strong Pearson's correlation coefficient between $\mathcal{E}$ and the load factor (0.92) is evident from Fig. 68.2a. When comparing the average value of tickets with $\mathcal{E}$, it is found anti-correlation (−0.89) until 2014 and strong correlation (0.96) after 2014 (Fig. 68.2b). As ticket prices fall, the load factor is expected to increase, and vice versa. The novelty here is that the Multiplex Efficiency Index followed the load factor. This points to the fact that the greater the diversification of routes, the higher the rate of aircraft seat utilization, and this is strongly related to ticket prices.

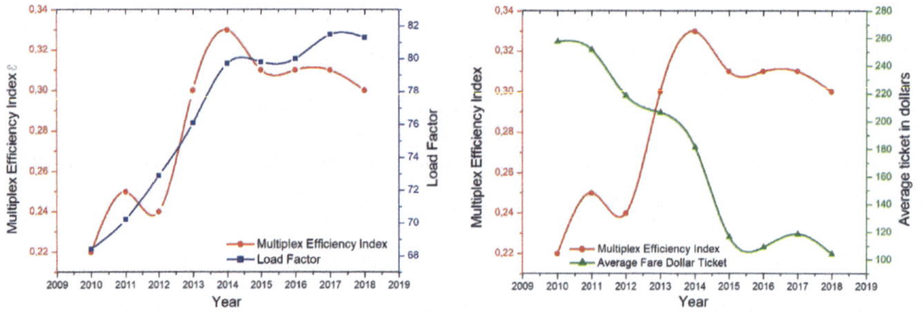

Fig. 68.2 Evolution of $\mathcal{E}$-index values and possible correlations with other BATMN quantities. **a** The left panel shows the $\mathcal{E}$ values and Load Factor. **b** The right panel displays the average value of BATMAN airline tickets (US dollar) and $\mathcal{E}$

By analyzing the node-level Multiplex Efficiency Index, it is possible to get the ranking of the most efficient airports on the network. In this article, the first 10 airports of 2010 and 2018, and also of 2014 are displayed, when the maximum value of $\mathcal{E}$ occurs. In addition to the top 10, airports that were privatized up to 2018 are highlighted [6]. It is possible to observe large variations in the ranking, caused by the merge of three airlines between 2010 and 2014. In 2013, begins a major expansion of AZU, gradually connecting many secondary airports and diversifying routes to other airlines. AZU became the airline contributing to the system from 2014, high diversity values. The same diversity order occurs over the last 5 years, as found by Oliveira et al., in [6]. Airports that had few connections, are more connected since this expansion. In addition, AZU operates more frequently at busiest airports in the country, connecting them to secondary airports. Depending on the demand for corporate flights and business niches, there may be more or less route overlap between airlines. Thus, changes in the connections of secondary airports and large hubs have been observed over the years (Table 68.1).

68.4 Conclusions

According to the construction of BATMN, it is noticed that there was a market concentration, reducing the number of airlines from 8 airlines in 2010 to only 4 airlines in 2019. The Multiplex Efficiency Index shows that as merges occur, the diversification of routes increases. Depending on the niche of each company, there was more or less diversification of routes from each airport, regardless of whether it is secondary or a network hub. Airports privatizations did not significantly influence the diversification of network links. With the exclusion of ONE, the network efficiency increases because many ONE's routes were overlapping the others. In other words, the remaining airlines have more diverse routes to each other. The growth in the network's

Table 68.1 Placement of privatized airports in the $\mathcal{E}$ ranking in 2010, 2014 and 2018 and other airports that were in the top 10

Year of privatization	ICAO Airport	Served City	2010 ranking	2014 ranking	2018 ranking
2011	SBSG[1]	Natal	17	15	21
2012	SBBR	Brasília	9	36	29
2012	SBKP	Campinas	4	43	4
2012	SBGR	Guarulhos	7	32	39
2013	SBCF	Confins (Belo Horizonte)	12	17	9
2013	SBGL	Rio de Janeiro (Galeão)	3	23	30
2017	SBFL	Florianópolis	6	12	19
2017	SBFZ	Fortaleza	1	18	13
2017	SBPA	Porto Alegre	10	9	7
2017	SBSV	Salvador	11	38	35
–	SBBH	Belo Horizonte (Pampulha)	60	19	3
–	SBCT	Curitiba	8	8	20
–	SBFI	Foz do Iguaçu	48	5	10
–	SBIL	Ilhéus	25	4	8
–	SBJP	João Pessoa	51	7	22
–	SBNF	Navegantes	31	39	6
–	SBPJ	Palmas	44	3	2
–	SBPV	Porto Velho	35	2	43
–	SBRF	Recife	5	16	23
–	SBRP	Ribeirão Preto	13	6	1
–	SBSP	São Paulo (Congonhas)	2	10	15
–	SBUL	Uberlândia	18	1	5

[1]SBSG is the new airport serving Natal city from 2014

Multiplex Efficiency Index means greater route diversification among airlines. Thus, the occupancy rate of the aircraft also increases and this fact influences the prices of airline tickets.

References

1. T.F. Costa, G. Lohmann, A.V. Oliveira, Res. Transp. Econ. **26**(1), 3 (2010)
2. Brazilian Government Publications. Ministério do Planejamento (2012). http://www.pac.gov.br/noticia/b1bbf26d

3. Brazilian Government Publications. ANAC (2010). https://www.anac.gov.br/A_Anac/institucional
4. Agência M Brasil, http://agenciambrasil.com.br/2019/02/sobre-o-transporte-ferroviario-de-passageiros-entre-sao-paulo-rio-de-janeiro-e-belo-horizonte (2019)
5. El País (2014). https://brasil.elpais.com/brasil/2014/07/15/economia/1405381188_257885.html
6. I.M. Oliveira, L.C. Carpi, A.P.F. Atman, The multiplex efficiency index: unveiling the Brazilian air transportation multiplex network—batmn Sci. Rep. **10**(1), 13339 (2020). https://doi.org/10.1038/s41598-020-69974-0
7. Brazilian Government Publications. ANAC (2016). http://www.anac.gov.br/assuntos/dados-e-estatisticas/dados-estatisticos/dados-estatisticos
8. L.C. Carpi, T.A. Schieber, P.M. Pardalos, G. Marfany, C. Masoller, A. Díaz-Guilera, M.G. Ravetti, Sci. Rep. **9**(1), 4511 (2019)
9. L.C. Carpi et al., Github **9**(1), 4511 (2019). https://github.com/tischieber/assessing_diversity_in_multiplex_networks/
10. I.M. Oliveira, L.C. Carpi, A.P.F. Atman, in *Añais ENMC 2019* (Encontro Nacional de Modelagem Computacional, Juiz de Fora, 2019), pp. 1959–1968. http://enmc.fadepe.org.br/anais
11. Brazilian Government Publications. ANAC (2019). https://www.anac.gov.br/noticias/2019/alocacao-de-slots-em-congonhas-e-definida
12. Portal de Notícias Uol (2017). https://airway.uol.com.br/ponte-aerea-rio-sao-paulo-e-rota-mais-movimentada-das-americas/

Chapter 69
Braess' Paradox in Networks with Microscopic Stochastic Dynamics and Traffic Information

Andreas Schadschneider and Stefan Bittihn

Abstract Braess' paradox describes the counterintuitive situation when the addition of a new road to a network leads to higher travel times for all users. Typically it occurs for selfish drivers that try to minimize their own travel times. Here we investigate whether the occurrence of Braess' paradox can be prevented if the drivers have access to more detailed traffic information, e.g. as provided by modern navigation apps.

69.1 Introduction

Nowadays commuters spend up to more than 100 h per year in congestions [1]. However, building new roads is not necessarily a solution to this problem! According to the *Fundamental Law of Road Congestion* [1] more roads will lead to more road users and thus the new roads will also be congested. But even without an increased traffic volume new roads do not necessarily improve the situation for the drivers. This is exemplified by the so-called *Braess paradox* [2]:

> In road networks of selfish users additional roads can lead to higher travel times for all users.

It can occur in transport networks [3–5] and other systems as well [6].

The paradox originates in situations where the user optimum (uo, also called Nash equilibrium), reflecting the driver's perspective, and the system optimum (so), reflecting a global perspective, are not identical. This can drive the network into a state, the user optimum, in which no driver can improve his/her travel time by choosing an alternative route. For certain network types the user optimum with an additional road may have higher travel times than that of the same network without the new road. Braess presented a mathematical model [2] based on simple networks with 4 and 5 links (where links represent roads), respectively. He showed that for adequately chosen travel time functions for the roads and selfish drivers, the paradoxical situation described above can occur.

A. Schadschneider (✉) · S. Bittihn
Institut für Theoretische Physik, Universität zu Köln, 50937 Köln, Germany
e-mail: as@thp.uni-koeln.de

I. Zuriguel et al. (eds.), *Traffic and Granular Flow 2019*,
Springer Proceedings in Physics 252,
https://doi.org/10.1007/978-3-030-55973-1_69

The paradox was since observed in several real world situations [7] where newly built roads led to a worse traffic situation or, inversely, where the closing of roads led to an improvement of the traffic situation. It is related to the concept of the *price of anarchy* which is a measure for the reduced efficiency of a system due to the selfish behavior of agents [8].

Here we extend our previous work [9, 10] in which we have shown that Braess' paradox occurs in networks of stochastic transport models in the following two cases: (1) routes are chosen stochastically by each driver. In this case, so-called mixed user optima (muo)/mixed Nash equilibria are observed [9], and (2) drivers use fixed strategies for their route choices. In this case, so-called pure user optima (puo)/pure Nash equilibria are observed [10]. We address the question whether Braess' paradox can be avoided by providing traffic information, i.e. with more realistic decisions by the drivers. Modern traffic information systems provide drivers with up-to-date information about the traffic state on different routes or even predictions of travel times. We have introduced such information into the model to study its effects on the occurrence of the paradox.

69.2 Model

Braess' original model is based on a macroscopic and deterministic traffic model for which the travel time depends only on the number of cars on the road. We have generalized this previously to more realistic microscopic stochastic dynamics [9, 10] without changing the structure of the original simple network (Fig. 69.1).

Link E_5 is the new link that is considered to be added to the network and the network without/with that link is from here on called the 4link/5link network (corresponding variable names are marked by the superscripts (4)/(5)). All drivers want to go from the start at j_1 to the finish at j_4. For this they can use 2 (3) different

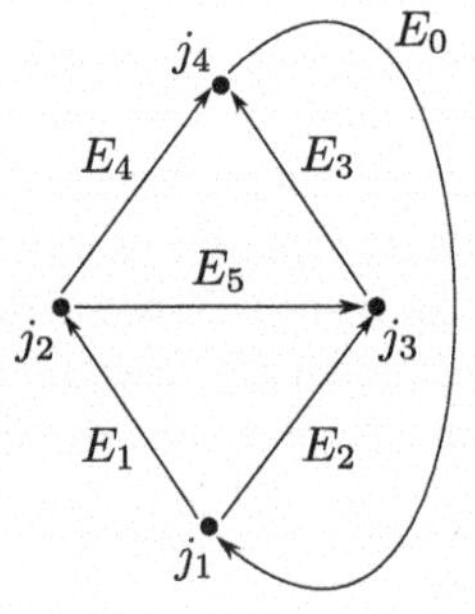

Fig. 69.1 Structure of the network. All particles are considered to move from the start at j_1 to the finish at j_4. In Braess' original network the link E_0, ensuring periodic boundary conditions, is missing. Without the link E_5 we call the network 4link, with link E_5 5link network. The dynamics on each link is modeled by TASEPs

Fig. 69.2 TASEP of length L with open boundary conditions. In the case of random sequential update rules a site is chosen randomly. If it is occupied, the particle can jump to the right neighbor site if this next site is empty. Particles are added (removed) at the left (right) boundary with rate α (β)

routes in the 4link (5link) network: routes 14, 23 and 153 (the latter is only available in the 5link network). Route 14 leads from j_1 to j_4 via E_1, j_2, E_4 (routes 23 and 153 accordingly). The traffic on the network is modeled by the paradigmatic totally asymmetric exclusion process (TASEP) (Fig. 69.2).

The TASEP is based on stochastic dynamics which allows including fluctuations and provides a mechanism for spontaneous formation of jams. It is the $v_{\max} = 1$ limit of the Nagel-Schreckenberg cellular automaton model of highway traffic. Note that we consider here the TASEP with random-sequential update.

Previously, we have considered two different cases in which the driver's route choices are tuned externally, i.e. each driver chooses routes (1) stochastically [9], and (2) using fixed strategies [10]. In the stochastic case, on junction j_1, particles turn left (i.e. onto E_1) with probability γ or right (i.e. onto E_2) with probability $1 - \gamma$. On junction j_2, particles turn left (i.e. onto E_4) with probability δ or right (i.e. onto E_5) with probability $1 - \delta$. In the case of fixed strategies, fixed subsets of N_{14} / N_{23} / N_{153} particles choose routes 14 / 23 /153. These three parameters can be encoded in the two variables $n_1^{(j_1)} = 1 - \frac{N_{23}}{M}$ and $n_1^{(j_2)} = \frac{N_{14}}{N_{14}+N_{153}}$ that denote the fractions of particles turning left at j_1 and j_2, respectively, with M being the total number of particles in the network. Mixed and pure user optima can be found by varying the γ, δ or the $n_1^{(j_1)}$, $n_1^{(j_2)}$, respectively.

Both model versions with externally tuned strategies show rich phase diagrams that include extended regions where the Braess paradox occurs (Fig. 69.3). The shown phase diagrams were obtained for a network with edges E_i of the lengths $L_0 = 1, L_1 = L_3 = 100, L_2 = L_4 = 500$. The length of a route i (i.e. the sum of the lengths of the TASEPs that make up the route and the number of junction sites on the route) is denoted by $\hat{L}_i$. The network was studied for different lengths of E_5 that lead to different lengths of the new route 153 ($\hat{L}_{153}$) compared to the lengths of the two old routes 14 and 23 ($\hat{L}_{14} = \hat{L}_{23}$), as denoted by the route length ratio $\hat{L}_{153}/\hat{L}_{14}$. The phase the system is in depends on $\hat{L}_{153}/\hat{L}_{14}$ and the global density ρ_{global}, which is M divided by the total number of sites (sum of all TASEP sites and junction sites).

Here we will study the effect of traffic information on the occurrence of Braess' paradox. Generically one can distinguish between *public information*, as e.g. provided by navigational systems, and *personal information* which could e.g. be based on personal experience. Furthermore, in networks one can distinguish (1) *historical information* based on existing data, (2) *current information* based on the up-to-date traffic state determined by measurements, and (3) *predictive information*, e.g. based on computer simulations.

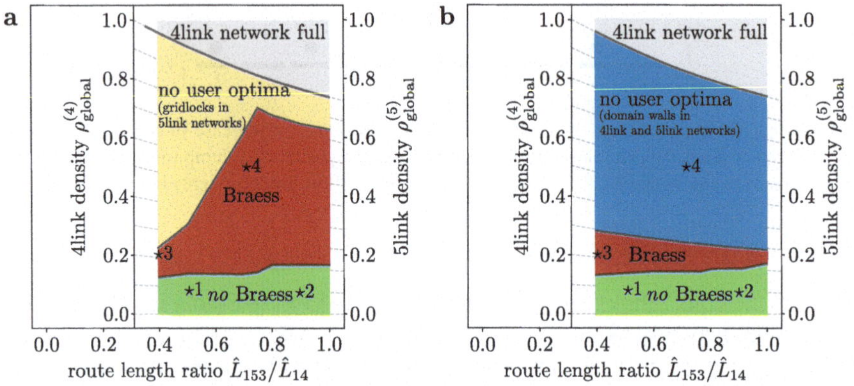

Fig. 69.3 Phase diagrams for externally tuned strategies via **a** fixed route choices [10] and **b** turning probabilities [9]. Sub-phases can be found in [9, 10]. The points ⋆1, ..., ⋆4 indicated have been studied for the influence of traffic information [12]

We have implemented two different route choice algorithms where all agents choose their routes intelligently based on

1. travel time predictions based upon the current network state, i.e. public predictive information,
2. their own memory of travel times, i.e. personal historical information.

Algorithm 1 captures the effects of modern navigation apps in a simple way whereas algorithm 2 corresponds to a commuter scenario where the agents drive each day from their home to their office relying only on their past experiences.

To find out whether user optima in the 4link and 5link networks that are obtained by externally tuning the particles' strategies are realized by intelligently deciding agents, we implemented the following route choice algorithm. First, M agents are placed randomly on the routes. Each agent is assigned an initial pure strategy (choosing either route 14, 23 or 153) randomly. After a relaxation procedure that depends on the type of information used [11, 12], all particles have information about the expected travel times $T_{14,\text{info}}$, $T_{23,\text{info}}$ and $T_{153,\text{info}}$ of all three routes 14, 23 and 153. After relaxation is complete, route choice decisions can occur at three points: before starting a new round (i.e. on edge E_0) and during a round on sites j_1 and j_2 when a jump to the desired target site is not possible. Before a new round, when sitting on E_0 before jumping to j_1, each particle generally chooses the route i with the lowest $T_{i,\text{info}}$. To make such a decision more realistic, it contains a stochastic component [11, 12].

69.3 Results

In order to reduce the required computational load we have chosen four interesting points from the phase diagrams determined previously for stochastic route choice [9] and fixed strategies [10] (see Fig. 69.3). Here we only present results for point $\star 3$, which is in the Braess phase in both previously studied situations, and the case of public predictive information. The parameters for point $\star 3$ in the phase diagram are the length of the new road, $L_5 = 37$, and a total of $M = 248$ particles in the system. This corresponds to a route length ratio of $\hat{L}_{153}/\hat{L}_{14} \approx 0.4$ and global densities $\rho^{(4)}_{\text{global}} \approx 0.21$ and $\rho^{(5)}_{\text{global}} \approx 0.20$ in the 4link and 5link systems. As detailed in [11, 12], the 5link system of state $\star 3$ with externally tuned parameters has two pure user optima (puo(i) and puo(ii)) and one mixed user optimum (muo). For a more detailed discussion of the other points and other types of traffic information, we refer to [11, 12].

Public predictive information is provided on the basis of the current positions of all cars in the network. It is a simplified version of the traffic information provided by navigation apps. The algorithm predicts travel times on the basis of the total number of vehicles on each link of the network, i.e. the global densities ρ_i on each link i. The predicted travel time for link i with length L_i is then

$$T_{i,\text{pred}} = \frac{L_i}{1 - \rho_i}, \tag{69.1}$$

which is based on the exact travel time of a particle in the stationary state of a TASEP with periodic boundary conditions at density ρ_i [9]. These $T_{i,\text{pred}}$ are then used as the $T_{i,\text{info}}$ for the route choice processes of the particles.

Figure 69.4 shows results for the case of public predictive information for the state $\star 3$ indicated in Fig. 69.3. The result of a single Monte Carlo simulation of the system is shown. The general behavior has been confirmed in numerous test runs. The shared x-axes of Fig. 69.4 show the system time. All times are measured in units of the number of performed Monte Carlo sweeps. Figure 69.4a shows the mean travel times $\bar{T}_i$ on the routes i, as realized by the intelligently deciding agents. The maximum travel times $T_{\max}$ in the user optima found by externally tuning the decisions are shown for comparison. Figure 69.4b shows which routes the particles take. The $m_1^{(j_1)}$ and $m_1^{(j_2)}$ show the fractions of agents turning left at junctions j_1 and j_2. The strategies of the accessible pure and mixed user optima realized by externally tuning the decisions are shown for comparison. We see that the user optimum of the 4link system is realized. Figure 69.4b shows that half of the agents use routes 14 and 23, respectively. Figure 69.4a indicates that the travel times of both routes equalize at the expected value. Interestingly, in the 5link system the route choice algorithm leads to strong fluctuations around the pure user optimum puo(ii) (Fig. 69.4b). In this state route 23 is not used and half the agents choose route 14 and the other half route 153. Due to the fluctuations around the user optimum the travel times of the two used routes 14 and 153 are close to each other but not exactly equal (Fig. 69.4a). All travel times are higher than those of the two routes in the 4link system. Thus indeed a Braess state is realized. Route 23, which is almost not used, has an even higher travel time, as expected.

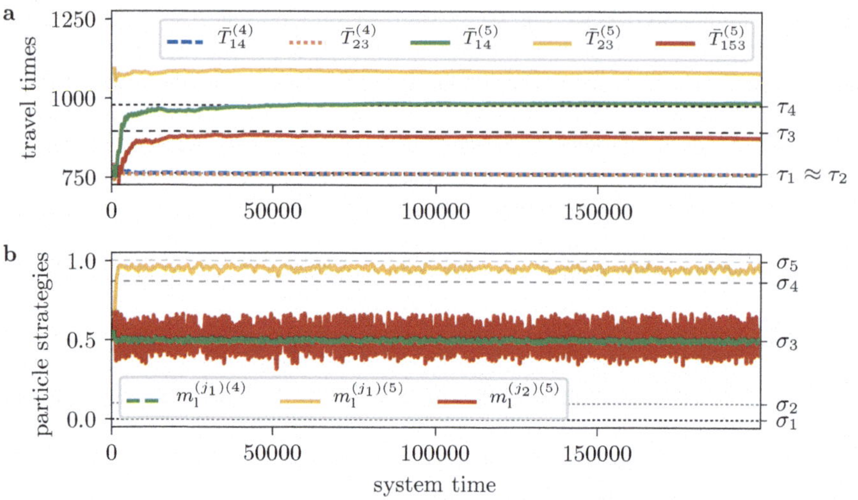

Fig. 69.4 Results for public predictive information in state $\star 3$. **a** Mean travel times realized by intelligently deciding agents (coloured lines). The expected travel times in the pure and mixed user optima found by externally tuning the strategies are shown by the dotted lines with $T^{(4)}_{\text{max,muo}} = \tau_1$, $T^{(4)}_{\text{max,puo}} = \tau_2$, $T^{(5)}_{\text{max,muo}} = \tau_3$, $T^{(5)}_{\text{max,puo(i)}} = T^{(5)}_{\text{max,puo(ii)}} = \tau_4$. **b** Particle strategies realized by the route choice algorithm (coloured lines). The strategies realizing the user optima by externally tuning the strategies are given for comparison with $n^{(j_1)(4)}_{1,\text{puo}} = \sigma_3$, $\gamma^{(4)}_{\text{muo}} = \sigma_3$, $\left(n^{(j_1)(5)}_{1,\text{puo(i)}}, n^{(j_2)(5)}_{1,\text{puo(i)}}\right) = (\sigma_3, \sigma_1)$, $\left(n^{(j_1)(5)}_{1,\text{puo(ii)}}, n^{(j_2)(5)}_{1,\text{puo(ii)}}\right) = (\sigma_5, \sigma_3)$, $\left(\gamma^{(5)}_{\text{muo}}, \delta^{(5)}_{\text{muo}}\right) = (\sigma_4, \sigma_2)$

69.4 Summary

We have studied the effects of traffic information on the occurrence of Braess' paradox in a network with microscopic, stochastic traffic dynamics. Modern traffic information systems provide drivers with up-to-date information about the traffic state on different routes or even predicted travel times. We have introduced such information into the model and find that Braess' paradox still occurs.

Acknowledgements Financial support by German Science Foundation (DFG) under grant SCHA 636/8-2, Bonn-Cologne Graduate School of Physics and Astronomy (BCGS) and the German Excellence Initiative through the University of Cologne Forum is gratefully acknowledged

References

1. G. Duranton, M.A. Turner, Am. Econ. Rev. **101**, 2616 (2011)
2. D. Braess, Unternehmensforschung **12**, 258 (1968); english translation: D. Braess, A. Nagurney, T. Wakolbinger, Transp. Sci. **39**, 446 (2005)
3. R. Steinberg, W.I. Zangwill, Transp. Sci. **17**, 301 (1983)

4. S. Dafermos, A. Nagurney, Transp. Res. B **18**, 101 (1984)
5. E.I. Pas, S.L. Principio, Transp. Res. B **31**, 265 (1997)
6. C.M. Penchina, L.J. Penchina, Am. J. Phys. **71**, 479 (2003)
7. G. Kolata, The New York Times (December 25, 1990), www.nytimes.com/1990/12/25/health/what-if-they-closed-42d-street-and-nobody-noticed.html
8. H. Youn, M.T. Gastner, H. Jeong, Phys. Rev. Lett. **101**, 128701 (2008)
9. S. Bittihn, A. Schadschneider, Phys. Rev. E **94**, 062312 (2016)
10. S. Bittihn, A. Schadschneider, Phys. A **507**, 133 (2018)
11. S. Bittihn, Doctoral dissertation. University of Cologne (2018), https://kups.ub.uni-koeln.de/9166/
12. S. Bittihn, A. Schadschneider, preprint (2020), https://arxiv.org/abs/2009.02158

Chapter 70
Robustness Analysis of Car-Following Models for Full Speed Range ACC Systems

Antoine Tordeux, Jean-Patrick Lebacque, and Sylvain Lassarre

Abstract Adaptive cruise control (ACC) systems are fundamental components of driving automation. At the upper control level, ACC systems are based on car-following models determining the acceleration rate of a vehicle according to the distance gap to the predecessor and the speed difference. The pursuit strategy consists in keeping a constant time gap with the predecessor, as recommended by industrial norms. The systematic active safety of the systems is tackled thanks to local and string stability analysis. In this contribution, we analyze and compare several classical constant time gap linear and non-linear car-following models. We critically evaluate the stability robustness of the models against latency, noise and measurement error, heterogeneity, or kinetic constraints operating in the dynamics at the lower control level. The results highlight that many factors can perturb or even break the stability and induce the formation of stop-and-go waves or collisions, for intrinsically stable car-following models.

70.1 Introduction

Road vehicles are becoming increasingly automated. Advanced driver assistance systems are common equipment of modern cars, while connected and automated vehicles are actively developed and tested by manufacturers and suppliers. One of the main arguments for the automation of the driving lies in safety aspects. Indeed, more than 90% of traffic accidents are due to human errors [1], that could be avoided with

A. Tordeux (✉)
Division for traffic safety and reliability, University of Wuppertal, Wuppertal, Germany
e-mail: tordeux@uni-wuppertal.de
URL: https://www.vzu.uni-wuppertal.de

J.-P. Lebacque · S. Lassarre
GRETTIA, COSYS, IFSTTAR, Gustave Eiffel University, Champs-sur-Marne, France
e-mail: jean-patrick.lebacque@univ-eiffel.fr

S. Lassarre
e-mail: sylvain.lassarre@univ-eiffel.fr

I. Zuriguel et al. (eds.), *Traffic and Granular Flow 2019*,
Springer Proceedings in Physics 252,
https://doi.org/10.1007/978-3-030-55973-1_70

safe automatized systems. Other arguments for the automation rely on performance, with platooning or optimal traffic assignment, and environment, by limiting jamming and traffic instability (i.e. stop-and-go waves), or by facilitating car sharing.

The automation of road vehicles is commonly classified in 6 levels [2]. Current automation levels are levels 1 and 2 for partial longitudinal and lateral controls in specific driving situations (mainly on highway) and under driver supervision. Nowadays, speculations on the levels 3, 4 and 5 of automation without supervision are going well [3]. Experiments in real condition have shown promising preliminary results for partial and full driving automation. However, rigorous proofs for the safety of automatized vehicles remain currently actively debated [4]. Even basic adaptive cruise control (ACC) systems currently available in the market have recently shown during an experiment unsafe behaviors [5].

In ACC systems, the automation is able thanks to sensors, cameras and algorithms, to detect and to track the predecessor and to measure in real-time the inter-vehicle distance and relative speed. Full speed range ACC systems (FACC) are supposed to regulate the speed even when the vehicle needs to stop. ACC and FACC systems are fundamental components of the automation of the driving. At upper levels, ACC and FACC systems are based on car-following models determining the acceleration of a vehicle according to the distance ahead and relative speed. The challenge consists in finding systematic models able to describe safe and comfortable longitudinal motions. The active safety of the models is tackled thanks to local (over-damped) and string (collective) stability analysis (see, e.g., [6, 7]). Many factors beyond the car-following model may perturb or even break the stability, e.g. noise, heterogeneity, delays or again kinetic constraints [8, 9]. Safe ACC systems should be robust to those perturbations.

The constant spacing policy, consisting in keeping a constant distance gap independently to the speed, does not describe string stable dynamic [10]. Instead, the current accepted pursuit policy supposes the distance gap proportional to the speed in order to maintain a constant time gap with the predecessor (see [6, 11] and Fig. 70.1). Actual ACC systems proposed by manufacturers offer a choice between pre-defined desired time gap settings, usually ranging from 0.8 to 2.2 s [11]. In this paper, several classical constant time gap linear and non-linear car-following models are compared.

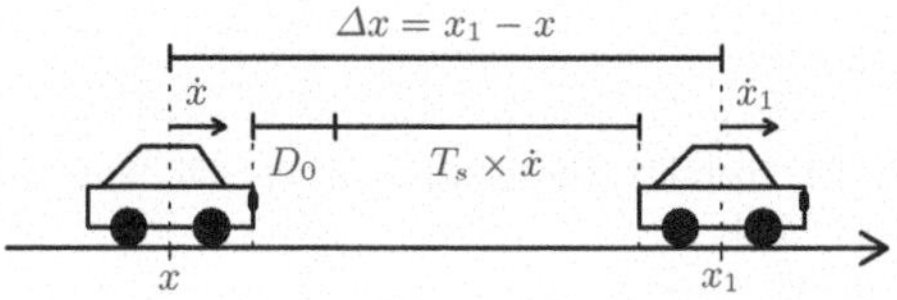

Fig. 70.1 Notation used. x is the position of the vehicle and x_1 is the position of the predecessor. $D_0 \geq 0$ is the gap during standstill while $T_s > 0$ is the desired time gap

We calculate the stability conditions and critically evaluate the stability robustness against delay, noise, heterogeneity, or kinetic constraints operating in the dynamics at lower control level.

70.2 Constant Time Gap Models for ACC Systems

In the following, we consider truncated car-following models with no maximal speed parameter. Indeed, our objective is to describe constant time gap interacting pursuit driving, and not a free driving. The models are second order ordinary differential equations describing the acceleration rate of a vehicle.

70.2.1 Linear Car-Following Models

The current accepted pursuit policy consists in keeping a constant time gap with the predecessor [6]. Indeed, many statistical analyses have shown that a driver tends to keep a constant time gap with the predecessor ranging, roughly speaking, from 1 to 3 s [12, 13]. The constant time gap strategy is notably the one recommended by the ISO 15622 norm for ACC systems [11]. Many car-following models allow to maintain a constant time gap with the predecessor. The simplest model is a linear version of the optimal velocity (OV) model [14]

$$\ddot{x}(t) = \frac{1}{T_r}\big[(\Delta x(t) - \ell)/T_s - \dot{x}(t)\big]. \tag{70.1}$$

Here, $\Delta x(t) = x_1(t) - x(t)$ is the spacing while $\Delta x(t) - \ell$ is the distance gap, $\ell \geq 0$ being the length of the vehicle plus the gap during standstill. $T_r > 0$ is the relaxation time parameter operating in unsteady state when the time gap is different from the desired time gap T_s. The OV model is solely based on the distance gap. A simple linear constant time gap model considering as well the speed difference term is the full velocity difference (FVD) model [15, 16]

$$\ddot{x}(t) = \frac{1}{T_r}\big[(\Delta x(t) - \ell)/T_s - \dot{x}(t)\big] + \frac{1}{\tilde{T}_r}\Delta\dot{x}(t). \tag{70.2}$$

The constant time gap (CTG) model is a FVD model for which $\tilde{T}_r = T_s$. The CTG model has a single relaxation time parameter T_r. It is frequently used for ACC systems at upper level control (see, e.g., [6]).

70.2.2 Non-linear Car-Following Models

OV, FVD and CTG models are linear models that can, even for a queue of vehicles, be explicitly solved. We consider in the following two non-linear models that are generally investigated by using numerical simulation. The truncated intelligent driver (ID) model [17] with no maximal speed is

$$\ddot{x}(t) = a\Big[1 - \Big(\dot{x}(t)\frac{T_s - \Delta\dot{x}(t)\,/\,2\sqrt{ab}}{\Delta x(t) - \ell}\Big)^2\Big]. \qquad (70.3)$$

The model has no relaxation parameters but instead two parameters $a > 0$ and $b > 0$ accounting for the maximal acceleration and the desired deceleration, respectivelly. The ID model has been extended to describe ACC systems with adaptive driving strategies in [18]. In the adaptive time gap (ATG) model [13], the time gap $T(t) = (\Delta x(t) - \ell)/\dot{x}(t)$ is relaxed to the desired time gap T_s, i.e. $\dot{T}(t) = \lambda\big[T_s - T(t)\big]$. The formulation of the model as an acceleration function is

$$\ddot{x}(t) = \lambda\dot{x}(t)\Big[1 - \frac{T_s}{T(t)}\Big] + \frac{1}{T(t)}\Delta\dot{x}(t). \qquad (70.4)$$

Here λ is a relaxation rate parameter. The ATG model is close to FVD and CTG models. The difference lies in the relaxation times, which are constant parameter in FVD and CTG models, while the relaxation time is the dynamical time gap variable in the ATG model. Note that the model is not defined when the speed or distance gap is zero. It can however be extended by bounding the time gap variable. The bounded ATG model corresponds then to the CTG model.

70.3 Linear Stability Analysis

Two types of linear stability are classically analysed in the literature: the local stability and the string stability (see, e.g., [19]). Local stability consists in analysing the dynamics of vehicles following a leader with assigned speed. Over-damped local stability conditions account for non-oscillating speed control. In string stability, a queue of vehicles, infinite or with periodic boundary conditions, is considered.

70.3.1 Perturbed Linear System

One denotes the car-following model as $\ddot{x}_n = F(\Delta x_n, \dot{x}_n, \dot{x}_{n+1})$, Δx_n and $\dot{x}_n$ being the spacing and speed of the vehicle n, and we suppose the existence of equilibrium spacing and speed solutions (s, v) for which $F(s, v, v) = 0$. The dynamics of the perturbations to the equilibrium state (s, v) denoted $\Delta y_n(t) = \Delta x_n(t) - s$ and $\dot{y}_n(t) = \dot{x}_n(t) - v$ are linearised to obtain

$$\ddot{y}_n(t) = a\Delta y_n(t) + b\,\dot{y}_n(t) + c\,\dot{y}_{n+1}(t), \tag{70.5}$$

with $F = F(x, y, z)$, $a = \frac{\partial F}{\partial x}(s, v, v)$, $b = \frac{\partial F}{\partial y}(s, v, v)$ and $c = \frac{\partial F}{\partial z}(s, v, v)$ the model's parameters. The uniform solutions (s, v) are linearly stable if $y_n(t) \to 0$ and $\dot{y}_n(t) \to 0$ as $t \to 0$. The predecessor speed v_{n+1} is given in the local stability analysis while it is coupled in the dynamics for the string stability. Indeed, string stability conditions are more restrictive than the local conditions by taking in consideration convective and advective perturbations as well [20].

70.3.2 Stability Condition

Several techniques are used in traffic engineering to analyse the string stability [21]. Transfer function of the perturbation Laplace transform allows to describe the convective or advective nature of the instability. String stability condition for a system with periodic boundaries can be obtained by spectrum analysis, while infinite systems are generally tackled using the exponential Ansatz $y_n(t) = \Re\big(Ae^{zt+in\theta}\big)$ with $z \in \mathbb{C}$. These two last cases describe the more restrictive stability conditions by considering as well stationary, advective and convective perturbations. The stability is hence related in the literature as *absolute string stability* [20]. The linear stability conditions are established by showing that the solution z of the characteristic equation $z^2 - z(b + c\,e^{i\theta}) + a(1 - e^{i\theta}) = 0$ of the linear differential system (70.5) are all defined is the strictly left-half plan of the complex number for all $\theta \in [0, \pi]$ excepted one solution $z_0 = 0$ for $\theta = 0$. The conditions for the locations of the zeros of polynomials with complex coefficients have been generalised in [22, Th. 3.2]. The conditions for the linear model (70.5) are respectively for the local over-damped stability and for the absolute string stability [20, 23]

$$a > 0,\ \ b^2/4 > a \qquad \text{and} \qquad a > 0,\ \ b + |c| < 0,\ \ b^2 - c^2 > 2a. \tag{70.6}$$

The conditions for the linear and non-linear constant time gap car-following models listed in Sect. 70.2 are given in Table 70.1. Parameters of OV, FVD and ID models are constraint by the stability conditions, while CTG and ATG models are systematically stable.

Table 70.1 Local over-damped and string linear stability conditions (70.6) for OV, FVD, CTG. ID and ATG models with parameters a, b, c

Model	a, b, c	Local over-damped	Absolute string
OV	$\frac{1}{T_sT_r}, \frac{-1}{T_r}, 0$	$T_s > 4T_r$	$T_s > 2T_r$
FVD	$\frac{1}{T_sT_r}, \frac{-1}{T_r} - \frac{1}{\tilde{T}_r}, \frac{1}{\tilde{T}_r}$	$T_s > \frac{4T_r}{(1+T_r/\tilde{T}_r)^2}$	$T_s > \frac{2T_r\tilde{T}_r}{2T_r+\tilde{T}_r}$
CTG	$\frac{1}{T_sT_r}, \frac{-1}{T_r} - \frac{1}{T_s}, \frac{1}{T_s}$	$T_s, T_r > 0$	$T_s, T_r > 0$
ID	$\frac{2a}{s-\ell}, \frac{-2aT_s}{s-\ell} - \frac{\sqrt{a}}{T\sqrt{b}}, \frac{\sqrt{a}}{T\sqrt{b}}$	$\frac{aT_s^2}{s-\ell} + \sqrt{\frac{a}{b}} + \frac{s-\ell}{4bT_s^2} > 2$	$\frac{aT_s^2}{s-\ell} + \sqrt{\frac{a}{b}} > 1$
ATG	$\frac{\lambda}{T_s}, \frac{-1}{T_s} - \lambda, \frac{1}{T_s}$	$T_s, \lambda > 0$	$T_s, \lambda > 0$

70.4 Stability Robustness

The experiment realised in [5] has shown that even recent ACC systems do not describe stable dynamics. Indeed, many factors operating at lower control level may perturb or even break the linear stability of the car-following planners [8, 9]. Examples are latency and delay in the dynamics, noise and measurement error, heterogeneity of the behaviors, or even kinetic bound for acceleration or jerk (as recommended by ISO 15622 norm [11]). The modelling challenge consists in finding models robust against such factors and to establish systematic stabilisation properties.

70.4.1 *Latency and Response Time*

Drivers have reaction time, but they can compensate for this delay by anticipation. Automated systems do not react instantaneously as well. The perception, computing and control require time and induce computational and mechanical latency in the dynamics. For ACC systems, the delay is estimated by 0.5–1 s [6]. The delayed linear car-following model is

$$\ddot{y}_n(t+\tau) = a\Delta y_n(t) + b\,\dot{y}_n(t) + c\,\dot{y}_{n+1}(t), \qquad \tau \geq 0. \tag{70.7}$$

The corresponding characteristic equation $\lambda^2 e^{\lambda\tau} - \lambda(b + c\,e^{i\theta}) + a(1 - e^{i\theta}) = 0$, with $\theta \in [0, \pi]$, is no more polynomial but exponential-polynomial. The generalized Hurwitz conditions do not hold. Yet, the location of the zeros can be obtained by sub-critical Hopf bifurcation [24], method consisting in investigating purely imaginary zeros and determining the locations by continuity, or by using Taylor approximations for the delayed quantities. For instance, string stability conditions are $T_s > 2T_r$ and $T_s > 4\tau(1 + \sqrt{1 - 2T^r/T_s})^{-1}$ for the delayed OV model, while they are $T_s > 2\tau$ for CTG and ATG models. The delay in the dynamics clearly results in stability breakdown. It can however be (partially) compensated by spatial or temporal anticipation mechanisms [8, 13, 23] (cooperative adaptive cruise control).

70.4.2 *Stochastic Noise*

ACC systems are based on the distance with the predecessor that is measured by sensors such as radar and Fourier signal processing [25]. The radar has a given precision threshold and external factors like the weather perturb the measurements. This noise may influence the dynamics and the stability as well, especially when it depends on the vehicle state [8, 26]. It is for instance reasonable to consider that the precision of the measurement decreases as the distance increases. A general perturbed (stochastic) car-following model is $\ddot{x} = F(x_1 - x, \dot{x}, \dot{x}_1) + \xi(\mathbf{x}, \dot{\mathbf{x}})$, where ξ is a noise that may depend on the system state. For instance, the CTG model with a Gaussian white noise is

$$\ddot{x}(t) = \frac{1}{T_r}\big[(\Delta x(t) - \ell)/T_s - \dot{x}(t)\big] - \frac{1}{T_s}\Delta\dot{x}(t) + \sigma\dot{W}(t), \tag{70.8}$$

$\dot{W}(t) = \mathrm{d}W(t)/\mathrm{d}t$ being the derivative of the Wiener process W in the stochastic sense. One may write such model as the coupled motion/noise first order models [27]

$$\begin{cases} \dot{x}(t) = \frac{1}{T_s}[\Delta x(t) - \ell] + \varepsilon(t) \\ \dot{\varepsilon}(t) = \frac{-1}{T_r}\varepsilon(t) + \sigma\dot{W}(t) \end{cases} \tag{70.9}$$

for which the noise $\varepsilon(t)$ is described by the Ornstein-Uhlenbeck process [28]. It turns out that the periodic perturbed system with periodic boundary is an Ornstein-Uhlenbeck process as well, whose invariant distribution is a centered Gaussian with explicit covariance matrix. Indeed, the noise being white in initial CTG, the stochastic linear stability is the deterministic one, i.e. stable for all $T_s, T_r > 0$. No deterministic phase transition is observed, however, noise-induced long wavelength stop-and-go phenomena are described. The stochastic waves can for instance be highlighted by analysing oscillation and periodicity in the vehicle speed autocorrelation.

70.4.3 Heterogeneity

The heterogeneity of the driving behaviors may also play a role in the collective dynamics. Indeed, it is observed that low penetration rate of "stable" ACC systems allows to homogenise the flows. For instance, a single automated vehicle suffices to stabilise 20 vehicles in a ring in [29], corresponding to a rate of 5 %. Further simulation experiments corroborate such estimate [30]. Yet, if single stable behaviors can stabilise the flow, it is not to exclude that unstable behaviors of certain vehicles may lead to stability breaking. The linear perturbed system for heterogeneous vehicles is

$$\frac{\mathrm{d}}{\mathrm{d}t}\bar{v}_n(t) = a_n\, \Delta\bar{x}_n(t) + b_n\, \bar{v}_n(t) + c_n\, \bar{v}_{n+1}(t), \tag{70.10}$$

where the underlying parameters for the desired time gap, desired speed, vehicle's length etc. are specific to each vehicle in a multi-level model. The corresponding characteristic equation is $\prod_{n=1}^{N}\left[\lambda^2 - b_n\lambda + a_n\right] - e^{i\theta N}\prod_{n=1}^{N}\left[c_n\lambda + a_n\right] = 0$, while the absolute string stability condition is [31]

$$\sum_{n=1}^{N}\frac{b_n^2 - c_n^2}{2a_n^2} > \sum_{n=1}^{N}\frac{1}{a_n}. \tag{70.11}$$

Note that the condition recovers the classical stability condition $b^2 - c^2 > 2a$ for homogeneous cases. The non-linear shape of the condition corroborates the fact that single vehicles may strongly influence the collective dynamics. Small a_n (i.e. large desired time gap) or large b_n (i.e. small reaction time) may bring over-weight allowing the condition to hold. On the other hand, large c_n (i.e. small time gap) for some vehicles lies in stability breaking.

70.4.4 Kinetic Constraints

The ISO 15622 norm recommends that the acceleration and jerk rates do not exceed safe and comfortable maximal and minimal thresholds. The bounds, depending on the speed, are dynamic. They range for instance from −5 to −3.5 m/s^2 as the speed increases for the deceleration rate, when acceleration bounds range from −4 to −2 m/s^2 [11]. These constraints in the dynamics may be the cause of instability or even collisions in the models, especially when the relaxation times are small. A basic bounded car-following model is

$$\ddot{x}(t) = \Big[F\big(x_1(t) - x, \dot{x}(t), \dot{x}_1(t)\big)\Big]_{a_{\min}(\dot{x}(t))}^{a_{\max}(\dot{x}(t))} \tag{70.12}$$

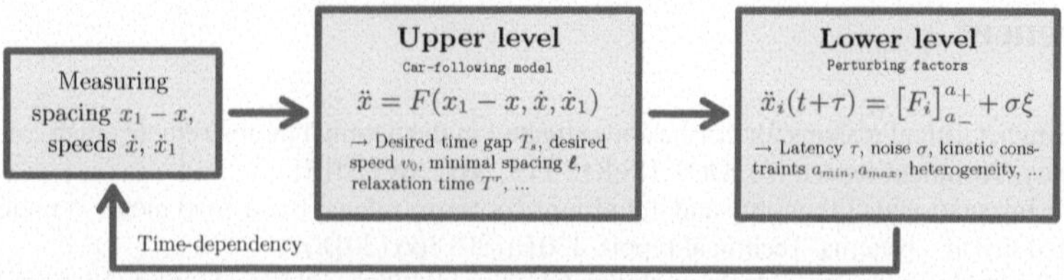

Fig. 70.2 Operational scheme for ACC systems. At upper level control, ACC systems are stable car-following models based on desired time gap, desired speed or relaxation time parameters. At lower level, factors such delay, noise and measurement error, heterogeneity and kinetic constraints affect the dynamics and may even break the stability

denoting $[x]_b^a = \min\{a, \max\{b, x\}\}$. A more regular model can be obtained by using smooth approximation functions, for instance by relaxing the process with anti-windup compensators [32]

$$\dddot{x}(t) = \frac{1}{\tau_a}\left(\left[F\big(x_1(t) - x, \dot{x}(t), \dot{x}_1(t)\big)\right]_{a_{\min}(\dot{x}(t))}^{a_{\max}(\dot{x}(t))} - \ddot{x}_n(t)\right), \quad \tau_a > 0. \qquad (70.13)$$

Such mechanism induces further inertia into the dynamics that potentially result in stability breaking. Conditions on τ_a are necessary to ensure stabilisation.

70.5 Summary

Local and string stability of the car-following models are necessary conditions to demonstrate operational safety properties of ACC systems. String stability is in particular fundamental in regard to collective dynamics and the formation of stop-and-go waves that are frequently observed in conventional traffic. Many factors operating at lower control level may perturb or even break the stability. Examples are time-delay in the dynamics, stochastic noise, heterogeneity, or kinetic constraints. A summary for the architecture of ACC systems and factors affecting the dynamics is proposed in Fig. 70.2. The modelling and safety analysis of ACC and FACC systems consist in determining systematic stability and stabilisation properties of systems of stochastic delayed differential equations under constraints. Special cases may be determined analytically (e.g. uniform model with periodic boundaries). The general problem has many parameters when we consider all factors acting and has to be investigated numerically.

References

1. S. Singh, Critical reasons for crashes investigated in the national motor vehicle crash causation survey. Technical report, No. DOT HS 812 115, NHTSA (2015)
2. SAE International, Taxonomy and definitions for terms related to on-road motor vehicle automated driving systems. Technical report, J3016_201806 (2018)
3. T. Litman, Autonomous vehicle implementation predictions. Technical report (Victoria Transport Policy Institute, 2018)
4. P. Koopman, M. Wagner, Challenges in autonomous vehicle testing and validation. SAE Int. J. Trans. Safety **4**, 15–24 (2016)
5. G. Gunter et al., String stability of commercial adaptive cruise control vehicles, in *Proceedings of the 10th ACM/IEEE International Conference on Cyber-Physical Systems* (ACM, New York, 2019), pp. 328–329
6. J. Zhou, H. Peng, Range policy of adaptive cruise control vehicles for improved flow stability and string stability. IEEE TITS **6**(2), 229–237 (2005)
7. B. Paden et al., A survey of motion planning and control techniques for self-driving urban vehicles. IEEE Trans. Intell. Veh. **1**(1), 33–55 (2016)
8. M. Treiber, A. Kesting, D. Helbing, Delays, inaccuracies and anticipation in microscopic traffic models. Phys. A **360**(1), 71–88 (2006)
9. A. Kesting, M. Treiber, How reaction time, update time, and adaptation time influence the stability of traffic flow. Comput. Aided Civil Infrastruct. Eng. **23**(2), 125–137 (2008)
10. S. Darbha, K.R. Rajagopal, Intelligent cruise control systems and traffic flow stability. Transp. Res. C **7**(6), 329–352 (1999)
11. Standard ISO 15622:2018. Technical report (2018)
12. J.H. Banks, Average time gaps in congested freeway flow. Transp. Res. A **37**(6), 539–554 (2003)
13. A. Tordeux, S. Lassarre, M. Roussignol, An adaptive time gap car-following model. Transp. Res. B **44**(8), 1115–1131 (2010)
14. M. Bando, K. Hasebe, A. Nakayama, A. Shibata, Y. Sugiyama, Dynamical model of traffic congestion and numerical simulation. Phys. Rev. E **51**, 1035–1042 (1995)
15. W. Helly., Simulation of bottlenecks in single lane traffic flow, in *ISTTT Proc.* (Elsevier, 1959), pp. 207–238
16. R. Jiang, Q. Wu, Z. Zhu, Full velocity difference model for a car-following theory. Phys. Rev. E **64**, 017101 (2001)
17. M. Treiber, A. Hennecke, D. Helbing, Congested traffic states in empirical observations and microscopic simulations. Phys. Rev. E **62**, 1805–1824 (2000)
18. A. Kesting, M. Treiber, M. Schönhof, D. Helbing, Extending adaptive cruise control to adaptive driving strategies. Transp. Res. Rec. **16–24**, 2007 (2000)
19. M. Treiber, A. Kesting, *Traffic Flow Dynamics: Data Models and Simulation* (Springer, 2013)
20. R.E. Wilson, J.A. Ward, Car-following models: fifty years of linear stability analysis—a mathematical perspective. Transport. Plan. Techn. **34**(1), 3–18 (2011)
21. G. Orosz, R.E. Wilson, G. Stépán, Traffic jams: dynamics and control. Phil. Trans. Roy. Soc. A **368**(1928), 4455–4479 (2010)
22. E. Frank, The location of the zeros of polynomials with complex coefficients. Bull. Amer. Math. Soc. **52**(2), 144–157 (1946)
23. A. Tordeux, M. Roussignol, S. Lassarre, Linear stability analysis of first-order delayed car-following models on a ring. Phys. Rev. E **86**, 036207 (2012)
24. G. Orosz, G. Stepan, Subcritical hopf bifurcations in a car-following model with reaction time delay. Proc. R. Soc. A **462**(2073), 2643–2670 (2006)
25. K. Reif, *Fahrstabilisierungssysteme und Fahrerassistenzsysteme* (Springer, 2010)
26. S.H. Hamdar, H.S. Mahmassani, M. Treiber, From behavioral psychology to acceleration modeling. Transp. Res. B **78**, 32–53 (2015)
27. A. Tordeux, A. Schadschneider, White and relaxed noises in OV models for pedestrian flow with stop-and-go waves. J. Phys. A **49**(18), 185101 (2016)

28. G.E. Uhlenbeck, L.S. Ornstein, On the theory of the brownian motion. Phys. Rev. **36**, 823–841 (1930)
29. R.E. Stern et al., Dissipation of stop-and-go waves via control of autonomous vehicles: field experiments. Transp. Res. C **89**, 205–221 (2018)
30. A. Kesting, M. Treiber, M. Schönhof, D. Helbing, Adaptive cruise control design for active congestion avoidance. Transp. Res. C **16**(6), 668–683 (2008)
31. D. Ngoduy, Effect of the car-following combinations on the instability of heterogeneous traffic flow. Transportmetrica B **3**(1), 44–58 (2015)
32. R. Mayr, *Regelungsstrategien für die automatische Fahrzeugführung* (Springer, 2001)

Chapter 71
Uniform Cluster Traffic Model on Closed Two-Contours System with Two Non-symmetrical Common Nodes

Marina V. Yashina and Alexander G. Tatashev

Abstract Spontaneous transitions of traffic flows, from the laminar state to the turbulent state, may occur. In 2008, the experimental results of traffic jam formation on a ring was published by Y. Sugiyama et al. Our paper studies a dynamical system which contains two closed contours with two common points (nodes). The nodes divide each contour into two parts with lengths d and $1 - d$. In each contour, there is a moving segment called a cluster. The clusters move in accordance with given rules. The system can be intepreted as a traffic model and belongs to the class of dynamical systems introduced by A.P. Buslaev. We study the system spectrum connected with the set of cyclic trajectories in the system state space. Some theorems regarding the spectrum have been proved.

71.1 Introduction

In well-known Nagel–Schreckenberg traffic model [1], particles move on a one-dimensional lattice in accordance with given rules. In Nagel–Schreckenberg model, the average velocity of movement is a flow density function which does not depend on the initial configuration of particles. It was supposed that traffic jams are induced by bottlenecks on highways. Experimental results of traffic jam formation on highways are represented in [2]. It is shown that a congested state occurs simultaneously if traffic flow density exceeds a critical value, and, in this case, no bottleneck is needed for the formation of a traffic jam. In more complicated versions of the model, particles move on a network structure. A lot of models were studied by computer simulation, and some analytical results have been obtained [3]. The main problem is to find

M. V. Yashina (✉)
Moscow Automobile and Road State Tech. Univ. (MADI) and Moscow Tech. Univ. of Communications and Informatics (MTUCI), Moscow, Russia
e-mail: mv.yashina@madi.ru

A. G. Tatashev
MADI and MTUCI, Moscow, Russia
e-mail: a-tatashev@yandex.ru

I. Zuriguel et al. (eds.), *Traffic and Granular Flow 2019*,
Springer Proceedings in Physics 252,
https://doi.org/10.1007/978-3-030-55973-1_71

the average velocity of particles taking into account the delays in the movement. A system state such that all particles move in the current moment and in the future is called a free movement state.

In [4–6], the movement of particles on a one-dimensional lattice is studied. It is assumed that any particle moves onto a one cell in the direction of movement at any step if the cell ahead is vacant. In accordance with results of [4–6], the system results in a state of free movement (all particles move at present time and in the future) from any initial state if and only if the number of particles is not greater than half the number of cells. Analytical results for some more general models was obtained in [7]. As it has been noted in [7], the Nagel—Shreckenberg traffic model provides an algorithm that is easy implemented on a computer, but too complicated for exact mathematical analysis. In [8], a concept of cluster movement in traffic model was introduced. The clusters are batches of neighboring particles (in the discrete version of the system) or segments (in the continuous version). The concept of Buslaev contour network was introduced in [9]. The contours of the network form a structure such that there are common points (nodes) of the neighboring contours. These delays are due to that particles cannot cross the node simultaneously. The concept of spectrum of a contour network was introduced in [10]. This concept regards the dependence of the system behavior on the initial configuration. Concepts and approaches of dynamical systems theory can be applied to the study contour networks. The results, regarding to the contour networks, can be applied to analyze traffic models with network structure.

The problems which are formulated in the study of contour networks are similar to problems appearing in the study of BML traffic model. The BML model [11, 12], is defined on two-dimensional toroidal lattice. Any cell of the lattice is occupied by a particle with probability p. There are two types of particles. Any particle belongs to each type equiprobably. The particles of the first type move downwards. The particles of the second type move rightwards. Any discrete moment, each particle moves onto one cell in the direction of movement if the cell ahead is vacant. If two particles tend to occupy the same cell, then only the particle of the first type move. In accordance with results of simulation, a hypothesis was formulated. In accordance with this hypothesis there exists a critical value p_c. The system results in a state of collapse, if $p > p_c$, or in a state of free movement, if $p < p_c$, asymptotically almost sure as $N \to \infty$. However, intermediate phases between free movement and collapse have been seen [12]. In terms of the spectrum, the problem can be formulated as follows. Does the measure of the spectral cycles, corresponding to intermediate phases, tend to zero as $N \to \infty$?

Analytical results were obtained for regular periodical structures, for example, in [13], for systems with two nodes [14, 15]. In [15], a discrete dynamical system with two contours is considered. There is a cluster of particles in each contour. The nodes divide each contour into two parts with different lengths. Two versions of the system are studied with one-directional movement and co-directional movement. It has been cleared out that the spectrum of the system is sufficiently different in these two cases.

The present paper studies a continuous counterpart of the system considered in [15]. We suppose that clusters are moving segments. There are two contours with nodes dividing the contours into to non-equal parts. We have obtained the spectrum of the system. There is no more than two possible values of the average cluster velocity for any fixed pair of the parameters.

71.2 Formulation of Problem

Suppose a dynamical system contains two closed *contours*, Fig. 71.1a. These contours are the contour 1 and the contour 2. The length of each contour is equal to 1. There is a moving segment in each contour. This segment is called a *cluster*. The cluster i is located in the contour i, $i = 1, 2$. The length of each cluster is equal to l ($l < 1$). The velocity of a cluster equals 1 if no delay occurs. There are two common points of the contours. These points are called *nodes*. The nodes divide each contour into the parts of lengths d ($d \le 1/2$) and $1 - d$. The clusters cannot move through the node simultaneously. If the leading point of a cluster is at the node at a time moment, and the other cluster moves through the node at this moment, then the cluster at the node waits for the node release. If the leading points the clusters are at the node, then the cluster 1 moves through the node first, Fig. 71.1b.

A *coordinate system* [0, 1) is given at each contour. One of the nodes is located at the point with the coordinate 0 in the contour 1 and with coordinate d in the contour 2. This node is called the node 1. The other node is located at the point with the coordinate 0 in the contour 2 and with coordinate d in the contour 1. This node is called the node 2. The direction of the coordinate system is the same as the direction of movement (by modulo 1). The state of the system at time $t \ge 0$ is the vector $\alpha(t) = (\alpha_1(t), \alpha_2(t))$, where $\alpha_i(t)$ is the coordinate of leading point of the cluster i, $i = 1, 2$, at time t. We say that *the cluster i is at the node i at time t* if $\alpha_i(t) = 0$, $i = 1, 2$; *the cluster i is at the node* $j \ne i$ *at time t* if $\alpha_i(t) = d$; *the cluster i covers the node i at time t* if $0 < \alpha_i(t) < l$; *the cluster i covers the node* $j \ne i$ *at time t* if $d < \alpha_i(t) < d + l$. If a delay of a cluster begins at time t, then this cluster is at a node at time t, and the other cluster covers the node at this time. The initial state $\alpha(0)$ is given.

Suppose each point of the cluster passes the total distance $S_i(T)$ in the time interval $(0, T)$, $i = 1, \dots, N$. The limit $v_i = \lim\limits_{T\to\infty} \frac{S_i(T)}{T}$ is called the average velocity of the cluster i, $i = 1, 2$. We shall prove that this limit exists for each cluster, any values of the system parameters and initial states, $v_1 = v_2$ for any values of the system

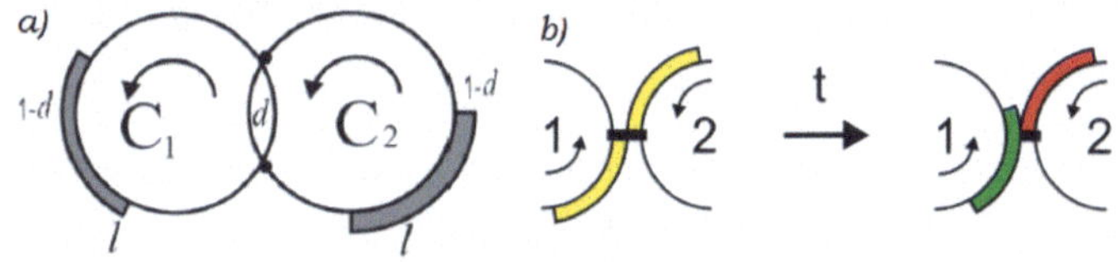

Fig. 71.1 **a** Two-contours system, **b** Left-priority competition resolution rule

parameters and initial states. *The system is in a state of free movement* at time t_0 if at any time $t \geq t_0$ the clusters move without delay. The system is at time t_0 in a state of *collapse* if at time t_0 no cluster moves. In this case, $v = 0$.

For each cluster and any values of the system parameters and initial states, there exist moments t_0, t_1 such that $\alpha(t_0) = \alpha(t_1)$. A cyclic trajectory of states is called *a spectral cycle.* The set of *spectral cycles* with values of velocities corresponding to these spectral cycles is called the *spectrum* of the system.

71.3 Behavior of Closed Two-Contours System with Two Non-symmetrical Common Nodes

Lemma 1 *For a state of collapse to exist it is necessary and sufficient to have $l > d$.*

Proof If $l > d$, then the state (d, d) is a state of collapse. If $l \leq d$ and a cluster is at the node, then the other cluster cannot be at a covered node, and therefore the system cannot be at the state of collapse. Lemma 1 has been proved.

Lemma 2 *If the system does not result in a state of free movement, and no collapse occurs, then there exists a moment t_0 such that, at this moment, the system is in one of the states $(0, d + l)$, $(d + l, 0)$, (d, l), (l, d).*

Proof Lemma 2 follows from that any delay of a cluster ends in one of the states $(0, l + d)$, $(l + d, 0)$, (d, l), (l, d).

Lemma 3 *If $l > 1 - d$, then the system results in the state of free movement from any initial state.*

Proof If $l > 1 - d$, then, at any time, each cluster covers at least of the node. Therefore, each node is covered at any time. From this, Lemma 3 follows.

Lemma 4 *If $l \leq \frac{1}{2} - d$, then the states $(0, l + d)$ and $(l + d, 0)$ are states of free movement.*

Proof Suppose $\alpha(t_0) = (0, l + d)$. Then, under the conditions of Lemma 4, we have $\alpha(t_0) = (0, l + d)$, $\alpha(t_0 + d) = (d, l + 2d)$, $\alpha(t_0 + 1 - l - d) = (1 - l - d, 0)$, $\alpha(t_0 + 1) = (0, l + d) = \alpha(t_0)$. Thus the state $(0, l + d)$ is a state of free movement. For the state $(l + d, 0)$, the the proof is analogous. Lemma 4 has been proved.

Lemma 5 *If $\frac{1}{2} - d < l < 1 - 2d$, then the states $(0, l + d)$ and $(l + d, 0)$ belong to a spectral cycle such that $v = \frac{1}{2(l+d)}$.*

Proof Suppose $\alpha(t_0) = (0, l + d)$. Then, under the conditions of Lemma 5, we have $\alpha(t_0) = (0, l + d)$, $\alpha(t_0 + d) = (d, l + 2d)$, $\alpha(t_0 + 1 - l - d) = (1 - l - d, 0)$, $\alpha(t_0 + l + d) = (l + d, 0), \alpha(t_0 + 2(l + d)) = (0, l + d) = \alpha(t_0)$. For the state $(l + d, 0)$, the proof is analogous. Lemma 5 has been proved.

Theorem 1 *For any l, d, and initial states, the velocities of clusters are equal to each other, i.e.,* $v_1 = v_2$.

Denote by v the value $v_1 = v_2$.

Theorem 2 *The following is true. The system results in a state of free movement from any initial state if the inequality* $l \leq d$ *and at least one of inequalities* $l \leq \frac{1}{2} - d$, $l \geq 1 - 2d$ *hold. The system results in a state of collapse from any initial state if and only if the inequalities* $l > d$, $l \geq 1 - 2d$ *hold. If* $(l \leq d)\,\&\,(\frac{1}{2} - d < l \geq 1 - 2d)$, *then, depending on the initial state, the system results in a state of free movement or the average velocity of each cluster is equal to* $v = \frac{1}{2(l+d)}$. *There exists a unique spectral cycle with the average velocity* $v = \frac{1}{2(l+d)}$. *This spectral cycle contains the states* $(0, l + d)$, $(l + d, 0)$. *If* $(l > d)\,\&\,(l \leq \frac{1}{2} - d)$, *then, depending on the initial state, the system results in a state of free movement or in a state of collapse. If* $(l > d)\,\&\,(\frac{1}{2} - d < l < 1 - 2d)$, *and* $\alpha(0) = (\alpha_{1,0}, \alpha_{2,0})$, *then the then the system results in a state of collapse or* $v = \frac{1}{2(l+d)}$. *There exists a unique spectral cycle with the average velocity* $v = \frac{1}{2(l+d)}$. *This spectral cycle contains the states* $(0, l + d)$, $(l + d, 0)$.

Thus the behavior of system may depend on the initial configuration of particles. We have shown that there exist five spectral types. For three of these types, the spectrum contains two values depending on the initial configuration of the system.

71.4 Conclusions

The paper studies a dynamical system which may interpreted as a traffic model and other applications of the results are possible. Exact analytical results have been found.

Acknowledgements This work has been supported by the RFBR, Grant No. 20-01-00222.

References

1. K. Nagel, M. Schreckenberg, Cellular automation models for freeway traffic. J. Phys. I. **2**, Nr. 12, S. 2221–2229 (1992). http://dx.doi.org/10.1051/jp1.1992277
2. Y. Sugiyama, M. Kikuchi, A.K. Husebe, M.K. Makayama, K. Nishinary, S.T. Minory, Y. Satochi, F. Minory, Traffic jam without bottlenecks—experimental evidence for physical mechanism of the formation of a jam. New J. Phys. **10**, S. 2221–2229 (2002). http://dx.doi.org/10.1088/1367-2630/10/303301
3. V.V. Kozlov, A.P. Buslaev, A.S. Bugaev, M.V. Yashina, A. Schadschneider, M. Schreckenberg, in Traffic and granular flow'11 (Springer, 2013)
4. M. Shreckenberg, K. Nagel, A. Schadshneide, Discrete stochastic models for traffic flow. In: In Phys. Rev. **51**, Nr. 4, S. 2939 (1995). https://doi.org/10.1103/PhysRevE.51.2939

5. M.L. Blank, Exact analysis of dynamical systems arising in models of traffic flow. Russ. Math. Surv. **55**, Nr. 3, S. 562–563 (2000). http://dx.doi.org/10.4213/rm95
6. V. Belitsky, P.A. Ferrari, Invariant measures and convergence properties for cellular automation 184 and related processes. J. Stat. Phys. **118**, Nr. 3, S. 589–623 (2005). http://dx.doi.org/10.1007/s10955-044-8822-4
7. L. Gray, D. Griffeath, The ergodic theory of traffic jams. J. Stat. Phys. **105**, Nr. 3/4, S. 413–452 (2001). http://dx.doi.org/10.1023/A:1012202706850
8. A.S. Bugaev, A.P. Buslaev, V.V. Kozlov, M.V. Yashina, Distributed problems of monitoring and modern approaches to traffic modeling. In: 14th international IEEE conference on intelligent transactions systems (ITSC 2011), Washington, USA, S. 477–481 (2011). http://dx.doi.org/10.1109/ITSC.2011.6082805
9. V.V. Kozlov, A.P. Buslaev, A.G. Tatashev, On synergy of totally connected flow on chainmails, in: *Proceedings of the 13 International Conference on Computational and Mathematical Methods in Science and Engineering, Almeria, Spain*, vol. 3, S. 861–874 (2013)
10. A.P. Buslaev, M.Y. Fomina, A.G. Tatashev, M.V. Yashina, On discrete flow networks model spectra: statement, simulation, hypotheses. J. Phys.: Conf. Ser. **1053**, Nr. 1, S. 012034 (2018). http://dx.doi.org/10.1088/1742/6596/1053/1/012034
11. O. Biham, A.A. Middleton, D. Levine, Self-organization and a dynamical transition in traffic-flow models. Phys. Rev. A. **71**, Nr. 6, S. R6124–R6127 (1992). http://dx.doi.org/10.1003/PhysRevA.46.R6124
12. T. Austin, I. Benjamini, For what number of cars must self organization occur in the Biham-Middleton-Levine traffic model from any possible starting configuration? (2006). arXiv:math/0607759
13. A.P. Buslaev, A.G. Tatashev, M.V. Yashina, Flows spectrum on closed trio of contours. Eur. J. Pure Appl. Math. **11**, Nr. 3, S. 893–897 (2018). http://dx.doi.org/10.29020/nybg.ejpam.v11i1.3201
14. A.G. Tatashev, M.V. Yashina, Behavior of continuous two-contours system. WSEAS Trans. Math. **18**, Nr. 5, S. 28–36 (2019)
15. M. Yashina, A. Tatashev, Spectral cycles and average velocity of clusters in discrete two-contours system with two nodes, in *Mathematical Methods in the Applied Sciences* First published 04 February 2020, S. 1–28 (2020). http://dx.doi.org/10.1088/1742-6596/1324/1/012001

Chapter 72
Paris-Gare-de-Lyon's DNA: Analysis of Passengers' Behaviors Through Wi-Fi Access Points

Yuji Yoshimura, Irene de la Torre - Arenas, Shinkyu Park, Paolo Santi, Stefan Seer, and Carlo Ratti

Abstract This paper proposes a novel analytical framework for passengers' spatio-temporal behaviors in a large-scale train station in Paris, France. We focus on extracting the passengers' behavioral patterns, considering their sequential movement between key locations, their length of stay, and the relationship between them. For this purpose, we employ the sequence alignment methods, applying them to a Wi-Fi access points data set. The sequence alignment methods were first introduced to analyze DNA or RNA strings of information in molecular biology. One of their significant aspects is their ability to consider the order (or sequence) of events, thus making them well suited for the analysis of passengers' behaviors in spatio-temporal aspects.

72.1 Introduction

This paper proposes a novel analytical framework for passengers' spatio-temporal behaviors at Paris-Gare-de-Lyon, one of the largest train stations in Paris, France. We focus on extracting the passengers' behavioral patterns, considering their sequential

Y. Yoshimura (✉) · I. de la Torre - Arenas · S. Park · P. Santi · S. Seer · C. Ratti
SENSEable City Laboratory, Massachusetts Institute of Technology, 77 Massachusetts Avenue, Cambridge, MA 02139, USA
e-mail: yyyoshimura@cd.t.u-tokyo.ac.jp

Y. Yoshimura
Research Center for Advanced Science and Technology, the University of Tokyo, 4-6-1 Komaba, Meguro-ku, Tokyo 153-8904, Japan

S. Park
Department of Mechanical and Aerospace Engineering, Princeton University, Princeton, NJ 08544, USA

P. Santi
Instituto di Informatica e Telematica, CNR, Via Giuseppe Moruzzi 1, 56124 Pisa, Italy

S. Seer
Austrian Institute of Technology, Giefinggasse 2, 1210 Vienna, Austria

I. Zuriguel et al. (eds.), *Traffic and Granular Flow 2019*,
Springer Proceedings in Physics 252,
https://doi.org/10.1007/978-3-030-55973-1_72

movement between key locations, their length of stay, and the relationship between them. For this purpose, we employ the sequence alignment methods (SAMs) [1], applying them to a Wi-Fi access points data set, obtained from all over the station.

Capturing pedestrians' presence and their sequential movement in the built environment is a well-established line of research in urban studies [2–14].

We can classify these studies into three types depending on the proposed data collection methodologies: manual, active, and passive (see [13] in detail). The first type uses direct observation, interviews, or questionnaires [15], and the second one relies on attaching digital tags to pedestrians, such as RFID [8], or requiring them to download the relevant application in advance [16]. The last one does not require any preparation for the passengers but rather detects their activities passively, for example, by capturing their media access control (MAC) address, which is a unique ID for each device [11]. In the context of the train station, manual data collection methods (e.g., interviews, questionnaires, or observations) are frequently employed to analyze the passengers' behaviors. They can be used to estimate the origin-destination matrix or the route choice, which works as an input for the pedestrian simulation. Recently, passengers' data collected through automatic fare collection systems has attracted much attention given its potentiality to enable the analysis of passengers' behaviors (i.e., the route choice for the transfer time, the dwelling time in the station) [17]. Asakura et al. [18] used multiple data sets such as travel surveys, pedestrian counts, and trajectories obtained from camera-based tracking technologies [19] and evaluated the level of service of pedestrian facilities in the large-scale train station.

This paper proposes the use of SAMs [20], which stem from biology to compare the region of similarity of two or more sequences (e.g., DNA, RNA, protein). Different from other quantitative analytical methods, the characteristics of SAMs include being able to determine the order of events [4]. This feature makes it possible to consider not only the structural variables embedded in activity travel patterns (e.g., transport mode) but also the sequential variability in activities and trips. Therefore, this method is also introduced in transportation and time use research [21]. Although the frequency of application of SAMs has significantly increased in travel behavior research [22, 23], most of these applications are based on the GPS data set. In the context of tourist behavioral analysis, [4] proposed to create typologies of tourists depending on their spatial and temporal factors and based on GPS data. Conversely, [6] used large-scale Bluetooth tracking data to analyze visitors' behaviors inside the building.

For the present study, we propose to use the Wi-Fi access points data set. The growth of smartphone usages enables us to collect human behavior data sets at more granular scales. In addition, many of recent smartphones have Wi-Fi communication enabled by default. People normally use Internet applications through Wi-Fi or cellular interface, but they prefer the former to reduce data charges if and when they are within Wi-Fi coverage. Thus, the mobile device is configured to scan Wi-Fi access points automatically by broadcasting the probe request periodically to seek nearby Wi-Fi access points. Our methodology utilizes Wi-Fi access points to register and accumulate the passengers' communication activities.

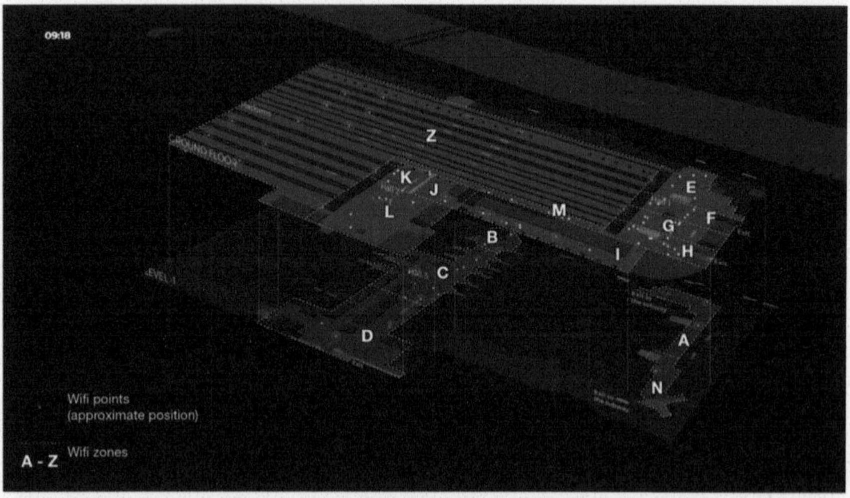

Fig. 72.1 Location of 15 Wi-Fi access points, indicating their approximate sensing range

72.2 Data Settings

72.2.1 Wi-Fi Access Points in the Station and the Definition of a Node

Figure 72.1 shows the location of 15 Wi-Fi access points deployed all over Paris-Gare-de-Lyon, one of the largest railway stations in Paris, which receives almost 90 million passengers per year. It contains a commercial area with more than 100 retail stores, restaurants, and cafes. The underground space is connected to the metro stations, bus stops, and parking.

Fifteen Wi-Fi access points cover almost all spaces continuously to offer Wi-Fi services to the passengers as seamlessly as possible. Node Z is the platform, which has 10 tracks. Nodes E, F, G, and H comprise the waiting node, where the ticket sales booths are located, while nodes B, I, J, K, L, and M form the commercial space in the station. Conversely, nodes A and N connect the station with the metro station, which is located at the underground level. Similarly, nodes B, C, and D are situated at the underground level. Within them, node D is connected to the parking area downstairs as well as the ticket sales desk and the large-scale waiting area.

72.2.2 A Subsection Sample

We collected data over 7 days—from March 26 to April 1, 2017. We selected data present from 6:00 a.m. to 23:59 p.m., which corresponds to the station's normal timetable. The raw data set contains 64,271 unique devices. We checked for possible synchronization issues caused by a lack of calibration between all Wi-Fi access points and then removed any inconsistencies (e.g., overlapping of the passenger's presence at many locations). To apply the SAMs, we transform our data set to express the passengers' spatial sequences with the time stamps as a string. For this purpose, we assign one letter to every 3 min of their length of stay in a location. For example, the original data of a passenger's sequential movement may be A-B-C: staying at node A for 12 min and 30 s, node B for 10 min and 10 s, and node C for 7 min and 4 s. In that case, the transformed data would be: AAAA-BBB-CC.

72.3 Results

72.3.1 Basic Statistics of Passengers' Behavior

Figure 72.2a shows the distribution of the number of passengers per length of stay in the station, binned for every 10 min. Although the maximum length of stay is around 4 h, 95% of the passengers' length of stay is less than 80 min. The distribution is skewed toward the lesser length of stay, with a peak of 10–20 min. Considering this is a train station, the "transient feature" may largely derive from its function. However, this plot shows its degree, that is, how fast the passengers leave the station. Conversely, Fig. 72.2b presents the distribution of the path sequence length. The maximum path sequence length is 62, but only 5% of passengers have extremely long path length, which is more than 22. It doesn't affect the statistical reliability; therefore, we aggregate them as the extreme case.

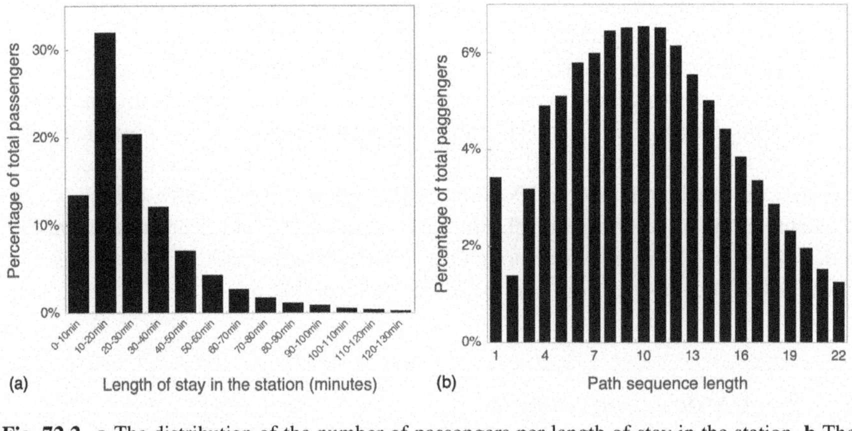

Fig. 72.2 **a** The distribution of the number of passengers per length of stay in the station. **b** The distribution of the path sequence length

72.3.2 *Path Analysis*

Based on the results of the basic statistics performed in the previous section, this section analyzes the passengers' paths and visualizes their sequences. We focus on examining the relationships among nodes rather than the individual node shown in the previous section. First, we analyze the distribution of arriving and leaving passengers–their next node to move to for the former and the node of origin for the latter. We define the arriving passengers as the ones who end their path in Z (e.g., D-E-G-Z) and the latter as the ones who start from Z (e.g., Z-E-G-H). Thus, the first and last two letters in their path provide information on whether they arrive at the station or leave from the station. We also define the types of passengers by their length of stay in the station. First, we sort all passengers by their total time spent in the station. By binning them into deciles, we obtain equally sized clusters of ~6.427 each. We refer to passengers found in the first decile as the "shorter-stay type" and the passengers of the tenth decile as the "longer-stay type".

Our analysis shows that 31.8% of passengers arrive at the station through node Z, while 51.4% leave the station from the same node, indicating that this station is largely oriented toward leaving rather than arriving. In addition, we found an uneven spatial distribution of the arriving and leaving passengers. For example, there is no significant difference between the arriving passengers' length of stay in each node, while the leaving passengers show a wider range of their length of stay among the nodes. To analyze the passengers' behaviors in more detail, we examine the frequently appearing paths in both groups: the longer-stay type and the shorter-stay type.

Table 72.1 in the supplementary materials presents the top five most frequently appearing paths in the longer-stay and shorter-stay types of passengers, and Fig. 72.3 is the visualization of those paths. As we can see, the longer-stay paths are concentrated in such regions as nodes C, D, L, and K. Conversely, the paths of shorter-stay passengers are likely to be dispersed and extended all over the station, such as nodes E, J, H, I, and M (see top left in Fig. 72.3). In the rest of the figures between those extreme cases, we can observe the transition of paths from the dispersing to the converging ones, that is, the dispersing ones gradually disappear when the passengers' length of stay becomes greater. Conversely, the converging paths gradually appear when their length of stay increases. This indicates that the shorter-stay passengers tend to explore much wider spatial dimensions than the longer-stay passengers. In addition,

Table 72.1 Top five frequently appearing paths from the longer- and shorter-stay types of passengers

Shorter-stay type; its frequency (%)	Longer-stay type; its frequency (%)
Z; 4.72	D; 1.49
Z-C; 3.96	Z; 0.84
Z-E; 1.07	C-Z; 0.54
C-Z; 1.05	L-D-L; 0.52
Z-C-B; 0.91	Z-C-Z; 0.38

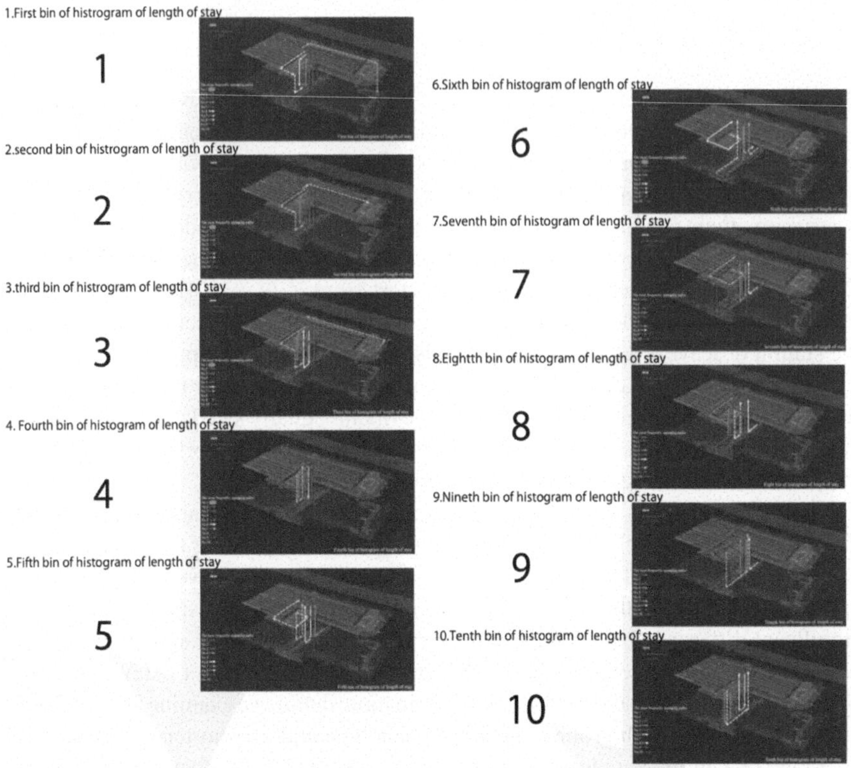

Fig. 72.3 Top 10 of the most frequently appearing paths in each group from the shorter-stay type to the longer-stay one

the difference between the path sequence length of the longer-stay and shorter-stay types is not significant; both of them have similar path sequence length, although their lengths of stay are significantly different.

72.4 Discussion and Conclusion

This paper analyzes passengers' behaviors in a large-scale train station in Paris, France. We examine their path sequence between key places in the station, their length of stay at each spot, and the relationship between these factors. For extracting frequently appearing patterns, we propose applying SAMs to the longitudinal Wi-Fi data set obtained from Wi-Fi access points dispersed all over the station.

Wi-Fi access points are primarily deployed to offer free Wi-Fi access to passengers in the station; thus, their usage in the analysis is for secondary purposes. This validates that we do not incur additional costs to prepare the infrastructure for data collection. More generally, although the methodology presented herein was applied to particular stations, it could easily be generalized to other stations or other types of buildings or public spaces. Also, we show the higher applicability of SAMs to the Wi-Fi access points data set. We successfully extract as meaningful knowledge the frequently appearing paths from the large-scale data set of the passengers' behaviors. Sequence alignment, which is a well-developed and established method in biology, can consider the order of events, suggesting the spatio-temporal aspects of human behaviors. This method can respond to the most critical issue in terms of the exponential increase of Wi-Fi access points data: how to deal with a large-scale data set to explore the results.

These findings are difficult and costly to obtain from traditionally used methodologies such as manual observations and surveys [15]. They provide us with passengers' inner thoughts and their attributions, but the smaller-scale samples make it impossible to analyze such detailed paths and lengths of stay in the station. The RFID-based methodology [8] enables us to provide all that information on a more granular spatial and temporal scale, but the sample to be collected tends to be smaller scale as well, resulting in impossibility to estimate the change of the passengers' behaviors depending on the day of the week or seasonal effects.

Overall, the method offers an effective means to extract the patterns of passengers' behaviors from Wi-Fi access points data and, by doing so, may also present new methods of analyzing such data. Our research considered the locations passengers visited frequently inside the station. These results demonstrate that the proposed methodology has great potential to uncover the passengers' spatial usage on a more granular scale in space and time.

For future works, we have just started to collect more sample data from longer periods. By comparing different periods and seasons, we would be able to detect seasonal effects. Also, based on the current results, we are preparing to analyze crowd effects for more efficient and more effective crowd management. Microscopic pedestrian simulation can be used to estimate those effects by putting an elaborated origin-destination matrix of pedestrian movement in the station.

References

1. J.B. Kruskal, An overview of sequence comparison: time warps, string edits, and macromolecules. SIAM Rev. **25**, 201–237 (1983)
2. B. Hillier, *Space is the Machine: a Configurational Theory of Architecture* (Cambridge University Press, Cambridge, 1996)
3. C. Ratti, R. Pulselli, S. Williams, D. Frenchman, Mobile Landscapes: using location data from cell phones for urban analysis. Environ. Plan. **33**(5), 727–748 (2006)

4. N. Shoval, B. McKercher, A. Birenboim, E. Ng, The application of a sequence alignment method to the creation of typologies of tourist activity in time and space. Environ. Plan. **42**(1), 76–94 (2015)
5. M. González, A. Hidalgo, A.L. Barabási, Understanding individual human mobility patterns. Nature **453**, 779–782 (2008)
6. M. Delafontaine, M. Versichele, T. Neutens, N. van de Weghe, Analysing spatiotemporal sequences in Bluetooth tracking data. Appl. Geogr. **34**, 659–668 (2012)
7. M. Versichele, T. Neutens, M. Delafontaine, N. Van de Weghe, The use of Bluetooth for analysing spatiotemporal dynamics of human movement at mass events: a case study of the Ghent festivities. Appl. Geogr. **32**, 208–220 (2011)
8. T. Kanda, M. Shiomi, L. Perrin, T. Nomura, H. Ishiguro, N. Hagita, Analysis of people trajectories with ubiquitous sensors in a science museum, in Proceedings 2007 IEEE International Conference on Robotics and Automation (ICRA'07), pp. 4846–4853 (2007)
9. S.K. Hui, E.T. Bradlow, P.S. Fader, Testing behavioral hypotheses using an integrated model of grocery store shopping path and purchase behavior. J. Consumer Res. **36**, 478–493 (2009)
10. Y. Yoshimura, F. Girardin, JP. Carrascal, C. Ratti, J. Blat, New tools for studying visitor behaviours in museums : a case study at the Louvre, in M. Fucks, F. Ricci, & L. Cantoni (Eds.), Information and Communication Technologies in Tourism 2012. Proceedings of the International Conference in Helsingborg (ENTER 2012), pp. 391–402. Mörlenback, Germany: Springer Wien New York (2012)
11. Y. Yoshimura, S. Sobolevsky, C. Ratti, F. Girardin, J.P. Carrascal, J. Blat, R. Sinatra, An analysis of visitors' behavior in the Louvre Museum: a study using Bluetooth data. Environ. Plan. **41**(6), 1113–1131 (2014)
12. Y. Yoshimura, A. Krebs, C. Ratti, Noninvasive Bluetooth monitoring of Visitors' length of stay in the Louvre. IEEE Pervasive Comput. **16**(2), 26–34 (2017)
13. Y. Yoshimura, S. Sobolevsky, J.N. Bautista Hobin, C. Ratti, J. Blat, Urban association rules: uncovering linked trips for shopping behavior. Environ. Plan. **45**(2), 367–385 (2018)
14. Y. Yoshimura, R. Sinatra, A. Krebs, C. Ratti, Analysis of visitors' mobility patterns through random walk in the Louvre Museum. J. Ambient Intell. Humaniz. Comput. (2019). https://doi.org/10.1007/s12652-019-01428-6
15. U. Flick, *An Introduction to Qualitative Research.* SAGE Publications Ltd (2009)
16. Y. Asakura, E. Hato, T. Maruyama, Behavioural data collection using mobile phones, in S. Rasouli & H. Timmermans (eds.), Mobile Technologies for Activity-Travel Data Collection and Analysis. IGI Global, pp. 17–35. New York (2014)
17. van den, J. Heuvel, JH. Hoogenraad, Monitoring the performance of the pedestrian transfer function of train stations using automatic fare collection data. Transp. Res. Procedia **2**, 642–650 (2014)
18. F.S. Hänseler, M. Bierlaire, R. Scarinci, Assessing the usage and level-of-service of pedestrian facilities in train stations: a Swiss case study. Transp. Res. Part A: Policy Pract. **89**, 106–123 (2016)
19. A. Alahi, V. Ramananthan, F. Li, Socially-aware Large-scale Crowd Forecasting, in IEEE Conference on Computer Vision and Pattern Recognition (CVPR). Columbus, OH, USA. June 24–27 (2014)
20. B. Chowdhury, G. Garai, A review on multiple sequence alignment from the perspective of genetic algorithm. Genomics **109**, 419–431 (2017)
21. W.C. Wilson, Activity pattern analysis by means of sequence-alignment methods. Environ. Plann. A **30**, 1017–1038 (1998)
22. J. Xianyu, S. Rasouli, HJP. Timmermans, Analysis of variability in multi-day GPS imputed activity-travel diaries using multi-dimensional sequence alignment and panel effects regression models. Transportation **44**(3), 533–553 (2017)
23. A. Moiseeva, HJP. Timmermans, J. Choi, CH. Joh, Sequence alignment analysis of variability in activity travel patterns through 8 weeks of diary data. Transp. Res. Record **2412**, 49–56 (2014)

Chapter 73
Departure Rates Optimization and Perimeter Control: Comparison and Cooperation in a Multi-region Urban Network

Kai Yuan, Victor L. Knoop, Boudewijn Zwaal, and Hans van Lint

Abstract With the renewed interest in the Macroscopic Fundamental Diagram (MFD) in the last decade, studies on network-level urban traffic control have increased in popularity. A popular urban traffic control approach is perimeter control, in which vehicle accumulation is kept below some critical accumulation value. An alternative control strategy is to optimize time series of departure rates as a means to limit inflows into the (sub)network. In this paper we test how these approaches compare in terms of minimizing total time spent (TTS), and whether network performance can be improved by combining these two approaches. To the best of the authors' knowledge, answers to these two questions are still missing. Our findings indicate that—for a particular multi-region network under a specific demand profile—optimizing departure rates outperforms perimeter control. Particularly, we find that the combination of perimeter control and departure rates optimization may even have adverse effects on minimizing TTS, compared to optimizing departure rates only. We also show that properly over-saturating part of a network could result in less TTS than under the application of a perimeter control, which keeps the accumulation under the critical accumulation.

K. Yuan (✉) · V. L. Knoop · B. Zwaal · H. van Lint
Department of Transport and Planning, Delft University of Technology, Delft, The Netherlands
e-mail: K.Yuan@tudelft.nl

V. L. Knoop
e-mail: V.L.Knoop@tudelft.nl

B. Zwaal
e-mail: boudewijn.zwaal@gmail.com

H. van Lint
e-mail: J.W.C.vanLint@tudelft.nl

I. Zuriguel et al. (eds.), *Traffic and Granular Flow 2019*,
Springer Proceedings in Physics 252,
https://doi.org/10.1007/978-3-030-55973-1_73

73.1 Introduction

The Macroscopic Fundamental Diagram (MFD), which describes an inverse-U shaped relationship between traffic flow and density on a network level, offers a parsimonious approach to model and study urban traffic dynamics and control in a large-scale network [1, 2].

A popular urban traffic control approach is perimeter control, in which vehicle accumulation is kept below some critical accumulation value. An alternative to perimeter control is to optimize departure rates, for example by means of pricing strategies [3]. In this paper we test how these approaches compare in terms of minimizing total time spent (TTS), and whether network performance can be improved by combining these two approaches. To the best of the authors' knowledge, answers to these two questions are still missing.

We formulate the research question as a comparison among four study cases. This research offers two main findings: (i) departure rates optimization outperforms perimeter control in minimizing TTS in a multi-region urban network; (ii) perimeter control may even have adverse effects on the performance of departure rates optimization when combining the two measures. The second finding also indicates that partial over-saturation could result in less TTS than fully under-saturation under the application of perimeter control. We believe our work contributes to the research of applying departure time control (by whatever means) in combination with perimeter control in more complex (multi-region) networks.

This paper is organized as follows: Sect. 73.2 describes the network transmission model and the genetic algorithm. Section 73.3 formulates the optimization problem, and conduct four case studies. Finally, we end this paper with conclusions in Sect. 73.4.

73.2 Methodology

This section describes methods for addressing the research questions. The traffic dynamics are described by a MFD-based model in Sect. 73.2.1. Section 73.2.2 describes the optimization of control measures.

73.2.1 Network Transmission Model: MFD-based Traffic Dynamics

This section introduces the MFD-based network transmission model. This model is used for describing the traffic dynamics on network level. The MFD function is expressed as $Q_i(N_i) = N_i \cdot v_{i,f} \cdot e^{-\frac{1}{2}(N_i/N_{i,c})^2}$ where $v_{i,f}$ and $N_{i,c}$ are the free-flow speed and the critical number of vehicles, $Q_i(N_i)$ is the potential outflow of zone i pertaining to N_i.

According to [4], the traffic dynamics of zone i are given by

$$N_i(t+1) = N_i(t) + [I_i(t) - O_i(t)] \cdot \Delta t \tag{73.1}$$

where

$$I_i(t) = \lambda_i(t) + \sum_{i \neq j, j \in \mathcal{U}_i} \min\left(\frac{N_{j\to i}}{N_j} Q_j(N_j), \frac{N_{i\leftarrow j}}{N_i} \tilde{Q}_i(N_i)\right) \tag{73.2a}$$

$$O_i(t) = \sum_{i \neq j, j \in \mathcal{D}_i} \min\left(\frac{N_{i\to j}}{N_i} Q_i(N_i), \frac{N_{j\leftarrow i}}{N_j} \tilde{Q}_j(N_j)\right) + \frac{N_{ii}}{N_i} Q_i(N_i) \tag{73.2b}$$

$I_i(t)$ is the number of vehicles entering region i per unit of time, while $O_i(t)$ is the number of vehicles leaving the region i per unit of time. $\mathcal{U}_i$ and $\mathcal{D}_i$ are the set of adjacent zones in the upstream and downstream of zone i, respectively. $N_{i\leftarrow j}$ means the number of vehicles in zone i that originating from zone j. $N_{i\to j}$ is the number of vehicles in zone i whose destination is zone j. $Q_i(N_i)$ represents the aggregated demand (shaped by the MFD function) of zone i when the vehicle accumulation is N_i, and $\tilde{Q}_i(N_i)$ is the aggregated supply of zone i whose accumulation is N_i. Remark here the term "aggregated" means all vehicle in the corresponding region are counted regardless their origins and destinations. $\lambda_i(t)$ indicates how many vehicles are generated per time unit within the region i at time t.

When perimeter control is off, the aggregated supply in one region is formulated as:

$$\tilde{Q}_i(N_i) = \begin{cases} Q_i(N_{i,c}) & \text{if } N_i \leqslant N_{i,c}, \\ Q_i(N_i) & \text{if } N_i > N_{i,c}. \end{cases} \tag{73.3}$$

When perimeter control is on, a feed-back controller is designed to limit the aggregated supply (73.3). The feed-back controller is expressed as:

$$\tilde{Q}_i(N_i(t+1)) = \tilde{Q}_i(N_i(t)) + K_1 \cdot (N_{i,c} - N_i(t+1))/L)) \tag{73.4}$$

where L is the network distance in every region. K_1 is a coefficient for the proportional term. When the perimeter controlled is on, the aggregated supply of each region will be the minimum value between (73.3) and (73.4).

73.2.2 *Optimization: A Genetic Algorithm*

We specify the demand profile along time as a vector over the monitored period. A genetic algorithm is used to find an optimal demand profile (or solution). Each solution corresponds to a cost pertaining to the total time spent (TTS), which is given by running simulations. We rank the solutions by the corresponding costs from low to high.

We repeatedly select two solutions (parents) to generate two new solutions (children) by crossover. In the selection, only a number of highly ranked solutions (named as the sample size) will be chosen to ensure the chosen 'parents' have the best 'gene'. From the top 75 out of 2000 solutions, a choice is made using a logit model. Before generating children from two parents, this crossover will take place for a random number of times. This study has a challenge to this with a constraint on the fixed total number of trips, that is, interchanging elements of any two solutions would immediately violate this constraint. Hence, after the crossover, the difference between the total number of trips in the solution and the desired total number of trips N will be added/subtracted from the very last element of the solution. If this value turns out negative after the crossover, we discard the solution and replace it with a new (random) one. The final step in the genetic algorithm is the mutation. We perform an iteration-dependent mutation by adding/subtracting a value from a random amount of elements of the parents. The repetition of generating two new solutions will end when the population size in the next generation is reached. The whole process is one iteration. In every iteration, the best solution in one generation will be recorded as the solution of the final generation. After finite iterations, the top ranked solution in the final generation would be the optimal solution.

73.3 Problem Formulation and Case Study

This section describes the studied urban network, and how we formulate the optimization problem. A particular case study is given.

Consider an urban network, which is divided into $R = 6$ regions; see Fig. 73.1. Each region ($i = 1, 2, ..., R$) is characterized by a well-defined MFD. In this network, during peak hours citizens in region 1 and 4 need to drive to region 3 and 6, respectively. This OD matrix structure illustrated in Fig. 73.1, where two directions of traffic streams intersect each other in region 2. The same MFD is applicable from region 1 to region 5; the MFD in the region 6 has a lower capacity and free-flow speed. In Fig. 73.2, the area of each region indicates its capacity.

In this study, the departure rates at every time window from region 4 was fixed; while the departure rate time series via region 1 is controllable by using some prospective traffic measures, e.g., time-specific trade-able peak permits [3]. This study uses $\lambda_i(t)$ to denote the expected departure rate via region i at time t per unit of time. As a result, region 6 can be a bottleneck for vehicles from region 4. Congestion could occur in region 5 due to the bottleneck. When the perimeter control is on, congestion will be prohibited from occurring in the region 5 by holding vehicles in its upstream.

The goal of the departure rates optimization is to minimize the cost pertaining to TTS, U. The cumulative number of vehicles entering and exiting the network at time t are denoted by $N_{\text{in}}(t)$ and $N_{\text{out}}(t)$. The cost function is formulated as:

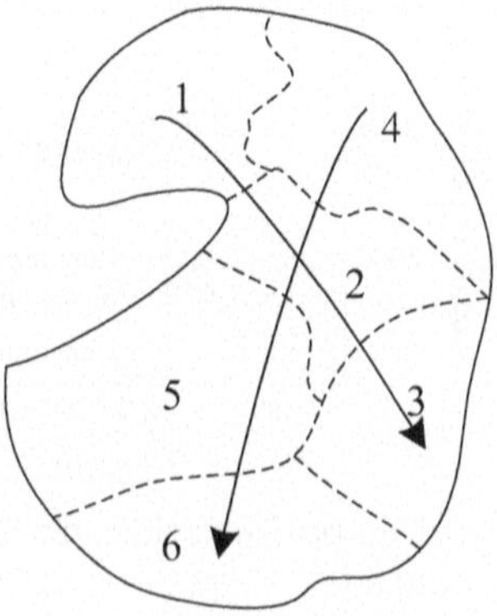

Fig. 73.1 Studied urban network structure

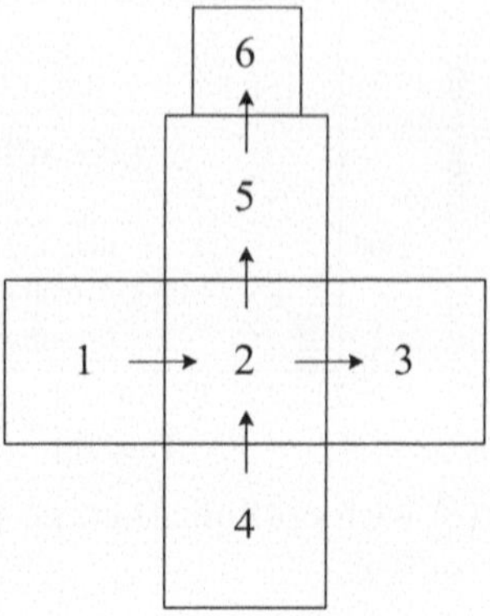

Fig. 73.2 A representation of the network

$U = \alpha \int_0^T ([N_{out}(t) - N_{in}(t)] \, dt$. Here, the total simulation time is denoted as T. α is the marginal cost of total time spent on roads.

For all simulations we consider a peak-period of three hours (from 7:00 am to 10:00 am) and a simulation time of six hours: three more hours following the peak duration for emptying the network.

As described in Sect. 73.3, two different exogenous flows enter our network: from zone 1–3; and from zone 4–6. Consider two demands: from 8:00 to 9:00, 1300 vehicles would like to take the path $1 \to 2 \to 3$; while 3900 vehicles take $4 \to 2 \to 5 \to 6$ during the 3-hour peak duration.

Four different cases are categorized regarding to whether the departure rate time series $\lambda_4(t)$ is optimized and whether perimeter control is active: : (Case 1) Without any control measure: The aggregated departure rate in every time window is the same. The perimeter control is off; (Case 2) Perimeter control: The aggregated departure rate in every time window is the same. The perimeter control is on; (Case 3) Departure rates optimization: The time series of departure rates is optimized through the genetic algorithm. The perimeter control is off; (Case 4) Combining perimeter control and the departure rates optimization: The time series of departure rates is optimized through the genetic algorithm. The perimeter control is on.

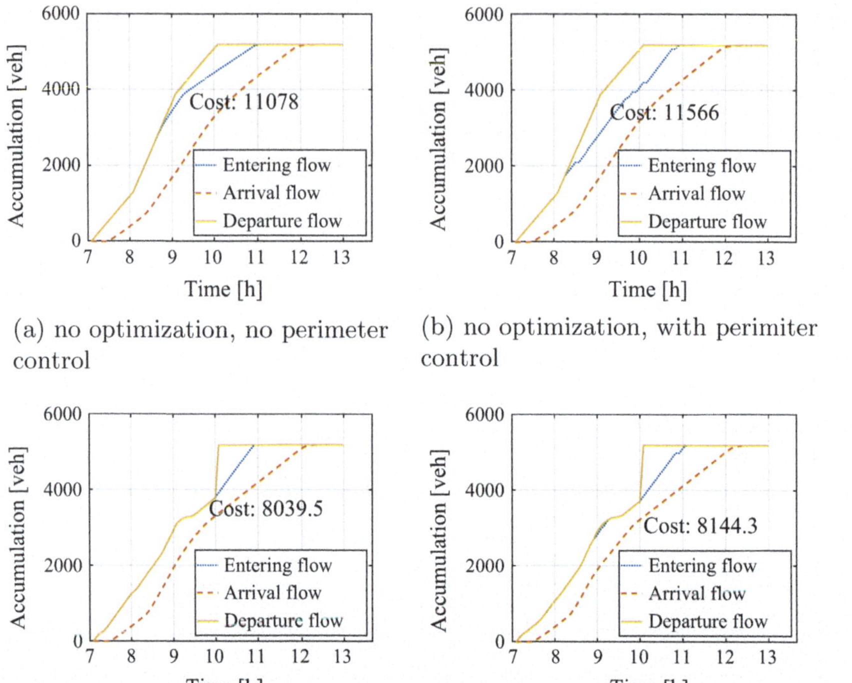

(a) no optimization, no perimeter control

(b) no optimization, with perimiter control

(c) with optimization, no perimeter control

(d) with optimization, with perimeter control

Fig. 73.3 Cumulative curves of departure, entering and arrival flow in four study cases. The cost in each case is put in each sub-figure

The departure rates tell how many vehicles depart in every time windows. Due to the traffic dynamics (e.g., low supply in the downstream zone) or perimeter control, some vehicles will be kept waiting out of the monitored network (before entering region 1 or 4). Hence, another term "entering rate" is used to indicate the number of vehicles entering into region 1 and 4 per unit of time.

Figure 73.3 shows the cumulative curves of departure rates, entering rates and arrival rates. The cost estimated using for each case is shown in the corresponding sub-figure. The cost is α times the area between the departure (solid line) and arrival (dashed line) cumulative curve. In Fig. 73.3, the area between the cumulative curve of entering and arrival rates indicates the total time spent in the monitored network, excluding the waiting time out of the network.

The case with departure rates optimization and without perimeter control has the lowest cost (8039) among four cases. In the case of combination of the departure rates optimization and perimeter control, the cost reaches 8144. Since the only difference in these two cases is whether the perimeter control is on, we can conclude that perimeter control adds *adverse* effects to the congestion mitigation performance of the departure rates optimization in this study. Comparing the cases without optimization, i.e., Case 1 and Case 2, we can find a slightly lower cost in the case without perimeter control. When the optimization is used in Case 3 and Case 4, the case with perimeter control has higher cost. All show that perimeter control will slightly increase total cost, and that departure rates optimization will decrease total cost.

73.4 Conclusions

In this paper we compared different combinations of perimeter control and departure rates optimization. To this end we used an MFD network traffic flow model and a genetic optimization algorithm to do the optimization.

This research offered two main findings: (i) departure rates optimization outperforms perimeter control in minimizing TTS in a multiregion urban network—given of course our assumptions on the supply dynamics; (ii) perimeter control may even have adverse effects on the performance of departure rates optimization when combining the two measures. The second finding also indicates that partial over-saturation could result in less TTS than fully under-saturation under the application of perimeter control. In the future, authors will explore whether the findings and conclusions can be generalised in a more complex but realistic city.

References

1. C.F. Daganzo, N. Geroliminis, An analytical approximation for the macroscopic fundamental diagram of urban traffic. Transp. Res. Part B Methodol. **42**(9), 771–781 (2008). http://www.sciencedirect.com/science/article/pii/S0191261508000799
2. S.P. Hoogendoorn, W. Daamen, V.L. Knoop, J. Steenbakkers, M. Sarvi, Macroscopic fundamental diagram for pedestrian networks: theory and applications. Transp. Res. Procedia **23**, 480–496 (2017). http://www.sciencedirect.com/science/article/pii/S2352146517303058. (papers Selected for the 22nd International Symposium on Transportation and Traffic Theory Chicago, Illinois, USA, 24–26 July, 2017)
3. E. Verhoef, P. Nijkamp, P. Rietveld, Tradeable permits: Their potential in the regulation of road transport externalities. Environ. Plan. B Plan. Design **24**(4), 527–548 (1997). https://doi.org/10.1068/b240527
4. V. Knoop, S. Hoogendoorn, An area-aggregated dynamic traffic simulation model. Eur. J. Transp. Infrastruct. Res. (EJTIR) (2015)

Author Index

I. Zuriguel et al. (eds.), *Traffic and Granular Flow 2019*,
Springer Proceedings in Physics 252,
https://doi.org/10.1007/978-3-030-55973-1

Subject Index

I. Zuriguel et al. (eds.), *Traffic and Granular Flow 2019*,
Springer Proceedings in Physics 252,
https://doi.org/10.1007/978-3-030-55973-1

The manufacturer's authorised representative in the EU is Springer Nature Customer Service Centre GmbH, Europaplatz 3, 69115 Heidelberg, Germany. If you have any concerns regarding our products, please contact ProductSafety@springernature.com

Printed and bound by CPI Group (UK) Ltd, Croydon, CR0 4YY
30/09/2026
02225689-0019